LIST OF ABBREVIATIONS

Handbook Series in Organic Electrochemistry

Vol. V

APPARATUS:

Cell			Oxygen removal			Data acquisition		
	0	two-electrode		A	inert gas		0	recorder
	1	iR correction		B	air		1	manual
	2	three-electrode		C	hydrogen		2	oscillographic
	3	agar bridge		D	sulfite		3	digital
	4	liquid junction (Kalousek)		E	hydrazine		4	other
	5	porous separator		F	other			
	6	salt bridge						
	7	plastic membrane						
	8	thin layer						
	9	stirred solution						

ALL OTHER:

A

A	(alone or as subscript) anodic
A=	area (cm^2)
ACET	acetate
AcNHME	N-methylacetamide
ads	adsorption
AE	abraded electrode
AM	ammonia
ANHYD	anhydride
Ap	approximate (value)
a.s.	anomalous shape
ATR	attenuated total reflectance
AuBE	gold-bead
AuDE	gold-disc
Aux	auxiliary (counter) electrode
Av(n)	average of n values

B

B	poorly separated
BARB	barbital
BB	Bates and Bower
BENZ	benzoic acid
BF	buffer
BOR	borate or borax
BPT,bp	boiling point
BR	Britton and Robinson

C

c	calculated value
C	(alone or as subscript) cathodic
C=	concentration
C=x-x	concentration range
C→	concentration up to
CA	convective chronoamperometry
calcd.	calculated
calcn.	calculation
CARB	carbonate
CBE	carbon-black
CC	convective chronocoulometry
CdSE	cadmium sulfate electrode
CE	calomel electrode
ce	chemical reaction followed by an electron-transfer step
CHA	carbon-hydrogen analysis
CHN	carbon-hydrogen-nitrogen analysis
CITR	citrate
CL	Clark and Lubs
ClACET	chloroacetic acid
corrd.	corrected
corrn.	correction
CP	controlled-potential electrolysis
CPDO	carbon-paste electrode containing dissolved organic substances
CPE	carbon-paste
CT	convective triangular-wave voltammetry
C_O	concentration of compound being reduced
C_R	concentration of compound being oxidized

D

D	diffusion coefficient ($10^{-6} cm^2 s^{-1}$)
d	diameter (cm)
Dcalc	calculated diffusion coefficient
D:D	comparison of diffusion coefficients
De	not energy-sufficient
detn	determination
DI	differential pulse polarography
dim	dimerization
DME	dropping mercury
DMF	N,N-dimethylformamide
DMSO	dimethyl sulfoxide
DP	derivative pulse polarography
Dplot	Delahay plot
dpn	disproportionation
DQ	incremental charge polarography
dr	drawn out

E

E	exponential (e.g., $1E4 = 1 \times 10^4$)
E=	potential
(E)	only potential listed
EC	programmed-current chronopotentiometry
EC=	electrode constant $m^{2/3} t^{1/6}$ ($mg^{2/3} s^{-1/2}$)
ec	electron-transfer step followed by a chemical reaction
ece	electron-transfer step followed by a chemical reaction, then a second electron-transfer step
ecl	electrochemiluminescence
ECM	electrocapillary maximum
EE	current-step chronopotentiometry
EF	ac chronopotentiometry
EK	alternating-voltage chronopotentiometry
EL	chronopotentiometry with linear current sweep
El	electrode
Elog	E vs. $\log[(i_\ell - i)/i]$
Elogi^a	E vs. $\log[(i_\ell - i)/i^a]$ (a = some numerical constant)
Elogv	E_p vs. $\log v$
ElogT	E vs. $\log[(\tau^{1/2} - t^{1/2})/t^{1/2}]$
Elogω	$E_{1/2}$ vs. $\log \omega$
EM	derivative chronopotentiometry
eqn	equation
ER	chronopotentiometry
ESR	electron-spin resonance
EV	chronopotentiometry with superimposed ac
EW	controlled-potential electrogravimetry
EY	cyclic chronopotentiometry
E_{app}	applied emf (V)
E_{disc}	potential of the disc (for a ring-disc electrode)
E_i	incision potential
E_j	liquid-junction potential
E_{max}	potential of a maximum
E_{min}	potential of a minimum
E_{MP}	potential of mercury pool
E_p	peak potential
$E_{p,A}$	anodic peak potential
$E_{p,C}$	cathodic peak potential
$E_{p/2}$	half-peak potential
E_{pi}	photoionization potential
E_{ref}	potential of a reference electrode (V)
E_{rf}	radio-frequency potential at which current changes sign
E_{ring}	potential of the ring (for a ring-disc electrode)
E_s	steric substituent constant
E_{su}	summit potential
E_t	tangent potential
E_λ	half-neutralization potential
$E_{\pi max}$	potential of peak emission (on scanning)
E_τ	switching potential
$E_{\tau/4}$	quarter-transition-time potential
$E_{1/2}$	half-wave potential
$E_{1/2}^0$	half-wave potential at pH=0
$E_{1/2}^\uparrow$	wave merges with initial current rise
$E_{1/4}$	quarter-wave potential

LIST OF ABBREVIATIONS

Handbook Series in Organic Electrochemistry
- Vol. V

ALL OTHER (cont.)

$E_{\frac{3}{4}}$	three-quarter-wave potential
$E_{0.22}$	0.22-transition-time potential
$E_{0.85}$	0.85-peak potential
E_o	standard potential
E_o'	formal potential
E_{2d}	second-harmonic null potential

F

F	from a figure
F( )	given non-linear functional dependence
f	frequency of a sinusoidal signal (Hz)
f( )	functional dependence, not given
Fc	ferrocene
FORM	formate
FP	demodulation polarography

G

GC	gas-phase chromatography
GCE	glassy carbon
GE	graphite
GEL	gelatin
GLC	gas-liquid chromatography
GLYC	glycine

H

H	reduction of hydrogen ion
h	height of the mercury column (cm)
Hg/HgAc	mercury-mercurous acetate electrode
HL	high-level faradaic rectification
HMDE	hanging mercury-drop
HMP	hexamethylphosphoramide
HN	electrochemiluminescence
HOMO	highest occupied molecular orbital
HQ	hydroquinone

I

I	diffusion current constant ($\mu A\ s^{1/2} mg^{-2/3} m\underline{M}^{-1}$)
i	current (μA)
(i)	only current listed
IB	amperometry with two indicator electrodes
ID	differential amperometry
I:I	comparison of diffusion current constants
i:i	comparison of (limiting) currents
IL	chronoamperometry with linear potential sweep
Ilk	from Ilkovič equation
IM	chronoamperometry with non-linear potential sweep
In	minimum current values measured
IO	amperometry
IP	Izmailov and Pivneva
IQ	stirred-pool-electrode chronocoulometry
IR	chronoamperometry
iR	ohmic potential as in iR drop
IRS, ir	infrared spectroscopy
IS	stirred-pool-electrode-chronoamperometry
i-t	potentiostatic measurement
IU	double potential-step chronoamperometry
i(u)	current in arbitrary units
IV	polarographic chrono-amperometry
i vs. E	log i vs. E
Ix	maximum currents measured
i_A	anodic current
i_a	adsorption current
i_{ac}	diffusion current constant for ac polarography (mmho $m\underline{M}^{-1}$)
i_C	cathodic current
i_c	capacity current
i_{cat}	catalytic current
i_d	diffusion current
i_{disc}	current at the disc of a ring-disc electrode
i_k	kinetic current
i_ℓ	limiting current (μA)
$i_{\ell,A}$	anodic limiting current
$i_{\ell,C}$	cathodic limiting current
i_{max}	current at the maximum
i_{min}	current at the minimum
i_p	peak current (μA)
$i_{p,A}$	anodic peak current
$i_{p,C}$	cathodic peak current
$\frac{1}{2}(i_{p1}+i_{p2})$	average of two peak currents
i_{ring}	current at the ring of a ring-disc electrode
i_{su}	summit current (μA)
i_{VDME}/i_{DME}	ratio of current at VDME to current at DME

J

j	current density ($\mu A\ cm^{-2}$)
j:j	comparison of current densities
j_ℓ	limiting current density ($\mu A\ cm^{-2}$)
j_p	peak current density ($\mu A\ cm^{-2}$)
$J\tau^{\frac{1}{2}}/C$	chronopotentiometric constant ($\mu A\ s^{1/2} cm^{-2} m\underline{M}^{-1}$)

K

k	heterogeneous rate constant ($cm\ s^{-1}$)
KHP	potassium biphthalate
Kplot	Koutecký plot
k_{app}	apparent rate constant
$k_{b,h}$	backward heterogeneous rate constant ($cm\ s^{-1}$)
$k^o_{b,h}$	the value of $k_{b,h}$ at 0 V vs. NHE
$k_{f,h}$	forward heterogeneous rate constant ($cm\ s^{-1}$)
$k^o_{f,h}$	the value of $k_{f,h}$ at 0 V vs. NHE
k^o_h	heterogeneous rate constant at 0 V vs. NHE
$k_{s,h}$	the common value of $k_{b,h}$ and $k_{f,h}$ at E^o
$k^{app}_{s,h}$	apparent value of $k_{s,h}$

L

L	light excluded
ℓ	thickness of thin-layer electrode gap (μm)
LLC	liquid-liquid chromatography
ℓp	limited proof
LSC	liquid-solid chromatography
LSCE	lithium-saturated calomel electrode
LSQ	least-squares value
LUMO	lowest unoccupied molecular orbital

M

M	merges with final current rise
m	rate of flow of Hg ($mg\ s^{-1}$)
MAS, MS	mass spectrometry
MB	McIlvaine
Mc( )	mistake corrected (column)
MeCN	acetonitrile
MG	1,2-dimethoxyethane (monoglyme)
Mn	minimum
MM	modulation polarography
MP	mercury-pool
MPCuD	mercury-coated copper disc
MPDE	mercury-coated disc
MPG	mercury-coated graphite
MPT, mp	melting point
MSE	mercury-mercurous sulfate electrode
MWT, mw	molecular weight
Mx	maximum

N

(n)	number of electrons only
NAgE	silver—normal silver ion electrode
NBS	National Bureau of Standards pH scale
NCE	normal calomel electrode
neg.	negative
NF	measurement of dynamic capacity
NHE	normal hydrogen electrode
NMR	nuclear magnetic resonance
N-S	αn_a or $k_{s,h}$ calculated from peak separation (Nicholson-Shain equation)
ns	not stated
n_a	number of electrons transferred through rate-determining step of a cathodic process

LIST OF ABBREVIATIONS

Handbook Series in Organic Electrochemistry

Vol. V

ALL OTHER (cont.)

n_{app}	an apparent number of electrons transferred
n_b	number of electrons transferred through rate-determining step of an anodic process
N_o	collection efficiency (ring-disc electrode)

O

0	no wave observed
(oc)	open circuit
oh–	number of hydroxide ions consumed through rate-determining step
OTE	optically transparent electrode (may be preceded by a chemical symbol)

P

P	merging with previous wave
p	preliminary
p=	number of protons consumed through rate-determining step
PA	triangular-wave polarography
PB	higher harmonic ac polarography with phase-sensitive rectification
PbDE	lead-disc
PBE	platinum-bead
PbSE	lead sulfide electrode
PC	current-scanning polarography
PCA	propylene carbonate
PD	derivative polarography
PDE	platinum-disc
PDME	mercury-coated platinum-disc
PDS	with periodic displacement of solution
PE	pretreated electrode
PF	differential polarography
PG	square-wave polarography
PGE	pyrolytic graphite
PH	higher harmonic ac polarography
PHEN	phenol
PHOS	phosphate
PHTH	phthalate
PI	alternating-voltage polarography
PK	Kalousek polarography
pK_d	–log(acidic dissociation constant)
pK'	pH at inflection point of polarographic dissociation curve
PL	intermodulation polarography
Pl	plateau
PM	multisweep polarography
PMC	mercury-coated platinum
PO	oscillopolarography
Po	postwave
pos.	positive
PP	pulse polarography
ppt	precipitate
PQ	polarographic coulometry
PR	cyclic triangular-wave polarography
PROC	propylene carbonate
Pr	prewave
PrCN	butyronitrile
PRW	Prideaux and Ward
PS	staircase polarography
PSCE	mercury-coated carbon
PSMC	mercury-coated platinum sphere
PSME	mercury-coated platinum sphere
PT	Tast polarography
PV	total ac polarography
PVC	poly(vinyl chloride)
PVI	in-phase ac polarography
PVQ	quadrature polarography
PW	single-sweep polarography
PX	double-tone polarography
PY	polarography
PYR	pyridine

Q

Q	in characteristic potential column, incision quotient; elsewhere, quantity of charge (μC)
QE	controlled-potential coulometry
QI	quasi-reversible
QP	controlled-current coulometry without a reagent precursor
Qp	integrated charge corresponding to the faradaic current (μC)
QR	chronocoulometry
QT	controlled-current coulometry with a reagent precursor
QU	double potential-step chronocoulometry
QW	electrogravimetry

R

R	reversible
r	reliable
RAuE	rotating gold
RCCE	rotating cylindrical carbon
RCGE	rotating glassy carbon cylinder
RDCP	rotating carbon paste
RDE	rotating disc
RDGC	rotating glassy carbon-disc
RDME	rotating dropping mercury
RGDE	rotating gold-disc
RhBE	rhodium bead
RMAuD	rotating mercury-coated gold-disc
RMCuD	rotating mercury-coated copper disc
RMDE	rotating mercury-coated ring-disc
RMPD	rotating mercury-coated platinum-disc
rms	root mean square
R_4NCE	mercury — mercurous chloride — tetraalkyl-ammonium chloride electrode
RP	radio-frequency polarography
RPDE	rotating platinum-disc
RPE	rotating platinum
RPGE	rotating pyrolytic graphite
RPMC	rotating mercury-coated platinum
RRDE	ring-disc
RSDE	rotating silver-disc
RWGE	rotating wax-impregnated graphite
rxn	reaction
r_d	radius of disc
$r_{r,i}$	inner radius of ring
$r_{r,o}$	outer radius of ring

S

S	merging with succeeding wave
satd	saturated
SCE	saturated calomel electrode
Se	energy-sufficient
SLiE	mercury — mercurous chloride — saturated lithium chloride electrode
SMDE	sessile mercury drop
SME	streaming mercury
SnDE	tin-disc
SnO	tin-oxide electrode
Sr	surface reaction
SRGE	silicone rubber-graphite
SSCE	mercury — mercurous chloride — saturated sodium chloride electrode
St	slope of tangent $E_{\frac{1}{2}}$
sttd	stated
SULF	sulfate
SULN	sulfolane
supp. elect.	supporting electrolyte

T

T	temperature
t	time (s)
t(c)	controlled time
$t(E_1)$	time of first step
t(oc)	drop time open circuit
Tc	temperature coefficient of the limiting current (% deg^{-1})
t(sc)	drop time short circuit
THF	tetrahydrofuran
THFA	tetrahydrofurfuryl alcohol
TLC	thin-layer chromatography
TLE	thin-layer electrode
TLIR	thin-layer chronoamperometry
TLQE	thin-layer coulometry
Tomeš	$E_{\frac{3}{4}} - E_{\frac{1}{4}}$ (mV)
TQ	thin-layer chronocoulometry
TRIS	tris(hydroxymethyl)aminomethane
TX100	Triton X-100

LIST OF ABBREVIATIONS

Handbook Series in Organic Electrochemistry

Vol. V

ALL OTHER (cont.)

T_{sw}	switching (reversal) time

U

(u)	arbitrary unit
UB	universal buffer
UVS, UV	ultraviolet spectroscopy

V

V	volts
v	scan rate (mV s^{-1})
VA	triangular-wave voltammetry
VC	current-scanning voltammetry
VD	derivative voltammetry
VDME	vibrating mercury drop
VF	differential voltammetry
VIS	visible spectroscopy
VM	multisweep voltammetry
VO	oscillovoltammetry
VPE	vibrating platinum
VR	cyclic triangular-wave voltammetry
Vr	volume reaction
VV	ac voltammetry
VY	hydrodynamic voltammetry

W

W	well defined
WGE	wax-impregnated graphite

X

X	ill defined
X preceding another symbol, see Introduction to Table I	
x→x(units)	continuous variation
x-x(units)	range

Y

Y	admittance (mmho)

Symbols:

→	becomes
≈ and QE	curve-fitting and controlled-potential electrolysis
⇍	irreversible
≠	unequal
∝̸	not proportional
∅	quantum yield efficiency
□	square wave
∧	triangular wave
/	ramp
α	cathodic charge-transfer coefficient
β	Hückel integral energy unit
β_a	anodic charge-transfer coefficient
Γ	surface excess (10^{-10} mol cm^{-2})
ΔE	pulse amplitude (mV)
Δe	amplitude of sinusoidal potential (mV)
ΔE_p	$E_{p,A} - E_{p,C}$
$\Delta E_{p/2}$	half-peak width
$\Delta E_{su/2}$	half-summit width
Δi	amplitude (peak-to-peak) of alternating current (μA)
$\Delta(1/\lambda)$	change in wavenumber (nm^{-1})
η	relative viscosity
λ	wavelength (nm)
λ_{max}	wavelength of maximum absorbance
μ	ionic strength (mol dm^{-3})
ν	frequency (= c/λ)
π_*	intensity of luminescence
π^*	excited π-orbital
π_{max}	maximum intensity of luminescence
ρ	Hammett reaction constant
ρ^*	Taft reaction constant
Σ	(except in equations) secondary source
σ, σ_x	Hammett substituent constant
σ^n	substituent constant for aromatic rings
σ^*, σ^*_x	Taft substituent constant
τ	transition time (s)
τ_A	anodic transition time (s)
τ_C	cathodic transition time (s)
τ_d	diffusion-controlled transition time (s)
τ_f	forward transition time (s)
τ_r	reverse transition time (s)
ϕ_2	potential difference across the diffuse double layer
ω	phase angle
φ^o	zero-charge potential (-ECM)
χ	for $A \underset{k_{-1}}{\overset{k_1}{\rightleftarrows}} B$, $\chi = 2k_1(k_1+k_{-1})^{\frac{1}{2}}t^{\frac{1}{2}}/k_{-1}$
ψ^o	ψ potential in double layer
ω	rotation rate (s^{-1})

LIST OF ABBREVIATIONS

Handbook Series in Organic Electrochemistry

Vol. V

ALL OTHER (cont.)

$E_{\frac{3}{4}}$	three-quarter-wave potential
$E_{0.22}$	0.22-transition-time potential
$E_{0.85}$	0.85-peak potential
E_o	standard potential
$E_o{}'$	formal potential
E_{2d}	second-harmonic null potential

F

F	from a figure
F( )	given non-linear functional dependence
f	frequency of a sinusoidal signal (Hz)
f( )	functional dependence, not given
Fc	ferrocene
FORM	formate
FP	demodulation polarography

G

GC	gas-phase chromatography
GCE	glassy carbon
GE	graphite
GEL	gelatin
GLC	gas-liquid chromatography
GLYC	glycine

H

H	reduction of hydrogen ion
h	height of the mercury column (cm)
Hg/HgAc	mercury-mercurous acetate electrode
HL	high-level faradaic rectification
HMDE	hanging mercury-drop
HMP	hexamethylphosphoramide
HN	electrochemiluminescence
HOMO	highest occupied molecular orbital
HQ	hydroquinone

I

I	diffusion current constant ($\mu A\ s^{1/2} mg^{-2/3} mM^{-1}$)
i	current (μA)
(i)	only current listed
IB	amperometry with two indicator electrodes
ID	differential amperometry
I:I	comparison of diffusion current constants
i:i	comparison of (limiting) currents
IL	chronoamperometry with linear potential sweep
Ilk	from Ilkovič equation
IM	chronoamperometry with non-linear potential sweep
In	minimum current values measured
IO	amperometry
IP	Izmailov and Pivneva
IQ	stirred-pool-electrode chronocoulometry
IR	chronoamperometry
iR	ohmic potential as in IR drop
IRS, ir	infrared spectroscopy
IS	stirred-pool-electrode chronoamperometry
i-t	potentiostatic measurement
IU	double potential-step chronoamperometry
i(u)	current in arbitrary units
IV	polarographic chronoamperometry
i vs. E	log i vs. E
Ix	maximum currents measured
i_A	anodic current
i_a	adsorption current
i_{ac}	diffusion current constant for ac polarography (mmho mM^{-1})
i_C	cathodic current
i_c	capacity current
i_{cat}	catalytic current
i_d	diffusion current
i_{disc}	current at the disc of a ring-disc electrode
i_k	kinetic current
i_ℓ	limiting current (μA)
$i_{\ell,A}$	anodic limiting current
$i_{\ell,C}$	cathodic limiting current
i_{max}	current at the maximum
i_{min}	current at the minimum
i_p	peak current (μA)
$i_{p,A}$	anodic peak current
$i_{p,C}$	cathodic peak current
$\frac{1}{2}(i_{p1}+i_{p2})$	average of two peak currents
i_{ring}	current at the ring of a ring-disc electrode
i_{su}	summit current (μA)
i_{VDME}/i_{DME}	ratio of current at VDME to current at DME

J

j	current density ($\mu A\ cm^{-2}$)
j:j	comparison of current densities
j_ℓ	limiting current density ($\mu A\ cm^{-2}$)
j_p	peak current density ($\mu A\ cm^{-2}$)
$J\tau^{\frac{1}{2}}/C$	chronopotentiometric constant ($\mu A\ s^{1/2} cm^{-2} mM^{-1}$)

K

k	heterogeneous rate constant (cm s^{-1})
KHP	potassium biphthalate
Kplot	Koutecký plot
k_{app}	apparent rate constant
$k_{b,h}$	backward heterogeneous rate constant (cm s^{-1})
$k_{b,h}^o$	the value of $k_{b,h}$ at 0 V vs. NHE
$k_{f,h}$	forward heterogeneous rate constant (cm s^{-1})
$k_{f,h}^o$	the value of $k_{f,h}$ at 0 V vs. NHE
k_h^o	heterogeneous rate constant at 0 V vs. NHE
$k_{s,h}$	the common value of $k_{b,h}$ and $k_{f,h}$ at E^o
$k_{s,h}^{app}$	apparent value of $k_{s,h}$

L

L	light excluded
ℓ	thickness of thin-layer electrode gap (μm)
LLC	liquid-liquid chromatography
ℓp	limited proof
LSC	liquid-solid chromatography
LSCE	lithium-saturated calomel electrode
LSQ	least-squares value
LUMO	lowest unoccupied molecular orbital

M

M	merges with final current rise
m	rate of flow of Hg (mg s^{-1})
MAS, MS	mass spectrometry
MB	McIlvaine
Mc( )	mistake corrected (column)
MeCN	acetonitrile
MG	1,2-dimethoxyethane (monoglyme)
Mn	minimum
MM	modulation polarography
MP	mercury-pool
MPCuD	mercury-coated copper disc
MPDE	mercury-coated disc
MPG	mercury-coated graphite
MPT, mp	melting point
MSE	mercury-mercurous sulfate electrode
MWT, mw	molecular weight
Mx	maximum

N

(n)	number of electrons only
NAgE	silver—normal silver ion electrode
NBS	National Bureau of Standards pH scale
NCE	normal calomel electrode
neg.	negative
NF	measurement of dynamic capacity
NHE	normal hydrogen electrode
NMR	nuclear magnetic resonance
N-S	αn_a or $k_{s,h}$ calculated from peak separation (Nicholson-Shain equation)
ns	not stated
n_a	number of electrons transferred through rate-determining step of a cathodic process

CRC Handbook Series in Organic Electrochemistry

Volume V

Authors:

Louis Meites
Petr Zuman
Professors of Chemistry
Department of Chemistry
Clarkson College of Technology

Elinore B. Rupp
Research Associate in Chemistry
Department of Chemistry
Clarkson College of Technology

and

Theodore L. Fenner
Ananthakrishnan Narayanan

With financial support from the National Science Foundation

CRC Press, Inc.
Boca Raton, Florida

Library of Congress Cataloging in Publication Data

Meites, Louis.
　Handbook of organic electrochemistry.

(CRC handbook series in organic electrochemistry; v. 5).
Bibliography: p.
　Includes indexes.
　1. Electrochemical analysis—Handbooks, manuals,
etc. 2. Chemistry, Organic—Handbooks, manuals,
etc. I. Zuman, Petr, joint author. II. Title.
III. Series.
QD272.E4M44　　　547'.1'37　　　77-24273
ISBN 0-8493-7220-8 (complete set)
ISBN 0-8493-7225-9 (v. 5)

　This Handbook was prepared with the support of the National Science Foundation Grants No. CHE74-19761 and CHE76-17435. However, any opinions, findings, conclusions, or recommendations expressed herein are those of the editors and do not necessarily reflect the view of NSF.

　This book represents information obtained from authentic and highly regarded sources. Reprinted material is quoted with permission, and sources are indicated. A wide variety of references are listed. Every reasonable effort has been made to give reliable data and information, but the author and the publisher cannot assume responsibility for the validity of all materials or for the consequences of their use.

　All rights reserved. This book, or any parts thereof, may not be reproduced in any form without written consent from the publisher.

　Direct all inquiries to CRC Press, Inc., 2000 Corporate Blvd., N.W., Boca Raton, Florida, 33431.

© 1982 by Clarkson College of Technology

International Standard Book Number 0-8493-7220-8 (Complete Set)
International Standard Book Number 0-8493-7225-9 (Volume V)

Library of Congress Card Number 77-24273
Printed in the United States

PREFACE

Sixty years have now gone by since Professor Jaroslav Heyrovský conducted the preliminary investigations that led to the development of polarography. In that time, nurtured by the efforts of a host of brilliant and dedicated scientists, it has grown from promising infancy to vigorous maturity. Mathematicians, electrochemists, analytical chemists, hydrodynamicists, biochemists, clinical chemists and analysts, inorganic chemists, physical organic and synthetic organic chemists, and others have contributed to its development and have conferred a wide spectrum of abilities on it. Analysts use it for detection, identification, and determination; organic chemists use it for studying the rates and equilibria of organic reactions, elucidating their mechanisms, making structural assignments, and designing or improving synthetic procedures; inorganic chemists use it for deducing the compositions of complexes and studying the thermodynamics and kinetics of their formations and dissociations; clinicians use it for diagnosis and for following the treatment of diseases. Many other kinds of uses have already been found for it, and many more can be foreseen for the future. Without doubt it is one of the most powerful and most widely applicable techniques of scientific measurement and investigation.

As polarography itself has grown, other techniques related to it have been revived or invented and have shared in its growth. Some of these, like ac polarography and oscillopolarography, involve changes of the natures of the exciting signal and the measured response. Some, like voltammetry, involve different kinds of mass-transfer processes and different indicator electrodes. Others, like controlled-potential electrolysis and coulometry, involve the same fundamental relationships among current, potential, composition, and time, but employ them in different ways and for different purposes.

The development and growth of this family of interrelated techniques has been recorded in a large number of publications. About 125 research papers dealing with polarography had been published by 1930, about 1000 by 1940, about 3500 by 1950, and about 10,000 by 1960. Professor Heyrovský's annual *Bibliography of Publications Dealing with the Polarographic Method* included a total of 17,306 citations through 1964, while the *Bibliografica Polarografica* published by the Consiglio Nazionale delle Richerche lists 21,798 through 1967. No more recent reliable figure is available, but an estimate of 50,000 through 1978 can hardly be very far from the mark.

Ready access to the information previously gathered is of course essential to further progress throughout the development of any field of science. The analytical chemist and the clinical chemist need data on the behaviors of the substances they wish to determine and of other substances that might accompany these so that they can design and use analytical methods that yield accurate and reliable results. Data on the behaviors of individual substances at different pH-values, ionic strengths, and temperatures and in different buffers and solvents are needed in correlating these with their chemical and biochemical properties; data obtained with different techniques and indicator electrodes are needed in clarifying electrochemical phenomena; data on the behaviors of different but related substances provide not only structure-reactivity correlations but also correlations with many other kinds of data from different fields, ranging from ultraviolet, infrared, and nuclear magnetic resonance spectroscopy to the abilities of polycyclic hydrocarbons to act as semiconductors; data on the behaviors of starting materials often enable the synthetic electrochemist to select conditions that will give the best possible yield of the purest possible product in the least possible time. Data compilations can also indicate which compounds or groups of compounds need further study, aid in the selection of systems that will repay study by newly devised techniques, and guide efforts to achieve uniformity in reporting the results of research.

A number of authors, including two of ourselves, strove to achieve these goals in compilations of electroanalytical data published between about 1950 and 1965. These differed widely in scope, length, and arrangement, and in the amounts of critical judgment they embodied. Some were limited to inorganic substances and others to organic ones. Some were intended only as representative samples of the available information, others were attempts to present it much more fully. Some were limited to half-wave potentials, to polarography alone, to data obtained in aqueous media, or in other ways; others, especially the more recent ones, attempted greater diversity. We shall not evaluate them in detail, for some of them were our own, all served a useful and important purpose, and we have striven to learn from the merits and defects of each. It is more appropriate to lament their increasing age, which renders them all less satisfactory and less useful than they were when they were published. The simple fact is that the volume of data available is already so large and is growing so rapidly that their collection, evaluation, and selection is no longer a feasible task for any single electrochemist.

That the need for such work was becoming more acute as the possibility of doing it decreased has been a matter of concern to Commission V.5 on Electroanalytical Chemistry of the International Union of Pure and Applied Chemistry for almost 20 years. A subcommission charged with responsibility for electroanalytical data and nomenclature, established as early as 1963 and having first one and then another of us as chairman, attempted for years to encourage, advise, and cooperate with everyone engaged in data compilation in this area. During this time there have been several groups, in as many different countries, that seriously considered engaging in such work, but the magnitudes of the task and of the expenditures of money and time required by work on a worthwhile scale have prevented some of these from actually beginning work and rendered significant achievement impossible for those who made the attempt.

When two of us came together on the staff of Clarkson College of Technology in 1970, we had published 12 different compilations of polarographic data: one of us had been involved in this work since 1950 and the other since 1955. In our common concern for continuing it we were fortunate enough to be able to secure financial support from the National Library of Medicine of the National Institutes of Health. That support enabled us to publish, in 1974, a volume containing data on 2015 organic compounds having 1 to 11 carbon atoms per molecule.

In mid-1974, support for the work was taken over by the National Science Foundation, and a new publisher had to be selected for reasons beyond our control. The first two volumes of the present series were published in 1977. They included the data contained in the 1974 volume, with various changes and corrections, and also gave data on 1848 new compounds having 12 or more carbon atoms per molecule. The third volume was published in 1978 and contained data on 1309 compounds, including 228 that had appeared in the first two volumes and for which additional data (obtained by different techniques, under different conditions, or in different solvents and supporting electrolytes) were given in Volume III. The fourth volume, published in 1980, followed the same pattern as Volume III. It contained data on 731 compounds, of which 190 had also appeared in Volumes I through III. The first four volumes thus dealt with a total of 5485 compounds.

This volume extends the series still further. It follows the same pattern as Volumes III and IV, and includes 528 compounds. There are newer or additional data for 196 compounds that also appeared in previous volumes.

Although these five volumes contain the largest collection of organic electrochemical data ever made, it is essential to describe their limitations because they include only a fraction of the data that have appeared in the original literature. We have excluded electrochemical techniques, such as conductometry, high-frequency conductometry, and dielectrometry*, in which neither the electrical double layer nor any electrode reaction need be considered. These furnish information so widely different from, and so rarely used in conjunction with, that provided by polarography and its congeners that combining them would be difficult and of little use. We have also excluded a number of other techniques, such as electrography and potentiometric, amperometric, and other titration techniques that, although of great analytical utility, do not in general provide fundamental data of lasting importance. Finally, we have excluded potentiometry itself to avoid duplicating the efforts of others who have prepared or are preparing compilations of standard and formal potentials. These exclusions leave a group of approximately 65 closely interrelated techniques, which include polarography, voltammetry, amperometry, controlled-potential coulometry, chronopotentiometry, chronoamperometry, controlled-potential coulometry, chronopotentiometry, chronoamperometry and chronocoulometry, and a host of variants on these such as ac and differential pulse polarography, triangular-wave voltammetry, and potential-step chronoamperometry and chronocoulometry. Table VIII lists the techniques that were used to obtain the data that appear in this volume.

This series of volumes is devoted to data on the behaviors of organic substances and organometallic compounds in which the metal is bound to a carbon atom. The importance of porphyrin and its derivatives to scientists in health-related fields led us to include hemes and related compounds in these volumes, and we also included some substitution-inert complexes in which the ligand is electroactive. A companion series, entitled *CRC Handbook Series of Inorganic Electrochemistry,* deals with inorganic substances, including the complexes of metal ions with organic ligands that are not electroactive, and data on complex compounds included in this volume are cross-referenced there.

Volumes I and II of the *CRC Handbook Series of Organic Electrochemistry* were still further restricted to information published in the 12-year period from 1960 through 1971. Volumes III and IV contained additional material from the same period, and also covered the later period beginning in 1972 and extending into 1976. Volume V follows the same pattern, covering additional information published between 1960 and 1974. Subsequent volumes will extend this coverage in both directions.

Within these restrictions the coverage of the literature in this volume and its predecessors is comprehensive but by no means complete. According to the estimates we gave above, there must have been roughly 30,000 to 35,000 publications in the field between the beginning of 1960 and the end of 1976, and we have not been able to inspect all of these. We have never relied on secondary sources such as *Chemical Abstracts,* and our coverage of the original publications has naturally been influ-

---

* The names used in this compilation are taken from Meites, L., Nürnberg, H. W., and Zuman, P., The classification and nomenclature of electroanalytical techniques, *Pure and Applied Chemistry,* 45, 81, 1976.

enced by the availability of journals and reprints. Although we have occasionally allowed our desire to record the fact that a compound has been found to be amenable to investigation by polarographic and related techniques, and our conviction that further study of its behavior would be worthwhile, to persuade us to include preliminary or dubious results (all of which we have scrupulously attempted to identify as such), we have rejected many more such results than we have included. Our attempt has been to select the best of the available data, the most useful, and the ones most likely to withstand the test of time.

To this end we and our collaborators have repeatedly scrutinized every bit of information recorded here. We have separately evaluated the conformity of the apparatus and experimental techniques to the best modern standards, the accuracy of the handling of the data obtained, the certainty with which products and intermediates were identified, the strength of the evidence given for each suggested reaction course or mechanism, and every other point that the published description of each research allowed us to check. In an effort to minimize inconsistency we have compared the data obtained under different conditions, by different techniques, and for different but related compounds. We have assayed alternative courses and mechanisms proposed by different authors and for related compounds. Despite all these precautions it would of course be idle to pretend that the result is wholly free from errors or contradictions. Some of these are in the original literature and have survived our best attempts to detect and expunge them; others have been introduced in transcription and typing. We can only hope that the merits and utility of this collection are sufficient to induce its users to forgive its defects, and urge those who find errors in the pages that follow to call them to our attention so that others may be apprised of them.

There is, however, one sort of defect for which we must disclaim responsibility: this is the result of the woeful lack of standardization in reporting data that runs rampant through the original literature and that we have been powerless to correct. In some circumstances this is trivial enough to be merely unaesthetic; in others it is an ineradicable blot on the meaningfulness or utility of the work that the data represent. In polarography some authors deduce ratios of n-values for processes that give rise to successive waves on a polarogram by comparing the heights of the waves, while others apply a correction for the dependence of $m^{2/3}t^{1/6}$ on potential by comparing diffusion current constants instead. In Table I, which constitutes the major portion of this volume, these two practices are represented by the entries "i:i" and "I:I", respectively, in the 21st column. Though the second of them is unarguably superior for diffusion-controlled waves, it would be so unlikely that the first would lead to incorrect values of n that the universal adoption of the better expedient would merely improve the beauty and legibility of a tabulation like this one. At the other extreme, there are still authors who report potentials that are referred to mercury-pool and other unpoised "reference" electrodes even though every reputable textbook or monograph published in the field for over a quarter of a century has inveighed against this archaic practice. There are actually even some who fail to describe the compositions of the solvents or supporting electrolytes they have used. Such defects becloud the significance of the data reported and render them far less useful than they might have been for identification, devising analytical methods, and many other purposes. In the present state of the art it has not been possible to exclude all of the data that are thus or otherwise blemished, but we strongly hope that these volumes will help to promote standardization and uniformity so that such blemishes will be less numerous in the raw material gathered for future ones.

We said above that this work owes its existence to the financial support we have received. That of the National Science Foundation is acknowledged in official terms on the title page. But in addition we are glad to have this opportunity to record our personal gratitude to Dr. A. F. Findeis and Dr. Janet Osteryoung at the Foundation: their sympathy with our aspirations and patience with our problems have given us moral support that has meant as much to us as the Foundation's financial support.

This work would also have been impossible without the advice, encouragement, and cooperation that we have received from many of our electrochemical colleagues. First among these are the ones whose names appear on the title page. Dr. Elinore B. Rupp joined the project early in 1975 and played a leading role in every stage of the preparation of this volume. Dr. Theodore L. Fenner joined us at the same time, and his care, diligence, and accuracy as an editor have contributed much to its quality. Dr. Ananthakrishnan Narayanan began work with us in mid-1976; his responsibility has been chiefly to the parallel *CRC Handbook Series in Inorganic Electrochemistry,* of which Volumes I and II have been published. He has also contributed substantially to the present volume. We also owe a continuing debt of gratitude to those who have worked on the project in prior years. These include Dr. William J. Scott, who shared in its birth pangs, and Dr. Bruce H. Campbell and Dr. Alex M. Kardos, who helped to prepare Volumes I and II and to develop most of the procedures employed in preparing this volume. It is a pleasure to be able to express our thanks to all of these for the talents, devotion, and enthusiasm that they brought to tasks that are both tedious and difficult. The bulk of this volume is a reproduction of a typescript prepared by Mrs. Helen Tyler. Her contributions to its appearance and legibility and to its accuracy and reliability are visible whereas her care, dedication, and patient good cheer are not, but all have been equally indispensable throughout our work. We are grateful to Professor Ernest I. Becker for his continued generous advice and help with problems of organic nomenclature. Dr. John J. Rupp assisted us in checking the ACS Registry Numbers, which appear in the present volume as a further aid to identifying the compounds listed. We have also had the benefit of consultation and help from our fellow members of Commission V.5 on Electroanalytical Chemistry of IUPAC, whose encouragement, support, and experience in data compilation in many diverse fields have been invaluable to us. Finally, it is a pleasure to acknowledge the advice and assistance that we have received at various stages from many eminent leaders of the electrochemical community.

Louis Meites
Petr Zuman
Elinore B. Rupp
Potsdam, New York
February, 1980

TABLE OF CONTENTS

General Introduction 1

Table I.	Electrochemical Data	5
Table II.	Structural Formulas	233
Table III.	Courses and Mechanisms of Half-Reactions	247
Table IV.	Compounds Included in Table I	367
Table V.	Functional-Group Index	389
Table VI.	Chemical Abstracts Service Registry Numbers	419
Table VII.	Index of Solvents Employed	425
Table VIII.	Index of Techniques Employed	429
Table IX.	Index of Indicator Electrodes Employed	435
Table X.	Key to Literature Citations	439
Table XI.	Author Index	447

Corrigenda 453

GENERAL INTRODUCTION

We have attempted in these volumes to give as much information as possible in as little space as possible. The information is arranged in ways that are often arbitrary, and it is presented with the aid of symbols and abbreviations whose meanings are often not obvious. This General Introduction, and the introductions to the individual tables that follow, are provided in the hope that they will enable the reader to make full use of the tables in the shortest possible time.

This volume of the *CRC Handbook Series in Organic Electrochemistry* comprises 11 tables. Of these, the longest and most important is Table I, which gives information about the electrochemical behaviors of 528 organic and organometallic compounds and about the experimental techniques and conditions employed in studying them. The other ten tables serve to supplement, cross-index, and interpret the contents of Table I.

Despite the physical bulk of this volume and its predecessors in this series and the stringency with which we applied the criteria, described in the Preface, for restricting their scopes, ruthless abbreviation and condensation have been necessary in compressing the information that is given here into the space available for it. The list of abbreviations and symbols is therefore essential to the correct decoding and interpretation of entries that would be unintelligible without it. One copy of this list appears inside the front and back covers of this volume, and there is a separate copy of it on unbound sheets that can be removed and used to follow the lines across the wide pages of Table I. In preparing this volume we added a few abbreviations to the lists that appeared in Volumes I through IV. Hence the current list can be used in conjunction with those volumes, whereas the earlier lists will occasionally fail to permit the complete interpretation of an entry appearing in this volume.

Each table is preceded by a description of its contents, its organization, and the manner in which it can be used. The present introductory discussion gives a more general description of the volume as a whole, stressing the purpose of each table and the ways in which different tables are related and supplement each other.

Table I contains the purely electrochemical data that have been obtained by polarography and the other techniques listed in Table VIII. There are so many columns that they extend across two facing pages. The division between pages is very nearly such that the left-hand pages identify the compounds for which data are given, the techniques by which the data were obtained, and the electrodes used, and also describe the solvent and supporting electrolyte, apparatus, and experimental conditions; the right-hand pages give the data and other information obtained and provide cross-references to additional information contained in other tables.

The compounds listed in Table I are arranged according to their empirical formulas, and this of course provides one mode of access to the information on any particular compound. Empirical formulas are written in the sequence C, H, and all other atoms in the alphabetical order of their symbols. These formulas are then arranged alphabetically and according to increasing number of atoms; thus, for example, $C_3H_7N_3O_3$ precedes $C_3H_{10}N_2$, and this in turn precedes $C_4H_2Cl_2O_4$. To permit the rapid location of any empirical formula, the empirical formula of the first compound that appears on each pair of facing pages is given in the upper left-hand corner of the left-hand page, and that of the last compound appearing on that pair of pages is given in the upper right-hand corner of the right-hand page. By virtue of these facts, chlorofumaric acid (*trans*-HOOCCCl:HCCOOH or $C_4H_3ClO_4$) can be located quickly on the pair of pages that contain information for compounds having formulas between $C_4H_2Cl_2O_4$ and $C_4H_4N_2$, and then further identified as the third compound on this pair of pages by means of either the empirical formulas listed in the second column or the names listed in the third. Different compounds having the same empirical formulas are listed in the alphabetical order of the names assigned to them.

Another mode of access to the same information is provided by Table IV. This is an index of names and a few common synonyms. It contains the entry "Chlorofumaric acid, EA43". The code number EA43 appears in both the first and last columns of Table I. Both the code number and the empirical formula of the first compound appearing on a pair of facing pages are given in the upper left-hand corner of the left-hand page, while those of the last compound appearing on that pair of pages are given in the upper right-hand corner of the right-hand page. Each code number consists of two letters followed by two digits; in Table I the code numbers appear in the order EA00, EA01, ..., EA99, EB00, EB01, ..., EF28. The code numbers serve not only to abbreviate cross-references between tables and in the tightly packed columns of Table I, but also to distinguish between different compounds that have identical empirical formulas. The first letter of the code number identifies the volume of this series: the letters A and B refer to compounds given in Table I of Volumes I and II, C to those in Volume III, D to those in Volume IV, and E to those in the present volume.

Table V is another index that makes it possible to locate compounds of interest in Table I: it classifies the compounds according to the electroactive (or possibly electroactive) functional groups they contain. One frequently wants to examine data on unsaturated compounds, on carboxylic acids, or on another more or less specific class of substances, and the purpose of Table V is to simplify searches of this sort. For example, chlorofumaric acid appears among the α,β-unsaturated acids and esters and halogen derivatives in Table V, and inspection of the compounds listed under this heading would quickly reveal any other closely related compound that appears in this volume. To minimize false leads in such searches, this index is selective: chlorofumaric acid does not appear among the aliphatic carboxylic acids, because its behavior, so far as it is known, is not representative of compounds bearing a carboxylic group. On the other hand, 4-nitroazobenzene would appear both among the nitro compounds and among the azo compounds because both the nitro group and the azo group are known (or can be assumed) to affect its electrochemical behavior.

Table VI lists the compounds under American Chemical Society Registry Numbers, arranged sequentially. These Numbers, which also appear in Column 3 of Table I, should enable the user to find a compound in Table I without having to be concerned with nomenclature.

Table VII is a list of the solvents that appear in the individual entries of Table I. It permits comparing data on different substances in a single solvent or solvent mixture. The typical entry

Acetone, EA12,EA28,EA62,EA94,EC42

provides rapid and easy access to all of the information that this volume includes about electrochemistry in this solvent. Water is omitted from this index because an entry for purely aqueous solutions would have contained 205 citations and would therefore have been far too long to be of any use.

Table VIII is an index to the techniques employed. For the sake of brevity classical dc polarography is excluded from it, as the dropping mercury electrode is from Table IX, but for each of the other 60-odd techniques that are included in these volumes an entry like

Chronopotentiometry(ER), EA19

provides not only access to all of the chronopotentiometric information that these volumes contain, but also an indication of how much chronopotentiometry has contributed to our knowledge of the electrochemical properties of organic substances during the years that these volumes cover. Since Table VIII lists some techniques for which no citations are given, it may be useful to state the policy we adopted in dealing with data obtained by rarely used techniques. There were a few of these that were rejected because they, or the interpretations based on them, disagreed radically with other information, but such cases were very few and we much more often found ourselves deciding to include an entry because it was obtained by a technique little represented in the data file. Such a selection procedure tends to yield overestimates of the utility and importance of rarely used techniques. The overwhelming preponderance of classical dc polarography is reflected by the omission of an entry for it from Table VIII because that would have consumed more space than we thought reasonable. If one had been included, its length would, if anything, have understated the predominating importance that dc polarography still has in obtaining fundamental electrochemical information.

Table IX is an index to the indicator electrodes employed. It provides access to all of the data obtained with mercury-pool, carbon-paste, rotating disc, and each of the numerous other electrode materials and configurations represented in Table I, with the single exception of the dropping mercury electrode, for which there would have been no less than 369 citations, and which is therefore omitted here as water is omitted from Table VII and polarography is omitted from Table VIII.

Tables IV to IX are thus indexes to Table I. Tables II, III, X, and XI, on the other hand, are supplements to Table I, providing additional information and used with Table I as a starting point. Inspection of these Tables is often necessary for the proper use of Table I. Table II gives, in the order of their code numbers, the structural formulas that could not conveniently be compressed into line form for inclusion in Column 4 of Table I. Table III gives equations that represent the courses and mechanisms of half-reactions, together with values of the rate and equilibrium constants for many of the homogeneous chemical steps involved in these. The equations are given in numerical order and keyed to entries in the fifteenth column (headed "C/M") of Table I; the appearance of a number in that column of Table I in the entry for any compound signifies that information about the course or mechanism of the half-reaction is given opposite the same number in Table III. Table X provides full literature references and is used together with the condensed citations that appear in the fourteenth column (headed "Ref.") of Table I. For each such citation, Table X gives the name of the journal, the volume number, the year of publication, the page number, and the names of the authors. Table XI is an author index which gives, in alphabetical order, the names of all of the authors listed in Table X and, for each, a reference to each citation of his work in Table X. Finally, there is a list of corrigenda giving all of the previously uncorrected errors in prior volumes of which we are aware.

This volume is dedicated to all those whose names appear in Table XI, in recognition of the labor that they have expended on assembling and bringing to its present state the enormous body of knowledge and understanding that is summarized here. Much remains to be discovered, and much of our understanding is still very fragmentary. We hand the volume over to its users in the hope that it will both ease and help to guide their work in the future, as well as the work of those who will follow them.

TABLE I.
ELECTROCHEMICAL DATA

This table consists of 28 columns divided between two facing pages. Here we shall first describe the contents, significance, and arrangement of these columns.

Columns 1 and 28: "Code No."

The contents of these columns are identical so that the user can more easily follow the lines of this very wide table across two pages. The columns give the four-character code number (see explanation given in the General Introduction) assigned to each compound and also indicate whether or not more information about a compound will be found on the following pair of pages.

In many instances the data for a single compound had to be divided between two, and occasionally even among three, pairs of facing pages. To avoid overlooking part of an entry thus divided, the user should observe the following points:

1a. When the last line is blank in Columns 1 and 28, the entry does *not* continue onto the next pair of pages, whereas
1b. An entry that *is* continued overleaf is identified by the symbol "CONT" on the last line in both Column 1 and Column 28.
2a. When the index in the upper left-hand corner of a left-hand page consists solely of a code number and an empirical formula (for example, "EA06, $C_2H_2Cl_2O_2$") the first entry on that page is *not* continued from the preceding pair of pages, whereas
2b. An entry that *is* being continued from the preceding pair of pages is identified by the appearance of "(CONT.)" between the code number and the empirical formula in the upper left-hand corner of the left-hand page (for example, "EA02 (CONT.) CH_2O_2").

For a number of these compounds additional data will be found in an earlier volume of this series. Each compound that also appears in an earlier volume is identified by the appearance, in both Column 1 and Column 28, of the code number assigned to it in each previous volume in which it appeared, but with small letters in place of the capitals used elsewhere. For formaldehyde the entry in these columns.

EA00
aa16
ca05

shows that additional data appear in Volumes I and II under the code number AA16 and in Volume III under the code number CA05. As was stated in the General Introduction, the first letter of a code number identifies the volume to which reference is made. The second letter and the two digits of the code number have only an ordinal significance.

Column 2: "Empirical Formula"

The empirical formulas in this column are given in the customary way, with carbon first, hydrogen second, and then the other elements in the alphabetical order of their chemical symbols, as described in the General Introduction. Deuterated compounds appear immediately after the corresponding hydrogen compounds.

Column 3: "Name and C.A. Number"

This column gives the name and, below it, the Chemical Abstracts Service Registry Number, introduced by the abbreviation C.A. for *Chemical Abstracts,* where available. Some of the names are the simplest and shortest trivial names for the compounds they denote, but most follow IUPAC or *Chemical Abstracts* nomenclature. The Registry Number is given only once for each compound; when an entry continues onto the following pair of pages the name of the compound is repeated but the Registry Number is not. Some compounds have not been assigned Registry Numbers, either because their structures are unknown or because Registry Numbers have not yet been published by *Chemical Abstracts*.

Wiswesser Line Notation (WLN) designations of the compounds were given in Volumes I through IV, but the Registry Numbers appear to be both more easily used and more widely accepted.

Column 4: "Structural Formula"

Only line structural formulas are given here. For many of these compounds, however, the structural formula cannot conveniently be given in line form, and is therefore shown in Table II. The entry "Table II" in Column 4 means that the structural formula may be found in Table II of this volume opposite the appropriate code number taken from Column 1. To avoid duplicating structural formulas in different volumes, the entry "Table II-2", which appears in Column 4, provides a cross-reference to Table II in Volume II, while the entry "Table II-3" is a cross-reference to Table II in Volume III. For the compound for which the entries in Columns 1 and 3 are

EA45 pyridazine
ac43 C.A. 289-80-5
da32

the reference "Table II-2" means that the structural formula may be found in Table II of Volume II with the code number (AC43) given for pyridazine in Column 1.

Column 5: "Solvent"

The symbols used in this column to denote non-aqueous solvents are given in the list of abbreviations. The entry "H_2O" denoted an aqueous solution; an entry like "MeCN" denotes a solution in the nominally pure (and anhydrous) solvent given; in an entry like "MeCN 50" the number denotes the percentage (by volume, unless otherwise stated) of the specified non-aqueous solvent in a solvent mixture, of which the balance is understood to be water. An occasional entry like "EtOH(aq)" reflects our inability to deduce the composition of the solvent mixture from the published information given; such entries appear only when the importance of the data seemed to us to override the laxity of the reporting. Of the entries

MeOH 50 and EtOH 50
C_6H_6 50 Me_2CO 25

the left-hand one denotes a nominally anhydrous mixture containing equal parts by volume of methanol and benzene. The right-hand one denotes a mixture containing 50% (V/V) of ethanol, 25% (V/V) of acetone, and 25% (V/V) of water; when the sum of the percentages given for the constituents of any solvent mixture is less than 100, the balance is always understood to be water.

Aqueous solutions are listed first, then alcohols, in order of increasing molecular weights, and finally other solvents in the order in which they appear on the list of abbreviations. Entries for mixtures of a non-aqueous solvent with water immediately precede those for the corresponding anhydrous solvent. A few data that seemed worth including were taken from papers that did not fully describe the composition of the solvent; for these data the symbol "ns" (for "not stated") is given in this column.

Column 6: "Technique"

The technique is identified by a two-letter symbol that can usually be interpreted by means of the list of abbreviations. Polarography ("PY") appears first in the data for any solvent, and the other techniques follow in the alphabetical order of the symbols assigned to them. Some of the information in this volume has been obtained by techniques as yet so rarely used that we have hesitated to assign symbols to them until further evidence of their utility appears. Each such technique is identified by the letters "XT" in Column 6, and Column 27 always gives its name on the same line in the form "XT = second-derivative chronoamperometry with linear potential sweep". In assigning such names we have followed the principles embodied in the 1975 IUPAC recommendations regarding the nomenclature of electroanalytical techniques (*Pure and Applied Chemistry*, **45**, 81, 1976).

Column 7: "Medium"

This column describes the composition of the supporting electrolyte. Concentrations are given in moles per cubic decimeter wherever possible: the entry "KOH 0.1" denotes 0.1 *M* potassium hydroxide. Britton-Robinson, McIlvaine, and other commonly used buffers, both mixed and simple, are identified by means of abbreviations. The word "buffer" means that the solution was said to be buffered at the pH quoted in Column 9, but that no information was given about the composition of the buffer employed. Maximum suppressors are identified in this column and their concentrations are given in weight/volume % (grams per 100 cm^3).

Column 8: "μ, M"

This gives the ionic strength of the solution, in moles per cubic decimeter, except where it is fully specified by the entry in Column 7. A dash in this column means that no information about the ionic strength is available in addition to that given in Column 7.

Column 9: "pH"

Values of the pH are generally given to the nearest 0.1 unit. An entry like "5.7→6.2" would signify that the pH-value changed over the range given during the course of a measurement; one

TABLE I. Electrochemical Data

like "5—7" means that all of the data in the columns that follow are independent of pH over the stated range. The entries for any given medium are usually arranged in order of increasing pH. A dash in this column means that no information about the acidity of the solution is available in addition to that in Column 7.

Column 10: "T, °C"

This column gives the temperature at which the data were obtained and, where possible, the precision with which it was maintained. A dash in this column means that no information about the temperature of the solution was given in the original publication.

Column 11: "Electrodes"

The indicator or working electrode is given first; then, after a solidus or virgule ("/") or on a second line, the reference electrode. In three-electrode configurations the nature of the auxiliary or counter electrode is given in Column 13.

Reference electrodes are divided into two groups. One comprises the saturated calomel electrode, its variants (such as the "lithium S.C.E.", $Hg/Hg_2Cl_2(s)$, LiCl(s), and others of the same ilk), and the normal hydrogen electrode. These are almost invariably prepared with water, so that their use with a non-aqueous solution entails a liquid-junction potential between the non-aqueous solution of the compound being studied and the aqueous solution in the reference electrode. Some workers have sought to circumvent this by preparing similar electrodes in the same solvents or solvent mixtures that contain the compounds they study; when this has been done, the symbol "(o)" (for "organic") follows the abbreviation that would denote the ordinary aqueous form of the reference electrode.

The other group comprises silver — silver halide electrodes, mercury pools, metal — metal-ion electrodes, and others normally prepared in the solvent used for the compound being studied (and often, indeed, employed as internal "reference" electrodes). For such an electrode, the abbreviation alone signifies that the solvent was the same throughout the cell, while the symbol "(aq)" following the abbreviation signifies that the reference electrode was prepared with water and used as an external reference electrode.

Some authors have used "pilot ions", whose half-wave potentials in non-aqueous solvents have been reliably established in prior work by themselves or others, to provide internal reference potentials and minimize problems like those that arise from variations of liquid-junction potentials with time. Potassium ion, rubidium ion, ferrocene, ferricinium ion, and several others have been used. This practice gives rise to entries of the form

$$\text{DME/MP} \atop \text{K}^+$$

which denotes the use of a dropping mercury electrode as the indicator electrode, of a mercury pool as the counter electrode, and of potassium ion as the added pilot ion.

A dash in place of the symbol for one electrode (e.g., "DME/-") means that the nature of that electrode was not given in the original paper.

Column 12: "App."

Some salient features of the apparatus are summarized in this column. The basic code is a three-position alphanumeric, consisting of one or more digits that describe the cell, one or more letters that indicate whether and how dissolved oxygen was removed, and a final digit that identifies the technique of data acquisition. By means of the list of abbreviations (in which the abbreviations used in this column appear at the very beginning), the typical entry "0A0" may be decoded to find that a two-electrode cell was used, that deaeration was accomplished with a stream of inert gas, and that the data were recorded on a pen-and-ink recorder. Inert gases are not identified because that would, in our judgment, consume more space than it would be worth, but they do not include hydrogen because hydrogen is sometimes not inert. More elaborate entries, such as "23CD3", are occasionally necessary; this one denotes the use of a three-electrode cell with an agar bridge, deaeration by both hydrogen and sulfite, and digital data-acquisition. When insufficient information was given to permit completing any part of this three-position code, a dash always appears in the appropriate position, as in the entry "0-0". This permits easy differentiation between, for example, "--0" and "0--".

Column 13: "Experimental Parameters"

Some of the information given in this column, such as the concentration C of the compound studied (which is always given in millimoles per cubic decimeter), would be significant no matter what technique was employed; some of it is governed by the nature of that technique. For polarographic data we give, wherever possible, values of the drop time t (in seconds), the rate of flow of mercury m (in milligrams per second), and the height h (in centimeters) of the column of mercury above the capillary tip.

The drop time for a dropping electrode with enforced mechanical detachment is denoted by the symbol "t(c)"; in the absence of the "(c)" (for "controlled") the value of t, like that of m, should be regarded only as an approximation, for there is little agreement about the conditions under which they are measured and the ways in which it is reported. An author chiefly interested in mass-transfer phenomena is apt to measure t at a potential on the plateau of the wave, one chiefly interested in charge-transfer phenomena is apt to measure it at the half-wave potential, and one studying a number of different compounds having different half-wave potentials is apt to measure it at the potential of the electrocapillary maximum (open circuit) or at the potential of the reference electrode employed (short circuit). Similarly, the value of m may be measured at the same potential as that of t, but often it is measured at the potential of the electrocapillary maximum while t is measured at a different potential and sometimes it is measured with the tip of the capillary immersed in mercury. In view of this diversity, exact specification would have been difficult and wasteful of space, and the difficulty would have been compounded in dealing with papers that report values of t and m without specifying how they were measured. Consequently we have usually foregone any attempt at greater exactness; the user who needs more information must seek it in the original literature. The value of h serves as a crude but useful indication of whether the capillary was a conventional one or not: an abnormally high or low value would signify that there was something unusual about the capillary, and the original literature should again be consulted for details.

Sometimes the value of $m^{2/3}t^{1/6}$ (abbreviated "EC") is given, always in $mg^{2/3}s^{-1/2}$, when this was stated in the paper cited in lieu of the value of t or m or both. What was said above about the difficulty of quoting, and often even of deducing, the potential at which a value of m or t was measured applies equally to the potential at which $m^{2/3}t^{1/6}$ was evaluated.

For other techniques other kinds of data are given instead. For chronopotentiometry these include the area of the indicator electrode and the current or current density; for stationary-electrode voltammetry they include the area of the indicator electrode and the scan rate; for cyclic voltammetry they include the starting and reversal potentials and the area of the indicator electrode, and so on. The list of abbreviations must always be consulted regarding the units of the quantities given in this column. In the space that was available for these purposes it is quite impossible to give a full description of the experimental conditions, but an attempt has been made to give an accurate idea of their nature.

This column is also used to give the identity of the auxiliary or counter electrode when a three-electrode configuration was used, and to indicate whether the indicator or working electrode was subjected to chemical, mechanical, or electrolytic pretreatment.

The symbol "ns" (for "not stated") is often used in this column to mean that information was not available about the concentration of the compound studied ("C=ns") or about important experimental parameters.

Column 14: "Ref."

Each entry in this column is a nine-character alphanumeric of which two letters and three digits appear on the first line while the remaining four digits appear on the second line. The first two letters denote the journal; these abbreviations do not appear in the list of abbreviations, but may be deciphered with the aid of Table X. The next three digits give the volume number, packed to three digits by the addition of leading zeros if necessary. The four digits on the second line give the page number, again with leading zeros added if necessary to yield a total of four digits. Since the letters "AA" denote *"Analytica Chimica Acta"*, the typical entries

AA072 and AA080
0169   0017

would refer to papers published in *Anal. Chim. Acta,* 72, 169 and *Anal. Chim. Acta,* 80, 17, respectively. The year is not given here but may be obtained by consulting Table X, which gives the names of the authors as well. In a reference to a journal that does not employ volume numbers, the two leading digits "19" are dropped from the number of the year, and the two remaining digits (e.g., 66 for a paper published in 1966) are followed by a letter. This is "O" if, like the *Bulletin de la Société Chimique de France,* the journal has no subsections; otherwise it denotes the subsection of a journal like the *Journal of the Chemical Society (London)*. Thus the entries

JL66B and BF63O
0103   2252

would denote *J. Chem. Soc., Sec. B,* 103 (1966) and *Bull. Soc. Chim. Fr.,* 2252 (1963), respectively.

TABLE I. Electrochemical Data

To the shame chiefly of the authors, but also to some extent of the referees, editors, and journals involved, we have found a number of cases in which nearly, and sometimes exactly, identical papers have been published in two or three different journals. This practice has been repeatedly criticized even in the formerly common situation where publication in a local or regional journal in the language of a small country was followed by publication in an international journal having a much wider circulation and in a more widely known language. Repeated publication of the same material in journals of comparable circulation and in the same language seems to us to defraud the scientific community. In such cases we have simply suppressed all of the citations save one; we prefer the risk of being criticized for having covered the literature a little less thoroughly than we have actually done to that of allowing the offenders to profit from actions that we must hope to have been ill-judged rather than something more reprehensible.

Columns 15 through 26

Information that belongs in any of these columns but that was not given in the original paper is represented by a dash on the first line of an entry to aid the user in following separate entries across the page.

Column 15: "C/M"

A number that appears in this column is a cross-reference to Table III. Opposite the same number in that table, equations are given that describe the course or mechanism of the half-reaction. Table III may also contain other information, such as the rate constants of homogeneous chemical steps in the overall mechanism; its introduction should be consulted for further details.

The numbers that appear in Column 15 are not in numerical order because virtually identical equations may often be written for the reductions or oxidations of a number of compounds that have widely different empirical formulas and are therefore widely scattered in Table I. Cross-references in Table III list the code numbers of other compounds to which the same mechanism, or a closely related one, is applicable. For further details see the introduction to Table III.

Columns 16—18: "Characteristic Potential"

This term denotes a potential whose nature depends on the technique used. Typical characteristic potentials are the half-wave potential in polarography, the quarter-transition-time potential in chronopotentiometry, and the peak or half-peak potential in linear-sweep (stationary-electrode) voltammetry. Regardless of its nature, the characteristic potential always depends on the identity of the electroactive substance, on the kinetics or thermodynamics of the electron-transfer process, and of course on the experimental conditions; for any particular technique and under any completely defined set of experimental conditions the value of any characteristic potential is a reproducible property of the electroactive substance.

Column 16 gives the symbol of the characteristic potential whose numerical value (in volts) appears in Column 17. Column 18 identifies the reference electrode to which that value is referred; this is not necessarily the same as the reference electrode used for the experimental measurements, which was identified in Column 11.

As is also true of the techniques employed, some of the characteristic potentials appearing here have not yet been assigned individual symbols. Each of these is identified by the letters "XE" in Column 16, and Column 27 always defines it on the same line in the form "XE = potential where $d^2i/dE^2 = 0$".

Some investigators report values of the half-wave potential obtained by triangular-wave voltammetry and other techniques that do not yield experimental values of the half-wave potential. These values are of dubious significance and are identified by the symbol "$E_{1/2}(?)$" in Column 16.

Columns 19 and 20. "Response Const."

This generic term may denote either the measured value of the independent variable under the experimental conditions employed or some function of that value: as with the characteristic potential, the nature of the response constant depends on the technique used, and in addition it depends on the behavior of the system being studied. For a diffusion-controlled process the preferred polarographic response constant is the diffusion current constant $I = i_l/Cm^{2/3}t^{1/6}$, but the ratio i_l/C is often given instead when a value of $m^{2/3}t^{1/6}$ could not be obtained from the original, and even values of the limiting current i_l alone are sometimes quoted. Values of i_l/C and i_l provide at least some indication of the relative heights of different waves. Wherever possible, values of i_l are accompanied (in Column 27) by an indication of whether the current is controlled by diffusion or some other process. For diffusion-controlled processes the preferred response constants in chronopotentiometry and stationary-electrode voltammetry are $i\tau^{1/2}/AC$ and $i_p/v^{1/2}AC$, respectively, given here in terms of current densities $j\tau^{1/2}/C$ and $j_p/v^{1/2}C$ for compactness. We have attempted to give all currents in microamperes, concentrations in millimoles per cubic decimeter, and areas in square centimeters, but as currents are sometimes reported in arbitrary units such as millimeters of recorder deflection we have occasionally had to quote these, using the symbol "(u)" to denote their limited significance.

For processes that are not diffusion-controlled, and even for diffusion-controlled processes when some techniques are used, closed-form descriptions of the preferred response constants may not be available, and this is another reason why currents, current densities, current- or current density-concentration ratios, and other incomplete response "constants" appear in these columns. There are even some authors who have reported diffusion coefficients in place of the response constants obtained experimentally; the justification for doing so is meager, but we have sometimes been forced to quote these values for want of anything with a more solid basis.

Column 19 identifies the response constant whose value is given in Column 20. Units are given in the list of abbreviations: a polarographic diffusion current constant identified by an "I" in Column 19 is always in $\mu A\ mmol^{-1} dM^3 Mg^{-2/3} S^{1/2}$, a calculated diffusion coefficient identified by a "D" in Column 20 is always 10^6 times the reported value in $cm^2 s^{-1}$, and so on.

Some rarely used response constants have not been assigned individual symbols but are identified by the letters "XI" in Column 19 and defined on the same line in Column 27.

Columns 21 and 22: "n"

Usually the value of n appearing in Column 22 is the total number of electrons involved in the overall half-reaction that consumes one molecule or ion of the electroactive substance, and the abbreviation appearing in Column 21 identifies the technique by which that value was obtained. Values obtained by controlled-potential coulometry ("QE" in Column 21), however, generally pertain to the experimentally determined ratio of the number of faradays consumed to the number of moles of electroactive substance taken or consumed; in the original literature this ratio is often denoted by a symbol like "n_{app}" to emphasize its origin. Nonintegral values of n obtained by controlled-potential coulometry and other techniques reflect the occurrence of coupled chemical reactions that produce or consume electroactive material during a measurement.

Columns 23—25: "Electrokinetic Data"

Information about charge-transfer kinetics is given in these columns in various ways whose diversity again reflects the lack of standardization in the literature. Some authors calculate and report values of the symmetry parameter αn_a (usually abbreviated here as αn), the heterogeneous rate constant at 0 V vs. N.H.E. ($k_{f,h}$) or at the standard or formal potential of the couple ($k_{s,h}$), or of other parameters similar to these; others report values of more directly accessible quantities, such as (in polarography) the slope of a plot of E vs. $\log[i/(i_d - i)]$ or $E_{3/4} - E_{1/4}$. Some calculate and report the number of hydrogen or hydroxide ions consumed in the rate-determining electron-transfer step and the steps that precede it; some report the slope of a plot of the characteristic potential against the pH. Column 23 gives the symbol of the parameter whose value is given in Column 24, while Column 25 identifies the experimental data from which that value was deduced.

Column 26: "Products and Identification"

This column indicates what final products were obtained and what techniques were used to isolate and identify them. Products are usually identified by names; only for unusual structures (e.g., some radical ions) are their formulas given. The percent yield of the product, when available, follows its name. Techniques used for isolation and identification are indicated by abbreviations given in the list of abbreviations. Additional information, such as the lifetime of a radical, is sometimes included.

Column 27: "Description and Remarks"

This column, in which abbreviations are very dense, indicates whether the wave or other signal is cathodic (reduction) or anodic (oxidation), whether it is well- or ill-defined, whether or not it shows a maximum or some other reproducible deviation from the idealized simple form, whether the response is controlled by mass-transfer or some other process such as adsorption, and gives other information not provided in the preceding columns, including definitions of the special symbols "XT", "XE", and "XI" where these appear in Columns 6, 16, or 19 respectively. Correlations of measured quantities, such as the half-wave potential, with structural parameters (such as the exchange integral β, the Hammett substituent constant σ, or the Taft polar constant σ^*) are often indicated here. Relationships between measured quantities, such as the current, and experimental variables, such as the concentration, are indicated together with the limits of their validities. Sometimes the nature of the electrochemical process is indicated by entries such as "redn. of H^+" (which means that the process consists of the direct or catalytic reduction of hydrogen ion rather than the reduction of the organic species) or "redn. of maleic acid", which for a maleate salt means that it is not the organic cation but the maleate anion that is responsible for the behavior observed. In addition, this column gives, very briefly, our assessment of the reliability of the data and interpre-

TABLE I. Electrochemical Data

tation. Usually we have confined ourselves to stating whether we believe the information and conclusions to be reliable and reasonably accurate and precise, of limited precision, or in need of further confirmation, but occasionally we have indicated doubts concerning individual components of the entry. The symbol "(?)" is used to signify that we doubt the correctness of the information it follows, while the symbol "(!)" usually means either that the information reported cannot be validly obtained in the manner specified, as for example when the slope of a "log plot" is adduced as evidence for an overall n-value, or that we are dubious of its significance.

When data reported in one of the papers abstracted for this volume were thought to be inferior to data reported in a previous volume for the same compound, only the name of the compound and the experimental conditions are given, and a cross-reference (in the form "see AG07") to the earlier data appears in this column.

EA00 CH_2O CRC Handbook Series in Organic Electrochemistry

Code No.	Empirical Formula	Name and C.A. Number	Structural Formula	Solvent	Tech.	Medium		μ, M	pH	T, °C	Electrodes	App.	Experimental Parameters
EA00 aa16 ca05	CH_2O	formaldehyde C.A. 50-00-0	HCHO	H_2O	PY	BOR		-	8.2	20	DME/SCE	0-0	C=4, m=0.806, t=2.88, h=68
					PA	BOR KCl	0.03 1.3	-	9.3	20	DME/SCE	0-2	-0.95 → -1.85 → -0.95 V, C=217, m=0.227, t=48, h=46, v=66.7
					VA	PHOS	0.1	-	6.8	25±2	SMDE SCE	2A-	C=50, A=2.5E-2, v=1E5, Pt Aux
						BOR	0.023		9.0				
					VY	PHOS	0.1	-	6.8	25±2	RMCuD SCE	2AO	C=20, A=3.431, v=10-300, ν=17-35, Pt Aux
						BOR	0.023		9.0				
						NaOH KCl	0.145 0.055		13				
				EtOH 2	PY	LiOH LiCl	0.1 0.4	-	-	25± 0.02	DME Hg/Hg_2Cl_2, $LiCl_2$	034A 0	C=0.2, m=2.59 ±0.37, t=3.55 ±0.45, h=65±5
EA01	CD_2O	formaldehyde-d_2 C.A. 1664-98-8	DCDO	D_2O	PY	BOR, deuterated		-	8.4	20	DME/SCE	0-0	C=4, m=0.806, t=2.88, h=68
						NaOD	ns		12.96				
									13.98				
									14.8				
					PA	BOR, deuterated KCl	0.03 1.3	-	-	20	DME/SCE	0-2	-0.95 → -1.85 → -0.95 V, C=555, m=0.227, t=48, h=46, v=66.7
													v=5.3E3
EA02 aa17	CH_2O_2	formic acid C.A. 64-18-6	HCOOH	H_2O	VA	H_2SO_4	0.5	-	-	-	Pt/NHE	2AO	0 → 0.76 → 0 V, C=1E3, v=143
													0 → 0.83 → 0 V, C=1E3, v=143
													0 → 0.89 → 0 V, C=1E3, v=143
						H_2SO_4	0.5	-	-	25± 0.1	PBE/NHE	24AO	0 → 1.5 → 0, C=100, v=140

CONT

TABLE I. Electrochemical Data CH_2O_2 (CONT.) EA02

Ref.	C/M	Charact. Potential Value	vs.	Response Const.	Value	n Tech.		Electrokinetic Data Parameter	Value	From	Products and Identification	Description and Remarks	Code No.
JE046 0051	32 b	-	-	i_ℓ	0.16F	-	-	-	-	-	-	C, i_k, max. i_ℓ at pH=13.2, similar results in D_2O, r	EA00 aa16 ca05
JE046 0051	32 b	$E_{\frac{1}{2}}$	-1.55F	SCE	i_ℓ	60F	-	-	-	-	-	-	C.a.s., $i_\ell=0.2 v^{\frac{1}{2}}+k_2$ for $v \geq 6E3$, W, r A, 0	
JE036 0109	32 f	E_p	-1.6F	SCE	i_p	138F	-	-	-	-	-	-	C, p A, 0	
			-1.74F			138F	-	-	-	-	-	-	C, p	
			-1.55F			87.5F	-	-	-	-	-	-	A	
JE036 0109	32 f	$E_{\frac{1}{2}}$	-1.44F	SCE	i_ℓ	750F	-	-	Elog	50	-	-	$C, i_\ell=kC$, r	
			-1.53F			2600F	-	-		46	-	-	$C, i_\ell=kC$, r	
	32 g		-1.84F			6.3E4F	-	-		50	-	-	C, r	
JE046 0323	32 bd	$E_{\frac{1}{2}}$	-1.72F	SCE	-	-	-	-	-	-	-	-	$C, \Delta E_{\frac{1}{2}}=0.2\sigma^*$, r	
JE046 0051	32 b	-	-	-	i_ℓ	0.12F	-	-	-	-	-	-	C, i_k, r	EA01
		-	-	-		1.6F	-	-	-	-	-	-	C, i_k, r	
		-	-	-		2.4F	-	-	-	-	-	-	C, i_k, r	
		-	-	-		0.74F	-	-	-	-	-	-	C, i_k, r	
JE046 0051	32 b	E_p	-1.62F	SCE	i_p	76.7F	-	-	-	-	-	-	C, r A, 0	
			-1.82F			410F	-	-	-	-	-	-	$C, i_p=1.3v^{\frac{1}{2}}+k_2$ for $v \geq 6E3$, r	
			-1.55F			small	-	-	-	-	-	-	A	
JE045 0205	-	E_p	0.51F 0.51F	NHE	i_p	89.2F 89.2F	-	-	-	-	-	-	A, p A, occurs in reverse scan	EA02 aa17
			0.51F -			104.6F -	-	-	-	-	-	-	A, p A, occurs in reverse scan	
			≈0.0F												
			0.48F			147.7F	-	-	-	-	-	-	A, occurs in reverse scan	
			0.51F -			106F -	-	-	-	-	-	-	A, p A, occurs in reverse scan	
			0.8F			212.3F	-	-	-	-	-	-		
			0.51F			212.3F	-	-	-	-	-	-	A, occurs in reverse scan	
JE044 0239	-	E_p	0.55F 0.9F 1.45F 0.61F	NHE	i_p	46.2F 83.1F 99.2F 169F	-	-	-	-	-	-	A, i_a, r A A A, occurs in reverse scan, i_a	
														CONT

EA02 (CONT.) CH_2O_2

Code No.	Empirical Formula	Name and C.A. Number	Structural Formula	Solvent	Tech.	Medium		μ, M	pH	T, °C	Electrodes	App.	Experimental Parameters
EA02 aa17	CH_2O_2	formic acid	HCOOH	H_2O	VA	H_2SO_4	0.5	-	-	25±0.1	RhBE NHE	24A0	$0 \to 1.0 \to 0$ V, C=100, v=70
													$0 \to 1.5 \to 0$ V, C=100, v=140
											PdWE NHE		$0 \to 1.5 \to 0$ V, C=100, v=70
													C=1E3
EA03	CH_3NO	formaldehyde oxime C.A. 75-17-2	$H_2C:NOH$	H_2O	PY	CITR GEL	0.1 0-0.01	-	1.4	-	DME/SCE	04A0	m=1.3, t=2.6, h=54
						AM		0.2	9.24				
									10.5				
						PYR			11.20				
						KOH	ns		12.10				
									13.15				
EA04 aa36	CH_6N_2	methylhydrazine C.A. 60-34-4	CH_3NHNH_2	H_2O	PY	NaOH NaClO$_4$		1	12.6	30±0.2	DME Hg/HgO NaOH 1	2-0	C=ns, m=2.79, t=3.0, h=28.2, Pt Aux
									13.8				
					VA	H_2SO_4	0.5	-	-	-	Au/MSE	2A0	$0 \to 1 \to 0$ V, C=6, A=0.17, v=1E2, PE
					VY	H_2SO_4	0.25	-	-	-	RAuE MSE	2A0	C=0.2, A=0.17, v=5, ω=135, PE
													C=4
EA05 aa41	$C_2HF_3O_2$	trifluoroacetic acid C.A. 76-05-1	CF_3COOH	DMF	PY	Et_4NClO_4	0.1	-	-	≈21	DME/SCE	236A0	C=1, m=2.28, t=3.5, h=50, MP Aux

14

TABLE I. Electrochemical Data $C_2HF_3O_2$ EA05

Ref.	C/M	Charact. Potential		Response Const.		n		Electrokinetic Data			Products and Identification	Description and Remarks	Code No.
		Value	vs.		Value		Tech.	Parameter	Value	From			
JE044 0239	–	E_p	0.1F 0.68F 0.22F	NHE	i_p	18.5F 87.7F 84.6F	–	–	–	–	–	–	A,r A A,occurs in reverse scan	EA02 aa17
			0.05F			15.7F	–	–	–	–	–	–	A,oxidn. of H_2, $i_p \downarrow$ as [HCOOH] $\uparrow$,r	
			0.77F 0.25F 0.04F			54.3F 51.4F 25.7F							A C C,$H^+ \rightarrow H_2$(ads), $i_p \downarrow$ as [HCOOH] $\uparrow$	
			0.33F 0.56F			166F 70F	–	–	–	–	–	–	A,r C,also occurs if [HCOOH]=0	
			0.33F			138.5F							A,occurs in reverse scan	
			0.36F 0.36F			1260F 815F	–	–	–	–	–	–	A,r A,occurs in reverse scan	
C0026 0052	–	$E_{\frac{1}{2}}$	–1.0	SCE	–	–	–	–	–	–	–	–	C,Mx,dr,$i_\ell=f(t)$,p	EA03
			–0.03		–	–	–	–	$dE_{\frac{1}{2}}/dpH$	54	–	–	A,$i_\ell=f(t)$,p	
			–0.08		–	–	–	–		54	–	–	A,$i_\ell=f(t)$,p	
			–0.12		–	–	–	–		54	–	–	A,$i_\ell=f(t)$,p	
			–0.18		–	–	–	–		54	–	–	A,$i_\ell=f(t)$,p	
			–0.24		–	–	–	–		54	–	–	A,$i_\ell=f(t)$,p	
JE034 0081	385 b	$E_{\frac{1}{2}}$	–0.29	SCE	–	–	–	–	$dE_{\frac{1}{2}}/dpH$	78 ± 2	plot	–	A,r	EA04 aa36
			–0.381F		D I	10.4±0.2 7.32	–	–		78 ± 2		–	A,$i_\ell=k_1C + k_2$ for C= 0.3-16, $E_{\frac{1}{2}}=k_1t + k_2$ for t=0.3-6, Tc=1.63 for T=5-40, i_d, r	
JE038 0381	385 a	E_p	0.22F	MSE	j_p	1750F	–	–	$d\log i_p/d\log v$ dE_p/dv	0.52 50	–	–	A,≠,p	
			0.5F			750F				–			A,occurs in reverse scan	
JE038 0381	385 a	$E_{\frac{1}{2}}$	0.11F	MSE	j_ℓ	620F	–	–	$d\log i/dE$ at foot of wave	10.0-11.5	sttd	–	A,$i_\ell \rightarrow 0$ for E > ~0.5 V,p	
			0.23F			5400F	–	–		–	–	–	A,$i_\ell \neq f(C,\omega)$, i_a, p	
			0.50F			4100F							A,$I_\ell \rightarrow 0$ for E > 0.7 V, $i_\ell = k_1C + k_2$, $i_\ell = k_1\omega^{\frac{1}{2}} + k_2$	
JE049 0085	–	$E_{\frac{1}{2}}$	–1.46	SCE	–	–	QE	1	–	–	–	CP → no F^-	C,i_d,p	EA05 aa41

Code No.	Empirical Formula	Name and C.A. Number	Structural Formula	Solvent	Tech.	Medium		μ, M	pH	T, °C	Electrodes	App.	Experimental Parameters
EA06 aa45	$C_2H_2Cl_2O_2$	dichloroacetic acid C.A. 79-43-6	$CHCl_2COOH$	DMF	PY	Et_4NClO_4	0.1	-	-	≈21	DME/SCE	236A0	C=ns, m=1.26, t(oc)=5.77, h=70, MP Aux
EA07	$C_2H_2F_2O_2$	difluoroacetic acid C.A. 381-73-7	CHF_2COOH	DMF	PY	Et_4NClO_4	0.1	-	-	≈21	DME/SCE	236A0	C=1, m=2.28, t=3.5, h=50, MP Aux
EA08 aa49 da05	$C_2H_2O_4$	oxalic acid C.A. 144-62-7	$(\cdot COOH)_2$	H_2O	IL	HOAc	1	-	2.3	25	PGE/SCE	-	-
EA09	$C_2H_3ClO_2$	chloroacetic acid C.A. 79-11-8	$CH_2ClCOOH$	DMF	PY	Et_4NClO_4	0.1	-	-	≈21	DME/SCE	236A0	C=0.5, m=1.26, t(oc)=5.77, h=70, MP Aux
EA10	$C_2H_3Cl_3N_2O$	trichloroacetamidoxime C.A. 2533-67-7	$CCl_3C(:NOH)NH_2$	EtOH 25	PY	BR		-	2.1	-	DME/SCE	04A0	C=0.5
									3.3				
									4.6				
EA11	$C_2H_3FO_2$	fluoroacetic acid C.A. 144-49-0	CH_2FCOOH	DMF	PY	Et_4NClO_4	0.1	-	-	≈21	DME/SCE	236A0	C=ns, m=2.28, t=3.5, h=70, MP Aux
EA12	$C_2H_3KOS_2$	potassium methyldithiocarbonate C.A. 2667-20-1	$CH_3OC(S)SK$	Me_2CO	PY	Et_4NClO_4	0.1	-	-	20±0.1	DME Ag/AgCl, LiCl 0.1	12A0	C=1, Pt Aux
						Et_4NClO_4 HCl	0.1 trace						
EA13	C_2H_3N	acetonitrile C.A. 75-05-8	CH_3CN	MeCN	CP	Bu_4NClO_4	0.1	-	-	20	Sn/SCE	-	E_{app} = -2.87

TABLE I. Electrochemical Data　　　　C_2H_3N　EA13

Ref.	C/M	Charact. Potential		Response Const.		Tech.	n	Electrokinetic Data			Products and Identification	Description and Remarks	Code No.
		Value	vs.		Value			Parameter	Value	From			
JE044 0025	377 b	$E_{\frac{1}{2}}$	-1.32	SCE	-	-	QE	1	-	-	-	$CP \to 0.9\ Cl^-/$ $CHCl_2COOH$	$C, E_{\frac{1}{2}}$ and $i_\ell \neq$ $f[H^+\ \text{donor}], r$	EA06 aa45
			-2.49					-					$C, E_{\frac{1}{2}}$ and $i_\ell \neq$ $f[H^+\ \text{donor}]$	
JE049 0085	-	$E_{\frac{1}{2}}$	-1.72	SCE	i_ℓ	0.5F	QE ⎫	≈1	-	-	-	$CP \to$ no F^-	$C, i_\ell \neq (h), i_k, H, p$	EA07
			-2.14			1.4F	⎭	-					$C, i_\ell = kh^{\frac{1}{2}},$ CHF_2COOH redn.	
JE055 0069	-	-	-	-	-	-	-	-	-	-	-	CO_2	see AA49, DA05	EA08 aa49 da05
JE044 0025	377 e	$E_{\frac{1}{2}}$	-1.93	SCE	i_p	1.3F	QE	1	-	-	-	$CP \to 1.0\ Cl^-/$ $CH_2ClCOOH$	$C, \neq, i_d, i_\ell = kC$ and $kh^{\frac{1}{2}}, i_\ell$ and $E_{\frac{1}{2}} \neq$ $f[H^+\ \text{donor}], r$	EA09
C0026 2438	-	$E_{\frac{1}{2}}$	-0.50F	SCE	-	-	sttd	2	-	-	-	-	C, i_d, p	EA10
			-					2					C	
			-					2					C	
			-0.56F		-	-		-				-	C, i_d, p	
			-					2					C	
			-					2					C	
			-0.62F		-	-		-				-	C, i_d, p	
			-					2					C	
			-					2					C	
JE049 0085	-	$E_{\frac{1}{2}}$	-2.27	SCE	-	-	QE	0.97	-	-	-	$CP \to$ no F^-	$C, i_\ell = kC$ and $kh^{\frac{1}{2}}, p$	EA11
JE048 0071	333 a	$E_{\frac{1}{2}}$	-0.354	Ag/AgCl, LiCl 0.1	-	-	-	-	Tomeš	63	-	-	$A, i_\ell = kh^{\frac{1}{2}}, i_d, i_\ell = kC$ for $C=0.5-10, E_{\frac{1}{2}} \to$ more neg. as $C \uparrow, r$	EA12
			0.09							76			$A, i_\ell = kh^{\frac{1}{2}}, i_d, i_\ell = kC$ for $C=0.5-10, E_{\frac{1}{2}} \to$ more pos. as $C \uparrow$	
			-							=			A, M	
			0.36F		$i_\ell(u)$	11F	=	=	=	=		=	A, p	
			0.2F			27F	=	=	=	=			A	
JE043 0308	-	-	-	-	-	-	-	-	-	-	-	tetramethyltin, GLC; butane	C, p	EA13

EA14 $C_2H_4N_2O_2$

Code No.	Empirical Formula	Name and C.A. Number	Structural Formula	Solvent	Tech.	Medium		μ, M	pH	T, °C	Electrodes	App.	Experimental Parameters
EA14 da06	$C_2H_4N_2O_2$	oxamide C.A. 471-46-5	$(H_2NCO\cdot)_2$	H_2O	PY	KCl GEL	0.2 0.01	-	3.2	25	DME/SCE	0A0	C=0.85,m=1.6, t=4.5,h=40
									5.65				
EA15	$C_2H_4N_2S_2$	dithiooximide C.A. 79-40-3	$[NH_2C(S)]_2$	H_2O	PY	KCl GEL	0.2 0.01	-	5.65	25	DME/SCE	0A0	C=8,m=1.6, t=4.5,h=40
EA16 aa74	C_2H_4O	acetaldehyde C.A. 75-07-0	CH_3CHO	EtOH 2	PY	BARB LiCl		0.5	8.0	25± 0.02	DME/SCE	034A0	C=0.2,m= 2.59±0.37, t=3.55±0.45, h=65±5
						BOR LiCl			9.8				
						LiOH LiCl	ns		11.4		DME Hg/ Hg_2Cl_2, LiCl		
						LiOH LiCl	0.1 0.4		-				
						LiOH	ns		14				
				EtOH 5	PY	BOR LiCl		0.5	8.4	2± 0.02	DME/SCE	034A0	C=0.05,m= 2.6±0.4,t= 3.6±0.5,h= 65±5
									10.75				
						LiOH LiCl	ns		12.8		DME Hg/ Hg_2Cl_2, LiCl 2		
									14.0				
EA17 aa77	$C_2H_4O_2$	glycolaldehyde C.A. 141-46-8	$HOCH_2CHO$	H_2O	PY	LiOH $LiClO_4$	0.05 1	-	-	3.6	DME/SCE	2A0	C=0.8,m= 1.026,t= 2.65,h=60, Pt Aux
						LiOH $LiClO_4$	ns 1	const	11.5				C=1.49
									12.75				
									13.5				
EA18 aa89	C_2H_5NO	acetaldehyde oxime C.A. 107-29-9	$CH_3CH:NOH$	H_2O	PY	KOH	0.1	0.2	-	-	DME/SCE	04A0	m=1.3,t= 2.6,h=54
		oxamide	$(H_2NCO\cdot)_2$										

TABLE I. Electrochemical Data C_2H_5NO EA18

Ref.	C/M	Charact. Potential Value	vs.	Response	Const. Value	Tech.	n	Electrokinetic Data Parameter	Value	From	Products and Identification	Description and Remarks	Code No.
JE043 0135	-	$E_{\frac{1}{2}}$	-1.50	SCE	i_ℓ	2.4F	-	-	-	-	-	-	$C,\ne,i_\ell=k_1[H^+]+k_2,r$	EA14 da06
			-1.70			2.2F							$C,\ne,i_\ell=-k_1[H^+]+k_2, k_1>0, \Sigma i_\ell=kC$	
			-1.68			3.96F	-	-	-	-	-	-	$C,\ne,E_{\frac{1}{2}}\ne f(C)$ for $C=0.08-1, i_\ell=kC,r$	
JE043 0135	-	$E_{\frac{1}{2}}$	-0.98 ± 0.02	SCE	i_ℓ	27.4F	-	-	-	-	-	-	$C,W,i_\ell\ne f(pH)$ for $pH=3.5-9, i_\ell=kC$ for $C=0.2-10,r$	EA15
JE046 0323	32 d	$E_{\frac{1}{2}}$	-1.7	SCE	i_ℓ	1.03	-	-	-	-	-	-	$C, Tc>2, i_\ell=k_1h^{\frac{1}{2}}+k_2, i_k, r$	EA16 aa74
			-1.77_5			1.7	-	-	-	-	-	-	C,r	
			-1.80_5			2.17	$i_\ell:i_\ell$ PQ	2 2	-	-	-	-	C,r	
			-1.81F			-	-	-	-	-	-	-	$C, \Delta E_{\frac{1}{2}}=0.2\sigma^*, r$	
	32 b		-1.89_5			2.07	-	-	-	-	-	-	$C, i_\ell=k_1h^{\frac{1}{2}}+k_2, i_d, r$	
JE046 0323	32 d	-	-	-	i_ℓ	0.65	-	-	-	-	-	-	C,r	
		-	-	-		1.5	-	-	-	-	-	-	C,r	
	32 b	-	-	-		2.5	-	-	-	-	-	-	C,r	
		-	-	-		1.95	-	-	-	-	-	-	C,r	
JE038 0191	137 c	$E_{\frac{1}{2}}$	-1.61F	SCE	i_ℓ	1.9F	QE	2.1 ± 0.1	$dE_{\frac{1}{2}}/dpH$ dlogi/dE	49 49 ± 1	- -	$CP\to$ acetaldehyde, GLC	C,W,i_d+i_k at higher temp, $i_\ell\downarrow$ with time, $dE_{\frac{1}{2}}/dpH$ for $pH=11.5-13.5, r$	EA17 aa77
			-1.8F			0.9F	-	-		-	-		$C,W,i_\ell=kC,i_d+i_k, dE_{\frac{1}{2}}/dpH\ne 0$ for $pH=11.5-13.5$	
		-	-	-		2.0F	- -	-	-	- -	-	-	C,r C	
		-	-	-		3.65F	- -	-	-	- -	-	-	C,r C	
		-	-	-		2.9F	- -	-	-	- -	-	-	C,r C	
C0026 0052	-	-	-	-	-	-	-	-	-	-	-	-	$A,0,p$	EA18 aa89
JE043 0135	-	$E_{\frac{1}{2}}$	-1.50	SCE	i_ℓ	2.4F	-	-	-	-	-	-	$C,\ne,i_\ell=k_1[H^+]+k_2,r$	EA14 da06

EA19 C_2H_5NS

Code No.	Empirical Formula	Name and C.A. Number	Structural Formula	Solvent	Tech.	Medium		μ, M	pH	T, °C	Electrodes	App.	Experimental Parameters
EA19	C_2H_5NS	thioacetamide C.A. 62-55-5	$CH_3C(S)NH_2$	H_2O	EE	buffer		-	4.22	-	MP/SCE	2AO	C=1E2, J= 1.5E2, Pt Aux
													C=1E3
									6.02				C=1E2
													C=1E3
									6.5				C=1E2
					ER	PHOS KNO_3	0.45	0.5	3	25±0.5	MP/SCE	2AO	C=1, J=5E2, Pt Aux
													C=9
EA20	$C_2H_6N_2O$	acetamide oxime C.A. 22059-22-9	$CH_3C(:NOH)NH_2$	EtOH 25	PY	BR		-	2.6	-	DME/SCE	O4AO	C=0.5
									4.6				
EA21	$C_2H_6N_4O_2$	oxamide bisoxime C.A. 2580-79-2	$[C(:NOH)NH_2]_2$	EtOH 25	PY	BR		-	2.1	-	DME/SCE	O4AO	C=0.5
									2.6				
									3.3				
EA22 ab08	$C_2H_8N_2$	1,1-dimethylhydrazine C.A. 57-14-7	$(CH_3)_2NNH_2$	MeCN	VA	Et_4NClO_4	0.1	-	-	-	PDE Ag/Ag+ 0.01	---	C=1, v=76.7
						CH_3COOH Et_4NClO_4	0.2 0.1						
					VY	Et_4NClO_4	0.1	-	-	-	RPDE Ag/Ag+ 0.01	---	C=1
EA23 ab39	$C_3H_3N_3O_2$	5-azauracil C.A. 71-33-0	Table II-2	H_2O	PO	H_2SO_4	1.0	-	-	-	DME/-	OB2	C=0.5
EA24 ca28 da15	C_3H_4O	acrolein C.A. 107-02-8	$CH_2:CHCHO$	H_2O	PY	MB		1.0±0.2	2	25±0.05	DME/SCE	OAO	ns
CONT		thioacetamide	$CH_3C(S)NH_2$										

TABLE I. Electrochemical Data C_3H_4O (CONT.) EA24

Ref.	C/M	Charact.	Value	Potential vs.	Response	Const. Value	Tech.	n	Electrokinetic Data Parameter	Value	From	Products and Identification	Description and Remarks	Code No.
JE040 0295	-	$E_{\tau/4}$	-0.06F	SCE	τ_f	24F	IR	1.8	-	-	-	-	A,R,r	EA19
		$E_{0.22}$	-0.06F		τ_r	7.6F		-		-			C,redn. of Hg complex	
		$E_{\tau/4}$	-0.18F		τ_f	24F	-	-	-	-	-	-	A,r	
		$E_{0.22}$	-0.18F		τ_r	7.6F		-		-			C,redn. of Hg complex	
		$E_{\tau/4}$	-0.06F		τ_f	23.7F	-	-	-	-	-	-	A,r	
		$E_{0.22}$	-0.06F		τ_r	0.9F		-		-			C,redn. of Hg complex	
		$E_{\tau/4}$	-0.17F		τ_f	23.7F	-	-	-	-	-	-	A,r	
		$E_{0.22}$	-0.17F		τ_r	7F		-		-			C,redn. of Hg complex	
		$E_{\tau/4}$	-0.07F		τ_f	24.9F	-	-	-	-	-	-	A,r	
			-			-		-		-			C,0	
JE040 0295	-	$E_{\tau/4}$	0.108	SCE	$i\tau^{\frac{1}{2}}/C$	65	-	-	-	-	-	-	A,r	
			0.076			60	-	-	-	-	-	-	A,r	
C0026 2438	-	$E_{\frac{1}{2}}$	-1.055F	SCE	-	-	-	-	-	-	-	-	C,i_d,p	EA20
			-1.09F		-	-	-	-	-	-	-	-	C,i_d,p	
C0026 2438	-	$E_{\frac{1}{2}}$	-0.99F	SCE	-	-	sttd	4	-	-	-	-	C,p	EA21
			-		-	-		-		-			C,i_{cat}	
			-1.05F		-	-		4	-	-	-	-	$C,pK'=2.96,p$	
			-		-	-		-		-			C,i_{cat}	
			-1.19F		-	-		4	-	-	-	-	$C,pK'=2.96,p$	
			-		-	-		-		-			C,i_{cat}	
JE040 006A	-	E_p	-0.08F	Ag/Ag$^+$ 0.01	i_p	35.8F	-	-	-	-	-	-	A,p	EA22 ab08
			0.17F			9.2F		-		-			A	
			0.1F			small		-		-			C	
			-1.0F			10F		-		-			C	
			0.25F			18.8F		-		-			A,p	
			-			-		-		-			C,0	
JE040 006A	-	$E_{\frac{1}{2}}$	-0.08	Ag/Ag$^+$ 0.01	i_ℓ	4.9F	-	-	Elog	110	sttd	CP→1,1,4,4-tetramethyl-2-tetrazene (85%), λ_{max}=280	$A,i_\ell(Ap)$	
			0.16			2.3F		-		-			$A,i_\ell(Ap),i_\ell,2/i_\ell,\neq f(\omega)$ for ω=10 33.3	
C0029 0182	-	Q	0.71	-	-	-	-	-	-	-	-	-	C,R,O at SME,p	EA23 ab39
			0.87			-		-		-			C,O at SME,p	
			0.71			-		-		-			A,R,p	
JE049 0433	-	$E_{\frac{1}{2}}$	-0.75c	SCE	I	1.14± 0.11	-	-	p dE$_{\frac{1}{2}}$/dpH	1.20± 0.23 163±6	Elog sttd	-	$C,E_{\frac{1}{2}}$ = -0.428-0.163 (±0.006)pH for pH= 2-5, $i_\ell=kh^{0.35}$,p	EA24 ca28 da15
			-1.36± 0.03			-	QE	1		0		CP→polymer,IRS,MAS	$C,E_{\frac{1}{2}}\neq f(pH)$ for pH= 2-10, i_k	
	-	$E_{\tau/4}$	-0.06F	SCE	τ_f	24F	IR	1.8						CONT

EA24 (CONT.) C_3H_4O

Code No.	Empirical Formula	Name and C.A. Number	Structural Formula	Solvent	Tech.	Medium		μ, M	pH	T, °C	Electrodes	App.	Experimental Parameters
EA24 ca28 da15	C_3H_4O	acrolein	$CH_2{:}CHCHO$	H_2O	PY	MB		1.0± 0.2	5	25± 0.05	DME/SCE	OAO	ns
						BOR KCl			9.3				
EA25 ca31	$C_3H_4S_3$	1,3-dithiolane-2-thione C.A. 822-38-8	Table II-3	MeCN	CP	$NaClO_4$	0.1	-	-	-	Pt/SCE	-	-
EA26 ca34	$C_3H_5FO_3S_4$	2-methylthio-1,3-dithietanium fluorosulfonate C.A. 56125-79-2	Table II-3	MeCN	VA	Et_4NClO_4	0.1	-	-	-	Pt Ag/ $AgNO_3$ 0.01, Et_4NClO_4 0.1	2A2	v < 1E3
EA27	$C_3H_5HgNO_4$	2-oxopropylmercury nitrate	$CH_3COCH_2HgNO_3$	H_2O	PY	KOH	1.0	-	-	-	DME/-	---	C=10(?),m= 2.15,t=3.68, h=100
EA28 da17	$C_3H_5KOS_2$	potassium ethyl-dithiocarbonate C.A. 140-89-6	$C_2H_5OC(S)SK$	H_2O	VA	NaF	0.1	-	-	25	XEl/SCE	2AO	-0.54 → -0.14 → -0.54 V,C= 0.08,A=1.09± 0.06,v=100, PE
													C=0.4,v=10
				Me_2CO	PY	Et_4NClO_4	0.1	-	-	20± 0.1	DME Ag/AgCl, LiCl 0.1	12AO	C=1, Pt Aux
													C=10
EA29 ab70	$C_3H_5N_3O_2$	5,6-dihydro-6-azauracil C.A. 18802-37-4	Table II-2	H_2O	PO	$HCOONH_4$	1.0	-	-	-	SME/-	OB2	C=0.5
EA24	C_3H_4O	acrolein	$CH_2{:}CHCHO$	H_2O	PY	MB							

TABLE I. Electrochemical Data $C_3H_5N_3O_2$ EA29

Ref.	C/M	Charact. Potential		Response Const.		Tech.	n	Electrokinetic Data			Products and Identification	Description and Remarks	Code No.
		Value	vs.		Value			Parameter	Value	From			
JE049 0433	-	$E_{\frac{1}{2}}$	-1.24c	SCE	-	-	-	-	$dE_{\frac{1}{2}}/dpH$	163±6	sttd	-	C,p	EA24 ca28 da15
			-1.36		-	-	-	-		0			C	
			-1.36		-	-	QE	0.78 ± 0.08		-	-		C,p	
			-1.76		-	-	-	-		-			$C, i_k, 0$ for pH < 9	
JE038 0245	337 c	-	-	-	-	-	-	-	-	-	-	CP → ethylene dithio-carbonate, IRS, MAS, NMR	see CA31	EA25 ca31
JE049 0105	338 b	E_p	-0.55	SCE	-	-	QE	0.2	$E_p-E_{p/2}$	50	-	2,2'-bis(methyl-thio)-2,2'-bis(1,3-dithietyl)	$C, \neq, E_p-E_{p/2}$ and $I_p/Cv^{\frac{1}{2}} \neq f(v),p$	EA26 ca34
			-			-	-			-			A,X	
C0028 0026	-	$E_{\frac{1}{2}}$	-0.23F	SCE	-	-	-	-	-	-	-	-	C,p	EA27
JE046 0411	-	E_p	-0.41F	SCE	i_p	15F	-	-	-	-	-	-	A,XEl=CuS disc(85% chalcocite,13% bornite),broad peak,p	EA28 da17
			-0.35F			25F	-			-			C	
			-0.47F			41.3F	-			-			C	
			-0.45F			8.3F	-			-			$A, i_p \neq f(C)$ for $C \geq 0.25, i_a, p$	
			-0.39F			7.6F	-			-			$A, i_p \neq f(C)$ for $C \geq 0.25, i_a$	
			-0.27F			23.6F	-			-			A	
			-0.48F			37.5F	-			-			C	
JE048 0071	333 a	$E_{\frac{1}{2}}$	-0.364	Ag/AgCl, LiCl 0.1	i_ℓ	3.72	-	-	Tomeš	60	-	-	$A, i_d, i_\ell=kh^{\frac{1}{2}}, i_\ell=kC$ for C=0.5-10,r	
			0.093			1.46	-			76			$A, i_d, i_\ell=kh^{\frac{1}{2}}, i_\ell=kC$ for C=0.5-10	
			-			-	-			-			A,M	
			-0.437			37	-	-	ElogX Tomeš	38 74	-	-	$A, ElogX=dE/dlog[i/(i_\ell-i)^3],r$	
			0.128			14.5			FlogX	84 30			$A, ElogX=dE/dlog[I^?/(I_\ell-I)^?]$	
			-			-				-			A,M	
C0029 0182	-	Q	0.73	-	-	-	-	-	-	-	-	-	C,p	EA29 ab70

EA30 C$_3$H$_5$N$_3$O$_3$

Code No.	Empirical Formula	Name and C.A. Number	Structural Formula	Solvent	Tech.	Medium	μ, M	pH	T, °C	Electrodes	App.	Experimental Parameters
EA30	C$_3$H$_5$N$_3$O$_3$	glyoxylic acid semicarbazone C.A. 928-73-4	CH(:NNHCONH$_2$)-COOH	H$_2$O	PY	MB	–	0.50	–	DME/SCE	OAO	C=0.5(?), m=2.62, t=2.8, h=55
								2.00				
								2.95				
								3.35				
								4.10				
								5.10				
								6.50				
								7.60				
								9.30				
								10.60				
								11.80				
EA31	C$_3$H$_6$	propylene C.A. 115-07-1	CH$_3$CH:CH$_2$	MeCN	VY	Et$_4$NBF$_4$	0.1	–	25	RPDE Ag/Ag$^+$ 0.01	2-0	C≈1, A=0.12, v=50, ω=16.3, Pt Aux
				CH$_2$Cl$_2$	VY	Bu$_4$NBF$_4$	0.2	–	25	RPDE Ag/Ag$^+$ 0.01	2-0	C≈1, A=0.12, v=50, ω=16.3, Pt Aux
				EtNO$_2$	VY	Et$_4$NBF$_4$	0.2	–	25	RPDE Ag/Ag$^+$ 0.01	2-0	C≈1, A=0.12, v=50, ω=16.3, Pt Aux
				MeNO$_2$	VY	Et$_4$NBF$_4$	0.1	–	25	RPDE Ag/Ag$^+$ 0.01	2-0	C≈1, A=0.12, v=50, ω=16.3, Pt Aux
				PCA	VY	Et$_4$NBF$_4$	0.1	–	25	RPDE Ag/Ag$^+$ 0.01	2-0	C≈1, A=0.12, v=50, ω=16.3, Pt Aux
				SULN	VY	Et$_4$NBF$_4$	0.2	–	25	RPDE Ag/Ag$^+$ 0.01	2-0	C≈1, A=0.12, v=50, ω=16.3, Pt Aux

TABLE I. Electrochemical Data C_3H_6 EA31

Ref.	C/M	Charact. Potential Value	vs.	Response Const. Value		n	Tech.	Electrokinetic Data Parameter	Value	From	Products and Identification	Description and Remarks	Code No.
C0036 0331	388 a	$E_{\frac{1}{2}}$ -0.50F	SCE	i_ℓ	7.6F	2	i:i, QE	-	-	-	-	C, i_d, p	EA30
		-0.66F			7.6F	2		-	-	-	-	C, i_d, Elog non-linear, p	
		-0.76F			7.3F	-	-	-	-	-	-	C, p	
		-0.79F			4.3F	-	-	-	-	-	-	C, p	
		-1.32			3.8F	-	-	-	-	-	-	C	
		-0.86F			1.0F	-	-	-	-	-	-	C, i_k, p	
		-1.36F			7.3F	-	-	-	-	-	-	C, i_d	
		-1.45F			7.4F	2	i:i	-	-	Elog	-	C, i_d, p	
		-1.52F			7.1F	-	-	-	-	-	-	C, p	
	388 b	-1.59F			1.2F	-	-	-	-	-	-	$C, i_k, Tc=3.6, p$	
		-1.77F			6.9F	2	i:i	-	-	-	-	C, i_d, p	
		-1.80F			6.4F	2		-	-	-	-	CP → glyoxylic acid semicarbazide	C, p
		-1.83F			6.2F	2		-	-	-	-	CP → glyoxylic acid semicarbazide	C, p
JE042 0133 JE036 0131	372 c	$E_{\frac{1}{2}}$ 2.52	Fc		-	-	-	-	-	-	-	QE at 2.8 V vs. Ag/Ag$^+$, C=0.03 → N-allylacetamide; 2,4,6-trimethyl-pyrimidine; 2,4,6-trimethyl-s-triazine; N-isopropyl-acetamide, GLC, CP with 2% HOAc at 2.6 V vs. Fc → N-allylacetamide (43%), 3-fluoropropene, trace	A, p	EA31
JE042 0133 JE043 0318	372 c	$E_{\frac{1}{2}}$ 2.35	Fc		-	-	-	-	-	-	-	CP at 2.5 V vs. Fc with 2% HOAc → 4-fluoro-2-propanone(30%), VPC, IRS, NMR, MAS; 3-fluoropropene, trace; at 3.05 V vs. Fc → 4-fluoro-2-propanone-(30%), GLC, IRS, 3-propenyl acetate (20%), GLC, IRS	A, p	
JE042 0133	372 c	$E_{\frac{1}{2}}$ 2.44	Fc		-	-	-	-	-	-	-	CP → polypropylene, IRS; CP at 2.5 V vs. Fc with 2% HOAc → 3-fluoropropene, trace	A, p	
JE042 0133	372 c	$E_{\frac{1}{2}}$ 2.51	Fc		-	-	-	-	-	-	-	"	A, p	
JE042 0133	372 c	$E_{\frac{1}{2}}$ 2.52	Fc		-	-	-	-	-	-	-	CP at 2.6 V vs. Fc with 2% HOAc → 4-fluoro-2-pentanone (4%), VPC, IRS, NMR, MAS; 3-fluoropropene, trace	A, p	
JE042 0133	372 c	$E_{\frac{1}{2}}$ 2.50	Fc		-	-	-	-	-	-	-	CP at 2.6 V vs. Fc with 2% HOAc → 3-fluoropropene, trace	A, p	
C0036 0331		$E_{\frac{1}{2}}$ -0.50F	SCE	i_ℓ	7.6F	2	i:i, QE	-	-	-	-	C, i_d, p	EA30

EA32 C$_3$H$_6$Br$_2$

Code No.	Empirical Formula	Name and C.A. Number	Structural Formula	Solvent	Tech.	Medium		μ, M	pH	T, °C	Electrodes	App.	Experimental Parameters
EA32 ca36	C$_3$H$_6$Br$_2$	1,3-dibromopropane C.A. 109-64-8	BrH$_2$CCH$_2$CH$_2$Br	MeCN	CT	Bu$_4$NClO$_4$	0.1	-	-	-	RMPD Ag/ AgClO$_4$ 0.01, Et$_4$NBr 0.1	2AO	C=10, A= 0.126, v=30, w=11.9, Pt Aux
EA33 ab82 ca39	C$_3$H$_6$O	acetone C.A. 67-64-1	CH$_3$COCH$_3$	H$_2$O	PY	KOH	0.4	-	-	-	DME/SCE	O4AO	C=10, m=2.15, t=3.68, h=100
						KOH	1.0						
						KOH	1.6						
					PW	KOH	1.0	-	-	-	HMDE SCE	O4AO	C=100, v=6.7
EA34	C$_3$H$_7$NO	2-propanone oxime C.A. 127-06-0	CH$_3$C(:NOH)CH$_3$	H$_2$O	PY	KOH	0.1	0.2	-	-	DME/SCE	O4AO	m=1.3, t=2.6, h=54
EA35	C$_3$H$_7$NO$_2$	L-alanine C.A. 56-41-7	CH$_3$CH(NH$_2$)COOH	H$_2$O	IL	NaOH	1	-	-	23	Cu/SCE	2AO	C=0.01, A= 3.14E-2, v= 7.07
													C=0.1
													C=0.5
EA36 ac06 da23	C$_3$H$_7$NO$_2$S	cysteine C.A. 52-90-4	HSCH$_2$CH(NH$_2$)- COOH	H$_2$O	PY	ACET Bu$_4$NCl 5E-6 CoCl$_2$ 3E-4		-	5.55	-	DME/SCE	---	-0.8→-2.0 V, C=0.2, m= 1.39, t=5.2, h=83
						ACET NiCl$_2$ 3E-4		-	6.1	-	DME/SCE	---	C=0.1, m= 1.89, t=4.78, h=59
EA37	C$_3$H$_7$N$_3$O$_2$	glycolaldehyde semi- carbazone C.A. 21205-25-4	CH$_2$(OH)CH:NNH- CONH$_2$	H$_2$O	PY	NH$_2$CONHNH$_2$ NH$_2$CONHNH$_3$Cl		-	-2.0	-	DME/SCE	OAO	C=0.5(?), m= 2.62, t=2.8, h=55
									-1.0				
									0.0				
CONT									2.0				

TABLE I. Electrochemical Data $C_3H_7N_3O_2$ (CONT.) EA37

Ref.	C/M	Charact. Potential		Response Const.		n	Electrokinetic Data			Products and Identification	Description and Remarks	Code No.
		Value	vs.		Value	Tech.	Parameter	Value	From			
JE043 0215	94 g	E_p -1.76F	Ag/AgClO$_4$ 0.01, Et$_4$NBr 0.1	i_p	1.4E3F	-	-	-	-	-	CP at -1.26 V → propane(61%),GLC;cyclopropane(30%), GLC	C, i_d for C < 1E-2, r	EA32 ca36
		-2.0F			1.9E3F	-	-	-	-	-	CP at -1.26 V → propane(5.4%); cyclopropane(11%); 1-bromopropane(3%); 1,6-dibromohexane (8%); all GLC	C	
		-			-	-	-	-	-	-	-	A,O	
C0028 0026	402 a	$E_{\frac{1}{2}}$ -0.23	SCE	i_ℓ	0.4F	-	-	-	-	-	2-oxo-1-propylmercury, PY	$A, i_\ell \neq kC, p$	EA33 ab82 ca39
		-0.23			1.9F	-	-	-	-	-	2-oxo-1-propylmercury, PY	$A, i_\ell \neq kC, p$	
		-0.23			3.7F	-	-	-	-	-	2-oxo-1-propylmercury, PY	$A, i_\ell \neq kC, p$	
C0028 0026	402 a	E_p -0.30	SCE	i_p	1.0	-	-	-	-	-	-	C,p	
		-0.47			0.5	-	-	-	-	-	-	C	
C0026 0052	-	-	-	-	-	-	-	-	-	-	A,O,p	EA34
JE034 0091	379 a	E_p -0.24F	NHE	i_p	14.8F	-	-	-	-	-	CP → acetonitrile, GLC	A, also occurs if C=0, p	EA35
		-0.05F			169.8F	-	-	-	-	-		A,S if C=0	
		0F			489F	-	-	-	-	-		A, also occurs if C=0	
		0.71F			69.2F	-	-	-	-	-		$A, i_p = k_1C + k_2$	
		-	-	-	5.6F	-	-	-	-	-	-	A,p	
					194.9F	-	-	-	-	-		A	
					309F	-	-	-	-	-		A	
					295F	-	-	-	-	-		A	
		-	-	-	3.8F	-	-	-	-	-	-	A,p	
					316.2F	-	-	-	-	-		A	
					208.9F	-	-	-	-	-		A	
					1000F	-	-	-	-	-		A	
JE055 0157	-	$E_{\frac{1}{2}}$ -0.97	SCE	i_ℓ	0.75F	-	-	-	-	-	-	$C, i_{cat}, Pr, i_\ell = f(C, [Co(II)]), p$	EA36 ac06 da23
		-1.16			3.8F	-	-	-	-	-	-	$C, Co(II)$ redn., $i_\ell = f(C, [Co(II)])$	
		-1.47			4.5F	-	-	-	-	-	-	C, H^+ Pr	
		-1.76			-	-	-	-	-	-	-	C,M,H	
JE047 0190	-	$E_{\frac{1}{2}}$ -0.6	SCE			-	-	-	-	-	-	C	
		-0.96			-	-	-	-	-	-	-	C	
		-1.27			-	-	-	-	-	-	-	C, Pr, i_{cat}, O with $[Ni^{2+}] = 0, i_\ell = f(pH, C)$	
		-1.73			-	-	-	-	-	-	-	C, i_{cat}	
C0036 0331	-	-	-	i_ℓ	8.4F	-	-	-	-	-	-	C,p	EA37
					8.4F	-	-	-	-	-	-	C,p	
		$E_{\frac{1}{2}}$ -0.64F	SCE	Σi_ℓ	-	-	-	-	-	-	-	C,p	
		-0.76F			12.2F	-	-	-	-	-	-	C	
		-0.84F			-	-	-	-	-	-	-	C,p	
		-0.95F			21.0F	-	-	-	-	-	-	C	CONT

27

Code No.	Empirical Formula	Name and C.A. Number	Structural Formula	Solvent	Tech.	Medium	μ, M	pH	T, °C	Electrodes	App.	Experimental Parameters
EA37	$C_3H_7N_3O_2$	glycolaldehyde semicarbazone	$CH_2(OH)CH:NNH-CONH_2$	H_2O	PY	$NH_2CONHNH_2$ $NH_2CONHNH_3Cl$	–	3.0	–	DME/SCE	OAO	C=0.5(?),m=2.62,t=2.8, h=55
								4.5				
								5.3				
EA38	$C_3H_7N_3O_3$	semicarbazidoacetic acid C.A. 138-07-8	$H_2NCONHNHCH_2COOH$	H_2O	PY	NaOH 0.5	–	–	–	DME/SCE	OAO	m=2.62,t=2.8,h=55
EA39	$C_3H_{10}N_2$	trimethylhydrazine C.A. 1741-01-1	$(CH_3)_2NNHCH_3$	MeCN	VY	Et_4NClO_4 0.1	–	–	–	RPDE Ag/Ag^+ 0.01	---	C=1
EA40	$C_4H_2Cl_2O_4$	dichlorofumaric acid C.A. 25144-43-8	trans-$(HOOCCCl:)_2$	H_2O	PY	HCl 2	–	–	25±0.1	DME/SCE	O3AO	C=ns,t=3.9±0.8,h=81, EC=1.5±0.2
						HCl KCl ns	0.5	0.5				
						CITR PHOS KCl		2.4				
								3				
EA41	$C_4H_2Cl_2O_4$	dichloromaleic acid C.A. 608-42-4	cis-$(HOOCCCl:)_2$	H_2O	PY	HCl 2	–	–	25±0.1	DME/SCE	O3AO	C=ns,t=3.9±0.8,h=81, EC=1.51±0.2
						HCl KCl ns	0.5	0.5				
								1.6				
						CITR PHOS KCl		2.4				
								3				
								4				
								4.8				

TABLE I. Electrochemical Data $C_4H_2Cl_2O_4$ (CONT.) EA41

Ref.	C/M		Charact. Potential		Response Const.		n		Electrokinetic Data			Products and Identification	Description and Remarks	Code No.
			Value	vs.		Value	Tech.		Parameter	Value	From			
C0036 0331	-	$E_{\frac{1}{2}}$	-0.95F -1.03F	SCE	Σi_ℓ	- 18.4F	-	-	-	-	-	-	C,p c	EA37
			-1.08F -1.20F			- 10.9F	-	-	-	-	-	-	C,p c	
			-1.15F -1.29F			- 5.0F	-	-	-	-	-	-	C,p c	
C0036 0331	388 ab	$E_{\frac{1}{2}}$	-0.33	SCE	-	-	QE (-0.30 V)	1.8	-	-	-	CP → glyoxylic acid semicarbazone	A,p	EA38
JE040 006A	-	$E_{\frac{1}{2}}$	-0.14F	Ag/Ag$^+$ 0.01	i_ℓ	6.7F	-	-	Elog	110	sttd	CP → N,N-dimethyl-N'-methylenehydrazine (95%), UVS	A,p	EA39
			0.76F			1.9F	-	-	-		-		$A, i_{\ell,2}/i_{\ell,1} \neq f(\omega)$ for $\omega = 10-33.3$	
JE038 0403	-	$E_{\frac{1}{2}}$	-0.5F	SCE	i_ℓ	1.5F	-	-	-	-	-	-	C, i_d, r	EA40
			-0.55F		I	7.5F	QE	4.2F	-	-	-	CP → 2-chloro-1,4-butanedioic acid (90%), IRS	C, i_d, r	
			-0.86F			9.0F		5.3F	-	-	-	CP → 2-chloro-1,4-butanedioic acid (65%), IRS	C, i_d, r	
			-0.98F			2.7F		5.4F	-	-	-	CP → 2-chloro-1,4-butanedioic acid (45%), IRS	C, i_k, r	
JE038 0403	-	$E_{\frac{1}{2}}$	-0.5F	SCE	i_ℓ	1.7F	-	-	-	-	-	-	C, 2 merging waves, r	EA41
			-0.46F		I	} 7.7F }	QE	4.08	-	-	-	CP → 2-chloro-1,4-butanedioic acid (90%), IRS; chlorofumaric acid (9%); chloromaleic acid (1%)	C, i_d, r	
			-0.55F			-			-		-		C, i_d	
			-0.52F -0.6F			} 7.9F }		4.6	-		-	-	C, i_d, r C, i_d	
			-0.7F			7.5F	-	-	-	-	-	CP → 2-chloro-1,4-butanedioic acid (65%), IRS; chlorofumaric acid (6%); 1,4-butynedioic acid (<1%)	C, 2 merging waves, i_d, r	
			-0.82F			7.5F	-	-	-	-	-	-	C, 2 merging waves, r	
			-0.92F			1.6F	QE	} 4.7	-	-	-	CP → 2-chlorobutanedioic acid (25%), IRS; chlorofumaric acid (5%); 1,4-butynedioic acid (15%)	C, 2 merging waves, i_k, r	
			-1.10F			5.6F			-		-	-	C, i_k	
			-			0.3F		} 5.1	-	-	-	-	C, 2 merging waves, i_k, r	
			-1.18F			5.9F							C, i_k	CONT

EA41 (CONT.) C₄H₂Cl₂O₄

Code No.	Empirical Formula	Name and C.A. Number	Structural Formula	Solvent	Tech.	Medium	μ, M	pH	T, °C	Electrodes	App.	Experimental Parameters
EA41	C₄H₂Cl₂O₄	dichloromaleic acid	cis-(HOOCCCl:)₂	H₂O	PY	CITR PHOS KCl	0.5	6.2	25±0.1	DME/SCE	O3AO	C=ns,t=3.9 ±0.8,h=81, EC=1.51±0.2
EA42	C₄H₂N₂O₄	tetraoxopiperazine C.A. 49715-78-8	Table II	H₂O	PY	HOAc	1	2.3	25	DME/SCE	2AO	C=1,v=10, Pt Aux
EA43	C₄H₃ClO₄	chlorofumaric acid C.A. 617-43-6	trans-HOOCCCl: CHCOOH	H₂O	PY	HCl	2	-	25±0.1	DME/SCE	O3AO	C=ns,t=3.9 ±0.8,h=81, EC=1.5±0.2
						HCl KCl	ns 0.5	0.5				
						CITR PHOS KCl		2.5				
								4.2				
								4.5				
								5				
						AM KCl	0.1	9				
EA44	C₄H₃ClO₄	chloromaleic acid C.A. 617-42-5	cis-HOOCCCl:CH-COOH	H₂O	PY	HCl	1	-	25±0.1	DME/SCE	O3AO	C=ns,t=3.9 ±0.8,h=81, EC=1.5±0.2
						HCl KCl	ns 0.5	0.5				
						CITR PHOS KCl		2.4				
								4.3				
								6				
								6.8				
						AM KCl	0.1	8.5				
EA45 ac43 da32	C₄H₄N₂	pyridazine C.A. 289-80-5	Table II-2	H₂O	PY	PHOS NaClO₄	1.0	1.39	25±0.1	DME Ag/AgCl, NaCl satd	2-0	C=3.64
CONT												

TABLE I. Electrochemical Data $C_4H_4N_2$ (CONT.) EA45

Ref.	C/M	Charact. Potential		Response	Const. Value	n Tech.	n	Electrokinetic Data			Products and Identification	Description and Remarks	Code No.
		Value	vs.					Parameter	Value	From			
JE038 0403	-	$E_{\frac{1}{2}}$	-1.24F	SCE	I	1.1F	-	-	-	-	-	-	C, i_k, r	EA41
JE055 0069	398 d	$E_{\frac{1}{2}}$	-0.42	SCE	-	-	-	-	-	-	-	-	C, cmpd. decomposes ($\rightarrow$ oxamide) at pH > 4, p	EA42
			-0.85										C	
			-1.05										C	
JE038 0403	-	$E_{\frac{1}{2}}$	-0.46F	SCE	i_ℓ	0.97F	-	-	-	-	-	-	C, i_d, r	EA43
			-0.52F		I	3.8F	QE	2.1 F	-	-	-	CP $\rightarrow$ 2-chloro-1,4-butanedioic acid (85%), IRS	C, i_d, r	
			-0.75F			4.8F		2.5 F	-	-	-	CP $\rightarrow$ 2-chloro-1,4-butanedioic acid (65%), IRS	C, i_d, r	
			-1.05F			5.8F		3.6	-	-	-	CP $\rightarrow$ 2-chloro-1,4-butanedioic acid (5%), IRS	C, 2 merging waves, r	
			-1.08F			3.7F	-	-					C, i_k, r	
			-1.48F			2.6F							C, i_a	
			-1.12F			0.4F	-	-					C, i_k, r	
			-1.49F			4.2F							C, i_a	
			-1.7F			7.6F	-	-	-	-	-	-	C, r	
JE038 0403	-	$E_{\frac{1}{2}}$	-0.55F	SCE	i_ℓ	1.3F	-	-	-	-	-	-	C, r	EA44
			-0.54F		I	4.6F	QE	2.1	-	-	-	CP $\rightarrow$ 2-chloro-1,4-butanedioic acid (85%), IRS	C, i_d, r	
			-0.68F			4.4F		-	-	-	-	CP $\rightarrow$ 2-chloro-1,4-butanedioic acid (70%), IRS	C, i_d, r	
			-0.89F			5.6F		3.2	-	-	-	CP $\rightarrow$ 2-chloro-1,4-butanedioic acid (45%), IRS	C, 2 merging waves, r	
			-1.07F			0.4F	-	-	-	-	-	-	C, i_k, r	
			-1.3F			5.0F							C, i_a	
			-1.36F			4.0F	-	-	-	-	-	CP $\rightarrow$ 2-chloro-1,4-butanedioic acid (15%), IRS	C, r	
			-1.44F			0.7F	-	-	-	-	-	CP $\rightarrow$ 2-chloro-1,4-butanedioic acid (5%), IRS	C, r	
JE041 0411	413 a	$E_{\frac{1}{2}}$	-0.58F	SCE	i_ℓ	40F	QE	2.2 ± 0.1 Tomeš	$dE_{\frac{1}{2}}/dpH$	64 26.5 ±2.5	-	4-hydrazinobut-3-enal, λ_{max}=220,295, UVS	$C, W, i_d, i_\ell \neq f(pH), i_\ell = k_1C + k_2$ for C=0.11-27.2, $E_{\frac{1}{2}} = -0.45 - 0.064pH$ for pH=2-4, $E_{\frac{1}{2}}=f(C), r$	EA45 ac43 da32
			-0.79F			5.6F		-		-			$C, X, \not\equiv, i_d, i_\ell=k/C, i_\ell \neq f(pH), 0$ for C > 5	
			-0.94F			4.0F							$C, X, i_\ell \neq f(pH)$	
		$E_{\frac{1}{2}}$	-1.24F											CONT

EA45 (CONT.) $C_4H_4N_2$ CRC Handbook Series in Organic Electrochemistry

Code No.	Empirical Formula	Name and C.A. Number	Structural Formula	Solvent	Tech.	Medium		μ, M	pH	T, °C	Electrodes	App.	Experimental Parameters
EA45 ac43 da32	$C_4H_4N_2$	pyridazine	Table II-2	H_2O	PY	PHOS $NaClO_4$		1.0	3	25± 0.1	DME Ag/AgCl, NaCl satd	2-O	C=1.13
									5.3				C=3.64
					VA	$HClO_4$	1	1.0	-	25± 0.1	HMDE Ag/AgCl, NaCl satd	2-O	C=7.33, v=100
													v=1E5
						PHOS $NaClO_4$			3.01				C=2.45, v=100
													v=1E5
									6.82				C=3.64, v=112
													v=167E5
EA46 ac44 ca55 da33	$C_4H_4N_2$	pyrimidine C.A. 289-95-2	Table II-2	H_2O	VY	FORM		0.5	3.56	25± 0.2	RDGC Ag/AgCl, KCl satd	OA2	-0.6 → -0.9 V, C=10
EA47	$C_4H_4N_2O_2$	4,5-dihydroxypyrimidine C.A. 15837-41-9	Table II	H_2O	IL	HOAc	1	-	2.3	25	PGE/SCE	2AO	C=1, v=10, Pt Aux
EA48	$C_4H_4N_2O_2$	uracil C.A. 66-22-8	Table II	H_2O	PO	NaOH	1.0	-	-	-	SME/-	OB2	C=0.5
EA49 ac57	$C_4H_4N_6O$	8-azaguanine C.A. 134-58-7	Table II-2	H_2O	PO	H_2SO_4	1.0	-	-	-	SME/-	OB2	C=0.5
EA50 ca60	$C_4H_5Cl_3O_2$	ethyl trichloroacetate C.A. 515-84-4	$CCl_3COOC_2H_5$	DMF	PY	Et_4NClO_4	0.1	-	-	≈21	DME/SCE	236AO	C=ns, m=1.26, t(oc)=5.77, h=70, MP Aux
EA51	$C_4H_5F_3O_2$	ethyl trifluoroacetate C.A. 383-63-1	$CF_3COOC_2H_5$	DMF	PY	Et_4NClO_4	0.1	-	-	≈21	DME/SCE	26AO	C=1, m=2.28, t=3.5, h=50, MP Aux

TABLE I. Electrochemical Data $C_4H_5F_3O_2$ EA51

Ref.	C/M	Charact.		Potential	Response	Const.	n		Electrokinetic Data			Products and Identification	Description and Remarks	Code No.
			Value	vs.		Value	Tech.		Parameter	Value	From			
JE041 0411	413 a	$E_{\frac{1}{2}}$	-0.69	SCE	i_ℓ	-	QE	2.2 ± 0.1	$dE_{\frac{1}{2}}/dpH$	64	-	-	C,r	EA45 ac43 da32
			-0.85F					3.15 ± 0.25		-			C	
			-1.07F					4.3					C	
			-			-		-		61	-	-	C,r	
			-1.04F			-		-		-			C	
JE041 0411	413 a	E_p	-0.546	SCE	-	-	-	-	$E_p-E_{p/2}$	30	sttd	-	$C,i_p/Cv^{\frac{1}{2}}\downarrow$ as $v\uparrow$,p	
			-										C?	
			-										A,O	
			-0.619		$i_p/Cv^{\frac{1}{2}}$	3.95	-	-		66	-	-	C,p	
			-			-				-			C,?	
			-0.480										A	
			-0.765			-	-	-		42	-	-	C,p	
			-1.03			-				-			C	
			-										A,O	
			-0.868			3.6	-	-		95	-	-	C,p	
			-			-				-			C?	
			-0.480										A	
			-1.06			-	-	-		68	-	-	C,r	
			-1.36							-			C	
			-										A,O	
			-1.313			3.04	-	-		116	-	-	C,p	
			-			-				-			C?	
			-0.811										A	
JE046 0089	75 de	$E_{\frac{1}{2}}$	-0.96F	SCE	$i_\ell(u)$	24F	-	-	-	-	-	-	C,p	EA46 ac44 ca55 da33
JE055 0069	-	E_p	0.72	SCE	-	-	QE	2.85 ± 0.15	-	-	-	$CP\to NH_3$(1.7 moles), CO_2(1.5 moles), formaldehyde(0.8 mole), formic acid (0.7 mole), urea (trace)	A,p	EA47
C0029 0182	-	Q	0.13 0.13	-	-	-	-	-	-	-	-	-	C,R,p A,R	EA48
C0029 0182	-	Q	0.82	-	-	-	-	-	-	-	-	-	C,p	EA49 ac57
JE044 0025	-	$E_{\frac{1}{2}}$	-0.49 -1.30 -2.3	SCE	-	-	sttd	≈3 -	-	-	-	-	$C,E_{\frac{1}{2}}$ corr.(E_j),p C C	EA50 ca60
JE049 0085	377 a	$E_{\frac{1}{2}}$	-2.34	SCE	-	-	QE	1	-	-	-	$CP\to F^-$,ppt as AgF	$C,E_{\frac{1}{2}}$ corr.(E_j),i_d, $i_\ell \uparrow$ on addition of H^+ donor,p	EA51
		$E_{\frac{1}{2}}$	-0.69	SCE	i_ℓ	-	QE	2.2	$dE_{\frac{1}{2}}/dpH$	64	-	-		

EA52 $C_4H_5N_3O_5$

Code No.	Empirical Formula	Name and C.A. Number	Structural Formula	Solvent	Tech.	Medium	μ, M	pH	T, °C	Electrodes	App.	Experimental Parameters
EA52	$C_4H_5N_3O_5$	mesoxalic acid semi-carbazone C.A. 27255-91-0	Table II	H_2O	PY	MB	-	1.35 2.0 3.3 4.0 5.0 6.0 7.0 7.8 9.6 12.2	-	DME/SCE	OAO	C=0.5(?), m=2.62, t=2.8, h=55
EA53	$C_4H_6Cl_2O_2$	ethyl dichloro-acetate C.A. 535-15-9	$CHCl_2COOC_2H_5$	DMF	PY	Et_4NClO_4 0.1 Et_4NClO_4 0.1 3,4-xylenol 3E-3	-	-	≈21	DME/SCE	236AO	C=1, m=1.26, t=5.77, h=70, MP Aux
EA54	$C_4H_6F_2O_2$	ethyl difluoro-acetate C.A. 454-31-9	$CHF_2COOC_2H_5$	DMF	PY	Et_4NClO_4 0.1 Et_4NClO_4 0.1 3,4-xylenol >1E-3	-	-	≈21	DME/SCE	236AO	C=1, m=2.28, t=3.5, h=70, MP Aux
EA55	$C_4H_6N_2O_2$	5,6-dihydrouracil C.A. 504-07-4	Table II	H_2O	PO	NaOH 1.0	-	-	-	SME/-	OB2	C=0.5
EA56 ac83 da38	C_4H_6O	crotonaldehyde C.A. 123-73-9	$CH_3CH:CHCHO$	H_2O	PY	BOR KCl		1–9.3	25.00 ±0.05	DME/SCE		
EA57	C_4H_6O	methacrolein C.A. 78-85-3	$H_2C:C(CH_3)CHO$	H_2O	PY	BOR KCl	1.0± 0.2	2 6 10	25.00 ±0.05	DME/SCE	OAO	ns

TABLE I. Electrochemical Data $\quad C_4H_8O \quad$ EA57

Ref.	C/M	Charact. Potential		Response Const.		n		Electrokinetic Data			Products and Identification	Description and Remarks	Code No.
		Value	vs.		Value		Tech.	Parameter	Value	From			
C0036 0331	388 a	$E_{\frac{1}{2}}$	-0.63F	SCE	-	-	i:i, QE	2	$2.3RT/\alpha n_a F$	130	Elog	-	C, i_d, p	EA52
			-0.68F		i_ℓ	7.4F		2		-		-	C, i_d, Elog non-linear, p	
			-0.72F			7.4F		2		-		-	C, i_d, p	
			-0.80F			7.4F		2		·		-	C, i_d, p	
			-1.01F			7.4F		2		67		-	C, i_d, p	
			-1.22F			7.3F		2		-		-	C, i_d, p	
	388 b		-1.29F			4.0F		-		-		-	C, p	
			-1.33F			1.2F		-		65		-	$C, i_k, T_c=4.1, p$	
			-1.70F			7.0F	i:i, QE	2		-		-	C, i_d, p	
			-1.70F			7.0F		2		-		-	C, i_d, p	
JE044 0025	377 cd	$E_{\frac{1}{2}}$	-1.31	SCE	i_p	3.5F	QE	1	-	-	-	CP → 1 Cl⁻/ CHCl₂COOC₂H₅	$C, E_{\frac{1}{2}} corr.(E_j), \not{\epsilon}, i_d, r$	EA53
			-1.31			5.4F	QE (-1.3 V)	1.38	-	-	-	CP at -1.3 V → 1.37 Cl⁻/CHCl₂COOC₂H₅	$C, E_{\frac{1}{2}} corr.(E_j), r$	
			-1.93			4.2F	(-2.0 V)	0.66	-	-	-	CP at -2.0 V → 1.83 Cl⁻/CHCl₂COOC₂H₅; CP at -2.5 V → 1.93 Cl⁻/CHCl₂COOC₂H₅ and n=0.39	C	
JE049 0085	377 a	$E_{\frac{1}{2}}$	-2.54	SCE	i_ℓ	2.3F	QE	≈1	-	-	-	CP → F⁻, pptn. as AgF	$C, E_{\frac{1}{2}} corr.(E_j), W, i_d, p$	EA54
			-2.54			4.3F	-	-	-	-	-	-	$C, E_{\frac{1}{2}} corr.(E_j), p$	
C0029 0182	-	Q	0.14	-	-	-	-	-	-	-	-	-	C, p	EA55
JE049 0433	-	-	-	-	-	-	-	-	-	-	-	-	see AC83, DA38	EA56 ac83 da38
JE049 0433	-	$E_{\frac{1}{2}}$	-0.89c	SCE	-	-	-	-	$dE_{\frac{1}{2}}/dpH$	56±6	sttd	-	$C, E_{\frac{1}{2}} = -0.777-0.056 (\pm 0.006)pH$ for pH=2-10, $i_\ell = kh^{0.95}, p$	EA57
			-1.11c		-	-	-	-		56±6		-	$C, i_\ell = kh^{0.95}, p$	
			-1.34c			2.66	QE	0.68 ± 0.05		56±6		-	$C, i_\ell = kh^{0.95}, n Av(2), p$	

EA58　C₄H₆O　　　　　　　　　　CRC Handbook Series in Organic Electrochemistry

Code No.	Empirical Formula	Name and C.A. Number	Structural Formula	Solvent	Tech.	Medium	μ, M	pH	T, °C	Electrodes	App.	Experimental Parameters
EA58 ac84 ca66	C_4H_6O	methyl vinyl ketone C.A. 78-94-4	$H_2C:CHCOCH_3$	MeOH 25	PY	UB	–	1 4.6 6 10	–	DME Hg/Hg_2Cl_2, LiCl 1	2A0	C=ns,m=4.9, t=1.9
EA59 ac95	$C_4H_7ClO_2$	ethyl chloroacetate C.A. 105-39-5	$CH_2ClCOOC_2H_5$	DMF	PY	Et_4NClO_4 0.1 Et_4NClO_4 0.1 3,4-xylenol 5E-4	–	–	≈ 21	DME/SCE	236A0	C=0.5,m= 1.26,t(oc)= 5.77,h=70, MP Aux
EA60 ca72	$C_4H_7ClO_4S_3$	2-(methylthio)-1,3-dithiolan-2-ylium perchlorate C.A. 2183-29-1	Table II-3	MeCN	VA	Et_4NClO_4 0.1	–	–	–	Pt Ag/$AgNO_3$ 0.01, Et_4NClO_4 0.1	2A2	v < 1E3
EA61	$C_4H_7FO_2$	ethyl fluoroacetate C.A. 459-72-3	$CH_2FCOOC_2H_5$	DMF	PY	Et_4NClO_4 0.1	–	–	≈ 21	DME/SCE	236A0	C>5,m=2.28, t=3.5,h=70, MP Aux
EA62	$C_4H_7KOS_2$	potassium O-isopropyl dithiocarbonate C.A. 140-92-1	$(CH_3)_2CHOC(S)SK$	Me_2CO	PY	Et_4NClO_4 0.1	–	–	20± 0.1	DME Ag/AgCl, LiCl 0.1	12A0	C=1,Pt Aux
EA63	$C_4H_7N_3O_3$	pyruvic acid semicarbazone C.A. 2704-30-5	$H_2NCONHN:C(CH_3)-COOH$	H_2O	PY	MB	–	0.6 2.7 3.5 4.1 4.5 5.3 6.75	–	DME/SCE	0A0	C=0.5(?),m= 2.62,t=2.8, h=55

TABLE I. Electrochemical Data $C_4H_7N_3O_3$ EA63

Ref.	C/M	Charact. Potential		Response Const.		Tech.	n	Electrokinetic Data			Products and Identification	Description and Remarks	Code No.
		Value	vs.		Value			Parameter	Value	From			
JE035 0381	156 i	$E_{\frac{1}{2}}$ -0.95F	Hg/ Hg_2Cl_2, LiCl 1	-	-	sttd	1	$dE_{\frac{1}{2}}/dpH$	69	plot	-	C,r	EA58 ac84 ca66
		-1.2F		-	-		1		69		-	C,r	
		-1.53F		-	-		1		0			C	
		-1.33F		-	-		-		69		-	C,r	
		-1.53F		-	-		-		0			C	
	156 j	-1.6F		-	-	-	-	-	-	-	-	C, $i_\ell \to 0$ with time for pH > 11,r	
JE044 0025	377 c	$E_{\frac{1}{2}}$ -1.84	SCE	i_p	1.6F	QE	1	-	-	-	CP → Cl⁻(100%)	C,$E_{\frac{1}{2}}$ corr.(E_j),≠, i_d,r	EA59 ac95
		-1.84			2.5F	-	-	-	-	-	-	C,$E_{\frac{1}{2}}$ corr.(E_j),i_ℓ ↑ as [3,4-xylenol] ↑ for [3,4-xylenol]/C < 1,r	
JE049 0105	338 b	E_p -0.5	SCE	-	-	QE	0.94	$E_p-E_{p/2}$	48	-	2,2'-bis(methyl-thio)-2,2'-bi(1,3-dithiolanyl),MAS,NMR	C,≠,$E_p-E_{p/2}$ and $j_p/Cv^{\frac{1}{2}} \ne f(v)$,p	EA60 ca72
								$dE_p/dlogv$	-19.3 ±0.6				
								$dE_p/dlogC$	19.8 ±2.3				
		-		-		-			-			A,X	
JE049 0085	-	$E_{\frac{1}{2}}$ -2.27	SCE	-	-	-	-	-	-	-	-	C,CH_2FCOOH redn.,p	EA61
JE048 0071	333 a	$E_{\frac{1}{2}}$ -0.358	Ag/AgCl, LiCl 0.1	-	-	-	-	Tomeš	59	-	-	A,$i_\ell=kh^{\frac{1}{2}}$,i_d,$i_\ell=kC$ for C=0.5-10,$E_{\frac{1}{2}} \to$ more neg. as C ↑,r	EA62
		0.094		-		-			76			A,$i_\ell=kh^{\frac{1}{2}}$,i_d,$i_\ell=kC$ for C=0.5-10,$E_{\frac{1}{2}} \to$ more pos. as C ↑	
		-		-					-			A,M	
C0036 0331	388 a	$E_{\frac{1}{2}}$ -0.55F	SCE	i_ℓ	9.6F	i:i	2.5	-	-	-	-	C,i_d,p	EA63
		-0.72F			9.4F		2.5	$2.3RT/\alpha n_a F$	74	Elog	-	C,i_d,p	
		-0.91F			8.3F		-	-	-	-	-	C,i_d,Elog non-linear,p	
		-1.00F			5.9F		-	-	-	-	-	C,p	
		-1.08F			3.4F		-	-	-	-	-	C,p	
		-1.11F			0.8F		-	-	-	-	-	C,i_k,Tc=5.6,p	
	388 b	-1.22F			-		-	-	-	-	-	C,i_k,p	

Code No.	Empirical Formula	Name and C.A. Number	Structural Formula	Solvent	Tech.	Medium		μ, M	pH	T, °C	Electrodes	App.	Experimental Parameters
EA64	C_4H_8	1-butene C.A. 106-98-9	$CH_2:CHCH_2CH_3$	MeCN	VY	Et_4NBF_4	0.1	–	–	25	RPDE Ag/Ag^+ 0.01	2-0	$C \approx 1, A=0.12$, $v=50, \omega=16.3$, Pt Aux
				CH_2Cl_2	VY	Bu_4NBF_4	0.2	–	–	25	RPDE Ag/Ag^+ 0.01	2-0	$C \approx 1, A=0.12$, $v=50, \omega=16.3$, Pt Aux
				$EtNO_2$	VY	Et_4NBF_4	0.2	–	–	25	RPDE Ag/Ag^+ 0.01	2-0	$C \approx 1, A=0.12$, $v=50, \omega=16.3$, Pt Aux
				$MeNO_2$	VY	Et_4NBF_4	0.1	–	–	25	RPDE Ag/Ag^+ 0.01	2-0	$C \approx 1, A=0.12$, $v=50, \omega=16.3$, Pt Aux
				PCA	VY	Et_4NBF_4	0.1	–	–	25	RPDE Ag/Ag^+ 0.01	2-0	$C \approx 1, A=0.12$, $v=50, \omega=16.3$, Pt Aux
				SULN	VY	Et_4NBF_4	0.2	–	–	25	RPDE Ag/Ag^+ 0.01	2-0	$C \approx 1, A=0.12$, $v=50, \omega=16.3$, Pt Aux
EA65	C_4H_8	2-butene C.A. 107-01-7	$CH_3CH:CHCH_3$	MeCN	VY	Et_4NBF_4	0.1	–	–	25	RPDE Ag/Ag^+ 0.01	2-0	$C \approx 1, A=0.12$, $v=50, \omega=16.3$, Pt Aux
				CH_2Cl_2	VY	Bu_4NBF_4	0.2	–	–	25	RPDE Ag/Ag^+ 0.01	2-0	$C \approx 1, A=0.12$, $v=50, \omega=16.3$, Pt Aux
				$EtNO_2$	VY	Et_4NBF_4	0.2	–	–	25	RPDE Ag/Ag^+ 0.01	2-0	$C \approx 1, A=0.12$, $v=50, \omega=16.3$, Pt Aux
				$MeNO_2$	VY	Et_4NBF_4	0.1	–	–	25	RPDE Ag/Ag^+ 0.01	2-0	$C \approx 1, A=0.12$, $v=50, \omega=16.3$, Pt Aux
				PCA	VY	Et_4NBF_4	0.1	–	–	25	RPDE Ag/Ag^+ 0.01	2-0	$C \approx 1, A=0.12$, $v=50, \omega=16.3$, Pt Aux
				SULN	VY	Et_4NBF_4	0.2	–	–	25	RPDE Ag/Ag^+ 0.01	2-0	$C \approx 1, A=0.12$, $v=50, \omega=16.3$, Pt Aux
EA66 ad01 ca74	$C_4H_8Br_2$	1,4-dibromobutane C.A. 110-52-1	$BrCH_2CH_2CH_2CH_2Br$	DMF	CP	Et_4NBr	0.1	–	–	–	Hg $Ag/AgClO_4$ 0.01, Et_4NBr 0.1	2A-	$E_{app} = -2.0$ V, $C=100$, Pt Aux
				MeCN	VR	Bu_4NClO_4	0.1	–	–	–	PMC $Ag/AgClO_4$ 0.01, Et_4NBr 0.1	2A0	$C=10, A=0.126, v=100$

TABLE I. Electrochemical Data $C_4H_8Br_2$ EA66

Ref.	C/M	Charact. Potential		Response Const.		n		Electrokinetic Data			Products and Identification	Description and Remarks	Code No.
		Value	vs.	Value		Tech.		Parameter	Value	From			
JE042 0133	-	$E_{\frac{1}{2}}$	2.48	Fc	-	-	-	-	-	-	-	-	A,p	EA64
JE042 0133	-	$E_{\frac{1}{2}}$	2.32	Fc	-	-	-	-	-	-	-	-	A,p	
JE042 0133	-	$E_{\frac{1}{2}}$	2.41	Fc	-	-	-	-	-	-	-	-	A,p	
JE042 0133	-	$E_{\frac{1}{2}}$	2.47	Fc	-	-	-	-	-	-	-	-	A,p	
JE042 0133	-	$E_{\frac{1}{2}}$	2.43	Fc	-	-	-	-	-	-	-	-	A,p	
JE042 0133	-	$E_{\frac{1}{2}}$	2.43	Fc	-	-	-	-	-	-	-	-	A,p	
JE042 0133	-	$E_{\frac{1}{2}}$	2.04	Fc	-	-	-	-	-	-	-	-	A,p	EA65
JE042 0133	-	$E_{\frac{1}{2}}$	1.88	Fc	-	-	-	-	-	-	-	-	A,p	
JE042 0133	-	$E_{\frac{1}{2}}$	1.95	Fc	-	-	-	-	-	-	-	-	A,p	
JE042 0133	-	$E_{\frac{1}{2}}$	2.04	Fc	-	-	-	-	-	-	-	-	A,p	
JE042 0133	-	$E_{\frac{1}{2}}$	2.03	Fc	-	-	-	-	-	-	-	-	A,p	
JE042 0133	-	$E_{\frac{1}{2}}$	2.0	Fc	-	-	-	-	-	-	-	-	A,p	
JE043 0215	94 f	-	-	-	-	-	-	-	-	-	-	partial CP at -1.6 V → 1-bromobutane (26%),GLC;butane (6.7%),GLC;buta-diene,MAS; $C_4H_9HgC_4H_8HgC_4H_9$, CHA,MAS;partial CP at -2.0 V → dibutyl-mercury(24%),GLC; butane(6%),GLC; butadiene(trace), MAS;CP at -2.0 V → dibutylmercury(24%), GLC;1-bromobutane (8.5%),GLC;butane (3.3%),GLC;buta-diene,MAS;no cyclo-butane	C,r	EA66 ad01 ca74
JE043 0215	94 f	E_p	-1.96F	Ag/ AgClO$_4$ 0.01, Et$_4$NBr 0.1	i_p	4.9E3F	-	-	-	-	-	CP at -2.31 V → di-butylmercury(60%), GLC;butane,GLC; butadiene,MAS; partial CP at -2.9 V → 1-bromooctane (10%),GLC;1,8-di-bromooctane(0.05%), GLC;1-bromobutane (18%),GLC;buta-diene(1.6%),GLC	C,$E_{\frac{1}{2}}$ and i_ℓ=f(v and number of scans),p	
			-			-							A,o	

Code No.	Empirical Formula	Name and C.A. Number	Structural Formula	Solvent	Tech.	Medium		μ, M	pH	T, °C	Electrodes	App.	Experimental Parameters
EA67 ad08	$C_4H_8N_2O_2$	dimethylglyoxime C.A. 95-45-4	Table II-2	EtOH 10	PY	H_3CITR NaClO$_4$	5E-3 0.1	-	-	-	DME/SCE	---	C=0.71,m= 0.72,t=4.07
						H_3CITR NaClO$_4$	1E-3 0.1						C=0.17
				EtOH 50	PY	HClO$_4$ NaClO$_4$	3E-3 0.1	-	-	-	DME/SCE	---	C=0.1,m= 0.72,t=4.07
EA68 ad10	C_4H_8O	2-methylpropion- aldehyde C.A. 78-84-2	$(CH_3)_2$CHCHO	EtOH 2	PY	BARB LiCl		0.5	6.8	25± 0.02	DME/SCE	034A0	C=0.2,m=2.6 ±0.4,t=3.6 ±0.5,h=65 ±5
						LiOH LiCl	0.1 0.4	-	-		DME Hg/ Hg$_2$Cl$_2$, LiCl 2		
				EtOH 5	PY	BARB LiCl		0.5	7.1	2± 0.02	DME/SCE	034A0	C=0.05,m= 2.6±0.4,t= 3.6±0.5,h= 65±5
						LiOH LiCl	ns		13		DME Hg/ Hg$_2$Cl$_2$, LiCl 2		
									14				
				EtOH	PY	LiOH LiCl	ns	0.5	11.7$_5$	25± 0.02	DME Hg/ Hg$_2$Cl$_2$, LiCl 2	034A0	C=0.2,m=2.6 ±0.4,t=3.6 ±0.5,h=65 ±5
									12.8				
EA69	$C_4H_8O_4$	D-erythrose C.A. 583-50-6	Table II	H_2O	PY	NH_3 NH_4Cl	0.1 0.5	-	8.5	-	DME/SCE	04A0	C=1.0,m= 2.23,t=4.5, h=50
EA70	$C_4H_8O_4$	L-glycero-tetrulose C.A. 533-49-3	Table II	H_2O	PY	NH_3 NH_4Cl	0.1 0.5	-	8.5	-	DME/SCE	04A0	C=1.0,m= 2.23,t=4.5, h=50
EA71	$C_4H_8O_4$	D-threose C.A. 95-43-2	Table II	H_2O	PY	NH_3 NH_4Cl	0.1 0.5	-	8.5	-	DME/SCE	04A0	C=1.0,m= 2.23,t=4.5, h=50
EA72	$C_4H_9N_3O_3$	dihydroxyacetone semicarbazone C.A. 21205-27-6	H_2NCONHN:C- $(CH_2OH)_2$	H_2O	PY	$NH_2CONHNH_2$ $NH_2CONHNH_3Cl$		-	2.0	-	DME/SCE	0A0	C=0.5(?),m= 2.62,t=2.8, h=55
									3.0				
									4.0				
									5.0				
									6.3				

TABLE I. Electrochemical Data $C_4H_9N_3O_3$ EA72

Ref.	C/M	Charact. Potential		Response Const.		Tech.	n	Electrokinetic Data			Products and Identification	Description and Remarks	Code No.
		Value	vs.		Value			Parameter	Value	From			
JE043 0095	-	$E_{\frac{1}{2}}$	-0.92F	SCE	-	-	-	-	-	-	-	-	C,r	EA67 ad08
			-		-	-	-	-	-	-	-	-	C	
			-0.96F		-	-	-	-	-	-	-	-	C,r	
			-		-	-	-	-	-	-	-	-	C	
JE043 0095	-	$E_{\frac{1}{2}}$	-1.048	SCE	D i_ℓ	1.6 0.865	-	-	Tomeš	80	-	-	C,r	
			-		-	-	-	-	-	-	-	-	C,H	
JE046 0323	32 d	-	-	-	i_ℓ	0.58	i:i, PQ	0.53	-	-	-	-	C,$i_\ell \neq f(h)$, Tc > 2, i_k,r	EA68 ad10
	32 b	$E_{\frac{1}{2}}$	-1.84F	SCE	-	-	-	-	-	-	-	-	C,$\Delta E_{\frac{1}{2}}=0.2\sigma^*$,r	
JE046 0323	32 d	-	-	-	i_ℓ	0.5	-	-	-	-	-	-	C,r	
	32 b	-	-	-		2.73	-	-	-	-	-	-	C,r	
			-			2.35	-	-	-	-	-	-	C,r	
JE046 0323	32 b	-	-	-	i_ℓ	1.7	i:i, PQ	1.74	-	-	-	-	C,r	
			-			1.98		1.82	-	-	-	-	C,r	
C0036 0114	-	$E_{\frac{1}{2}}$	-1.44	SCE	-	-	-	-	-	-	-	-	C,i_k,$E_{\frac{1}{2}}$ and $i_\ell =$ f(pH),p	EA69
C0036 0114	267 c	$E_{\frac{1}{2}}$	-1.42	SCE	-	-	sttd	2	-	-	-	-	C,i_d,$E_{\frac{1}{2}}$ and $i_\ell \neq$ f(pH),p	EA70
			-1.62					< 2					C	
C0036 0114	-	$E_{\frac{1}{2}}$	-1.44	SCE	-	-	-	-	-	-	-	-	C,i_k,$E_{\frac{1}{2}}$ and $i_\ell =$ f(pH),p	EA71
C0036 0331	-	$E_{\frac{1}{2}}$	-0.90F	SCE	Σi_ℓ	-	-	-	-	-	-	-	C,$i_\ell = f[NH_2CONHNH_2]$,p	EA72
			-1.11F			8.0F							C	
			-1.00F			-	-	-	-	-	-	-	C,i_k,$i_\ell =$ f$[NH_2CONHNH_2]$,p	
			-1.16F			22F							C	
			-1.08F			-							C,$i=f[NH_2CONHNH_2]$,p	
			-1.26F			28F							C	
			-1.16F			-	-	-	-	-	-	-	C,$i=f[NH_2CONHNH_2]$,p	
			-1.37F			2.5F							C	
			-1.30F			-	-	-	-	-	-	-	C,$i=f[NH_2CONHNH_2]$,p	
			-1.47F			5.5F							C	

Code No.	Empirical Formula	Name and C.A. Number	Structural Formula	Solvent	Tech.	Medium		μ, M	pH	T, °C	Electrodes	App.	Experimental Parameters
EA73	$C_4H_9N_3O_3$	glyceraldehyde semicarbazone C.A. 21205-26-5	$H_2NCONHN:CH-CHOHCH_2OH$	H_2O	PY	$NH_2CONHNH_2$ $NH_2CONHNH_3Cl$		–	-2.0	–	DME/SCE	OAO	C=0.5(?), m=2.62, t=2.8, h=55
									0.0				
									1.0				
									2.0				
									3.0				
									4.5				
									5.3				
EA74	$C_4H_9N_3O_3$	2-semicarbazido-propionic acid C.A. 55846-35-0	$H_2NCONHNH-CH(CH_3)COOH$	H_2O	PY	NaOH	0.5	–	–	–	DME/SCE	OAO	m=2.62, t=2.8, h=55
EA75	$C_4H_{10}KO_2PS_2$	potassium 0,0-diethyl dithiophosphate C.A. 3454-66-8	$(C_2H_5O)_2P(S)SK$	H_2O	VA	ACET		–	3.8	–	Pt/SCE	2AO	C=0.1, A=0.917, v=50, PE
						PHOS			7.1				
													v=2.97
						BOR	0.025		9.3		Cu/SCE		C=0.5, A=2.85, v=50, PE
													C=2, A=2.85, v=0.5
													v=5
						BOR			11		Pt/SCE		C=0.1, A=0.917, v=50, PE
CONT		glyceraldehyde semi-	$H_2NCONHN:CH-CHOHCH_2OH$	H_2O	PY	$NH_2CONHNH_2$ $NH_2CONHNH_3Cl$			-2.0		DME/SCE		C=0.5(?),

TABLE I. Electrochemical Data $C_4H_{10}KO_2PS_2$ (CONT.) EA75

Ref.	C/M	Charact. Potential		Response Const.		n Tech.	n	Electrokinetic Data			Products and Identification	Description and Remarks	Code No.
		Value	vs.		Value			Parameter	Value	From			
C0036 0331	-	-	-	i_ℓ	7.2F	-	-	-	-	-	-	C,p	EA73
	$E_{\frac{1}{2}}$	-0.70F	SCE		7.2F	-	-	-	-	-	-	C,p	
		-0.80F			11.3F	-	-	-	-	-	-	C,p	
		-0.88F		Σi_ℓ	-	-	-	-	-	-	-	C,p	
		-1.04F			16.0F	-	-	-	-	-	-	C	
		-0.97F			-	-	-	-	-	-	-	C,p	
		-1.12F			21.3F	-	-	-	-	-	-	C	
		-1.04F			-	-	-	-	-	-	-	C,p	
		-1.18F			15.0F	-	-	-	-	-	-	C	
		-		i_ℓ	7.5F	-	-	-	-	-	-	C,p	
C0036 0331	388 ab	$E_{\frac{1}{2}}$ -0.25	SCE	-	-	sttd	2	-	-	-	2-oxopropanoic acid semicarbazone	A,p	EA74
JE056 0217	408 a	E_p 1.24	NHE	i_p	240	-	-	$E_p-E_{p/2}$	130	-	-	A,≠,p	EA75
								βn_b	0.37	N-S			
								dE_p/dpH	60	-			
		0.44F			-				60			C,PtO redn., $F_p \neq f(v)$	
		≈ -0.3F							-			C,M,bis(diethoxy-thiophosphoryl)di-sulfide redn.	
		1.03			227.7F	-	-	$E_p-E_{p/2}$	130	-		A,p	
								βn_b	0.37	N-S			
								dE_p/dpH	60	-			
		0.29F			-		-		60			C	
		-							-			C	
		0.91F			10F	-	-	$dE_p/d\log v$	60	-	-	A,E_p and $i_p=k_1\log v + k_2$,p	
		-			-				-			C	
	408 b	0.09		D i_p	7.4 170	-	-	βn_b	0.38	N-S		A,$E_p \neq f(C)$ for C= 0.2-2, $i_p=kC$,p	
		-0.7			260				-			C,$E_p \neq f(C)$ for C= 0.2-2, $i_p=f(C)$	
		-0.8			310							C,$E_p \neq f(C)$ for C= 0.2-2, $i_p=f(C)$	
		-0.2F		i_p	46.2F	-	-	-	-	-	-	A,$i_p=k_1v + k_2$,p	
		-0.37F			43F							C	
		-0.64F			26.9F							C	
		-0.07F		D i_p	4.6 160F	-	-	$di_p/dv^{\frac{1}{2}}$	72.7	-		A,$i_p=k_1v + k_2$,D from Dplot,p	
								βn_b	0.4	Dplot			
		-0.57F			110F				-			C	
		-0.7F			85F							C	
	408 a	-0.75		i_p	80F	-	-	$E_p-E_{p/2}$	220	-		A,p	
								βn_b	0.22				
								dE_p/dpH	60				
		0F			-				60			C	
		-							-			C	

CONT

EA75 (CONT.) $C_4H_{10}KO_2PS_2$

Code No.	Empirical Formula	Name and C.A. Number	Structural Formula	Solvent	Tech.	Medium	μ, M	pH	T, °C	Electrodes	App.	Experimental Parameters
EA75	$C_4H_{10}KO_2PS_2$	potassium O,O-diethyl dithiophosphate	$(C_2H_5O)_2P(S)SK$	H_2O	VA	BOR	-	11.7	-	Cu/SCE	2AO	C=2, A=2.85, v=50, PE
EA76	$C_4H_{10}N_2$	3-ethyl-3-methyl-diaziridine C.A. 4901-75-1	Table II	MeOH 20	PY	BR	-	2.4	20	DME Ag/AgCl	OAO	C=0.66, m=1.65, t=0.16
								3.5				
								4.2				
								5.0				
								5.3				
								6.0				
								7.0				
								7.6				
								9.0				
EA77	$C_4H_{12}ClNS$	2-mercaptoethyl-N,N-dimethylammonium chloride C.A. 13242-44-9	$[(CH_3)_2NHCH_2CH_2SH]^+ Cl^-$	H_2O	PY	BR KNO_3 0.1	-	4.4	26	DME/SCE	OAO	C=1, m=1.427, t=3.62, EC=1.50
								9.6				
								10.8				
				MeOH 50	PY	BR KNO_3 0.1	-	6.48	26	DME/SCE	OAO	C=1, m=1.427, t=3.62, EC=1.57
								10.0				
								11.6				
				EtOH 50	PY	BR KNO_3 0.1	-	5.42	26	DME/SCE	OAO	C=1, m=1.427, t=3.62, EC=1.57
								10				
								12				
				MeCN 25	PY	BR KNO_3 0.1	-	7.2	26	DME/SCE	OAO	C=1, m=1.427, t=3.62, EC=1.57
								9.2				
								11.6				
EA78	$C_4H_{22}B_{18}Ni$	bis[(7,8,9,10,11-η)-undecahydro-1,7-dicarbaundecaborato-(11)$^{2-}$]nickelate(1-) C.A. 51850-05-6	$[(\pi-1,7-B_9C_2H_{11})_2Ni]^-$	MeCN	PY	Bu_4NPF_6 0.3	-	-	25	DME/SCE	2AO	ns
CONT												

TABLE I. Electrochemical Data $C_4H_{22}B_{18}Ni$ (CONT.) EA78

Ref.	C/M	Charact. Potential		Response Const.		n		Electrokinetic Data			Products and Identification	Description and Remarks	Code No.
		Value	vs.		Value	Tech.		Parameter	Value	From			
JE056 0217	408 b	E_p	0.02 -0.65 -0.75	NHE	i_p	170 170 200	- - -	- - -	-	- - -	-	-	A,p C C	EA75
C0030 4178	295 c	$E_{\frac{1}{2}}$	-0.50F -1.31F	Ag/AgCl	i_ℓ	2.4F 4.9F	-	-	-	-	-	-	C,r C	EA76
			-0.55F -1.31F			2.4F 4.7F	-	-	-	-	-	-	C,r C	
			-0.62F -1.31F			2.5F 3.9F	-	-	-	-	-	-	C,r C	
			-0.80F -1.31F			2.1F 3.5F	-	-	-	-	-	-	C,r C	
			-0.96F -1.33F			1.8F 3.5F	-	-	-	-	-	-	C,r C	
			-1.00F -1.34F			1.4F 2.9F	-	-	-	-	-	-	C,r C	
			-1.01F -1.34F			0.6F 1.7F	-	-	-	-	-	-	C,r C	
			-1.03F -1.34F			0.3F 1.0F	-	-	-	-	-	-	C,r C	
			-1.03F -1.39F			0.1F 0.2F	-	-	-	-	-	-	C,r C	
JE036 0249	-	$E_{\frac{1}{2}}$	-0.22F	SCE	D	2.62	-	-	-	-	-	-	A,p	EA77
			-0.5F			-	-	-	-	-	-	-	A,p	
			-0.56F			-	-	-	-	-	-	-	A, $E_{\frac{1}{2}} \neq f(pH)$ for pH $\geq$ 10.8,p	
JE036 0249	-	$E_{\frac{1}{2}}$	-0.27F	SCE	D	2.89	-	-	-	-	-	-	A,p	
			-0.6F			-	-	-	-	-	-	-	A,p	
			-0.63F			-	-	-	-	-	-	-	A, $E_{\frac{1}{2}} \neq f(pH)$ for pH $\geq$ 11.6,p	
JE036 0249	-	$E_{\frac{1}{2}}$	-0.3F	SCE	D	0.74	-	-	-	-	-	-	A,p	
			-0.62F			-	-	-	-	-	-	-	A,p	
			-0.66F			-	-	-	-	-	-	-	A,p	
JE036 0249	-	$E_{\frac{1}{2}}$	-0.4F	SCE	D	2.89	-	-	-	-	-	-	A,p	
			-0.6F			-	-	-	-	-	-	-	A,p	
			-0.66F			-	-	-	-	-	-	-	A,p	
JE050 0031	-	$E_{\frac{1}{2}}$	- -0.92 -	SCE	D	- 30 -	sttd	- 1 -	Elog	- ≈ 59 -	-	$[(\pi-1,7-B_9C_2H_{11})_2Ni]^{2-}$	A,M,r C,R C	EA78
														CONT

EA78 (CONT.) $C_4H_{22}B_{18}Ni$

Code No.	Empirical Formula	Name and C.A. Number	Structural Formula	Solvent	Tech.	Medium		μ, M	pH	T, °C	Electrodes	App.	Experimental Parameters
EA78	$C_4H_{22}B_{18}Ni$	bis[(7,8,9,10,11-η)-undecahydro-1,7-dicarbaundecaborato-(11)$^{2-}$]nickelate(1-)	[(π-1,7-B_9C_2-H_{11})$_2$Ni]$^-$	MeCN	PVI	Bu_4NPF_6	0.3	-	-	25	DME/SCE	2AO	ns
EA79	$C_4H_{22}B_{18}Ni$	bis[(7,8,9,10,11-η)-undecahydro-7,8-dicarbaundecaborato-(11)$^{2-}$]nickelate(1-) C.A. 51159-91-2	[(π-1,2-B_9C_2-H_{11})$_2$Ni]$^-$	MeCN	PY	Bu_4NPF_6	0.3	-	-	25	DME/SCE	2AO	ns
					PVI	Bu_4NPF_6	0.3	-	-	25	DME/SCE	2AO	ns
EA80 ae03	$C_5H_4N_4S$	6-mercaptopurine C.A. 50-44-2	Table 11-2	H_2O	PO	NaOH	1.0	-	-	-	SME/-	OB2	C=0.5
EA81	$C_5H_5NO_2$	N-methylmaleimide C.A. 930-88-1	Table 11	EtOH 50	PY	BR		-	7.25	-	DME/SCE Tl+	O4AO	C=1.0, m=3.3, t=3.5, h=40
EA82 ae13	$C_5H_5N_5$	adenine C.A. 73-24-5	Table 11-2	H_2O	PY	MB		-	3.4	0	DME/SCE	12AO	C=0.1, m=1.7-2.2, t=3-4, Pt Aux
													C=0.5
									4.5				
									5.0				C=0.1
					PO	H_2SO_4	1.0	-	-	-	SME/-	OB2	C=0.5
					PVI	MB		0.5	3.4	0	DME/SCE	12AO	C=0.1, m=1.0, t(c)=3, f=50, Δe=3.54(rms), Pt Aux
													C=0.5
									4.5				
									5.0				C=0.1

TABLE I. Electrochemical Data $C_5H_5N_5$ EA82

Ref.	C/M	Charact. Potential Value	vs.	Response Const.	Value	Tech.	n	Electrokinetic Data Parameter	Value	From	Products and Identification	Description and Remarks	Code No.
JE050 0031	-	-	-	-	-	-	-	α	-	eqn	-	A,r	EA78
								$k_{s,h}^{app}$	-				
								$k_{s,h}$	-				
		-			-				0.47			C	
									1.6				
									11				
		-			-				-			C	
JE050 0031	-	$E_\frac{1}{2}$	0.25	SCE	D	34	sttd	1	Elog	≈59	-	$(\pi-1,2-B_9C_2H_{11})_2Ni$	A,R,r	EA79
			-0.57			34		1		≈59		$[(\pi-1,2-B_9C_2H_{11})_2Ni]^{2-}$	C,R
			-			-		-		-			C
JE050 0031	-	-	-	-	-	-	-	α	0.5	eqn	-	A,r	
								$k_{s,h}^{app}$	0.43				
								$k_{s,h}$	0.14				
		-			-				0.5			C	
									0.66				
									1.9				
		-			-				-			C	
C0029 0182	-	Q	0.35	-	-	-	-	-	-	-	-	-	C,p	EA80 ae03
C0026 2749	-	$E_\frac{1}{2}$	-0.782	SCE	-	-	sttd	1	-	-	-	-	C, i_ℓ=f(t),p	EA81
			-1.218					1		-			C, i_ℓ=f(t)
JA095 8495	113 mn	$E_\frac{1}{2}$	-1.231	SCE	i_ℓ I D	1.2 8.7 12.8	1:1 QE	4 6.3	Elog	50	sttd	-	C, i_d,D(||k),≠,r	EA82 ae13
			-1.274			6.4 9.5 15.3				76		-	C,i_d,≠,$E_\frac{1}{2}$→more neg. as C↑,r
			-1.344		i_ℓ I	6.8 10.2	-	-		84		-	C,i_d,≠,r
			-1.372			1.3 9.9	-	-		69		-	C,i_d,≠,r
C0029 0182	-	Q	0.80	-	-	-	-	-	-	-	-	-	C,μ	
JA095 8495	113 n	E_{su}	1.30	SCE	i_{su}/C	1.53	-	-	-	-	-	-	r,i_d,≠,r
			-1.32			1.12	-	-	-	-	-	-	C,≠,E_{su}→more neg. and Δi_{su}/C↓ as C ↑,r
			-1.38			1.24	-	-	-	-	-	-	C,≠,r
			-1.43			1.30	-	-	-	-	-	-	C,≠,r

Code No.	Empirical Formula	Name and C.A. Number	Structural Formula	Solvent	Tech.	Medium	μ, M	pH	T, °C	Electrodes	App.	Experimental Parameters
EA83	$C_5H_6N_2OS$	4-methylthiouracil C.A. 35551-31-6	Table II	H_2O	PY	BR KCl	0.2	1	25	DME/SCE	OAO	C=0.5,m=2.54, t=3.3,h=60
								2.5				
								6				
								6.8				
								12				
					PV	BR KCl	0.2	4.3	25	DME/SCE	OAO	C=0.5,m=2.81, t=4.8,h=55, Δe=20,f=78
								10.5				
					VA	BR KCl	0.2	5	25	HMDE SCE	OAO	C=ns,v=25
								8				
EA84	$C_5H_6N_2S_2$	5-methyl-2,4-dithiouracil C.A. 6217-61-4	Table II	H_2O	PO	NaOH	1.0	-	-	SME/-	OB2	C=0.5
EA85	$C_5H_6O_3$	4-oxo-2-pentenoic acid C.A. 4743-82-2	$CH_2COCH:CHCOOH$	H_2O	PY	BR(?)	-	4.45	-	DME/SCE	O4AO	C=1.0(?),m= 4.0,t=2.1, h=65
								5.5				
								5.85				
								6.3				
CONT		4-methylthiouracil		H_2O	PY	BR		1		DME/SCE	OAO	

TABLE I. Electrochemical Data $C_5H_6O_3$ (CONT.) EA85

Ref.	C/M	Charact. Potential		Response Const.		n		Electrokinetic Data			Products and Identification	Description and Remarks	Code No.
		Value	vs.	Value		Tech.		Parameter	Value	From			
JE048 0433	89 e	$E_{\frac{1}{2}}$	-0.82	SCE	-	-	-	-	-	-	-	-	C, Pr, i_a, r	EA83
			-0.83F						$dE_{\frac{1}{2}}/dpH$	5			$C, E_{\frac{1}{2}} = -0.830-0.005pH,$ $i_d, i_\ell = kh^{\frac{1}{2}}, i_\ell = kC$ for $C=0.1-1, Tc=2.6$	
			-1.24F							40			$C, E_{\frac{1}{2}} = -1.240-0.040pH,$ i_{cat} for $pH < 3, X$	
			-0.83	-	-	-	-	-	-	-	-	-	C, Pr, i_a, r	
			-0.84F		2.1F				5			C, i_d	
			-1.29F		6.9F				40			C, i_{cat}, M	
			-1.14		-	QE	-		-			C, Pr, i_a, r	
			-1.15F		1.88F		2.2		100		$HSCH_3$	$C, E_{\frac{1}{2}} = -0.840-0.100pH$ $(?), i_d, pK'=2.9$	
			-1.43F		2.13F		-		40		6,6'-bis(3,6-dihydropyrimidone-2), UVS; 3,6-dihydropyrimidone-2, UVS	$C, i_d + i_k, i_\ell = kh^{\frac{1}{2}}, M,$ $Tc=1.70$	
			-1.3F		4.5	-	-		60		6,6'-bis(3,6-dihydropyrimidone), UVS; 3,6-dihydropyrimidone-2, UVS	$C, E_{\frac{1}{2}} = -1.100-0.060pH$ $(?), 2$ merging waves, r	
			-1.58F		4F	-	-		60			C, i_d for $pH > 9.5$, $Tc=1.8, i_\ell = kC$ for $C=0.1-1.2, r$	
JE048 0433	89 e	E_{su}	-1.19F	SCE	-	-	-	-	dE_{su}/dpH	80	sttd	-	$C, E_{su} = -0.850$ $-0.080pH$ for $pH=$ $1.8-7, E_{su} \neq f(C), i_{su} =$ kC, p	
			-1.38F		-		-			-			$C, E_{su} \neq f(pH)$ for $pH=$ $1.8-7, E_{su} \to$ more neg. as $C \uparrow$	
			-1.6F		-	-	-			65		-	$C, E_{su} = -1.400$ $-0.065pH(?), p$	
JE048 0433	89 e	E_p	-0.9	SCE	-	-	-	-	-	-	-	-	C, i_a, X, p	
			-1.1										C	
			-										A, i_a	
			-0.2										A, X	
			-1.3										C, p	
			-0.37										A	
C0029 0182	-	U	0.29	-	-	-	-	-	-	-	-	-	C, R, p A, R	EA84
			0.29											
C0027 2717	-	$E_{\frac{1}{2}}$	-0.45F	SCE	i_ℓ	31F	-	-	-	-	-	-	C, p	EA85
			-0.55F		29F								C, p	
			-0.7F		2F								C	
			-0.6F		23F								C, p	
			-0.75F		10F								C	
			-0.65F	-	17F	-	-	-	-	-	-	-	C, p	
			-0.8F		15								C	
			-1.1F		2.5								C	
														CONT

EA85 (CONT.) $C_5H_8O_3$

Code No.	Empirical Formula	Name and C.A. Number	Structural Formula	Solvent	Tech.	Medium		μ, M	pH	T, °C	Electrodes	App.	Experimental Parameters
EA85	$C_5H_8O_3$	4-oxo-2-pentenoic acid	$CH_3COCH:CHCOOH$	H_2O	PY	BR(?)		-	6.9	-	DME/SCE	O4AO	C=1.0(?),m=4.0,t=2.1,h=65
									7.2				
									7.75				
									8.35				
									8.75				
									9.1				
									9.8				
									10.35				
									11.4				
						NaOH	0.05		-				
						NaOH	0.3						
EA86	$C_5H_7N_3O_2$	3,5-dimethyl-6-aza-uracil	Table II	H_2O	PY	BR		-	2.5	25	DME/SCE Tl$^+$	O4AO	C=0.24,h=90
									3.9				
									5.6				
									7.7				
									10.7				
EA87 ae48	C_5H_8O	2-methyl-2-butenal C.A. 1115-11-3	$CH_3CH:C(CH_3)CHO$	H_2O	PY	buffer KCl		1.0 ±0.2	2	25.00 ±0.05	DME/SCE	OAO	ns
									6				
						BOR KCl			10				
EA88	C_5H_8O	3-methyl-3-buten-2-one C.A. 814-78-8	$H_2C:CCH_3COCH_3$	MeOH 25	PY	UB		-	1.0	-	DME Hg/ Hg_2Cl_2, LiCl 1.0	2AO	C=ns,m=4.9,t=1.9
									4.6				
									10.5				
				MeOH	PY	$LiClO_4$ $C_6H_5SO_3H$	0.1 2E-4	-	-	-	DME Ag/AgCl, LiCl satd	2AO	C=1,m=4.9,t=1.9
EA85	$C_5H_8O_3$	4-oxo-2-pentenoic acid	$CH_3COCH:CHCOOH$										

TABLE I. Electrochemical Data C_5H_8O EA88

Ref.	C/M	Charact. Potential		Response Const.		Tech.	n	Electrokinetic Data			Products and Identification	Description and Remarks	Code No.
		Value	vs.		Value			Parameter	Value	From			
C0027 2717	-	$E_{\frac{1}{2}}$	-0.7	SCE	i_ℓ	5	-	-	-	-	-	-	C, i_k, p	EA85
			-0.85			24							C	
			-1.05			4							C	
			-0.7F			1F	-	-	-	-	-	-	C, i_k, p	
			-0.87F			21F							C	
			-1.02F			9F							C	
			-0.87F			15F							C,p	
			-1.00F			18F							C	
			-0.87F			6F							C, i_k, p	
			-1.00F			27F							C	
			-0.87F			1							C, i_k, p	
			-1.05F			28							C	
			-1.05F			29F							C,p	
			-1.35F			1.5F							C	
			-1.05F			24F							C,p	
			-1.37F			5F							C	
			-1.05F			19F							C,p	
			-1.38F			10F							C	
			-1.05F			19F							C,p	
			-1.38F			10F							C	
			-1.05F			15F							C,p	
			-1.38F			14F							C	
			-1.05F			26F							$C, i_\ell = f([Na^+]), p$	
			-1.4F			12F							$C, M, i_\ell = f([Na^+])?$	
C0027 0546	-	$E_{\frac{1}{2}}$	-0.70F	SCE	i	1.35F	sttd	2	-	-	-	-	C,p	EA86
			-0.80F			1.25F		2					C,p	
			-1.30F			1.30F		2					C,X,p	
			-1.40F			1.30F		2					C,p	
			-1.58F			0.60F		1					C,p	
JE049 0433	-	$E_{\frac{1}{2}}$	-1.03c	SCE	-	-	-	-	$dE_{\frac{1}{2}}/dpH$	71±6	sttd	-	$C, E_{\frac{1}{2}} = -0.886 - 0.071$ (±0.006)pH for pH= 2-10, p	EA87 ae48
			-1.31c		-	-				71±6		-	C,p	
	-		-1.60c		I	2.64	QE	1.03 ± 0.07		71±6		CP→polymer, IRS, MAS	$C, i_\ell = kh^{0.56}, n\,Av(5), p$	
JE035 0381	156 I	$E_{\frac{1}{2}}$	-1.02F	Hg/ Hg_2Cl_2, LiCl 1.0	-	-	sttd	1	$dE_{\frac{1}{2}}/dpH$	69	plot	-	C,r	EA88
			-1.27F		-	-		1		69	-	-	C,r	
			-1.55F					1		0			C	
	156 J		-1.68F		-	-	-	-		0	-	-	$C, i_\ell \to 0$ on standing at pH > 11, $E_{\frac{1}{2}} \neq f(pH)$ for pH=10.5-13, r	
JE035 0381	156 J	$E_{\frac{1}{2}}$	-0.95F	Ag/AgCl, LiCl satd	i_ℓ	0.94F	-	-	-	-	-	-	C,r	
			-1.55F			4.8F							C	
			-1.74F			2.5F							C	

Code No.	Empirical Formula	Name and C.A. Number	Structural Formula	Solvent	Tech.	Medium	μ, M	pH	T, °C	Electrodes	App.	Experimental Parameters
EA89	C_5H_8OS	3-oxothiane C.A. 19090-03-0	Table II	H_2O	PY	BR	-	3	20	DME/SCE	2-0	C=0.5
								4.5				
								7.5				
								10				
EA90	C_5H_8OS	4-oxothiane C.A. 1072-72-6	Table II	H_2O	PY	BR	-	8	20	DME/SCE	2-0	C=0.5
								10.5				
EA91	$C_5H_8O_2$	4-oxooxane C.A. 29943-42-8	Table II	H_2O	PY	BR	-	8	20	DME/SCE	2-0	C=0.5
								12				
EA92	$C_5H_9BF_4S_3$	2-ethylthio-1,3-dithiolanium tetrafluoroborate C.A. 51823-96-2	Table II	MeCN	VA	Et_4NClO_4 0.1	-	-	-	Pt Ag/$AgNO_3$ 0.01, Et_4NClO_4 0.1	2A2	v < 1E3
EA93	$C_5H_9ClO_4S_3$	2-methylthio-1,3-dithianium perchlorate C.A. 51348-33-5	Table II	MeCN	VA	Et_4NClO_4 0.1	-	-	-	Pt Ag/$AgNO_3$ 0.01, Et_4NClO_4 0.1	2A2	v < 1E3
EA94	$C_5H_9KOS_2$	potassium O-butyl-dithiocarbonate C.A. 871-58-9	$CH_3CH_2CH_2CH_2$-$OC(S)SK$	Me_2CO	PY	Et_4NClO_4 0.1	-	-	20±0.1	DME Ag/AgCl, LiCl 0.1	12A0	C=1, Pt Aux
					PV	Et_4NClO_4 0.1	-	-	20±0.1	DME Ag/AgCl, LiCl 0.1	12A0	C=10, Δe=10, f=50, Pt Aux

TABLE I. Electrochemical Data $\quad$ $C_5H_9KOS_2$ $\quad$ EA94

Ref.	C/M	Charact. Potential		Response Const.		n Tech.	n	Electrokinetic Data			Products and Identification	Description and Remarks	Code No.
		Value	vs.		Value			Parameter	Value	From			
JE039 0195	400 a	$E_{\frac{1}{2}}$	-1.08F	SCE	i_ℓ	4.2F	QE	1	-	-	-	CP at MP → 3,3'-di-hydroxy-3,3'-bi-thiane, MAS	C, compare $i_\ell=f(pH)$ with $n=f(pH)$, p	EA89
			-1.11F			1.2F		-	-	-	-	CP at MP → 3,3'-di-hydroxy-3,3'-bi-thiane, MAS	C,0 at pH > 5, p	
	400 b		-1.36F			2.4F		-	-	-	-	CP at MP → 3-hydroxythiane, MAS	C, p	
			-1.4F			2.4F		2	-	-	-	CP at MP → 3-hydroxythiane, MAS	C, p	
JE039 0195	400 ab	$E_{\frac{1}{2}}$	-1.5F	SCE	-	-	-	-	-	-	-	-	C, p	EA90
			-1.59F		-	-	QE	2	-	-	-	CP → 4-hydroxy-thiane, MAS	C, p	
JE039 0195	400 ab	$E_{\frac{1}{2}}$	-1.53F	SCE	-	-	-	-	$dE_{\frac{1}{2}}/dpH$	5	plot	-	C, p	EA91
			-1.55F		-	-	-	-		5		CP at pH=10 → 4-hydroxyoxane, MAS	C, p	
JE049 0105	338 c	E_p	-0.48	SCE	-	-	QE	1.0	$E_p-E_{p/2}$	50	-	2,2',5,5'-tetra-thiabicyclopentyl-idene, MAS, NMR, VA; dimethyldisulfane, MAS, NMR	C,≠,$E_p-E_{p/2}$ and $J_p/Cv^{\frac{1}{2}}\neq f(v)$, p	EA92
			-					-		-			A, X	
JE049 0105	338 b	E_p	-0.43	SCE	-	-	QE	0.94	$E_p-E_{p/2}$	43	-	2,2'-bis(methyl-thio)-2,2'-bi(1,3-dithianyl), MAS, NMR	C,≠,$E_p-E_{p/2}$ and $J_p/Cv^{\frac{1}{2}}\neq f(v)$, p	EA93
									$dE_p/d\log v$	-18.7 ±0.9	-			
									$dE_p/d\log C$	24.7 ±2.0	-			
			-					-					A	
JE048 0071	333 a	$E_{\frac{1}{2}}$	-0.36	Ag/AgCl, LiCl 0.1	-	-	-	-	Elogx	33	-	-	A, Elogx=E vs. $\log[i/(i_\ell-i)^3]$, $i_\ell = kh^{\frac{1}{2}}$, i_d, $i_\ell=kC$ for C= 0.5-10, $E_{\frac{1}{2}}$ → more neg. as C ↑, r	EA94
									Tomeš	60				
			0.091						Elogy	29			A, Elogy=E vs. $\log[i^3/(i_\ell-i)^2]$, $i_\ell = kh^{\frac{1}{2}}$, i_d, $i_\ell=kC$ for C= 0.5-10, $E_{\frac{1}{2}}$ → more pos. as C ↑	
									Tomeš	72				
			-							-		-	A, M	
JE048 0071	333 a	E_{su}	-0.42F	Ag/AgCl, LiCl 0.1	i_{su}	33.3F	-	-	-	-	-	-	A, r	
			-0.1F			22.7F							A	
			-			-							A	

Code No.	Empirical Formula	Name and C.A. Number	Structural Formula	Solvent	Tech.	Medium		μ, M	pH	T, °C	Electrodes	App.	Experimental Parameters
EA95 ae65 cb28 da74	$C_5H_{10}NNaS_2$	sodium diethyldithiocarbamate C.A. 148-18-5	$(C_2H_5)_2NC(S)SNa$	H_2O	DI	BOR		—	9.2	25	DME Ag/AgCl, NaCl 1.0	26AO	$C=1E-3, t=2, v=2, \Delta e=25$, Pt Aux $\Delta e=100$
				MeCN	VY	$NaClO_4$	0.4	—	—	—	RPDE Ag/Ag$^+$ 0.01	---	$C=0.053$
													$C=0.42$
													$C=5.9$
EA96	$C_5H_{10}N_6O_2$	methylglyoxal bis-semicarbazone C.A. 10200-47-2	$CH_3C(:NNHCONH_2)-$ $CH:NNHCONH_2$	H_2O	PY	MB		—	0.35	—	DME/SCE	OAO	$C=0.5, m=2.62, t=2.8, h=55$
									1.65				
									2.40				
									3.70				
									4.00				
									4.90				
									6.20				
									7.00				
									8.00				
									11.80				
EA97	$C_5H_{10}N_6O_3$	hydroxypyruvaldehyde semicarbazone C.A. 27255-88-5	$CH_2(OH)C(:NNH-$ $CONH_2)CH:NNH-$ $CONH_2$	H_2O	PY	HCl	2.0	—	—	—	DME/SCE	OAO	$C=0.5, m=2.62, t=2.8, h=55$
						HCl	0.5						
						MB			1.0				
									1.8				
									2.9				
CONT			$(C_2H_5)_2NC(S)SNa$	H_2O	DI	BOR			9.2				

TABLE I. Electrochemical Data $C_5H_{10}N_8O_3$ (CONT.) EA97

Ref.	C/M	Charact. Potential		Response Const.		n		Electrokinetic Data			Products and Identification	Description and Remarks	Code No.
		Value	vs.		Value	Tech.		Parameter	Value	From			
JE044 0291	–	E_{su}	$-0.65F$	Ag/AgCl, NaCl 1.0	–	–	–	–	$\Delta E_{su/2}$	120	sttd	–	A,r	EA95 ae65 cb28 da74
			-0.691		i_{su}	$0.5F$	–	–		158		–	$A, i_{su}=k_1C+k_2$ for $C=2.5E-3-0.02, i_{su}=f(\Delta e),r$	
JE043 0205	415 a	$E_{\frac{1}{2}}$	-0.201	Ag/Ag$^+$ 0.01	–	–	–	–	$dE_{\frac{1}{2}}/d\log C$	17.3	plot	–	$A, dE_{\frac{1}{2}}/d\log C$ for $C=0.05-3, r$	
										–			A	
			-0.215		–	–	–	–		17.3		–	A,r	
			–							–			A	
			-0.230		$i_\ell(u)$	1	QE	1		17.3 $\approx 3E-2$	eqn	$CP \to (C_2H_5)_2N-CS-S-S-CS-N(C_2H_5)_2(90\%)$, MAS, IRS	$A, QI, i_d, i_\ell=k\omega$ for $\omega=60-600, i_\ell=kC$ for $C=0.053-11, E_{\frac{1}{2}} \to -0.195\pm 0.005$ V as $C \to 0, r$	
			0.9			1	–						A, i_a	
C0036 0331	–	$E_{\frac{1}{2}}$	$-0.59F$	SCE	i_ℓ	3.3F	i:i	2.5	–	–	–	–	C,p	EA96
			$-0.76F$			} 4.3F		–					C	
			$-0.88F$										C	
			$-0.71F$			3.3F		2.5					C,p	
			$-0.88F$			} 8.0F		–					C	
			$-1.00F$										C	
			$-0.76F$			3.3F		2.5					C,p	
			$-0.94F$			} 10.0F		≥6.0					C	
			$-1.06F$										C	
			$-0.82F$			3.3F		–					C,p	
			$-1.06F$			} 10.5F		–					C	
			$-1.19F$										C	
			$-0.84F$			3.1F		–					C,p	
			$-1.08F$			} 9.8F		–					C	
			$-1.21F$										C	
			$-0.90F$			3.0F		2.0					C,p	
			$-1.15F$			} 6.9F		–					C	
			$-1.29F$										C	
			$-0.98F$			2.8F		2.0					C,p	
			$-1.28F$			} 1.4F		–					C	
			$-1.43F$										C	
			$-1.02F$			2.4		–	–	–	–	–	$C, i_k, pK'=7.0, p$	
			$-1.50F$			0.9							C	
			$-1.50F$			3.0F		2.0	–	–	–	–	C,p	
			$-1.50F$			3.0F		2.0	–	–	–	–	C,p	
C0036 0331	–	$E_{\frac{1}{2}}$	–	SCE	i_ℓ	–	i:i	2.5	–	–	–	–	C,p	EA97
			$-0.68F$			2.25F		1.6					C	
			–			–		2.5					C,p	
			$-0.78F$			3.0F		2.0					C	
			$-0.60F$			4.9F		2.5					C,p	
			$-0.83F$			} 4.0F		–					C	
			$-0.96F$										C	
			$-0.65F$			5.0F		2.5					C,p	
			$-0.88F$			} 5.9F		–					C	
			$-1.00F$										C	
			$-0.70F$			4.7F	i:i, QE	–					C,p	
			$-0.95F$			} 6.2F		3.5					C	
			$-1.05F$										C	CONT

EA97 (CONT.) $C_5H_{10}N_6O_3$

Code No.	Empirical Formula	Name and C.A. Number	Structural Formula	Solvent	Tech.	Medium		μ, M	pH	T, °C	Electrodes	App.	Experimental Parameters
EA97	$C_5H_{10}N_6O_3$	hydroxypyruvaldehyde semicarbazone	$CH_2(OH)C(:NNH-CONH_2)CH:NNH-CONH_2$	H_2O	PY	MB		–	3.8	–	DME/SCE	OAO	C=0.5, m=2.62, t=2.8, h=55
									5.0				
									6.0				
									6.9				
									9.3				
									11.6				
EA98	$C_5H_{11}N_3O_2$	3-hydroxy-2-butanone semicarbazone C.A. 5551-23-5	$CH_3CH(OH)C(:N-NHCONH_2)CH_3$	H_2O	PY	MB		–	2.75	–	DME/SCE	OAO	C=0.5, m=2.62, t=2.8, h=55
									4.0				
									5.0				
									6.5				
					ACET	0.2		4.2				
					ACET $NH_2NHCONH_2$	0.19 0.006		4.2				
					ACET $NH_2NHCONH_2$	0.16 0.02		4.2				
					$NH_2NHCONH_2$	0.10		4.2				
EA99	C_5H_{12}	pentane C.A. 109-66-0	$CH_3(CH_2)_3CH_3$	CF_3COOH	IL	FSO_3H	2.5	–	–	–	Pt/SCE	25-2	C=30, v=50, Pt Aux
EB00	$C_5H_{14}INS$	thiocholine iodide C.A. 625-00-3	$HSCH_2CH_2N(CH_3)_3^+ I^-$	H_2O	PY	ACET		–	3.65	–	DME/MSE	OAO	C=0.6, m=3.2, t=2.6
						PHOS			6.7				
						BOR			8.1				
						NaOH	0.01		12.0				
EB01 af09	$C_6Cl_4O_2$	chloranil C.A. 118-75-2	Table II-2	EtOH 50	PY	BR		–	5.3	–	DME/SCE	O4AO	C=0.4, m=2.72, t=3.75, h=50
									6.8				

TABLE I. Electrochemical Data $C_6Cl_4O_2$ EB01

Ref.	C/M	Charact. Potential		Response Const.		Tech.	n	Electrokinetic Data			Products and Identification	Description and Remarks	Code No.
		Value	vs.		Value			Parameter	Value	From			
C0036 0331	–	$E_{\frac{1}{2}}$	-0.75F -1.00F -1.17F	SCE	i_ℓ }	4.2F 5.2F	i:i	2.0 –	–	–	–	–	C,p C C	EA97
			-0.81F			4.2F		2.0	–	–	–	CP (-1.0 V) → n=3.7	C,p	
			-1.10F -1.20F		}	2.7F		–	–	–	–	CP (-1.25 V) → n=4.0, 2,3-diaminopropanol, urea (reaction with xanthydrol)	C C	
			-0.90F -1.15F -1.27F		}	4.2F 2.0F		2.0 –	–	–	–	–	C,p C C	
			-0.95F			0.2F	–	–	–	–	–	–	C,i_k,pK'=7.2,p	
			-1.31F			1.0F	–	–	–	–	–	–	C	
			-1.31F			4.0F		2	–	–	–	–	C,p	
			-1.32F			4.0F		2	–	–	–	–	C,n=2.15 from QE at -1.5 V in NaOH 0.1,p	
C0036 0331	–	$E_{\frac{1}{2}}$	-1.10F	SCE	i_ℓ	2.7F	sttd	4	–	–	–	–	C,p	EA98
			-1.15F -1.40F		}	10.0F 7.5F		4	–	–	–	CP → 3-amino-2-butanol,urea	C,p C	
			-1.22F -1.40F		}	8.5F 7.5F		4	–	–	–	CP → 3-amino-2-butanol,urea	C,p C,Mx	
			-1.32F -1.40F			2.0F 2.0F	–	–	–	–	–	–	C,p C,Mx	
			-1.13			6.0	–	–	–	–	–	–	C,p	
			-1.10 -1.19		Σi_ℓ	– 16.5	sttd	4	–	–	–	–	C,p C	
			-1.13 -1.20			– 24.5		4	–	–	–	–	C,p C	
			-1.13 -1.23			– 35.0		4	–	–	–	–	C,p C	
JE054 0181	–	$E_{p/2}$	2.44	SCE	–	–	–	–	–	–	–	–	A,≠,p	EA99
C0027 0693	–	$E_{\frac{1}{2}}$	-0.15F	MSE	–							–	A,p	EB00
			-0.35F		–	–	–	–	–	–	–	–	A,p	
			-0.41F		–	–	–	–	–	–	–	–	A,p	
			-0.41F		–	–	–	–	–	–	–	–	A,p	
C0028 2163	–	$E_{\frac{1}{2}}$	0.19	SCE	–	–	–	–	–	–	–	–	C,p	EB01 af09
			0.08		–	–	–	–	–	–	–	–	C,p	

EB02 $C_6HCl_3O_2$

Code No.	Empirical Formula	Name and C.A. Number	Structural Formula	Solvent	Tech.	Medium	μ, M	pH	T, °C	Electrodes	App.	Experimental Parameters
EB02 af12	$C_6HCl_3O_2$	2,3,5-trichloro-benzoquinone C.A. 634-85-5	Table 11-2	H_2O	PY	BR	-	5.3 6.8	-	DME/SCE	04A0	C=0.4, m=2.72, t=3.75, h=15
EB03	$C_6HCl_3O_3$	2,3,5-trichloro-6-hydroxybenzoquinone C.A. 877-13-4	Table 11	H_2O	PY	BR	-	6.8	-	DME/SCE	04A0	m=2.72, t=3.75, h=50
EB04 af15	$C_6H_2Cl_2O_2$	2,6-dichlorobenzoquinone C.A. 697-91-6	Table 11-2	H_2O	PY	BR	-	5.3 6.8	-	DME/SCE	04A0	C=1.5, m=2.72, t=3.75, h=50
EB05	$C_6H_2Cl_2O_3$	3,5-dichloro-2-hydroxybenzoquinone	Table 11	H_2O	PY	BR	-	6.8	-	DME/SCE	04A0	m=2.72, t=3.75, h=50
EB06 af16	$C_6H_2Cl_2O_4$	3,6-dichloro-2,5-dihydroxybenzoquinone C.A. 87-88-7	Table 11-2	H_2O	PY	BR	-	4.6 5.3 6.8 7.9 10.0 11.6	-	DME/SCE	04A0	C=0.4, m=2.72, t=3.75, h=50
EB07 af24	$C_6H_3ClO_2$	chlorobenzoquinone C.A. 695-99-8	Table 11-2	H_2O	PY	BR	-	3.3 5.3 6.8 8.0 9.2	-	DME/SCE	04A0	C=1.5, m=2.72, t=3.75, h=50
EB08	$C_6H_3ClO_3$	2-chloro-5-hydroxy-benzoquinone	Table 11	H_2O	PY	BR	-	6.8	-	DME/SCE	04A0	m=2.72, t=3.75, h=50
EB09 af69	$C_6H_4Cl_2$	1,4-dichlorobenzene C.A. 106-46-7	$4-ClC_6H_4Cl$	HF	VY	KF 0.1	-	-	0	RPDE Cu/CuF$_2$ KF 0.2	2-0	C=satd, A=5E-3, v=100, ω=1, Pt Aux
EB10 af90 cb63	$C_6H_4N_2$	2-cyanopyridine C.A. 100-70-9	Table 11-2	H_2O	PY	buffer	-	9.8	-	DME/SCE	0-0	ns

TABLE I. Electrochemical Data $C_6H_4N_2$ EB10

Ref.	C/M	Charact. Potential		Response Const.		n Tech.	n	Electrokinetic Data			Products and Identification	Description and Remarks	Code No.
		Value	vs.		Value	Tech.		Parameter	Value	From			
C0028 2163	-	$E_{\frac{1}{2}}$	0.17	SCE	-	-	-	-	-	-	-	-	C,p	EB02 af12
			0.10		-	-	-	-	-	-	-	-	C,p	
C0028 2163	-	$E_{\frac{1}{2}}$	-0.12	SCE	-	-	-	-	-	-	-	-	C,p	EB03
C0028 2163	-	$E_{\frac{1}{2}}$	0.14	SCE	-	-	-	-	-	-	-	-	C,p	EB04 af15
			0.05		-	-	-	-	-	-	-	-	C,p	
C0028 2163	-	$E_{\frac{1}{2}}$	-0.22	SCE	-	-	-	-	-	-	-	-	C,p	EB05
C0028 2163	-	$E_{\frac{1}{2}}$	-0.20	SCE	-	-	-	-	-	-	-	-	C,i_d,p	EB06 af16
			-0.08F		-	-	-	-	-	-	-	-	C,i_a,p	
			-0.26			-							C	
			-0.08F		-	-	-	-	-	-	-	-	C,i_a+i_d,p	
			-0.41			-							C,i_d	
			-0.08F		-	-	-	-	-	-	-	-	C,i_a,p	
			-0.50F										C,i_d	
			-0.89F										C,i_k	
			-0.08F		-	-	-	-	-	-	-	-	C,i_a,p	
			-0.68F										C,i_k	
			-0.89F										C,i_k	
			-0.08F		-	-	-	-	-	-	-	-	C,i_a,p	
			-0.89F										C,i_k	
C0028 2163	-	$E_{\frac{1}{2}}$	0.26F	SCE	i_ℓ	4.5F	-	-	-	-	-	-	C,p	EB07 af24
			0.10			4.5F	-	-	-	-	-	-	C,p	
			0.06			4.5F	-	-	-	-	-	-	C,p	
			0.0F			4.3F	-	-	-	-	-	-	C,i=f(t),p	
			-0.06F			3.4F	-	-	-	-	-	-	C,i=f(t),p	
C0028 2163	-	$E_{\frac{1}{2}}$	-0.16	SCE	-	-	-	-	-	-	-	-	C,p	EB08
JE054 0232	-	$E_{\frac{1}{2}}$	1.341	NHE	i_ℓ	3.7F	-	-	Elog	54	-	-	A,p	EB09 af69
JE036 0383	1 fg	$E_{\frac{1}{2}}$	-1.70	SCE	-	-	-	-	-	-	-	-	$C,\propto$ LUMO,p	EB10 af90 cb63

EB11 C$_6$H$_4$N$_2$

Code No.	Empirical Formula	Name and C.A. Number	Structural Formula	Solvent	Tech.	Medium	μ, M	pH	T, °C	Electrodes	App.	Experimental Parameters
EB11 af91 cb64	C$_6$H$_4$N$_2$	3-cyanopyridine C.A. 100-54-9	Table 11-2	H$_2$O	PY	buffer	-	9.8	-	DME/SCE	0-0	ns
EB12 af99	C$_6$H$_4$N$_2$O$_4$	1,2-dinitrobenzene C.A. 528-29-0	2-O$_2$NC$_6$H$_4$NO$_2$	DMF	XT	Et$_4$NClO$_4$ 0.1	0.1	-	0	HMDE Ag/AgI, Bu$_4$NI 0.1, Et$_4$NClO$_4$ 0.1	2-2	v=(1.5-5.5)E3
EB13 ag00	C$_6$H$_4$N$_2$O$_4$	1,3-dinitrobenzene C.A. 99-65-0	3-O$_2$NC$_6$H$_4$NO$_2$	DMF	XT	Et$_4$NClO$_4$ 0.1	-	-	0	HMDE Ag/AgI, Bu$_4$NI 0.1, Et$_4$NClO$_4$ 0.1	2-2	v=(1.5-5.5)E3
EB14 ag01	C$_6$H$_4$N$_2$O$_4$	1,4-dinitrobenzene C.A. 100-25-4	Table 11-2	MeCN	IL	Et$_4$NClO$_4$ 0.1	-	-	-30 ±1	HMDE Ag/AgClO$_4$ 0.01	2-2	v=2E3
				DMF	XT	Et$_4$NClO$_4$ 0.1	-	-	-	HMDE Ag/AgI, Bu$_4$NI 0.1, Et$_4$NClO$_4$ 0.1	2-2	v=(1.5-5.5)E3
EB15	C$_6$H$_4$N$_2$O$_4$	2,4-dinitrosoresorcinol C.A. 118-02-5	Table 11	MeOH 10	PY	buffer GEL 8E-3	-	1.9 7.33 10.5	25±1	DME/SCE	0-0	C=1,m=2.04, t(oc)=3.3
				MeOH	PY	LiCl 0.1 C$_6$H$_5$COOH 6E-3 LiCl 0.1 NaOH 5E-3	-	-	25±1	DME/SCE	0-0	C=1,m=2.04, t(oc)=3.3
EB16 ag05	C$_6$H$_4$N$_4$	pteridine C.A. 91-18-9	Table 11-2	H$_2$O	PY	BR	-	2.0 3.0 4.0 5.0	-	DME/SCE	04A0	C=0.43,L
CONT												

TABLE I. Electrochemical Data $C_6H_4N_4$ (CONT.) EB16

Ref.	C/M	Charact. Potential		Response Const.		n		Electrokinetic Data			Products and Identification	Description and Remarks	Code No.
		Value	vs.		Value	Tech.		Parameter	Value	From			
JE036 0383	1 h	$E_{\frac{1}{2}}$	-1.76	SCE	-	-	-	-	-	-	-	-	$C, \propto LUMO, p$	EB11 ab91 cb64
JE057 0027	57 c	XE	0.334	Ag/AgI, Bu_4NI 0.1 Et_4NClO_4 0.1	-	-	-	-	-	-	-	-	C,XT=semiintegral chronoamperometry with linear potential sweep, XE=difference between standard potentials for the two steps, see reference for dependence of XE on $[Li^+],[H_2O]$, and T,p C	EB12 af99
			-		-		-		-					
JE057 0027	124 1	XE	0.398	Ag/AgI, Bu_4NI 0.1 Et_4NClO_4 0.1	-	-	-	-	-	-	-	-	C,XT=semiintegral chronoamperometry with linear potential sweep, XE=difference between standard potentials for the two steps, see reference for dependence of XE on $[Li^+],[H_2O]$, and T,p C	EB13 ag00
			-		-		-		-					
JE047 0215	-	E_p	-0.9F	Ag/ $AgClO_4$ 0.01	i_p	29.2F	-	-	-	-	-	-	C,p	EB14 ag01
			-1.05F			16.2F	-	-	-	-	-	-	C	
JE057 0027	59 b	XE	0.305	Ag/AgI, Bu_4NI 0.1 Et_4NClO_4 0.1	-	-	-	-	-	-	-	-	C,XT=semiintegral chronoamperometry with linear potential sweep, XE=difference between standard potentials for the two steps, see reference for dependence of XE on $[Li^+],[H_2O]$, and T,p C	
			-		-		-		-					
JE041 0105	-	$E_{\frac{1}{2}}$	0.08F	SCE	i_ℓ	14.7F	-	-	-	-	-	-	C, i_d, r	EB15
			-0.17F			7.1F	i:i	2	Elog	60	-	-	C, i_d, r	
			-0.34F			7.6F		2		65		-	C, i_d	
			-0.55F			13.2F	-	-	-	-	-	-	C, i_d, r	
JE041 0105	-	$E_{\frac{1}{2}}$	-0.07F -0.29F -1.4	SCE	i_ℓ	9.7F 9.7F 3.8F	-	-	-	-	-	-	C, i C C,H	
			-1.0F			17.1F	-	-	-	-	-	-	C,p	
C0027 0199	-	$E_{\frac{1}{2}}$	-0.49F	SCE	i_ℓ/i_d at t=0	1.00F	i:i	2	-	-	-	-	$C, i_d, i_\ell=f(t), r$	EB16 ag05
			-0.09F -0.53F			0.05F 0.95F	}	2	-	-	-	-	$C, i_d, i_\ell=f(t), r$ C	
			-0.17F -0.62F			0.51F 0.49F	}	2	-	-	-	-	$C, i_d, i_\ell=f(t), r$ C, i_d	
			-0.27F -0.67F			0.87F 0.13F	}	2	-	-	-	-	$C, i_d, i_\ell=f(t), r$ C, i_k	CONT

EB16 (CONT.) C₆H₄N₄

Code No.	Empirical Formula	Name and C.A. Number	Structural Formula	Solvent	Tech.	Medium	μ, M	pH	T, °C	Electrodes	App.	Experimental Parameters
EB16 ag05	$C_6H_4N_4$	pteridine	Table II-2	H_2O	PY	BR	–	6.0	–	DME/SCE	O4AO	C=0.43, L
								8.0				
								9.0				
								10.0				
								11.0				
								12.0				
						NaOH	0.1	–				
EB17 ag06	$C_6H_4N_4O$	4-hydroxypteridine C.A. 700-47-0	Table II-2	H_2O	PY	HCl	–	0.0	–	DME/SCE	O4AO	C=0.19, h=50, L
								2.0				
								4.0				
								6.0				
								8.0				
								10.0				
								11.6				
						NaOH	1.0	13.9				
EB18 cb67	$C_6H_4N_4O$	6-hydroxypteridine C.A. 2432-26-0	Table II-3	H_2O	PY	BOR	–	7.1	25.0 ±0.1	DME/SCE	2AO	C=ns, t(c)=2, Pt Aux
								8.1				C=1
								8.8				C=ns
								9.8				C=0.72
												C=2
												C=5
						Na(?)OH Na(?)Cl	ns	12				C=ns
CONT		pteridine	Table II-2	H_2O	PY	BR					O4AO	

TABLE I. Electrochemical Data $C_6H_4N_4O$ (CONT.) EB18

Ref.	C/M	Charact. Potential Value	vs.	Response Const. Value		Tech.	n	Electrokinetic Data Parameter	Value	From	Products and Identification	Description and Remarks	Code No.
C0027 0199	-	$E_{\frac{1}{2}}$	-0.34F -0.73F	SCE	i_ℓ/i_d at t=0	0.99F 0.01F	i:i	2	-	-	-	-	$C,i_d,i_\ell=f(t),r$ C,i_k	EB16 ag05
			-0.49F -0.88F			0.99F 0.01F	}	2	-	-	-	-	$C,i_d,i_\ell=f(t),r$ C,i_k	
			-0.54F -0.94F			0.85F 0.19F	}	2	-	-	-	-	$C,i_d,i_\ell=f(t),r$ C,i_k	
			-0.59F -1.08F -1.34F			0.68F 0.29F 0.14F	}	2	-	-	-	-	$C,i_d,i_\ell=f(t),r$ C C,i_k	
			-0.62F -1.16F -1.44F			0.50F 0.38F 0.25F	}	2	-	-	-	-	$C,i_d,i_\ell=f(t),r$ C C	
			-0.68F -1.24F -1.47F			0.40F 0.36F 0.29F	}	2	-	-	-	-	$C,i_\ell=f(t),r$ C C	
			-0.72F -1.34F			0.14 -	}	2	-	-	-	-	$C,i_\ell=f(t),r$ C	
C0027 0199	-	$E_{\frac{1}{2}}$	-0.09F	SCE	-	-	i:i	2	-	-	-	-	C,i_d,p	EB17 ag06
			-0.22F		-	-		2	-	-	-	-	C,i_d,p	
			-0.37F		-	-		2	-	-	-	-	C,i_d,p	
			-0.53F		-	-		2	-	-	-	-	C,i_d,p	
			-0.67F		-	-		2	-	-	-	-	C,i_d,p	
			-0.82F		-	-		2	-	-	-	-	C,i_d,p	
			-1.00F		-	-		2	-	-	-	-	C,2 merging waves,p	
			-1.03F		-	-		2	-	-	-	-	C,i_d,Mx,p	
JE047 0479	398 a	$E_{\frac{1}{2}}$	-0.61F	SCE	I	0.18F	-	-	$dE_{\frac{1}{2}}/dpH$	64	sttd	-	$C,E_{\frac{1}{2}}=-0.162-0.064pH,$ r	EB18 cb67
			-0.64F		I i_ℓ	0.65F 1.8F	-	-		64		-	C,r	
			-0.84F			0.5F				-			$C,i_\ell \neq f(C)$ for C= 0.5-4	
			-0.75F		I	1.50F	QE	1.83		64		CP → 7,8-dihydro-6-hydroxypteridine, UVS,VA	C,r	
			-0.7F		i_ℓ	1.2F	-	-		-		-	$C,Pr,i_a,i_\ell \downarrow$ as C $\uparrow$, r	
			-0.77F			3.00F				64			C	
			-0.67F -0.86F			small 8.5F	-	-		-	-	-	C,Pr,i_a,r C	
			-0.9F			22.9F	QE	1.97		-	-	CP → 7,8-dihydro-6-hydroxypteridine, UVS,VA	C,Mx,r	
			-0.93F		I	3.3F		1.89		64		QE (-1.1V) in NaOH 1 → 7,8-dihydro-6-hydroxypteridine, UVS,VA	C,r	

Ref.	C/M	Charact. Potential Value	vs.	Response Const. Value		Tech.	n	Electrokinetic Data Parameter	Value	From	Products and Identification	Description and Remarks	Code No.
C0027		$E_{\frac{1}{2}}$												CONT

Code No.	Empirical Formula	Name and C.A. Number	Structural Formula	Solvent	Tech.	Medium	μ, M	pH	T, °C	Electrodes	App.	Experimental Parameters
EB18 cb67	$C_6H_4N_4O$	6-hydroxypteridine	Table II-3	H_2O	IL	MB	-	3	25	PGE/SCE	2AO	C=1,v=10, Pt Aux
								5.6				
								7				
								10				
					PV	MB	-	5.6	25±0.1	DME/SCE	2AO	C=0.1-0.4, t(c)=2,Δe= 10,f=100, Pt Aux
					VA	BOR	0.5	7.1	25±0.1	PGE/SCE	2AO	C=ns,Pt Aux
						Na(?)OH Na(?)Cl	ns	12				
					VR	BOR	0.5	11.1	25±0.1	PGE/SCE	2AO	0→0.75→ -1.5→0.75 →0 V,C= 0.5,v=50, Pt Aux
EB19 cb68	$C_6H_4N_4O$	7-hydroxypteridine C.A. 2432-27-1	Table II-3	H_2O	PY	MB	-	0	25±0.1	DME/SCE	2AO	C=ns,t(c)= 2,Pt Aux
								1				
								2.8				
						BOR		8.8				
CONT												

TABLE I. Electrochemical Data $C_6H_4N_4O$ (CONT.) EB19

Ref.	C/M	Charact. Potential Value	vs.	Response Const. Value		Tech.	n	Electrokinetic Data Parameter	Value	From	Products and Identification	Description and Remarks	Code No.
JE055 0069	398 b	-	-	$J_p/Cv^{\frac{1}{2}}$	425F	QE	1.9 ± 0.1	dE_p/dpH	71	sttd	$CP \to 6,7$-dihydroxy-pteridine, $\lambda_{max}=302$, PY and IL at PGE	$A, E_p=1.13-0.071$pH for pH=2.2-10, $i_p/C \downarrow$ as $C \uparrow$ for $C=0.1-2$, $J_p/Cv^{\frac{1}{2}} \uparrow$ as $v \uparrow$, r	EA18 cb67
		-	-	-	-				64		-	$A, E_p=1.4-0.064$pH for pH=2.2-9, $J_p/Cv^{\frac{1}{2}} \uparrow$ as $v \uparrow$, $J_p/C \downarrow$ as $C \uparrow$ for $C=0.1-2$	
		-	-	-	1150F		1.85 ± 0.09		71		$CP \to 6,7$-dihydroxy-pteridine (95-100%), $\lambda_{max}=302$, PY and IL at PGE	$A, J_p/Cv^{\frac{1}{2}} \uparrow$ as $v \uparrow$, $J_p/C \downarrow$ as $C \uparrow$ for $C=0.1-2$, r	
		-	-	-	-		-		-		-	$A, J_p/Cv^{\frac{1}{2}} \uparrow$ as $v \uparrow$, $J_p/C \downarrow$ as $C \uparrow$ for $C=0.1-2$	
		-	-	-	1475F		1.98 ± 0.1		71		$CP \to 6,7$-dihydroxy-pteridine, $\lambda_{max}=302$, PY and IL at PGE	$A, J_p/Cv^{\frac{1}{2}} \uparrow$ as $v \uparrow$, $J_p/C \downarrow$ as $C \uparrow$ for $C=0.1-2$, r	
		-	-	-	-		-		64		-	$A, J_p/Cv^{\frac{1}{2}} \uparrow$ as $v \uparrow$, $J_p/C \downarrow$ as $C \uparrow$ for $C=0.1-2$	
		-	-	-	500F		-		71		-	$A, J_p/Cv^{\frac{1}{2}} \uparrow$ as $v \uparrow$, $J_p/C \downarrow$ as $C \uparrow$ for $C=0.1-2$, r	
		-	-	-	-		-		-		-	A	
JE047 0479	-	-	-	-	-	-	-	-	-	-	-	$C, Mn \to$ wider as $C \uparrow, p$	
JE047 0479	398 a	E_p	-0.67F	SCE	-	-	QE	2.0 ± 0.1	dE_p/dpH	64	-	$CP \to 7,8$-dihydro-6-hydroxypteridine, UVS, VA	$C, E_p=-0.232-0.064$pH, r	
			0.75F		-	-		-		100		-	$A, E_p=1.45-0.10$pH	
			-0.99F		-	-		1.96		64		$CP \to 7,8$-dihydro-6-hydroxypteridine, UVS, VA	C, r	
			0.26F		-	-		-		100		-	A	
JE047 0479	398 a	E_p	-0.8F	SCE	i_p	11.2F	-	-	-	-	-		$C, J_p/v^{\frac{1}{2}} \uparrow$ as $v \uparrow, i_a$, Pr, $J_p/C \downarrow$ as $C \uparrow$ for $C=0.4-0.6$, $J_p/C \neq f(C)$ for $C > 0.6$, r	
			-1.05F			27.2F							$C, J_p/v^{\frac{1}{2}} \neq f(v^{\frac{1}{2}})$, $J_p/C \neq f(C)$	
			0.4F			35.2F							A	
JE047 0479	-	$E_{\frac{1}{2}}$	-0.19F	SCE	I	2.64F	-	-	$dE_{\frac{1}{2}}/dpH$	70	sttd	$CP \to 5,6$-dihydro-7-hydroxypteridine, UVS, VA	$C, W, E_{\frac{1}{2}}=-0.23-0.07$pH, r	EB19 cb68
			-0.27F			2.56F	-	-		70		$CP \to 5,6$-dihydro-7-hydroxypteridine, UVS, VA	C, W, r	
			-0.42F			3.32F	QE	1.8		70		$CP \to 5,6$-dihydro-7-hydroxypteridine, UVS, VA	C, W, r	
			-0.86F			3.32F		1.9		70		$CP \to 5,6$-dihydro-7-hydroxypteridine, UVS, VA	C, W, r	CONT

EB19 (CONT.) $C_6H_4N_4O$

Code No.	Empirical Formula	Name and C.A. Number	Structural Formula	Solvent	Tech.	Medium		μ, M	pH	T, °C	Electrodes	App.	Experimental Parameters
EB19 cb68	$C_6H_4N_4O$	7-hydroxypteridine	Table 11-3	H_2O	PY	Na(?)OH Na(?)Cl	ns	-	11.8	25±0.1	DME/SCE	2AO	C=ns, t(c)=2, Pt Aux
					IL	H_2SO_4	2	-	-	25	PGE/SCE	2AO	C > 1, v=10, Pt Aux
					VR	MB		-	7.0	25±0.1	PGE/SCE	2AO	0→0.9→-1.5→0.9→0 V, C=ns, v=50, Pt Aux
EB20 cb69	$C_6H_4N_4O_2$	6,7-dihydroxypteridine C.A. 3947-46-4	Table 11-3	H_2O	IL	HOAc	1	-	2.3	25	PGE/SCE	2AO	C=1, v=10, Pt Aux
						MB			7				
EB21	$C_6H_4N_4O_4$	2,4,6,7-tetra-hydroxypteridine C.A. 2817-14-3	Table 11	H_2O	IL	HOAc	1	-	2.3	25	PGE/SCE	2AO	C=1, v=10, Pt Aux
EB22 ag08 cb70 da86 dh59	$C_6H_4O_2$	1,4-benzoquinone C.A. 106-51-4	Table 11-2	H_2O	PY	BR		-	5.3	-	DME/SCE	O4AO	C=1.0, m=2.72, t=3.25, h=50
				DMF	PY	$Ba(ClO_4)_2$	0.05	-	-	-	DME/SCE	26AO	C=1, m=3.38, t=3.05
						$KClO_4$	0.1						
						$LiClO_4$	0.1						
						Et_4NClO_4	0.1						
						$NaClO_4$	0.1						
						$Sr(ClO_4)_2$	0.05						
CONT		7-hydroxypteridine											

TABLE I. Electrochemical Data $C_6H_4O_2$ (CONT.) EB22

Ref.	C/M		Charact. Potential		Response Const.		Tech.	n	Electrokinetic Data			Products and Identification	Description and Remarks	Code No.
			Value	vs.		Value			Parameter	Value	From			
JE047 0479	-	$E_{\frac{1}{2}}$	-1.02F	SCE	I	2.56F	QE	1.9	$dE_{\frac{1}{2}}/dpH$	70	sttd	$CP \rightarrow$ 5,6-dihydro-7-hydroxypteridine, UVS,VA	C,W,r	EB19 cb68
JE055 0069	-	E_p	1.14	SCE	$J_p/Cv^{\frac{1}{2}}$	≈911	QE	1.95 ± 0.04	-	-	-	$CP \rightarrow$ 6,7-dihydroxy-pteridine,(95-100%) λ_{max}=307,223, and 207, PY	$A, J_p/Cv^{\frac{1}{2}} \uparrow$ as v↑, $J_p/C \downarrow$ as C↑ for C=0.1-2,E_p=1.08 for C=1 and $[H_2SO_4]$=1,r	
			1.52			-		-				-	$A,M,J_p/Cv^{\frac{1}{2}} \uparrow$ as v↑	
JE047 0479	-	E_p	-	SCE	i_p	-	QE	-	-	-	-	-	C,X,Pr,i_a,r	
			-0.86F			41.6F		1.9	$dE_{\frac{1}{2}}/dpH$	72	sttd	5,6-dihydro-7-hydroxypteridine	C,E_p=-0.2-0.072pH	
			-0.4F			84.8F		-		66		-	A,E_p=0.88-0.066pH, 0 on first scan	
JE055 0069	398 c	E_p	1.23c	SCE	-	-	QE	5.5 ± 0.3	dE_p/dpH	56	sttd	$CP \rightarrow NH_3$(1.7 mole); CO_2(1.93 mole); formaldehyde(0.36 mole);formic acid (0.44 mole);urea (0.24 mole),TLC; oxamide(0.4 mole), MAS,IRS;tetraketo-piperazine(0.29 mole),TLC,PY	A,E_p=1.36-0.056pH for pH=2.3-9,r	EB20 cb69
			0.97c		$J_p/Cv^{\frac{1}{2}}$	1500		5.9 ± 0.3		56		-	A,r	
JE055 0069	-	E_p	0.73	SCE	-	-	QE	1.9 ± 0.2	-	-	-	$CP \rightarrow CO_2$(0.5 mole); NH_3(1.2 mole); oxamide(0.33 mole)	A,r	EB21
C0028 2163	-	$E_{\frac{1}{2}}$	0.04	SCE	-	-	-	-	-	-	-	-	C,p	EB22 ag08 cb70 da86 dh59
JE055 0277	-	$E_{\frac{1}{2}}$	-0.55	SCE	i_ℓ	4.2	-	-	Elog	63	-		$C,R,E_{\frac{1}{2}} \neq f(t)$ for t= 1.02-5.5,r	
			-			-				-			$C,\neq$	
			-0.57			4.1				61			$C,R,E_{\frac{1}{2}} \neq f(t)$ for t= 1.02-5.5,r	
			-			-				-			$C,\neq$	
			-0.56			4.3				65			$C,R,E_{\frac{1}{2}} \neq f(t)$ for t= 1.02-5.5,r	
			-			-				-			$C,\neq$	
			-0.57			4.1				63			$C,R,E_{\frac{1}{2}} \neq f(t)$ for t= 1.02-5.5,r	
			-			-				-			$C,\neq$	
			-0.57			4.3				65			$C,R,E_{\frac{1}{2}} \neq f(t)$ for t= 1.02-5.5,r	
			-			-				-			$C,\neq$	
			-0.54			4.2				62			$C,R,E_{\frac{1}{2}} \neq f(t)$ for t= 1.02-5.5,r	
			-			-				-			$C,\neq$	
JE047	-	$E_{\frac{1}{2}}$	-1.02F			2.56F	QE	1.9	$dE_{\frac{1}{2}}/dpH$	70	sttd	$CP \rightarrow$ 5,6-dihydro-7-hydroxypteridine,	C,W,r	CONT

EB22 (CONT.) $C_6H_4O_2$

Code No.	Empirical Formula	Name and C.A. Number	Structural Formula	Solvent	Tech.	Medium		μ, M	pH	T, °C	Electrodes	App.	Experimental Parameters
EB22 ag08 cb70 da86 dh59	$C_6H_4O_2$	1,4-benzoquinone	Table II-2	DMF	VA	Et_4NClO_4	0.1	-	-	25±1	Au Ag/ $AgClO_4$ 0.01, Et_4NClO_4 0.1	-45A 0	C=5, A≈0.16, v=170, PE
											MPAu Ag/ $AgClO_4$ 0.01, Et_4NClO_4 0.1		
											Ir Ag/ $AgClO_4$ 0.01, Et_4NClO_4 0.1		
											Pd Ag/ $AgClO_4$ 0.01, Et_4NClO_4 0.1		
											Pt Ag/ AgCl 0.01, Et_4NClO_4 0.1		
											Rh Ag/ $AgClO_4$ 0.01, Et_4NClO_4 0.1		
					VR	Bu_4NClO_4	1.0	-	-	20.0 ±0.3	PDE Ag/AgCl	2A2	C=0.4-1.0, Pt Aux
EB23	$C_6H_5BrO_2Se$	3-bromobenzene-seleninic acid C.A. 33350-65-1	$3-BrC_6H_4SeO_2H$	H_2O	VY	$HClO_4$ $NaClO_4$	0.05 0.2	-	-	25	Pt PDS MSE	126A 0	C=1, v=5.6, MSE Aux, PE
EB24	$C_6H_5BrO_2Se$	4-bromobenzene-seleninic acid C.A. 20825-08-5	$4-BrC_6H_4SeO_2H$	H_2O	VY	$HClO_4$ $NaClO_4$	0.05 0.2	-	-	25	Pt PDS MSE	126A 0	C=1, v=5.6, MSE Aux, PE
EB25	$C_6H_5BrO_3Ru$	bromotricarbonyl-(η^3-2-propenyl)-ruthenium C.A. 31781-74-5	$(\pi-C_3H_5)Ru-(CO)_3Br$	MeCN	PY	Et_4NClO_4	0.1	-	-	-	DME Ag/Ag$^+$	125A 0	Pt Aux
EB26	$C_6H_5ClO_2Se$	3-chlorobenzene-seleninic acid C.A. 33350-63-9	$3-ClC_6H_4SeO_2H$	H_2O	VY	$HClO_4$ $NaClO_4$	0.05 0.2	-	-	25	Pt PDS MSE	126A 0	C=1, v=5.6, MSE Aux, PE
EB27	$C_6H_5ClO_2Se$	4-chlorobenzene-seleninic acid C.A. 20753-53-1	$4-ClC_6H_4SeO_2H$	H_2O	VY	$HClO_4$ $NaClO_4$	0.05 0.2	-	-	25	Pt PDS MSE	126A 0	C=1, v=5.6, MSE Aux, PE

TABLE I. Electrochemical Data $C_6H_5ClO_2Se$ EB27

Ref.	C/M	Charact. Potential		Response Const.		n Tech.	n	Electrokinetic Data			Products and Identification	Description and Remarks	Code No.
		Value	vs.		Value			Parameter	Value	From			
JE046 0215	-	E_p	-0.15F	NHE	i_p	44F	-	-	k	(5.3 ± 0.8)E-3	N-S	-	C,p	EB22 ag08
			-0.99F			18F							C	cb70
			-0.87F			small							A	da86
			-0.09F			32F							A	dh59
			-0.15F			60F				(5.2 ± 0.6)E-3		-	C,p	
			-0.99F			20F							C	
			-0.87F			small							A	
			-0.09F			44F							A	
			-0.15F			48F				(6.2 ± 0.8)E-3		-	C,p	
			-0.99F			22F							C	
			-0.87F			small							A	
			-0.09F			36F							A	
			-0.15F			44F				(5.7 ± 1)E-3		-	C,p	
			-0.99F			18F							C	
			-0.87F			small							A	
			-0.09F			38F							A	
			-0.15F			54F				(5.2 ± 0.7)E-3		-	$C, i_p=kv^{\frac{1}{2}},p$	
			-0.99F			26F							C	
			-0.87F			small							A	
			-0.09F			40F							A	
			-0.15F			60F				(3.7 ± 0.6)E-3		-	C,p	
			-0.99F			24F							C	
			-0.87F			small							A	
			-0.09F			44F							A	
JA096 0249	50 a	XE	-0.52	Ag/AgCl	-	-	sttd	1	-	-	-	radical anion	$C, XE=E_{p,A}+E_{p,C}/2$ $\rightarrow\rightarrow (v=0), R$	
			-1.40					1				dianion	C,R	
JE040 0339	404 a	$E_{\frac{1}{2}}$	0.44F	SCE	-	-	-	-	-	-	-	-	$C, \neq, \Delta E_{\frac{1}{2}}\neq 0.10\sigma, p$	EB23
JE040 0339	404 a	$E_{\frac{1}{2}}$	-0.36F	SCE	-	-	-	-	-	-	-	-	$C, \neq, \Delta E_{\frac{1}{2}}=0.10\sigma, p$	EB24
JE042 0151	-	$E_{\frac{1}{2}}$	-1.70	Ag/Ag$^+$	-	-	QE	1.0 ± 0.1	-	-	-	$[(\pi-C_3H_5)(CO)_3Ru]_2$, IRS	$C, \neq, p$	EB25
JE040 0339	404 a	$E_{\frac{1}{2}}$	-0.38F	SCE	-	-	-	-	-	-	-	-	$C, \neq, \Delta E_{\frac{1}{2}}=0.10\sigma, p$	EB26
JE040 0339	404 a	$E_{\frac{1}{2}}$	-0.33F	SCE	-	-	-	-	-	-	-	-	$C, \neq, \Delta E_{\frac{1}{2}}=0.10\sigma, p$	EB27
JE046 0215	-	E_p	-0.15F	NHE	i_p	44F	-	-	k	(5.3 ± 0.8)E-3	N-S	-	C,p	EB22

EB28 C_6H_5NO

Code No.	Empirical Formula	Name and C.A. Number	Structural Formula	Solvent	Tech.	Medium		μ, M	pH	T, °C	Electrodes	App.	Experimental Parameters
EB28 ag41 da89	C_6H_5NO	nitrosobenzene C.A. 586-96-9	C_6H_5NO	EtOH 20	VA	NaOH $NaNO_3$	<0.01	1.0	12.06	25	GCE/SCE	28A0	C=1,v=10, Pt Aux
				MeCN	IR	Et_4NClO_4	0.1	-	-	22.5 ±0.5	PBE/SCE	256A FO	-0.6→-1.2 V, C=4.82,A= 0.25,t=0.01, Pt Aux
													t=0.5
													t=2.24
					VR	Et_4NClO_4	0.1	-	-	22.5 ±0.5	PBE/SCE	256A FO	0→-1.6→0 V, C=4.82,A= 0.25,v=100, Pt Aux
				DMF	PY	$KClO_4$	0.01	-	-	-	DME/SCE	25A0	C=0.5,m= 2.74,t=2.5
						$KClO_4$	0.1						
						$LiClO_4$	0.01						
						$LiClO_4$	0.1						
						$Mg(ClO_4)_2$	0.01						
						$Mg(ClO_4)_2$	0.1						
						$NaClO_4$	0.01						
						$NaClO_4$	0.1						
						NH_4ClO_4	0.01						
						NH_4ClO_4	0.1						
						Et_4NClO_4	0.01						
						Et_4NClO_4	0.1						
						Pr_4NI	0.01						
						Pr_4NI	0.1						
CONT	C_6H_5NO	nitrosobenzene C.A. 586-96-9	C_6H_5NO	EtOH 20	VA	NaOH $NaNO_3$	<0.01	1.0	12.06	25	GCE/SCE	28A0	C=1,v=10, Pt Aux

TABLE I. Electrochemical Data C_6H_5NO (CONT.) EB28

Ref.	C/M	Charact. Potential		Response Const.		n		Electrokinetic Data			Products and Identification	Description and Remarks	Code No.
		Value	vs.		Value	Tech.		Parameter	Value	From			
JE046 0421	124 k	E_p	-0.42F -0.36F	SCE	i_p	6.3F 4.8F	- -	- -	- -	- -	- -	-	C,r A	EB28 ag41 da89
JE057 0179	86 b	-	-	-	$it^{\frac{1}{2}}/C$	63.6F	-	1	-	-	-	CP (-1.20 V) → azoxybenzene (82%), GLC, n=0.28; with 2.2E-3$\underline{M}$ Et$_4$NOH and C=2.2, CP (-1.20 V) → azoxybenzene (69%), λ_{max}=320 nm	C,r	
		-			44.8F	-	-	-	-	-	-	C,r	
		-			17.6F	-	-	-	-	-	-	C,r	
JE057 0179	86 b	E_p	-0.27F	SCE	i_p	14.3F	QE	-	-	-	-	-	C,0 on first scan and for v>3E4, r	
			-0.96F			164.3F		≈0.28				CP → azoxybenzene (75%), GLC, UVS	C,i_p ↓ on following scans, C_6H_5NO redn.	
			-1.47F			100F	-						C,0 for v>3E4, azoxybenzene redn.	
			-1.37F			50F							A,0 for v>3E4	
			-0.88F			50F							A	
			-0.17F			28.6F							A,0 for v>3E4	
JE057 0339	86 f	$E_{\frac{1}{2}}$	-0.950	SCE	i_ℓ	3F	-	-	Elog dE/dlog[K$^+$]	82 50	- -	-	C,r	
			-1.40			0.67F				-			C	
			-1.65			2F							C	
			-0.900		-	-	-	-		81	-	-	C,r	
			-1.53							-			C	
			-0.905		-	-	-	-	dE/dlog[Li$^+$]	87(?)	-	-	C,r	
			-							-			C,X	
			-0.82		-	-	-	-	Elog	81	-	-	C,r	
			-							-			C,X	
			-0.730		-	-	-	-	Elog dE/dlog[Mg^{2+}]	98 116(?)	- -	-	C,r	
			-0.62		i_p	4.38F	-	-	Elog	52	-	-	C,r	
			-0.920		-	2.6F	-	-	Elog dE/dlog[Na$^+$]	83 59(?)	- -	-	C,r	
			-1.51			2.3F				-			C	
			-0.865		-	-	-	-	Elog	101	-	-	C,r	
			-1.4							-			C	
			-0.300		-	-	-	-	Elog dE/dlog[NH$_4^+$]	109 86(?)	- -	-	C,H?,r	
			-0.220		i_ℓ	4.25F	-	-	Elog	98	-	-	C,H?,p	
			-0.940		-	-	-	-	Elog dE/dlog[Et$_4$N$^+$]	83 40(?)	- -	-	C,r	
			-1.41							-			r	
			-1.81							-			r	
			-0.905		i_ℓ	2.4F	-	-	Elog	79	-	-	C,r	
			-1.4			0.6F				-			C	
			-1.8			1.6F							C	
			-0.930		-	-	-	-	Elog dE/dlog[Pr$_4$N$^+$]	69 39	- -	-	C,r	
			-1.35			-	-	-		-			C	
			-1.85			-	-	-		-			C	
			-2.14			-	-	-		-			C	
			-0.89						Elog	75	-	-	C,r	
			-1.28							-			C	
			-1.7										C	
			-1.93										C	
														CONT

EB28 (CONT.) C_6H_5NO

Code No.	Empirical Formula	Name and C.A. Number	Structural Formula	Solvent	Tech.	Medium	μ, M	pH	T, °C	Electrodes	App.	Experimental Parameters
EB28 ag41 da89	C_6H_5NO	nitrosobenzene	C_6H_5NO	DMF	PY	Bu_4NClO_4 0.01	-	-	-	DME/SCE	25AO	C=0.5, m=2.74, t=2.5
						Bu_4NClO_4 0.1						
						tri-<u>iso</u>-amyl-n-butyl-ammonium iodide 0.01						
						tri-<u>iso</u>-amyl-n-butyl-ammonium iodide 0.1						
					IR	Et_4NClO_4 0.1	-	-	22.5 ±0.5	Pt/SCE	256A FO	E_{app}= -1.2 V, C=ns, t= 0.004-4, Pt Aux
					VA	$KClO_4$ 0.01	-	-	-	HMDE SCE	25AO	C=0.5, d=0.068
						$Mg(ClO_4)_2$ 0.5						
						NH_4ClO_4 0.1						
						Et_4NClO_4 0.1						
						Et_4NClO_4 0.1	-	-	22.5 ±0.5	Pt/SCE	256A FO	C=3.68, v=100, Pt Aux
						$NaClO_4$ 0.01	-	-	-	HMDE SCE	25AO	C=0.5, d=0.068
EB29 ag47 cb73 da91	$C_6H_5NO_2$	nitrobenzene C.A. 98-95-3	$C_6H_5NO_2$	MeOH 10	PY	buffer GEL 7E-3	-	11	-	DME/SCE	04AO	C=0.5
				EtOH 55	PY	GLYC	-	0	-	DME/SCE	04AO	C=0.1, m=4.0, t=3.5
				DMF	PY	Et_4NClO_4 0.1	-	-	-	DME/SCE	24AO	m=2.74, t=2.5
						Et_4NClO_4 0.1	-	-	-	DME/SCE	26-0	C=0.5, m=1.61, t=3.6
CONT												

TABLE I. Electrochemical Data $C_6H_5NO_2$ (CONT.) EB29

Ref.	C/M	Charact. Potential		Response Const.		n	Electrokinetic Data			Products and Identification	Description and Remarks	Code No.
		Value	vs.		Value	Tech.	Parameter	Value	From			
JE057 0339	86 f	$E_{\frac{1}{2}}$ -0.950	SCE	-	-	-	-	$dE_{\frac{1}{2}}/$ $dlog[Bu_4N^+]$	- 60(?)	- -	-	C,r C C	EB28 ag41 da89
		-1.35 -2.24											
		-0.900 -1.35 -2.00		-	-	-	-	-	-	-	-	C,r C C	
		-0.910		-	-	-	-	Elog dE/dlog $[RR_3^!N^+]$	86 32(?)	- -	-	C,r	
		-1.35 -2.00 -2.29		-	-	-	-	-	-	-	-	C C C	
		-0.880 -1.28 -2.00		-	-	-	-	Elog	73 -	-	-	C,r C C	
JE057 0179	86 b	-	-	$it^{\frac{1}{2}}/C$	42.5±1	-	1	-	-	-	with C=5.54,CP (-1.2 V) → radical anion,VA;azoxyben-zene(76%),GLC;with C=5.54,CP (-1.2 V) → azoxybenzene(93%), λ_{max}=320 nm;with Et_4NOH 1.54E-3 and C=1.54,CP(-1.20 V) → azoxybenzene(73%)	C,i_d,r	
JE057 0339	86 f	E_p -0.97F -1.44F -1.68F -1.37F -0.88F -0.64F	SCE	i_p	3F 0.5F 1.17F 1F 0.5F 1.5F	-	-	-	-	-	-	C C C A A A	
		-0.66F -			2.17F -	-	-	-	-	-	-	C,r A,O	
		-0.25F -0.08F			1.92F 0.92F	-	-	-	-	-	-	C,r A	
		-1.0F -1.44F -1.72F -1.92F -1.28F -0.84F			3F 1.5F 2F 2F - 2F	-	-	-	-	-	-	C,r C C C A A	
JE057 0179	86 b	E_p -0.92F -1.4F -1.36F -0.86F	SCE	$i_p(u)$	18F small small 18F	- - - -	- - - -	-	-	-	-	C,r C A A	
JE057 0339	86 f	E_p -0.76F -0.94F -1.42F -1.3F -0.84F -0.55F	SCE	i_p	3.3F 0.67F 1F 0.83F 0.83F 3F	- - - - - -	- - - - - -	-	-	-	-	C C C A A A	
JE056 0223	-	$E_{\frac{1}{2}}$ ≈0.0	SCE	-	-	-	-	-	-	-	-	C;$dE_{\frac{1}{2}}$/dpH=60 for pH= 1-10.8,0 for pH > 11;p	EB29 ag29 cb73 da91
C0026 1733	-	$E_{\frac{1}{2}}$ -0.16	SCE	-	-	sttd	4	$dE_{\frac{1}{2}}/dpH$	77	sttd	-	C,p	
JE042 0261	130 l	$E_{\frac{1}{2}}$ -1.125	SCE	-	-	-	-	Elog $dE_{\frac{1}{2}}/$ dlog[	≈65 51	- plot	-	C,r	
JE039 0395	-	$E_{\frac{1}{2}}$ -1.1 -1.77	SCE	I	2.2 4.4	-	-	Elog	67 -	-	-	C,R,r C,r	

CONT

EB29 (CONT.) C$_6$H$_5$NO$_2$

Code No.	Empirical Formula	Name and C.A. Number	Structural Formula	Solvent	Tech.	Medium		μ, M	pH	T, °C	Electrodes	App.	Experimental Parameters
EB29 ag47 cb73 da91	C$_6$H$_5$NO$_2$	nitrobenzene	C$_6$H$_5$NO$_2$	DMF	PY	CsClO$_4$	0.1	-	-	-	DME/SCE	24A0	m=2.74,t=2.5
						LiClO$_4$	0.1						
						KClO$_4$	0.1	-	-	-	DME/SCE	26-0	C=0.5,m= 1.61,t=3.6
						KClO$_4$	0.1	-	-	-	DME/SCE	24A0	m=2.74,t=2.5
						NaClO$_4$ H$_2$O	0.01 <1E-3	-	-	-	DME/SCE	256A0	C=0.5,m= 2.74,t=2.5
						NaClO$_4$	0.1	-	-	-	DME/SCE	24A0	m=2.74,t=2.5
						NaClO$_4$ H$_2$O	0.13 <1E-3	-	-	-	DME/SCE	256A0	C=0.5,m= 2.74,t=2.5
						NaClO$_4$ H$_2$O	0.008 1.66						
						NaClO$_4$ H$_2$O	0.1 1.66						
						NaClO$_4$ H$_2$O	0.19 1.66						
						NaClO$_4$ H$_2$O	0.71 1.66						
						NaClO$_4$ H$_2$O	0.013 16.6						
						NaClO$_4$ H$_2$O	0.16 16.6						
				HMP	IL	Et$_4$NClO$_4$	0.1	-	-	-	GCE Li/LiCl satd	25A0	d=0.3
						LiCl	0.3						
EB30 ag49 da92	C$_6$H$_5$NO$_2$	4-nitrosophenol C.A. 104-91-6	4-ONC$_6$H$_4$OH	DMSO	PY	Bu$_4$NClO$_4$	0.1	-	-	-	DME/MP	0-0	C=1
EB31 ag52	C$_6$H$_5$NO$_3$	2-nitrophenol C.A. 88-75-5	2-O$_2$NC$_6$H$_4$OH	MeOH 10	PY	buffer GEL	7E-3	-	12.5	-	DME/SCE	04A0	C=0.5
EB32 ag53	C$_6$H$_5$NO$_3$	3-nitrophenol C.A. 554-84-7	3-O$_2$NC$_6$H$_4$OH	MeOH 10	PY	buffer GEL	7E-3	-	11.5	-	DME/SCE	04A0	C=0.5
				DMSO	PY	Bu$_4$NClO$_4$	0.1	-	-	-	DME/MP	-	-

TABLE I. Electrochemical Data $C_6H_5NO_3$ EB32

Ref.	C/M	Charact. Potential		Response Const.		n Tech.		Electrokinetic Data			Products and Identification	Description and Remarks	Code No.
		Value	vs.		Value			Parameter	Value	From			
JE042 0261	130 1	$E_{\frac{1}{2}}$	-1.075	SCE	-	-	-	-	$E\log$ $dE_{\frac{1}{2}}/$ $d\log[Cs^+]$	≈65 64	- plot	-	C,r	EB29 ag47 cb73 da91
			-0.967		-	-	-	-	$E\log$ $dE_{\frac{1}{2}}/$ $d\log[Li^+]$	≈65 153	- plot	-	C,r	
JE039 0395	-	$E_{\frac{1}{2}}$	-1.065	SCE	I	2.3	-	-	$E\log$	69	-	-	C, $E_{\frac{1}{2}}$=-1.026 + 0.0088σ, $E_{\frac{1}{2}}$=0.665+ 1.088 LUMO,r	
			-1.38			5.4				-			C	
JE042 0261	130 1	$E_{\frac{1}{2}}$	-1.058	SCE	-	-	-	-	$E\log$ $dE_{\frac{1}{2}}/$ $d\log[K^+]$	≈65 70	- plot	-	C,r	
JE049 0017	-	$E_{\frac{1}{2}}$	-1.11F	SCE	-	-	-	-	$dE_{\frac{1}{2}}/$ $d\log[Na^+]$	96	-	-	C,p	
			-		-		-			-			C	
JE042 0261	130 1	$E_{\frac{1}{2}}$	-1.005	SCE	-	-	-	-	$E\log$ $dE_{\frac{1}{2}}/$ $d\log[Na^+]$	≈65 96	- plot	-	C,r	
JE049 0017	-	$E_{\frac{1}{2}}$	-0.99F	SCE	-	-	-	-	$dE_{\frac{1}{2}}/$ $d\log[Na^+]$	96	-	-	C,p	
			-		-		-			-			C	
			-1.05F -		-	-	-	-		65(?) -	-	-	C,p C	
			-0.98F -		-	-	-	-		65(?) -	-	-	C,p C	
			-0.96F -		-	-	-	-		105 -	-	-	C,p C	
			-0.91F -		-	-	-	-		105 -	-	-	C,p C	
			-0.87 -		-	-	-	-		36 -	-	-	C,p C	
			-0.83F -		-	-	-	-		36 -	-	-	C,p C	
JE041 0405	-	E_p	1.93 1.18	Li/satd LiCl					-	-		-	C,p C	
			1.99 1.20										C,p C	
JE057 0191	123 k	$E_{\frac{1}{2}}$	-0.5F 1.48Γ -1.84F	MP	$i_\ell(u)$	1.5F 8F 32.5F	-	-		-	-	-	C,p C C	EB30 ag49 da92
JE036 0223	-	$E_{\frac{1}{2}}$	-0.90	SCE	-	-	-	-	-	-	-	-	C,$dE_{\frac{1}{2}}/dpH$=52 for pH= 1-4.25, 72 for pH= 4.25-11.9, 0 for pH > 12;r	EB31 ag52
JE036 0223	-	$E_{\frac{1}{2}}$	-0.77	SCE	-	-	-	-	-	-	-	-	C;$dE_{\frac{1}{2}}/dpH$=55 for pH= 1-8.4, 95 for pH=8.4 -11.1, 0 for pH > 11.5;r	EB32 ag53
JE057 0191 (JE050 0073)	-	-	-	-	-	-	-	-	-	-	-	-	see AG53	

Code No.	Empirical Formula	Name and C.A. Number	Structural Formula	Solvent	Tech.	Medium		μ, M	pH	T, °C	Electrodes	App.	Experimental Parameters
EB33 ag54 da94	$C_6H_5NO_3$	4-nitrophenol C.A. 100-02-7	$4-O_2NC_6H_4OH$	MeOH 10	PY	buffer GEL	7E-3	-	1.1 7 8 10 12.5	-	DME/SCE	O4AO	C=0.5
				DMSO	PY, VA	Bu_4NClO_4	0.1	-	-	-	DME/MP	-	-
EB34	$C_6H_5NO_4Se$	3-nitrobenzene-seleninic acid C.A. 33439-12-2	$3-O_2NC_6H_4SeO_2H$	H_2O	VY	$HClO_4$ $NaClO_4$	0.05 0.2	-	-	25	Pt PDS MSE	126AO	C=1, v=5.6, MSE Aux, PE
EB35	$C_6H_5NO_4Se$	4-nitrobenzene-seleninic acid C.A. 3350-70-8	$4-O_2NC_6H_4SeO_2H$	H_2O	VY	$HClO_4$ $NaClO_4$	0.05 0.2	-	-	25	Pt PDS MSE	126AO	C=1, v=5.6, MSE Aux, PE
EB36 ag61	$C_6H_5N_5$	2-aminopteridine C.A. 700-81-2	Table II-2	H_2O	PY	HCl BR	0.1	-	1.0 2.0 3.0 4.0 5.0 6.0 8.0 10.0 11.0 12.0 13.0	-	DME/SCE	O4AO	C=0.21, h=70, L
EB37	C_6H_5NaS	sodium thiophenolate C.A. 930-69-8	C_6H_5SNa	DMSO	PY	$KClO_4$	0.1	-	-	25±0.1	DME Hg/ Hg_2Cl_2, LiCl	256AO 1	C=0.05, h=48.4, EC=2.06, Pt Aux C=0.88

TABLE I. Electrochemical Data C_6H_5NaS EB37

Ref.	C/M	Charact. Potential		Response Const.		n		Electrokinetic Data			Products and Identification	Description and Remarks	Code No.
		Value	vs.	Value		Tech.		Parameter	Value	From			
JE036 0223	-	$E_{\frac{1}{2}}$ -0.204F	SCE	-	-	-	-	$dE_{\frac{1}{2}}/dpH$	63	-	-	$C, E_{\frac{1}{2}}=-0.135-0.063pH$ for pH=1-7.6,p	EB33
		-0.54F		-	-	-	-		63	-	-	C,p	
		-0.633F		-	-	-	-		130	-	-	$C, E_{\frac{1}{2}}=-0.498-0.130pH$ for pH=7.6-10.2,p	
		-0.86F		-	-	-	-		130	-	-	C,p	
		-0.91F		-	-	-	-		0	-	-	$C, E_{\frac{1}{2}} \neq f(pH)$ for pH $\geq$ 12.0,p	
JE057 0191	123 k	-	-	-	-	-	-	-	-	-	-	see AG54, DA94	
JE040 0339	404 a	$E_{\frac{1}{2}}$ -0.18F	SCE	-	-	-	-	-	-	-	-	$C, \not\models, \Delta E_{\frac{1}{2}}=0.10\sigma,p$	EB34
JE040 0339	404 a	$E_{\frac{1}{2}}$ -0.23F	SCE	-	-	-	-	-	-	-	-	$C, \not\models, \Delta E_{\frac{1}{2}}=0.10\sigma,p$	EB35
C0027 0199	-	$E_{\frac{1}{2}}$ -0.40F	SCE	i_ℓ/i_d	1.00F	i:i	2	-	-	-	-	C, i_d, p	EB36 ag61
		-0.48F			0.95F		2	-	-	-	-	C, i_d, p	
		-0.10F			0.06F		2	-	-	-	-	C, i_k, p	
		-0.56F			0.92F		2	-	-	-	-	C, i_d	
		-0.20F			0.29F		2	-	-	-	-	C, i_d, p	
		-0.64F			0.75F			-	-	-	-	C, i_d	
		-0.29F			0.87F		2	-	-	-	-	C, i_d, p	
		-0.74F			0.13F			-	-	-	-	C, i_k	
		-0.39F			0.98F		2	-	-	-	-	C, i_d, p	
		-0.80F			0.02F			-	-	-	-	C, i_k	
		-0.59			1.00F		2	-	-	-	-	C, i_d, p	
		-0.72F			0.93F		2	-	-	-	-	C, i_d, p	
		-1.10F			0.05F			-	-	-	-	C, i_k	
		-0.77F			0.84F		2	-	-	-	-	C, i_d, p	
		-1.16F			0.10F			-	-	-	-	C, i_k	
		-			0.03F			-	-	-	-	C, i_d	
		-0.81F			0.75F		2	-	-	-	-	C, i_d, p	
		-1.22F			0.23F			-	-	-	-	$C, i_k(?)$	
		-			0.04F			-	-	-	-	C, i_d	
		-0.84F			0.56F		2	-	-	-	-	C, i_d, p	
		-1.27F			0.28			-	-	-	-	$C, i_k(?)$	
		-			0.20			-	-	-	-	$C, i_d(?)$	
JE056 0373	-	$E_{\frac{1}{2}}$ -0.69	SCE	i_ℓ	0.1	QE	0.67	-	-	-	-	$A, i_d, i_\ell/C \neq f(C)$ for C= 0.05-0.7, r	EB37
		-0.51			0.03		0.32	-	-	-	CP→Hg thiophenolate, PY	$A, i_\ell/C \neq f(C)$ for C= 0.05-0.9, i_d	
		-0.76			1.5		-	-	-	-	-	A, i_d, r	
		-0.39			0.55			-	-	-	-	A, i_d	

Code No.	Empirical Formula	Name and C.A. Number	Structural Formula	Solvent	Tech.	Medium		μ, M	pH	T, °C	Electrodes	App.	Experimental Parameters
EB38 ag84 cb78	$C_6H_6N_2O$	isonicotinamide C.A. 1453-82-3	Table 11-2	H_2O	PY	HCl	6	-	-	-	DME/SCE	0-0	C=0.33,m= 2.45,t(oc)= 3.7,h=48.5
						HCl	2.5						
						HCl KCl			0.40				
									0.70				
									0.90				
									1.50				
						GLYC			2.10				
						CITR			2.60				
									3.05				
						ACET			4.25				
						succinate			5.05				
						PHOS			6.15				
									6.60				
									7.10				
						BOR			8.25				
									9.00				
									9.70				
						PHOS			11.05				
						KOH	0.1		13.0				
EB39 ag85 cb79 db04	$C_6H_6N_2O$	nicotinamide C.A. 98-92-0	Table 11-2	H_2O	PY	HCl	0.5	-	0.60	25.0 ±0.2	DME Ag/AgCl, KCl satd	2AO	C=0.1,m=0.63, t(c)=0.5,h= 35,v=4.2
						PHOS	0.5		1.96				C=0.01,m= 1.86±0.04, t(c)=0.5,h= 35
													C=0.1
													C=1
													C=10
						GLYC	0.5		2.4				C=1
						FORM	0.5		3.6				C=0.1
CONT		isonicotinamide	Table 11-2	H_2O	PY	HCl	6				DME/SCE		

TABLE I. Electrochemical Data $C_6H_8N_2O$ (CONT.) EB39

| Ref. | C/M | Charact. | Potential | Response | Const. | Tech. | n | Electrokinetic Data | | | Products and | Description and | Code |
		Value	vs.		Value			Parameter	Value	From	Identification	Remarks	No.
AS017 1077	387 a	$E_{\frac{1}{2}}$	-0.57	SCE	i_ℓ	3.30	-	-	-	-	-	-	C,r	EB38 ag84 cb78
			-0.58			3.68	-	-	-	-	-	-	C,r	
			-0.605			3.75	-	-	-	-	-	-	C,r	
	-		-0.63			3.75	-	-	-	-	-	-	C,r	
			-0.64			3.80	-	-	-	-	-	-	C,r	
			-0.665			4.30	-	-	-	-	-	-	C,r	
			-0.705			5.00	-	-	-	-	-	-	C,r	
			-0.72			5.35	-	-	-	-	-	-	C,r	
			-0.75			5.55	-	-	-	-	-	-	C,r	
			-0.86			6.65	-	-	-	-	-	-	C,r	
			-0.93			6.70	-	-	-	-	-	-	C,r	
			-1.03			6.60	-	-	-	-	-	-	C,r	
			-1.07			6.80	-	-	-	-	-	-	C,r	
			-1.105			5.70	-	-	-	-	-	-	C,r	
			-1.17			3.90	-	-	-	-	-	-	C,r	
			-1.215			3.80	-	-	-	-	-	-	C,r	
			-1.24			3.70	-	-	-	-	-	-	C,r	
			-1.30			3.70	-	-	-	-	-	-	C,r	
			-1.345			3.90	-	-	-	-	-	-	C,r	
JE040 0197	-	$E_{\frac{1}{2}}$	-0.76F	NHE	-	-	-	-	$dE_{\frac{1}{2}}/dpH$	75	-	-	$C, E_{\frac{1}{2}} = -0.69-0.075pH \pm 0.02$ for pH=0-10, $dE_{\frac{1}{2}}/dpH=f(C), i_a, p$	EB39 ag85 cb79 db04
			-1.0F							30			$C, E_{\frac{1}{2}} = -0.97-0.03pH \pm 0.01$ for pH=0-4, $dE_{\frac{1}{2}}/dpH=f(C)$	
			-0.82F		i_ℓ/C	5.2F	-	-	Tomeš	46.5	-	-	C, i_a, p	
			-			-				-			C,0	
			-0.84F			4.8F	-	-	Tomeš	46.5	-	-	$C, dE_{\frac{1}{2}}/dpH=f(C), i_a, p$	
									$dE_{\frac{1}{2}}/dpH$	75	-			
			-1.04F			10.5F			Tomeš	35			$C, dE_{\frac{1}{2}}/dpH=f(C), i_\ell \downarrow$ as C ↑	
									$dE_{\frac{1}{2}}/dpH$	30				
			-0.86F			4.8F	-	-	Tomeš	70	-	-	$C, E_{\frac{1}{2}} \neq f(pH), dE_{\frac{1}{2}}/dpH= f(C), i_a, p$	
			-1.06F			6F				42.5			$C, i_\ell \downarrow$ as C ↑	
			-1.22F			45F				42.5			C	
			-0.79F			4F	-	-		30	-	-	C, i_a, p	
			-0.96F			4F				25			$C, E_{\frac{1}{2}} = -0.86 \ 0.056pH \pm 0.02$ for pH=0-7, $i_\ell \downarrow$ as C ↑	
									$dE_{\frac{1}{2}}/dpH$	56				
			-1.31F			large			Tomeš	100			C	
			-0.84± 0.01			-	-	-	$dE_{\frac{1}{2}}/dpH$	0	-	-	C, i_a, p	
			-0.98F			-		-		63			$C, E_{\frac{1}{2}} = -0.8-0.063pH \pm 0.015$ for pH=2-10	
			-1.08F							-			C	
			-0.97F			-	-	-		75	-	-	C,p	
			-1.1F							30			C	
AS017 1077	387 a	$E_{\frac{1}{2}}$		SCE										CONT

Code No.	Empirical Formula	Name and C.A. Number	Structural Formula	Solvent	Tech.	Medium		μ, M	pH	T, °C	Electrodes	App.	Experimental Parameters
EB39 ag85 cb79 db04	$C_6H_6N_2O$	nicotinamide	Table II-2	H_2O	PY	ACET	0.5	-	4.8	25.0 ±0.2	DME Ag/AgCl, KCl satd	2AO	C=0.01, t(c)=0.5, h=35, v=4.2
													C=0.1
													C=1.0
													C=10
						TRIS H$^+$ Cl$^-$	0.5		8.2				C=0.1
						BOR	0.5		9.6				C=10
EB40	$C_6H_6N_2O_2$	2-nitroaniline C.A. 88-74-4	$2\text{-}O_2NC_6H_4NH_2$	MeOH 10	PY	buffer GEL	7E-3	-	12	-	DME/SCE	04AO	C=0.5
EB41 ag92 cb81 db05	$C_6H_6N_2O_2$	3-nitroaniline C.A. 99-09-2	$3\text{-}O_2NC_6H_4NH_2$	MeOH 10	PY	buffer GEL	7E-3	-	12	-	DME/SCE	04AO	C=0.5
EB42 ag93 cb82 db06	$C_6H_6N_2O_2$	4-nitroaniline C.A. 100-01-6	$4\text{-}O_2NC_6H_4NH_2$	MeOH 10	PY	buffer GEL	7E-3	-	0.4	-	DME/SCE	04AO	C=0.5
									1.6				
									2.6				
									7.5				
									8.5				
									10.2				
									11				

TABLE I. Electrochemical Data $C_6H_6N_2O_2$ EB42

Ref.	C/M	Charact. Potential Value	vs.	Response Const.	Value	n	Tech.	Electrokinetic Data Parameter	Value	From	Products and Identification	Description and Remarks	Code No.
JE040 0197	-	$E_{\frac{1}{2}}$	-1.03F	NHE	i_ℓ/C	5.8F	-	-	Tomeš	32.5	-		C,i_a,p	EB39 ag85 cb79 db04
			-		-					-			C	
			-1.03F			5.5F	-	-	$dE_{\frac{1}{2}}/dpH$	75	-		C,i_a,p	
									Tomeš	42.5				
			-1.45F		i_{Max}/C	300F				32.5			C,i_{cat}	
			-0.94F		i_ℓ/C	1.2F	-	-		30	-	-	$C,E_{\frac{1}{2}}=-0.635-0.064pH$	
									$dE_{\frac{1}{2}}/dpH$	64			± 0.004 for pH=3.2-7, $dE_{\frac{1}{2}}/dpH=f(C),i_a,p$	
			-1.09F			4.2F			Tomeš	30	-		$C,E_{\frac{1}{2}}=-0.800-0.063pH$	
									$dE_{\frac{1}{2}}/dpH$	63			± 0.015 for pH=2-10, $dE_{\frac{1}{2}}/dpH=f(C)$	
			-1.5F		i_{Max}/C	263		-	Tomeš	65			C,i_{cat}	
			-0.93F		i_ℓ/C	1.2F	-	-		55	-	-	$C,E_{\frac{1}{2}}=-0.668-0.054pH$	
									$dE_{\frac{1}{2}}/dpH$	54			± 0.01 for pH=3.2-7, $dE_{\frac{1}{2}}/dpH=f(C),i_a,p$	
			-1.13F			3F	-	-	Tomeš	42.5			C	
									$dE_{\frac{1}{2}}/dpH$	56				
			-1.3F		-	-	-	-		75	sttd	-	C,p	
			-1.27F		-	-	-	-		47	-	-	$C,E_{\frac{1}{2}}=-0.862-0.047pH$	
													± 0.015 for pH=7-10, $dE_{\frac{1}{2}}/dpH=f(C),p$	
			-1.46F		-	-	-	-		87	-	-	$C,E_{\frac{1}{2}}=-0.66-0.087pH$	
													± 0.025 for pH=7-10, $dE_{\frac{1}{2}}/dpH=f(C)$	
JE036 0223	-	$E_{\frac{1}{2}}$	-0.81	SCE	-	-	-	-	-	-	-	-	$C; E_{\frac{1}{2}} \neq f(pH)$ for pH $\geq$ 12; $dE_{\frac{1}{2}}/dpH=88$ for pH=0.5-2.3, 60 for pH=2.3-8.3, and 50 for pH=8.3-11;r	EB40
JE036 0223	-	$E_{\frac{1}{2}}$	-0.72	SCE	-	-	-	-	-	-	-	-	$C; E_{\frac{1}{2}} \neq f(pH)$ for pH $\geq$ 12; $dE_{\frac{1}{2}}/dpH=73$ for pH=0.5-1.6, 61 for pH=1.6-7.8, and 52 for pH=7.8-11.2;r	EB41 ag92 cb81 db05
JE036 0223	-	$E_{\frac{1}{2}}$	-0.15F	SCE	-	-	-	-	$dE_{\frac{1}{2}}/dpH$	120	-	-	$C,E_{\frac{1}{2}}=-0.104-0.120pH$ for pH=0.4-1.6,r	EB42 ag93 cb82 db06
			-0.29F		-	-	-	-		120	-		C,r	
			-0.37F		-	-	-	-		65	-		$C,E_{\frac{1}{2}}=-0.20-0.065pH$ for pH=1.6-7.6,r	
			-0.69F		-	-	-	-		65	-		C,r	
			-0.73F		-	-	-	-		49	-		$C,E_{\frac{1}{2}}=-0.31-0.049pH$ for pH=7.6-10.7,r	
			-0.83F		-	-	-	-		49	-		C,r	
			-0.84F		-	-	-	-		0	-		$C,E_{\frac{1}{2}} \neq f(pH)$ for pH $\geq$ 11, r	

Code No.	Empirical Formula	Name and C.A. Number	Structural Formula	Solvent	Tech.	Medium		μ, M	pH	T, °C	Electrodes	App.	Experimental Parameters
EB43	$C_6H_8N_2O_3$	4-nitrophenyl-hydroxylamine C.A. 16169-16-7	$4-O_2NC_6H_4NHOH$	MeOH 10	PY	buffer GEL	7E-3	-	13.5	-	DME/SCE	04A0	C=0.5
EB44	$C_6H_8N_2O_4S$ (see also EE76)	4-nitrobenzene-sulfonamide C.A. 6325-93-5	$4-O_2NC_6H_4SO_2NH_2$	DMF	IR	Et_4NClO_4	0.1	-	-	22.5 ±0.5	Pt/SCE	256A F0	E_{app}= -1.05 V, A= 0.25, t=0.01-10, Pt Aux
													E_{app}= -1.3 V, t=0.01
													t=0.34
													t=3.9
													E_{app}=-1.6 V, t=0.01-10
					VA	Et_4NClO_4	0.1	-	-	22.5 ±0.5	Pt/SCE	256A F0	0→-1.06→ 0 V,C=0.63, A=0.25,v= 100,Pt Aux
					VR	Et_4NClO_4	0.1	-	-	22.5 ±0.5	Pt/SCE	256A 0	0→-1.26→ 0 V,C=0.63, A=0.25,v= 100,Pt Aux
													0→-1.62→ 0 V
EB45	$C_6H_8N_4O$	5,6-dihydro-7-hydroxypteridine C.A. 6742-34-3	Table II	H_2O	VR	MB		-	7.0	25± 0.1	PGE/SCE	2A0	0→-1.5→ 0.9→-1.5→ 0 V,C=ns, v=50,Pt Aux
EB46	$C_6H_8N_4O$	7,8-dihydro-6-hydroxypteridine C.A. 51036-16-9	Table II	H_2O	VR	BOR		0.5	11.1	25± 0.1	PGE/SCE	2A0	0→-1.5→ 0.75→-1.5 →0 V,C=ns, v=50,Pt Aux
EB47 ah06 cb85 db07	$C_6H_6O_2$	hydroquinone C.A. 123-31-9	$4-HOC_6H_4OH$	H_2O	VA	HCl KCl	0.1	-	1	22±1	XEl Ag/AgCl	--0	C=1,A=0.28, v=8.3
											XEl Ag/AgCl		
						ACET	0.2		4.85		XEl Ag/AgCl		
CONT		4-nitrophenyl-hydroxylamine	$4-O_2NC_6H_4NHOH$	MeOH 10	PY								

TABLE I. Electrochemical Data $C_6H_6O_2$ (CONT.) EB47

Ref.	C/M	Charact.	Potential Value	vs.	Response	Const. Value	Tech.	n	Electrokinetic Data Parameter	Value	From	Products and Identification	Description and Remarks	Code No.
JE036 0223	-	$E_{\frac{1}{2}}$	-1.07	SCE	-	-	-	-	-	-	-	-	$C, dE_{\frac{1}{2}}/dpH=55$ for pH=1-8.9, 100 for pH=8.9-13.1, and 0 for pH>13.5, r	EB43
JE053 0293	412 b	-	-	-	$it^{\frac{1}{2}}/C$	27	-	1	-	-	-	-	C, i_d, r	EB44
		-	-	-	-	-	-	1.08	-	-	-	-	C, i_k, r	
		-	-	-	-	-	-	1.46	-	-	-	-	C, i_k, r	
		-	-	-	-	-	-	1.99	-	-	-	-	C, i_k, r	
		-	-	-	-	54	-	2	-	-	-	-	C, i_d, r	
JE053 0293	412 b	E_p	-0.93	SCE	i_p	25.7F	-	-	-	-	-	-	C, ec, r	
			-0.87			10F							$A, i_{p,A}/i_{p,C} \uparrow$ as $v \uparrow$ or $C \downarrow$	
JE053 0293	412 b	E_p	-0.93	SCE	i_p	27.1F	-	-	-	-	-	-	$C, i_p \downarrow$ on later scans, r	
			-1.14F			2.9F							$C, i_p \uparrow$ on later scans	
			-1.06F			small							A	
			-0.86F			8.6F							A	
			-0.93			25.7F	QE	0.95 ± 0.05	-	-	-	$4-O_2NC_6H_4SO_2NH^-$, TLC	$C, i_p \downarrow$ on later scans, r	
			-1.16			2.9F		-					$C, i_p \uparrow$ on later scans	
			-1.44			2.9F		-					$C, \mathrel{\updownarrow}, i_p \uparrow$ as $v \uparrow$ or $C \downarrow$	
			-1.08F			12.1F		-					A	
			-0.87			small		-					A, present only in first scan	
JE047 0479	-	E_p	0.45F	SCE	-	-	-	-	-	-	-	7-hydroxypteridine	A, r	EB45
			-0.86F									-	C	
JE047 0479	-	E_p	0.4F	SCE	i_p	16F	-	-	-	-	-	6-hydroxypteridine, VA	A, r	EB46
			-1.0F			60.0F		-				-	C	
JE052 0037	-	$E_{p/2}$	0.311	SCE	i_p/C	13.5	-	-	$\Delta E_{p/2}$	158	sttd	-	$C, XEl=1$-bromonaphthalene CPE, $E_0'=0.390, p$	EB47 ah06 cb85 db07
			0.469			26.2							A	
			0.409			22.0		-		4			$C, XEl=$silicone fluid(MS200) CPE, $E_0'=0.411, p$	
			0.413			24.6				-			A	
			0.111			23.7		-		112	-		$C, XEl=1$-bromonaphthalene, CPE, $E_0'=0.167$ V, p	
			0.223			23.0							A	CONT

83

EB47 (CONT.) $C_6H_6O_2$ CRC Handbook Series in Organic Electrochemistry

Code No.	Empirical Formula	Name and C.A. Number	Structural Formula	Solvent	Tech.	Medium		μ, M	pH	T, °C	Electrodes	App.	Experimental Parameters
EB47 ah06 cb85 db07	$C_6H_6O_2$	hydroquinone	4-HOC_6H_4OH	H_2O	VA	ACET	0.2	–	4.85	22±1	XE1 Ag/AgCl	--0	C=1, A=0.28, v=8.3
						BOR	0.1		8.9		XE1 Ag/AgCl		
											XE1 Ag/AgCl		
				MeCN	VR	$KClO_4$	0.03	–	–	22	PDE/SCE	04A0	C=0.45, A=0.0018, v=0.066
				HF	VA	NaF	1	–	–	0	GCE Pd/H_2, NaF 1	24-0	C=3.3, v=5
EB48	$C_6H_6O_2Se$	benzeneseleninic acid C.A. 6996-92-5	$C_6H_5SeO_2H$	H_2O	VY	$HClO_4$ $NaClO_4$	0.05 0.2	–	–	25	Pt PDS MSE	126A0	C=1, v=5.6, MSE Aux, PE
EB49 db09	C_6H_6S	benzenethiol C.A. 108-98-5	C_6H_5SH	DMSO	PY	$KClO_4$	0.1	–	–	25±0.1	DME Hg/Hg_2Cl_2, LiCl 1	256A0	C=0.5, h=48.4, EC=2.06, Pt Aux
EB50 ah15 cb88	C_6H_7N	aniline C.A. 62-53-3	$C_6H_5NH_2$	H_2O	IL	ACET PHOS		–	0.8 5.3 12	25±0.1	CPE/MSE	2-0	C=1, A=3.8E-2
				MeCN	VY	$NaClO_4$	0.1	–	–	25	RPDE SSCE	2-0	C=0.3, ω=12.5, Pt Aux
EB51 ah19	C_6H_7NO	2-aminophenol C.A. 95-55-6	2-$HOC_6H_4NH_2$	MeCN	VR	$KClO_4$	0.03	–	–	22	PDE/SCE	04A0	C=0.6, A=0.0018, v=0.066
EB52 ah20	C_6H_7NO	3-aminophenol C.A. 41736-65-6	Table II-2	MeCN	VR	$KClO_4$	0.03	–	–	22	PDE/SCE	04A0	C=0.5, A=0.0018, v=0.066
EB53 ah21 cb89 db15	C_6H_7NO	4-aminophenol C.A. 123-30-8	4-$H_2NC_6H_4OH$	H_2O	VR	H_2SO_4	1	–	–	23±0.5	Pt TLE SCE	---	C=2, v=3
					VY	H_2SO_4	0.5	–	–	23±0.5	XE1/SCE	---	C=1
				MeCN	VR	$KClO_4$	0.03	–	–	22	PDE/SCE	04A0	C=0.5, A=0.0018, v=0.066

TABLE I. Electrochemical Data C_6H_7NO EB53

Ref.	C/M	Charact. Potential		Response Const.		n Tech.	n	Electrokinetic Data			Products and Identification	Description and Remarks	Code No.
		Value	vs.		Value			Parameter	Value	From			
JE052 0037	-	$E_{p/2}$ 0.175	SCE	i_p/C	26.4	-	-	$\Delta E_{p/2}$	13	-	-	C,XEl=silicone fluid (MS200) CPE,E_O'= 0.169,p	EB47 ah06 cb85 db07
		0.162			25.2		-		-				
		-0.079			24.6		-		4	-		C,XEl=1-bromonaph- thalene CPE,E_O'= -0.077 V,p A	
		-0.075			21.9		-		-				
		-0.069			23.4		-		8	-		C,XEl=silicone fluid (MS200) CPE,E_O'= -0.073,p A	
		-0.077			21.9		-		-				
C0036 1644	-	$E_{p/2}$ 0.901	SCE	-	-	i:i	2(?)	$E_p-E_{p/2}$	90	-	-	A,$\neq$,p	
JE051 0456	-	E_p 0.53	Pd/1 M NaF,H_2	$i_p(u)$	1	-	-	-	-	-	-	A,$i_p/v^{\frac{1}{2}}\neq f(v)$,R,p	
		0.47			1							C,$i_p/v^{\frac{1}{2}}\neq f(v)$,R	
JE040 0339	404 a	$E_{\frac{1}{2}}$ -0.41F	SCE	$i_\ell(u)$	28F	-	-	Elog $dE_{\frac{1}{2}}/dpH$	120 128±2	-	-	C,$\neq$,i_ℓ=kC for C= 0.05-5,Tc=1.54,i_d, $E^o_{\frac{1}{2}}$=0.178±0.004, $\Delta E_{\frac{1}{2}}$=0.10σ,p	EB48
JE056 0373	-	$E_{\frac{1}{2}}$ -0.25	SCE	i_ℓ	1.6F	QE	1	-	-	-	-	A,r	EB49 db09
JE043 0387	-	E_p 0.95F	SCE	-	-	-	-	-	-	-	-	A,p	EB50 ah15 cb88
		0.72F		-	-	-	-	-	-	-	-	A,p	
		0.62F		-	-	-	-	-	-	-	-	A,p	
JE043 0267	38 r	$E_{\frac{1}{2}}$ 0.88F	SSCE	D(?) i_ℓ	34.2 35F	-	-	-	-	-	CP (1.3 V) → 4-amino- diphenylamine,VY; CP at (1.8 V)→ aniline black	A,i_ℓ=kω$^{\frac{1}{2}}$,r	
C0036 1644	-	$E_{p/2}$ 0.70	SCE	-	-	-	-	$E_p-E_{p/2}$	100	sttd	-	A,$\neq$,i_p ↓ on later scans,p	EB51 ah19
C0036 1644	383 a	$E_{p/2}$ 0.75	-	-	-	-	-	$E_p-E_{p/2}$	140	sttd	-	A,$\neq$,i_p ↓ on later scans,p	EB52 ah20
JE044 0275	-	E_p 0.54F	SCE	i_p	12F	-	-	-	-	-	4-benzoquinoneimine	A,p	EB53 ah21 cb09 db15
		0.52F			9F						4-aminophenol	C,4-benzoquinone- imine redn.	
		0.44F			4.5F						hydroquinone	C,4-benzoquinone redn.	
		0.48F			4.5F						-	A,hydroquinone oxidn.,0 on first scan	
JE044 0275	-	$E_{\frac{1}{2}}$ 0.52F	SCE	i_ℓ	68.8F	-	-	-	-	-	-	A,XEl=twin thin- layer Pt electrode, p	
C0036 1644	105 e	$E_{p/2}$ 0.387	SCE	-	-	sttd	1	$E_p-E_{p/2}$	50	sttd	-	A,R,p	
		0.65					1		90			A,$\neq$	
		0.25					1		20			C,R	

Code No.	Empirical Formula	Name and C.A. Number	Structural Formula	Solvent	Tech.	Medium	μ, M	pH	T, °C	Electrodes	App.	Experimental Parameters
EB54 ah25 cb90	C₆H₇NO	phenylhydroxylamine C.A. 100-65-2	C₆H₅NHOH	H₂O	PY	CITR	-	3.15	-	DME/SCE	O4AO	C=0.5
								3.85				
								4.65				
				MeOH 10	PY	CITR	-	2.0	-	DME/SCE	O4AO	C=0.5
								3.8				
								4.75				
EB55	C₆H₇NO₂	N-ethylmaleimide	Table II	EtOH 50	PY	BR	-	7.25	-	DME/SCE Tl⁺ SCE	O4AO	C=1.0, t=3.5, m=3.3, h=40
EB56	C₆H₇NO₃S	metanilic acid C.A. 121-47-1	3-H₂NC₆H₄SO₃H	H₂O	IL	ACET PHOS	-	0.6	25±0.1	CPE/MSE	2-0	C=1, A=3.8E-2
								3.3				
								6				
								9.4				
								12.5				
					VR	ACET PHOS	-	2.5	25±0.1	CPE/MSE	2-0	-0.35→1.25 →-0.35→ 0.9V, C=1, A= 3.8E-2, v=5
					VY	ACET PHOS	-	1.6	25±0.1	Pt PDS MSE	2-0	C=0.5, A= 6.45E-2, v= 0.92, PE
								3.6				
								5.6				
								7.0				
								10.0				
EB57	C₆H₇NO₃S	orthanilic acid C.A. 88-21-1	2-H₂NC₆H₄SO₃H	H₂O	IL	ACET PHOS	-	0.6	25±0.1	CPE/MSE	2-0	C=1, A=3.8E-2
								3.2				
								5.5				
								8.9				
								12.5				
CONT		phenylhydroxylamine	C₆H₅NHOH	H₂O	PY	CITR						

TABLE I. Electrochemical Data $C_6H_7NO_3S$ (CONT.) EB57

Ref.	C/M	Charact. Potential		Response Const.		n Tech.	Electrokinetic Data			Products and Identification	Description and Remarks	Code No.
		Value	vs.		Value		Parameter	Value	From			
JE043 0311	103 c	$E_\frac{1}{2}$	-0.58F -0.95F	SCE	i_ℓ	2.0F 0.4F	-	-	-	-	-	C,r C	EB54 ah25 cb90
			-0.67F -0.98F			1.7F 0.7F	-	-	-	-	-	C,r C	
			-0.77F -1.18F			1.0F 1.4F	-	-	-	-	-	C,Tc<2.1,i_k,r C,M	
JE043 0311	103 b	$E_\frac{1}{2}$	-0.5F -0.92F	SCE	i_ℓ	1.9F 0.13F	-	-	-	-	-	C,r C	
	103 c		-0.62F -1.15F			≈1.7F 0.6F	-	-	-	-	-	C,rounded Mx,r C	
			-0.71F -1.27F			≈0.9F 1.4F	-	-	-	-	-	C,rounded Mx,r C,M	
C0026 2749	-	$E_\frac{1}{2}$	-0.791 -1.240	SCE	-	-	sttd	1 1	-	-	-	-	C,i_ℓ=f(t),p C,i_ℓ=f(t)	EB55
JE043 0387	38 g	E_p	1.13F	SCE	-	-	-	-	-	-	-	A,r	EB56
			0.94F		-	-	-	-	-	-	-	A,r	
			0.9F		-	-	-	-	-	-	CP→azobenzenedi-sulfonic acid(20%), PY,VA;N-phenyl-quinonediimine(37%), PY,VA	A,r	
	38 p		0.8F		-	-	-	-	-	-	CP→azobenzenedi-sulfonic acid(42%), PY,VA;N-phenyl-quinonediimine(32%), PY,VA	A,r	
			0.78F		-	-	-	-	-	-	-	A,r	
JE043 0387	38 g	E_p	0.95F	SCE	$i_p(u)$	46F	-	-	-	-	4-aminodiphenyl-amine-2-sulfonic acid,VA	A,r	
			0.38F 0.40F			7F 4F	-	-	-	-		C A,O on first scan	
JE043 0387	38 g	$E_\frac{1}{2}$	0.99F	SCE	-	-	QE	β_a	2.0 ±0.1	1.1	Elog	-	A,i_ℓ=kC for C=0.05-1,i_d,Tc=1.9,r	
			0.91		-	-			2.0 ±0.1	-		-	A,i_ℓ=kC for C=0.05-1,i_d,Tc=1.9,r	
			0.88F		-	-			2.0 ±0.1	1.07		-	A,i_ℓ=kC for C=0.05-1,i_d,r	
	38 p		0.85F		-	-			2.0 ±0.1	1.08		CP→azobenzenedi-sulfonic acid(40%), PY,VA;N-phenyl-quinonediimine(33%), PY,VA	A,i_ℓ=kC for C=0.05-1,i_d,Tc=1.9,r	
			0.78F		-	-			2.0 ±0.1	0.8		CP→azobenzenedi-sulfonic acid(42%), PY,VA,N=phenyl-quinonediimine(32%), PY,VA	A,i_ℓ=kC for C=0.05-1,i_d,	
JE043 0387	38 g	E_p	1.1F	SCE	-	-	-	-	-	-	-	A,r	EB57
			0.97F		-	-	-	-	-	-	-	A,r	
			0.94F		-	-	-	-	-	-	-	A,r	
	38 p		0.83F		-	-	-	-	-	-	CP→azobenzenedi-sulfonic acid(3%), PY,VA;N-phenyl-quinonediimine(68%), PY,VA	A,r	
			0.79F		-	-	-	-	-	-	-	A,r	

CONT

Code No.	Empirical Formula	Name and C.A. Number	Structural Formula	Solvent	Tech.	Medium	μ, M	pH	T, °C	Electrodes	App.	Experimental Parameters
EB57	$C_6H_7NO_3S$	orthanilic acid	$2-H_2NC_6H_4SO_3H$	H_2O	VY	ACET PHOS	-	0.7	25±0.1	Pt PDS MSE	2-0	C=0.5, A=6.45E-2, v=0.92, PE
								3.3				
								5.6				
								7.6				
								10.4				
EB58	$C_6H_7NO_3S$	sulfanilic acid C.A. 121-57-3	$4-H_2NC_6H_4SO_3H$	H_2O	IL	ACET PHOS	-	0.9	25±0.1	CPE/MSE	2-0	C=1, A=3.8E-2
								3.0				
								5.8				
								9.4				
								12.5				
					VY	ACET PHOS	-	0.7	25±0.1	Pt PDS MSE	2-0	C=0.5, A=6.45E-2, v=0.92, PE
								3.2				
								5.9				
								7.1				
								10				
EB59 ah35 cb92	$C_6H_7N_3O$	isonicotinic acid hydrazide C.A. 54-85-3	Table II-2	H_2O	PY	HCl	6	6	-	-	DME/SCE	--0	C=0.3, m=2.45, t=3.7, h=48.5
						HCl	2.5	2.5				
						HCl KCl	-	0.70				
								1.50				
						glycine		2.10				
CONT												

TABLE I. Electrochemical Data $C_6H_7N_3O$ (CONT.) EB59

Ref.	C/M	Charact. Potential		Response Const.		n Tech.	n	Electrokinetic Data			Products and Identification	Description and Remarks	Code No.
		Value	vs.		Value			Parameter	Value	From			
JE043 0387	38 g	$E_{\frac{1}{2}}$	0.98F	SCE	-	-	QE	2.0 ±0.1	β_a	1.05	Elog	-	$A,i_\ell=kC$ for C=0.05-1, i_d, Tc=1.8, r	EB57
			0.94F	-	-	-	QE	2.0 ±0.1		1.03		-	$A,i_\ell=kC$ for C=0.05-1, i_d, Tc=1.3, r	
			0.93F	-	-	-	QE	2.0 ±0.1		1.04		-	$A,i_\ell=kC$ for C=0.05-1, i_d, r	
	38 p		0.88F	-	-	-	QE	2.0 ±0.1		0.9		-	$A,i_\ell=kC$ for C=0.05-1, i_d, Tc=1.5, r	
			0.84F	-	-	-	QE	2.0 ±0.1		0.7		CP→azobenzenedi-sulfonic acid(3%), PY,VA;N-phenyl-quinonediimine(68%), PY,VA	$A,i_\ell=kC$ for C=0.05-1, i_d, r	
JE043 0387	38 g	E_p	1.03F	SCE	-	-	-	-	-	-	-	-	A,r	EB58
			0.95F	-	-	-	-	-	-	-	-	-	A,r	
			0.91F	-	-	-	-	-	-	-	-	CP→azobenzenedi-sulfonic acid(21%), PY,VA;N-phenyl-quinonediimine(46%), PY,VA	A,r	
	38 p		0.82F	-	-	-	-	-	-	-	-	CP→azobenzenedi-sulfonic acid(32%), PY,VA;N-phenyl-quinonediimine(39%), PY,VA	A,r	
			0.78F	-	-	-	-	-	-	-	-	-	A,r	
JE043 0387	38 g	$E_{\frac{1}{2}}$	1.0F	SCE	-	-	QE	2.0 ±0.1	β_a	1.06	Elog	-	$A,i_\ell=kC$ for C=0.05-1, i_d, Tc=1.8, r	
			0.93F	-	-	-	QE	2.0 ±0.1		1.04		-	$A,i_\ell=kC$ for C=0.05-1, i_d, Tc=1.9, r	
			0.89F	-	-	-	QE	2.0 ±0.1		-		CP→azobenzenedi-sulfonic acid(21%), PY,VA;N-phenyl-quinonediimine(46%), PY,VA	$A,i_\ell=kC$ for C=0.05-1, i_d, r	
	38 p		0.85F	-	-	-	QE	2.0 ±0.1		1.02		CP→azobenzenedi-sulfonic acid(31%), PY,VA;N-phenyl-quinonediimine(43%), PY,VA	$A,i_\ell=kC$ for C=0.05-1, i_d, Tc=2.0, r	
			0.79F	-	-	-	QE	2.0 ±0.1		0.79		CP→azobenzenedi-sulfonic acid(32%), PY,VA;N-phenyl-quinonediimine(39%), PY,VA	$A,i_\ell=kC$ for C=0.05-1, i_d, r	
AS017 1077	387 a	$E_{\frac{1}{2}}$	-0.50	SCE	i_ℓ	2.64	i:i	2	$dE_{\frac{1}{2}}/dpH$	0	-	-	$C,i_\ell=kC$ for C < 10	EB59 ah35 ab92
			-0.57			2.42	sttd	2		55			C	
			-0.475			2.70	i:i	2		0	-	-	$C,i_\ell=kC$ for C < 10, r	
			-0.58			2.88	sttd	2		55			C	
			-0.475			2.70	i:i	2		0	-	QE(-0.50V,C=21.5, pH=0.4)→isonico-tinamide(74%), MPT, IRS	$C,i_\ell=kC$ for C < 10, r	
			-0.625			2.75	sttd	2		55			C	
			-0.48			2.70	i:i	2		0	-	-	$C,i_\ell=kC$ for C < 10, r	
			-0.67			3.50	sttd	2		55			C	
			-0.53			2.70	i:i	2		95	sttd	-	C,r	
			-0.705			4.00	-	-		55			C	

CONT

EB59 (CONT.) $C_6H_7N_3O$

Code No.	Empirical Formula	Name and C.A. Number	Structural Formula	Solvent	Tech.	Medium		μ, M	pH	T, °C	Electrodes	App.	Experimental Parameters
EB59 ah35 cb92	$C_6H_7N_3O$	isonicotinic acid hydrazide	Table 11-2	H_2O	PY	CITR		-	2.6	-	DME/SCE	--0	C=0.3, m=2.45, t=3.7, h=48.5
									3.05				
						ACET			4.25				
						succinate			5.05				
						PHOS			6.15				
									6.60				
									7.10				
						BOR			8.25				
									9.00				
									9.70				
						PHOS			11.05				
						KOH		0.02	12.5				
						KOH		0.1	13.0				
						KOH		0.5	13.5				
EB60 ah37 cb96	$C_6H_7N_3O_2$	4-nitrophenyl-hydrazine C.A. 100-16-3	$4-O_2NC_6H_4NHNH_2$	MeOH 10	PY	buffer GEL		- 7E-3	10.5	-	DME/SCE	04A0	C=0.5
EB61 ah47	$C_6H_8N_2$	1,4-diaminobenzene C.A. 106-50-3	$4-H_2NC_6H_4NH_2$	MeCN	VR	$KClO_4$		0.03	-	22	PDE/SCE	04A0	C=0.45, A=0.0018, v=0.066
EB62	$C_6H_8N_2OS$	1-methyl-4-methyl-thiouracil C.A. 49844-94-2	Table 11	H_2O	PY	BR KCl		0.2	1	25	DME/SCE	0A0	C=0.5, m=2.54, t=3.3, h=60
CONT			Table 11-2										

TABLE I. Electrochemical Data $C_6H_8N_2OS$ (CONT.) EB62

Ref.	C/M	Charact. Potential		Response Const.		n Tech.	Electrokinetic Data			Products and Identification	Description and Remarks	Code No.
		Value	vs.		Value		Parameter	Value	From			
AS017 1077	-	$E_{\frac{1}{2}}$	-0.59	SCE	i_ℓ	2.55	i:i 2	$dE_{\frac{1}{2}}/dpH$	95	sttd	-	C,r	EB59 ah35
			-0.73			4.30	-		55			C	cb92
			-0.61			2.60	2		95	sttd	-	C,r	
			-0.755			4.80	4		85			C	
			-0.725			2.70	2		95		-	C,r	
			-0.865			5.35	4		85			C	
			-0.795			2.70	2		95		-	C,r	
			-0.94			5.40	4		85			C	
	387 b		-0.93			2.80	2		95		-	C,r	
			-1.035			5.20	4		85			C	
			-0.965			2.70	2		95		-	C,r	
			-1.07			5.45	4		85			C	
			-1.01			2.80	2		-		-	C,r	
			-1.105			4.20	-					C	
	-		0.075			5.0	4		77		-	A,r	
			-1.125			2.70	2		-			C,S	
			-1.185			2.70	2					C,P	
			0.02			5.3	4		77		-	A,i_ℓ,$A/i_{\ell,C}$=1.11 for C=0.15, 1.05 for C=1.5, 0.2 for C=15;r	
			-1.20			5.50	4		57			C	
			-0.02			5.5	4		77		-	A,r	
			-1.23			5.3	4		57			C	
	387 b		-0.12			4.9	4		77		QE(-0.50 V) → NH_3;QE (0.00V,j=2mA/cm^2,T= 0-5) → 1,2-diiso- nicotinoylhydra- zine(90%),MPT,IRS; pyridine-4-alde- hyde(5%),PY;2,4-di- nitrophenylhydra- zone,n=2	A;i_ℓ,$A/i_{\ell,C}$=1.08 for C=0.15, 0.95 for C=1.5, 0.56 for C=15;r	
			-1.305			5.05	4		57			C	
	387 c		-0.24			4.9	4		77		-	A;i_ℓ,$A/i_{\ell,C}$=1.90 for C=0.15-1.5, 1.63 for C=15;r	
			-1.40			3.40	-		-			C	
			-0.28			5.0	4		77		QE(KOH 0.2,KCl 1) → n=2.9,isonicotinic acid(45%);1,2-di isonicotinoylhydra- zine(55%);QE(0.00 V, same medium, T=0-5) → same pro- ducts,same yields	A,r	
			-1.45			2.8	-		-			C	
			-0.30			5.0	4		77		-	A,r	
			-1.46			2.8	-		-			C	
JE036 0223	-	$E_{\frac{1}{2}}$	-0.76	SCE	-	-	-	-	-	-	-	C;$dE_{\frac{1}{2}}/dpH$=60 for pH 1-10.2, 0 for pH > 10.5,pKa=12.8,r	EB60 ah37 cb96
C0036 1644	-	$E_{p/2}$	0.245	SCE	-	-	-	$E_p-E_{p/2}$	115	-	-	A,R,p	EB61 ah47
			0.805			-	-		85			A,R	
			0.19F			-	-		-			C,R	
			0.72F			-	-					C,R	
JE048 0433	89 e	$E_{\frac{1}{2}}$	-	SCE	-	-	-	-	-	-	-	C,i_a,Pr,r	EB62
			-0.82									C,E,$\neq$f(pH) for pH=1-$\frac{1}{2}$	

CONT

Code No.	Empirical Formula	Name and C.A. Number	Structural Formula	Solvent	Tech.	Medium	μ, M	pH	T, °C	Electrodes	App.	Experimental Parameters
EB62	$C_6H_8N_2OS$	1-methyl-4-methyl-thiouracil	Table II	H_2O	PY	BR KCl	0.2	3	25	DME/SCE	OAO	C=0.5, m=2.54, t=3.3, h=60
								7				
								8.5				
								10				
					VA	BR KCl	0.2	5	25	HMDE SCE	OAO	C=ns, v=25
								8				
EB63	$C_6H_8N_4O_2$	2-methyl-4-hydroxy-pyrimidine-5-carbo-hydrazide	Table II	H_2O	PY	CITR	–	4.5	28±0.2	DME/SCE	OAO	C=0.1, EC=1.3
EB64 ah52	C_6H_8O	2,4-hexadienal C.A. 142-83-6	$CH_3CH:CHCH:CH-CHO$	H_2O	PY	MB	1.0±0.2	5	25±0.05	DME/SCE	OAO	ns
						BOR KCl		10				
EB65 ah54 db13	$C_6H_8O_6$	L-ascorbic acid C.A. 50-81-7	Table II-2	H_2O	PY	ACET GEL 0.01	–	5.47	25±0.1	DME/SCE	2-0	C=2, m=2.07, t(c)=1
					IL	H_2SO_4 0.5	–	–	22	Pt/NHE	01AO	C=10, v=10
					VY	H_2SO_4 0.5	–	–	22	RPDE NHE	01AO	ns

TABLE I. Electrochemical Data $C_6H_8O_6$ EB65

Ref.	C/M	Charact. Potential		Response Const.		n Tech.	Electrokinetic Data			Products and Identification	Description and Remarks	Code No.
		Value	vs.		Value		Parameter	Value	From			
JE048 0433	89 e	$E_{\frac{1}{2}}$	−	SCE	−	−	QE	−	−	−	−	C,i_a,Pr,r	EB62
		−1.08c				2.2	$dE_{\frac{1}{2}}/dpH$	85	sttd	methyl mercaptan	$C, E_{\frac{1}{2}} = -0.82-0.085$pH for pH=2-7.5	
		−1.5c				1.0		30		6,6'-bis-(3,6-di-hydro-1-methyl-pyrimidone-2), λ_{max}=248,MAS,1-methyl-3,6-dihydro-pyrimidone-2	$C, E_{\frac{1}{2}} = -1.41-0.030$pH for pH=2.5-7.5	
		−		−	−	−		−	−	−	C,i_a,r	
		−1.42c						85			C	
		−1.62c						30			C	
		−1.74		−	−	−		40		−	$C, E_{\frac{1}{2}} = -1.4-0.040$pH for pH=8.5-12,r	
		−1.8c		−	−	−		40		6,6'-bis-(3,6-di-hydro)-1-methyl-pyrimid-2-one,MAS; 1-methyl-3,6-di-hydropyrimid-2-one	C,r	
JE048 0433	89 e	E_p	−0.97	SCE	−	−	−	−	−	−	−	C,W,i_a,p	
			−1.10									C	
			−									A,i_a	
			−0.2									A,X	
			−		−	−	−	−	−			C,i_a,p	
			−1.3		−	−	−		−			C	
			−0.37		−	−	−		−			A	
JE055 0445	−	$E_{\frac{1}{2}}$	−1.25	SCE	−	−	i;i	2	−	−	−	$C,W,i_d,Tc-1.6,E_{\frac{1}{2}}\neq f(pH)$ for pH=2-10,p	EB63
JE049 0433	−	$E_{\frac{1}{2}}$	−0.91c	SCE	I	−	−	−	$dE_{\frac{1}{2}}/dpH$	74±3	−	$CP \rightarrow$ polymer,IRS,MAS	$C, E_{\frac{1}{2}} = -0.544-0.074 (\pm 0.003)$pH for pH=2-10, $i_\ell = kh^{0.45}, \neq, p$	EB64 ah52
			−1.44± 0.05			0.47	−		0			$C, E_{\frac{1}{2}} \neq f(pH)$ for pH=5-10, i_k	
			−1.53c			0.53	−		80±60			$C, E_{\frac{1}{2}} = -1.105-0.085 (\pm 0.063)$pH for pH=5-10, $i_\ell = kh^{-0.59}$	
			−1.3c			0.32	−	−		74±3		$C, i_\ell \downarrow$ as $t \uparrow, p$	
			−1.44			0	QE	0.57 ± 0.01		0		C, n Av(2), $i_\ell \downarrow$ as $t \uparrow$	
			−1.95c			0.23	−		80±60			$C, i_\ell \downarrow$ as $t \uparrow$	
JE041 0127	−	$E_{\frac{1}{2}}$	0.04F	SCE	−	−	−	−	$dE_{\frac{1}{2}}/dpH$	43.5	sttd	CP at 0.58 V → radical,ESR,relatively stable	$A,W, E_{\frac{1}{2}} = 0.268-0.0435$ pH for pH=2.5-6.0,p	EB65 ah54
JE040 0013	−	E_p	−0.88F	NHE	j_p	0.77F	−	αn_a	≈0.4	$dE_{p/2}/dv$	−	A,r	
JE040 0013	−	−	−	−	D	6.8	sttd	2	αn_a	0.35	Elog	−	$A, \neq, i_\ell = k\omega^{\frac{1}{2}}, p$	

Code No.	Empirical Formula	Name and C.A. Number	Structural Formula	Solvent	Tech.	Medium	μ, M	pH	T, °C	Electrodes	App.	Experimental Parameters
EB66	$C_6H_8S_4$	2,2',5,5'-tetra-thiabicyclopentylidene C.A. 24719-68-4	Table II	$AlCl_3$ 63 NaCl 37	VA	-	-	-	140	Pt Al/Al^{3+}	25-0	C=10, A=0.1, v=100
										W Al/Al^{3+}		
EB67	$C_6H_9N_3O_2$	1,3,5-trimethyl-6-azauracil C.A. 53400-11-6	Table II	H_2O	PY	BR	-	2.5 3.9 5.6 7.7 10.7	25	DME/SCE Tl^+	O4AO	C=0.24, h=90
EB68 ah63	C_6H_{10}	cyclohexene C.A. 110-83-8	Table II-2	MeCN	VY	Et_4NBF_4 0.1	-	-	25	RPDE Ag/Ag^+ 0.01	2-0	C≈1, A=0.12, v=50, ω=16.3, Pt Aux
				CH_2Cl_2	VY	Bu_4NBF_4 0.2	-	-	25	RPDE Ag/Ag^+ 0.01	2-0	C≈1, A=0.12, v=50, ω=16.3, Pt Aux
				$EtNO_2$	VY	Et_4NBF_4 0.1	-	-	25	RPDE Ag/Ag^+ 0.01	2-0	C≈1, A=0.12, v=50, ω=16.3, Pt Aux
				$MeNO_2$	VY	Et_4NBF_4 0.2	-	-	25	RPDE Ag/Ag^+ 0.01	2-0	C≈1, A=0.12, v=50, ω=16.3, Pt Aux
				PCA	VY	Et_4NBF_4 0.1	-	-	25	RPDE Ag/Ag^+ 0.01	2-0	C≈1, A=0.12, v=50, ω=16.3, Pt Aux
				SULN	VY	Et_4NBF_4 0.2	-	-	25	RPDE Ag/Ag^+ 0.01	2-0	C≈1, A=0.12, v=50, ω=16.3, Pt Aux
EB69	$C_6H_{10}O$	2-ethyl-2-butenal C.A. 19780-25-7	$CH_3CH:C(CH_2CH_3)$-CHO	H_2O	PY	MB	1.0 ±0.2	2	25.00 ±0.05	DME/SCE	OAO	ns
						BOR KCl	1.0± 0.2	10	25± 0.05			
EB70	$C_6H_{10}O$	2-methyl-2-pentenal C.A. 623-36-9	$CH_3CH_2CH:C(CH_3)CHO$	H_2O	PY	MB	1.0 ±0.2	2	25.00 ±0.05	DME/SCE	OAO	ns
						BOR KCl	1.0± 0.2	10	25± 0.05			
EB71 ah74	$C_6H_{10}O$	4-methyl-3-penten-2-one C.A. 141-79-7	$(H_3C)_2C:CHCOCH_3$	MeOH 25	PY	buffer	-	1 4.3 6 8 9.3	-	DME Hg/Hg_2Cl_2, LiCl 1	2AO	C=ns, m=4.9, t=1.9
CONT												

TABLE I. Electrochemical Data $C_8H_{10}O$ (CONT.) EB71

| Ref. | C/M | Charact. Potential | | Response | Const. | n | Electrokinetic Data | | | Products and | Description and | Code |
		Value	vs.	Tech.	Value		Parameter	Value	From	Identification	Remarks	No.
JE047 0081	338 d	E_p	0.63 0.78 1.22 $\approx$1.65 1.13 0.57	Al/Al^{3+}	i_p	82F small 63F - 56F 106F	-	-	-	-	-	A,r A A A C C	EB66
			0.6F			50F	-	-	-	-	radical cation	A, E_p and $i_p/v^{\frac{1}{2}} \neq f(v)$ for $v < 2E4, R, r$	
			1.21F			50F					dication	A, E_p and $i_p/v^{\frac{1}{2}} \neq f(v)$ for $v < 2E4, R$	
			$\approx$1.65			-					-	A,0 for $v > 1E3$	
			1.14F 0.54F			50F 50F						C,R C,R	
C0027 0546	-	$E_{\frac{1}{2}}$	-0.73F	SCE	i_ℓ	1.50F	sttd	2	-	-	-	-	C,r	EB67
			-0.85F			1.37F		2	-	-	-	-	C,r	
			-1.03F			0.94F		2	-	-	-	-	C,X,r	
			-1.40F			1.35F		2	-	-	-	-	C,r	
			-1.49F			1.30F		1	-	-	-	-	C,r	
JE042 0133	-	$E_{\frac{1}{2}}$	1.93	Fc	-	-	-	-	-	-	-	-	A,p	EB68 ah63
JE042 0133	-	$E_{\frac{1}{2}}$	1.70	Fc	-	-	-	-	-	-	-	-	A,p	
JE042 0133	-	$E_{\frac{1}{2}}$	1.75	Fc	-	-	-	-	-	-	-	-	A,p	
JE042 0133	-	$E_{\frac{1}{2}}$	1.83	Fc	-	-	-	-	-	-	-	-	A,p	
JE042 0133	-	$E_{\frac{1}{2}}$	1.86	Fc	-	-	-	-	-	-	-	-	A,p	
JE042 0133	-	$E_{\frac{1}{2}}$	1.82	Fc	-	-	-	-	-	-	-	-	A,p	
JE049 0433	-	$E_{\frac{1}{2}}$	-1.03c	SCE	I	0.97	-	-	$dE_{\frac{1}{2}}/dpH$	71.3 ±6.0	sttd	-	$C, E_{\frac{1}{2}} = -0.886-0.071$ (± 0.006)pH for pH= 2-10,p	EB69
			-1.60c			2.45	QE	0.98 ± 0.01		71.3 ±6.0		-	$C, i_\ell = kh^{0.62}, n$ Av(2), p	
JE049 0433	-	$E_{\frac{1}{2}}$	-1.03c	SCE	I	1.65	-	-	$dE_{\frac{1}{2}}/dpH$	71.3 ±6.0	sttd	-	$C, E_{\frac{1}{2}} = -0.886-0.071$ (± 0.006)pH for pH= 2-10,p	EB70
		$E_{\frac{1}{2}}$	-1.60c			2.94	QE	1.02 ± 0.12	p =	1.2 ± 0.23	Elog	-	$C, i_\ell = kh^{0.73}, n$ Av(4), p	
JE035 0381	156 i	$E_{\frac{1}{2}}$	-1.12F	Hg/ Hg_2Cl_2, LiCl 1	-	-	sttd	1	$dE_{\frac{1}{2}}/dpH$	86	plot	-	$C, E_{\frac{1}{2}} = -1.03-0.083$pH, r	EB71 ah74
			-1.38F		-	-	-	-		86		-	C,r	
			-1.74F		-	-	-	-		$\approx$10		-	C,r	
			-1.76F		-	-	-	-		$\approx$10		-	C,r	
			-1.85F		-	-	-	-		-	-	-	C,r	CONT

EB71 (CONT.) $C_6H_{10}O$

Code No.	Empirical Formula	Name and C.A. Number	Structural Formula	Solvent	Tech.	Medium		μ, M	pH	T, °C	Electrodes	App.	Experimental Parameters
EB71 ah74	$C_6H_{10}O$	4-methyl-3-penten-2-one	$(H_3C)_2C:CHCOCH_3$	MeOH 25	PY	buffer		-	10.5	-	DME Hg/ Hg_2Cl_2, LiCl 1	2AO	C=ns, m=4.9, t=1.9
				MeOH	PY	$LiClO_4$	0.1	-	-	-	DME Ag/Ag^+ 0.1	---	C=1
						$LiClO_4$ $C_6H_5SO_3H$	0.1 4E-4						
						$LiClO_4$ $C_6H_5SO_3H$	0.1 1.0E-3						
						$LiClO_4$ $C_6H_5SO_3H$	0.1 1.0E-2						
				EtOH 50	PY	HCl KCl	0.05	-	2.27	-	DME Ag/AgCl	04AO	C=0.2
						ACET			4.6				
						PHOS			7.8 9.4				
					PVI	ACET		-	4.6	-	DME Ag/AgCl	23AO	C=0.6, Δe=50, f=20, v=ns, Hg Aux
					PV	ACET		-	4.6	-	DME Ag/AgCl	23AO	C=0.2, Δe=50, f=20, v=2, Hg Aux v=5E2
					PW	ACET		-	4.6	-	DME Ag/AgCl	04AO	C=0.2, v=500, Hg Aux
EB72	$C_6H_{10}OS$	3-thiepanone C.A. 25057-68-5	Table II	H_2O	PY	Et_4NClO_4	0.1	-	-	20	DME/SCE	2-0	C=0.5
				EtOH(?) 75	PY	Et_4NClO_4	0.1	-	-	-	DME/SCE	2-0	C=1
EB73	$C_6H_{10}OS$	4-thiepanone C.A. 22072-22-6	Table II	H_2O	PY	Et_4NClO_4	0.1	-	-	20	DME/SCE	2-0	C=0.5
				EtOH(?) 75	PY	Et_4NClO_4	0.1	-	-	-	DME/SCE	2-0	C=1
EB74	$C_6H_{10}O_3S$	3-thiepanone 1,1-dioxide C.A. 36165-01-2	Table II	EtOH(?) 75	PY	Et_4NClO_4	0.1	-	-	-	DME/SCE	2-0	C=1
EB75	$C_6H_{10}O_3S$	4-thiepanone 1,1-dioxide C.A. 41511-88-0	Table II	EtOH(?) 75	PY	Et_4NClO_4	0.1	-	-	-	DME/SCE	2-0	C=1

TABLE I. Electrochemical Data $C_6H_{10}O_3S$ EB75

Ref	C/M	Charact	Potential Value	vs.	Response	Const Value	Tech.	n	Electrokinetic Data Parameter	Value	From	Products and Identification	Description and Remarks	Code No.
JE035 0381	156 j	$E_{\frac{1}{2}}$	-1.84F	Hg Hg_2Cl_2, LiCl 1	-	-	-	-	-	-	-	-	$C, E_{\frac{1}{2}} \neq f(pH)$ for pH= 10.5-13, r	EB71 ah74
JE035 0381	156 j	$E_{\frac{1}{2}}$	-1.69F	Ag/Ag$^+$ 0.1	i_ℓ	4.5F	-	-	-	-	-	-	C,p	
			-0.98F -1.72F			1.9F 3.1F							C,p C	
			-0.97F -			4.4F -							C,p C,M	
			-0.97F -1.21F			5.4F 2.1F							C,p C	
JE043 0349	-	$E_{\frac{1}{2}}$	-1.12F	Ag/AgCl	i_ℓ	3.6F	i:i	1	-	-	-	-	C,r	
			-1.24			3.6F	sttd	0.93	-	-	-	-	$C, \log i_\ell = k_1 \log C + k_2$ for C=0.01-1, i_a for $C \geq 1$, r	
			-1.35F			0.92F	-	-	-	-	-	-	C,r	
			-1.42F			small	-	-	-	-	-	-	C,r	
JE043 0349	-	E_{su}	-1.28F	Ag/AgCl	i_{su}	1.4F	sttd	0.68	$dE_{\frac{1}{2}}/d\log X$	177	-	-	$C, d\log X = d\log[(\frac{i_{su}}{i})^{\frac{1}{2}} \pm (\frac{i_{su}-i}{i})^{\frac{1}{2}}], i_{su} = k_1 \log C + k_2$ for C= 0.001-1, r	
JE043 0349	-	E_{su}	-1.31F	Ag/AgCl	i_{su}	7F	sttd	0.87	-	-	-	-	$C, \log i_{su} = k_1 \log C + k_2$ for C=0.25-1, r	
			-1.33F			2.8F		0.84					$C, \log i_{su} = k_1 \log C + k_2$ for C=0.1-1, r	
JE043 0349	-	E_p	-1.31F	Ag/AgCl	i_p	2.2F	sttd	0.90	-	-	-	-	$C, \log i_\ell = k_1 \log C + k_2$ for C=0.0032-0.4, r	
JE039 0195	400 ab	$E_{\frac{1}{2}}$	-1.82	SCE	-	-	-	-	-	-	-	-	C,p	EB72
JE035 0363	-	$E_{\frac{1}{2}}$	-1.82	SCE	-	-	-	-	-	-	-	-	C,p	
JE039 0195	400 ab	$E_{\frac{1}{2}}$	-2.0	SCE	-	-	-	-	-	-	-	-	C,p	EB73
JE035 0363	-	$E_{\frac{1}{2}}$	-2.0	SCE	-	-	-	-	-	-	-	-	C,p	
JE035 0363	-	$E_{\frac{1}{2}}$	-1.54	SCE	-	-	-	-	-	-	-	-	C,p	EB74
JE035 0363	-	$E_{\frac{1}{2}}$	-1.81	SCE	-	-	-	-	-	-	-	-	C,p	EB75

Code No.	Empirical Formula	Name and C.A. Number	Structural Formula	Solvent	Tech.	Medium		μ, M	pH	T, °C	Electrodes	App.	Experimental Parameters
EB76	$C_6H_{11}BF_4S_3$	2-(ethylthio)-1,3-dithian-2-ylium tetrafluoroborate(1-) C.A. 51823-98-4	Table II	MeCN	VA	Et_4NClO_4	0.1	-	-	-	Pt Ag/ $AgNO_3$ 0.01 Et_4NClO_4 0.1	2A2	v < 1E3
EB77	$C_6H_{11}N_3O_3$	ethyl pyruvate semi-carbazone C.A. 14923-66-1	$CH_3C(:NNHCONH_2)$ $COOCH_2CH_3$	H_2O	PY	MB		-	1.4	-	DME/SCE	OAO	C=0.5(?),m= 2.62,t=2.8, h=55
									2.7				
									4.0				
									4.7				
									5.4				
									6.0				
									6.75				
									10.55				
EB78 ah86	C_6H_{12}	cyclohexane C.A. 110-82-7	Table II-2	CF_3COOH	IL	FSO_3H	2.5	-	-	-	Pt/SCE	25-2	C=30,v=50, Pt Aux
					VA	CF_3COONa FSO_3H	0.5 3	-	-	-	Pt/SCE	25-2	C=18.5,v= 10,Pt Aux
EB79 cc18	$C_6H_{12}N_2$	1,2-diazaspiro[2.5]-octane C.A. 185-79-5	Table II-3	MeOH 20	PY	BR		-	1.5	20	DME Ag/AgCl	OAO	C=0.66,m= 1.65,t=0.16
									2.9				
									3.5				
									4.0				
									5.0				
									5.2				
									5.4				
									6.1				
									6.8				
									7.5				
									8.0				
									8.9				
									9.5				

TABLE I. Electrochemical Data $C_6H_{12}N_2$ EB79

Ref	C/M	Charact.	Potential		Response Const.		n		Electrokinetic Data			Products and	Description and	Code
			Value	vs.		Value	Tech.		Parameter	Value	From	Identification	Remarks	No.
JE049 0105	338 c	E_p	-0.42	SCE	-	-	QE	1.0	$E_p-E_{p/2}$	50	-	2,2'6,6'-tetrathia-bicyclohexylidene, MAS,NMR,VA;tetra-thioethylene,MAS,NMR	$C,\not=,E_p-E_{p/2}$ and $j_p/Cv^{\frac{1}{2}}\not=f(v),p$	EB76
			-			-		-		-			A,X	
C0036 0331	-	$E_{\frac{1}{2}}$	-0.61F	SCE	i_ℓ	9.2F	i:i	2	-	-	-	-	C,i_d,p	EB77
			-0.70F			9.2F		2	-	-	-	-	C,i_d,p	
			-0.83F			8.3F		-	-	-	-	-	C,p	
			-0.91F			4.5F		-	-	-	-	-	C,p	
			-1.50F			5.0F		-	-	-	-	-	C	
			-0.94F			1.8F		-	-	-	-	-	C,i_k,p	
			-1.50F			6.8F		-	-	-	-	-	C	
			-0.97F			0.4F		-	-	-	-	-	C,i_k,p	
			-1.50F			7.4F		2	-	-	-	-	C,i_d	
			-1.50F			7.8F		2	-	-	-	-	C,i_d,p	
			-1.50F			7.8F		2	-	-	-	CP→ethyl pyruvate semicarbazide	C,i_d,p	
JE054 0181	-	$E_{p/2}$	2.06	SCE	-	-	-	-	-	-	-	CP→cyclohexyltri-fluoroacetate	$A,\not=,E_{p/2}$ → more neg. as $[FSO_3H]\uparrow,i_p=kC,p$	EB78 ah86
JE054 0181	-	E_p	2.16F	SCE	$i_p/v^{\frac{1}{2}}$	17.6F	-	-	-	-	-	-	$A,\not=,i_p/v^{\frac{1}{2}}\downarrow$ as $v\uparrow$, ce,p	
C0030 4178	295 c	$E_{\frac{1}{2}}$	-0.41F	Ag/AgCl	i_ℓ	2.4F	-	-	Elog	69	lower part of curve	-	C,p	EB79 cc18
										119	upper part of curve			
			-1.35F			3.0F							C	
			-0.42F			2.4F	-	-				-	C,p	
			-1.35F			4.3F							C	
			-0.42F			2.4F	-	-				-	C,p	
			-1.35F			4.0F							C	
			-0.45F			2.4F	-	-				-	C,p	
			-1.35F			2.9F							C	
			-0.51F			2.4F	-	-		160		-	C,p	
			-1.35F			2.4F				-			C	
			-0.58F			2.4F	-	-		-		-	C,p	
			-1.37F			2.4F							C	
			-0.71F			2.4F	-	-		-		-	C,p	
			-1.40F			2.3F							C	
			-0.87F			2.4F	-	-		-		-	C,p	
			-1.41F			2.2F							C	
			-0.93F			2.4F	-	-		-		-	C,p	
			-1.42F			2.1F							C	
			-0.98F			2.4F	-	-		-		-	C,p	
			-1.42F			1.8F							C	
			-1.01F			1.4F	-	-		-		-	C,p	
			-1.42F			0.8F							C	
			-1.05F			0.7F	-	-		-		-	C,p	
			-1.45F			0.4F							C	
			-1.06F			0.4F	-	-		-		-	C,p	
			-1.46F			0.4F							C	

Code No.	Empirical Formula	Name and C.A. Number	Structural Formula	Solvent	Tech.	Medium		μ, M	pH	T, °C	Electrodes	App.	Experimental Parameters
EB80 ah94 cc20	$C_6H_{12}N_2O_4S_2$	cystine C.A. 56-89-3	$(SCH_2CH[NH_2]-COOH)_2$	H_2O	PA	AM $Cd(NO_3)_2$	0.1 3E-5	-	10.45	-	DME Ag/AgCl, KCl? 1	26A2	$-0.1 \to -1.2 \to -0.1$ V, C= 0.03, Pt Aux
													$-0.2 \to -1.2 \to -0.2$ V
													$-0.3 \to -1.2 \to -0.3$ V
					PR	HNO_3 $NaNO_3$	1E-2 9E-2	-	2	-	DME Ag/AgCl, KCl? 1	26A2	C=0.025, v= 2E3
													$0 \to -0.50 \to 0$ V, C=0.5, v=0.025
													$0.05 \to -0.45 \to 0.05$ V
													$-0.05 \to -0.55 \to -0.05$ V
					PW	HNO_3 $NaNO_3$	1E-2 9E-2	-	2	-	DME Ag/AgCl, KCl? 1	26A2	C=0.025, v=2E3
													$0 \to -1.0$ V
						HNO_3 $NaNO_3$ $Cu(NO_3)_2$	0.01 0.09 3E-5						$0 \to -1.0$ V, C= 0.025
						HNO_3 $NaNO_3$ $Cu(NO_3)_2$	0.01 0.09 1E-4						$0 \to -1.0$ V, C= 0.5
						HNO_3 $NaNO_3$ $Cu(NO_3)_2$	0.01 0.09 4E-4						
						ACET	0.25		4.95				C=0.025
													$-0.2 \to -1.2$ V
													$-0.35 \to -1.35$ V
						AM	0.1		10.45				$-0.05 \to -1.05$ V
													$-0.3 \to -1.3$ V
													$-0.5 \to -1.5$ V
EB81 ah95	$C_6H_{12}N_2O_4Se_2$	selenocystine C.A. 1464-43-3	$(SeCH_2CH[NH_2]-COOH)_2$	H_2O	PY	ACET $NiCl_2$ GEL	4E-4 5E-3	-	6	-	DME/SCE	2-0	C=0.02, m= 2.64, t(oc)= 3.65

TABLE I. Electrochemical Data $C_6H_{12}N_2O_4Se_2$ EB81

Ref	C/M	Charact. Potential Value	vs.	Response Const.	Value	n Tech.	n	Electrokinetic Data Parameter	Value	From	Products and Identification	Description and Remarks	Code No.
JE036 0157	141 k	E_p -0.62F	Ag/AgCl, KCl? 1	i_p	4F	-	-	-	-	-	-	C,p	EB80 ah94 cc20
		-0.97F			0.6F							C	
		-0.74F			1.4F							A	
		-0.62F			2.7F	-	-	-	-	-	-	C,p	
		-0.82F			small							C	
		-1.0F			small							C	
		-0.74F			1.2F							A	
		-0.61F			2F	-	-	-	-	-	-	C,p	
		-0.82F			1.4F							C	
		-1.0F			small							C	
		-0.74F			1.2F							A	
JE036 0157	141 k	E_p -0.23F	Ag/AgCl, KCl? 1	i_p	1F	-	-	-	-	-	-	C,R,r	
		-0.67F			small							C,$i_p \downarrow$ on later scans	
		-0.45F			small							A	
		-0.15F			small							A	
		-0.29F			12F							C,r	
		-0.25F			8F							A	
		-0.28F			10F							C,r	
		-0.25F			6F							A	
		-0.29F			15F							C,r	
		-0.25F			10F							A	
JE036 0157	141 k	E_p -0.23F	Ag/AgCl, KCl? 1	i_p	0.4F	-	-	-	-	-	-	C,R,Hg-cysteinate redn.,r	
		-0.67F			1.0F						cysteine	C,$\neq$,cystine redn.	
		-0.22F			0.9F	-	-	-	-	-	-	C,p	
		-0.31F			1.1F							C	
		-0.7F			small							C	
		-0.24F			0.7F	-	-	-	-	-	-	C,$i_p \downarrow$ as $[Cu^{2+}] \uparrow$,p	
		-0.34F			1.0F							C,$i_p \uparrow$ as $[Cu^{2+}] \uparrow$	
		-0.6F			small							C	
		-0.22F			10F	-	-	-	-	-	-	C,p	
		-0.31F			10F							C	
		-0.52F			6F							C	
		-0.7F			small							C	
		-0.17F			6F	-	-	-	-	-	-	C,p	
		-0.36F			9.5F							C	
		-0.49F			13F							C	
		-0.7F			small							C	
		-			0.5F	-	-	-	-	-	-	C,p	
		-			-							C	
		-			1.0F	-	-	-	-	-	-	C,p	
		-			-							C	
		-			0.6F	-	-	-	-	-	-	C,p	
		-			-							C	
		-			6.0F	-	-	-	-	-	-	C,p	
		-			-							C	
		-			3.5F	-	-	-	-	-	-	C,p	
		-			-							C	
		-			1.1F	-	-	-	-	-	-	C,p	
		-			-							C	
JE043 0257	-	$E_{\frac{1}{2}}$ -0.58	SCE	i_ℓ	1.3F	-	-	-	-	-	-	C,i_{cat},Pr,Mx,i_ℓ= $f(C)$,$i_\ell \uparrow$ as pH $\uparrow$,r	EB81 ah95
		-0.94			1.5F							C,i_d,Ni(II) redn.	
		-1.14			2.7F							C,i_{cat},Pr,H,$i_\ell \neq f(h)$, $i_\ell=f([Ni(II)],C,pH)$	
		-1.67			6F							C,H,i_d	

EB82 $C_6H_{12}O$

Code No.	Empirical Formula	Name and C.A. Number	Structural Formula	Solvent	Tech.	Medium	μ, M	pH	T, °C	Electrodes	App.	Experimental Parameters
EB82	$C_6H_{12}O$	2-ethylbutanal C.A. 97-96-1	$(C_2H_5)_2CHCHO$	EtOH 2	PY	BOR LiCl	0.5	9.45	25.00 ± 0.02	DME/SCE	034AO	C=0.2, m=2.6, ±0.4, t=3.6± 0.5, h=65±5
						LiOH LiCl		11.4				
								12.0				
								13				
								13.5				
				EtOH 5	PY	BOR LiCl	0.5	8.4	25.00 ± 0.02	DME/SCE	034AO	C=0.05, m=2.6, ±0.4, t=3.6± 0.5, h=65±5
								9.45				
								10.15				
								10.75				
								11.75				
								12.8				
								13.5				
								14.3				
EB83 ai09 db37	$C_6H_{12}O_6$	D-glucose C.A. 50-99-7	Table II-2	H_2O	PP	PHOS KCl	0.1	—	25.0 ±0.1	DME/SCE	---	—
EB84 ai16	$C_6H_{13}NO_2$	isoleucine C.A. 73-32-5	$CH_3CH_2CH(CH_3)-CH(NH_2)COOH$	H_2O	CP	NaOH	1	—	23	Cu/SCE	---	—
EB85 ai17	$C_6H_{13}NO_2$	leucine C.A. 61-90-5	$(CH_3)_2CHCH_2CH-(NH_2)COOH$	H_2O	CP	NaOH	1	—	23	Cu/SCE	---	—
EB86 ai18	$C_6H_{13}NO_2$	norleucine C.A. 327-57-1	$CH_3(CH_2)_3CH-(NH_2)COOH$	H_2O	CP	NaOH	1	—	23	Cu/SCE	---	—
EB87	C_6H_{14}	hexane C.A. 110-54-3	$CH_3(CH_2)_4CH_3$	CF_3COOH	IL	FSO_3H	2.5	—	—	Pt/SCE	25-2	C=30, v=50, Pt Aux
EB88	C_6H_{14}	3-methylpentane C.A. 96-14-0	$CH_3CH_2CH(CH_3)-CH_2CH_3$	MeCN	VY	Et_4NBF_4	0.1	—	25	RPDE Ag/Ag^+ 0.01	2-0	C≈1, A=0.12, v=50, ω=16.3, Pt Aux
				CH_2Cl_2	VY	Bu_4NBF_4	0.2	—	25	RPDE Ag/Ag^+ 0.01	2-0	C≈1, A=0.12, v=50, ω=16.3, Pt Aux
				$MeNO_2$	VY	Et_4NBF_4	0.1	—	25	RPDE Ag/Ag^+ 0.01	2-0	C≈1, A=0.12, v=50, ω=16.3, Pt Aux
				$EtNO_2$	VY	Et_4NBF_4	0.2	—	25	RPDE Ag/Ag^+ 0.01	2-0	C≈1, A=0.12, v=50, ω=16.3, Pt Aux
EB89	$C_7H_3Cl_3O_2$	2,3,5-trichloro-6-methylbenzoquinone C.A. 4592-97-6	Table II	H_2O	PY	BR	—	5.3	—	DME/SCE	04AO	C=1.0, m=2.72, t=3.75, h=50
								6.8				

TABLE I. Electrochemical Data $C_7H_3Cl_3O_2$ EB89

Ref.	C/M	Charact. Potential		Response Const.		Tech.	n	Electrokinetic Data			Products and Identification	Description and Remarks	Code No.
		Value	vs.		Value			Parameter	Value	From			
JE046 0323	32 d	-	-	i_ℓ	0.9	i:i, PQ	0.89	-	-	-	-	$C, i_\ell \neq f(pH)$ for pH= 9.45-11, Tc > 2, $i_\ell = k_1 h^{\frac{1}{2}} + k_2, i_k, r$	EB82
		-	-		0.93		0.92	-	-	-	-	C,r	
	32 b	-	-		0.95		-	-	-	-	-	C,r	
		-	-		1.05		-	-	-	-	-	C,r	
		$E_{\frac{1}{2}}$	-1.86	SCE	1.15		1.14	-	-	-	-	$C, \Delta E_{\frac{1}{2}} = 0.2\sigma^*, r$	
JE046 0323	32 d	-	-	i_ℓ	0.5	-	-	-	-	-	-	C,r	
		-	-		0.68	-	-	-	-	-	-	C,r	
		-	-		0.85	-	-	-	-	-	-	C,r	
		-	-		1.0	-	-	-	-	-	-	C,r	
		-	-		1.30	-	-	-	-	-	-	C,r	
	32 b	-	-		1.50	-	-	-	-	-	-	C,r	
		-	-		1.58	-	-	-	-	-	-	C,r	
		-	-		1.63	-	-	-	-	-	-	C,r	
JE036 0479	-	-	-	-	-	-	-	-	-	-	-	see A109, DB37	EB83 ai09 db37
JE034 0091	379 a	-	-	-	-	-	-	-	-	-	2-methylbutylnitrile, GLC	see A116	EB84 ai16
JE034 0091	379 a	-	-	-	-	-	-	-	-	-	isovaleronitrile	see A117	EB85 ai17
JE034 0091	379 a	-	-	-	-	-	-	-	-	-	valeronitrile	see A118	EB86 ai18
JE054 0181	-	$E_{p/2}$	2.37	SCE	-	-	-	-	-	-	-	$A, \neq, p$	EB87
JE042 0133	-	$E_{\frac{1}{2}}$	2.74	Fc	-	-	-	-	-	-	-	A,p	EB88
JE042 0133	-	$E_{\frac{1}{2}}$	2.55	Fc	-	-	-	-	-	-	-	A,p	
JE042 0133	-	$E_{\frac{1}{2}}$	2.70	Fc	-	-	-	-	-	-	-	A,p	
JE042 0133	-	$E_{\frac{1}{2}}$	2.61	Fc	-	-	-	-	-	-	-	A,p	
C0028 2163	-	$E_{\frac{1}{2}}$	0.16	SCE	-	-	-	-	-	-	-	C,p	EB89
			0.11		-	-	-	-	-	-	-	C,p	

103

Code No.	Empirical Formula	Name and C.A. Number	Structural Formula	Solvent	Tech.	Medium	μ, M	pH	T, °C	Electrodes	App.	Experimental Parameters
EB90	$C_7H_3N_3$	2,3-dicyanopyridine C.A. 17132-78-4	Table 11	H_2O	PY	BOR(?)	-	9.8	-	DME/SCE	0-0	ns
EB91	$C_7H_3N_3$	2,4-dicyanopyridine C.A. 29181-50-8	Table 11	H_2O	PY	BOR(?)	-	9.8	-	DME/SCE	0-0	ns
EB92	$C_7H_3N_3$	2,5-dicyanopyridine C.A. 20730-07-8	Table 11	H_2O	PY	BOR(?)	-	9.8	-	DME/SCE	0-0	ns
EB93	$C_7H_3N_3$	2,6-dicyanopyridine C.A. 2893-33-6	Table 11	H_2O	PY	BOR(?)	-	9.8	-	DME/SCE	0-0	ns
EB94	$C_7H_3N_3$	3,4-dicyanopyridine C.A. 1633-44-9	Table 11	H_2O	PY	BOR(?)	-	9.8	-	DME/SCE	0-0	ns
EB95	$C_7H_3N_3$	3,5-dicyanopyridine C.A. 1195-58-0	Table 11	H_2O	PY	BOR(?)	-	9.8	-	DME/SCE	0-0	ns
EB96	C_7H_4FN	2-fluorobenzonitrile C.A. 394-47-8	$2-FC_6H_4CN$	DMF	IR	Et_4NClO_4 0.1	-	-	-	Pt/SCE	125A F02	C=0.67-3.77
					IU	Et_4NClO_4 0.1	-	-	-	Pt/SCE	125A F02	C=1.08-3.82
					VR	Et_4NClO_4 0.1	-	-	-	Pt/SCE	125A F02	0→-2.4→0, C=1.4, A=3.1E-3, v=1E3, Pt Aux
												0→-2.7→0, C=2.64, A=0.25, v=80.6, Pt Aux
EB97	C_7H_4FN	3-fluorobenzonitrile C.A. 403-54-3	$3-FC_6H_4CN$	DMF	IR	Et_4NClO_4 0.1	-	-	-	Pt/SCE	125A F02	E_{app}=-2.5, A=0.25, t=(2E-6)-8, Pt Aux
					VR	Et_4NClO_4 0.1	-	-	-	Pt/SCE	125A F02	C=2.22, v=80.6, A=0.25, Pt Aux
CONT												

TABLE I. Electrochemical Data C_7H_4FN (CONT.) EB97

Ref.	C/M	Charact. Potential		Response Const.		n Tech.		Electrokinetic Data			Products and Identification	Description and Remarks	Code No.
		Value	vs.		Value			Parameter	Value	From			
JE036 0383	-	$E_{\frac{1}{2}}$	-1.18	SCE	-	-	-	-	-	-	-	-	$C, E_{\frac{1}{2}} \propto LUMO, p$	EB90
JE036 0383	-	$E_{\frac{1}{2}}$	-1.09	SCE	-	-	-	-	-	-	-	-	$C, E_{\frac{1}{2}} \propto LUMO, p$	EB91
JE036 0383	-	$E_{\frac{1}{2}}$	-1.19	SCE	-	-	-	-	-	-	-	-	$C, E_{\frac{1}{2}} \neq LUMO, p$	EB92
JE036 0383	-	$E_{\frac{1}{2}}$	-1.31	SCE	-	-	-	-	-	-	-	-	$C, E_{\frac{1}{2}} \propto LUMO, p$	EB93
JE036 0383	-	$E_{\frac{1}{2}}$	-0.99	SCE	-	-	-	-	-	-	-	-	$C, E_{\frac{1}{2}} \propto LUMO, p$	EB94
JE036 0383	-	$E_{\frac{1}{2}}$	-1.35	SCE	-	-	-	-	-	-	-	-	$C, E_{\frac{1}{2}} \propto LUMO, p$	EB95
JA095 6033	107 k	-	-	-	-	-	-	-	-	-	-	-	$1 < [n_{app}=(it^{\frac{1}{2}})/(it^{\frac{1}{2}})_{k=0}=f(t)] < 2$, $k=1.0E-2\ s^{-1}$	EB96
JA095 6033	107 k	-	-	-	-	-	QE at -2.10V	2.02 ± 0.04	-	-	-	benzonitrile (101 ± 3%), VR, GLC	$i_A/i_C = f(\tau-C)$ for $t_A/\tau = 0.2-0.5, r$	
JA095 6033	107 k	E_p	-2.04F	SCE	i_p	2.75F	sttd	1	-	-	-	-	$C, R,$ for $v > 1E5$, $i_{p,A}/i_{p,C}=1$	
			-1.96F			1.5F							$A, R, i_p \downarrow$ on later scans	
			-0.88F			0.5F							$A, \not\models, i_p \uparrow$ on later scans	
			-2.17F			175F	-	-					C	
			-2.32			5F							C	
			-2.25F			15F							A	
			-2.05F			35F							A, i_k	
			-1.03F			47F							$A, \not\models, i_k$	
JA095 6033	107 i	-	-	-	$it^{\frac{1}{2}}/C$	44	IR	1	-	-	-	-	C, i_d	EB97
JA095 6033	107 i	E_p	-1.68F	-	i_p	12F	QE at -2.4V	1.05	-	-	-	fluorobenzene(45±2%), VR, GLC	C, Pr	
			-1.84F			19F		-					C, Pr	
			-2.19			62F							$C, i_p \downarrow$ on later scans	
			-2.86			238F							$C, i_\ell \uparrow$ on later scans, C_6H_5F redn.	CONT

Code No.	Empirical Formula	Name and C.A. Number	Structural Formula	Solvent	Tech.	Medium		μ, M	pH	T, °C	Electrodes	App.	Experimental Parameters
EB97	C_7H_4FN	3-fluorobenzonitrile	$3\text{-}FC_6H_4CN$	DMF	VR	Et_4NClO_4	0.1	-	-	-	Pt/SCE	125A F02	C=2.22,v=1E5, A=0.25, Pt Aux
EB98 ai73	C_7H_4FN	4-fluorobenzonitrile C.A. 1194-02-1	$4\text{-}FC_6H_4CN$	DMF	IR	Et_4NClO_4	0.1	-	-	-	Pt/SCE	125A F02	C=0.25
													C=5-10, t < 0.01
					VR	Et_4NClO_4	0.1	-	-	-	Pt/SCE	125A F02	0→-2.8→0, C=0.25,A= 0.25,v=100, Pt Aux
													C=0.25,A= 0.25,v=1E5, Pt Aux
													C=2.5,A= 3.1E-3,v= 3E4
EB99 cc43	C_7H_5ClO	3-chlorobenzaldehyde C.A. 587-04-2	$3\text{-}ClC_6H_4CHO$	EtOH 10	PY	KOH	1	-	-	-	DME/MSE	2-0	C=1,m=2.73, t=3.55,v= 3.3
EC00 ai93 cc44	C_7H_5ClO	4-chlorobenzaldehyde C.A. 104-88-1	$4\text{-}ClC_6H_4CHO$	EtOH 10	PY	BR		-	3.1	-	DME/SCE	2AO	C=0.03,t= 0.25
													C=0.1
													C=0.3
													C=1.0
													C=2.8
						KOH	0.04	-	-	-	DME/MSE	2-0	C=1,m=2.73, t=3.55,v= 3.3
						KOH	1						
						KOH	1.8						
				N-methyl-pyrrol-idene	PY	Et_4NClO_4	0.1	-	-	-	DME Hg/Hg^+ 0.02, $HClO_4$ 0.1	2AO	C=1-2
					VY	Et_4NClO_4	0.1	-	-	-	MP Hg/Hg^+ 0.02, $HClO_4$ 0.1	2AO	C=1-2, A=0.5

TABLE I. Electrochemical Data C_7H_5ClO EC00

Ref.	C/M	Charact. Potential Value	vs.	Response Const.	Value	Tech.	n	Electrokinetic Data Parameter	Value	From	Products and Identification	Description and Remarks	Code No.
JA095 6033	107 i	E_p -1.20	SCE	-	-	-	-	-	-	-	-	C, Pr at $v=1E4, i_a, \not\models, r$	EB97
		-1.68										$C, i_a, \not\models$	
		-1.84										$C, i_a, \not\models$	
		-2.10										C, minor peak, $\not\models$	
JA095 6033	107 l	-	-	-	-	-	-	-	-	-	-	$1 < [n_{app}=(it^{\frac{1}{2}})/(it^{\frac{1}{2}})_{k=0}=f(t)] < 2.2, k=11.2\ s^{-1}, r$	EB98 ai73
	107 m	-	-	-	-	IR	<1.3	-	-	-	-	$C, k_d=3E3\ M^{-1}\ s^{-1}, r$	
JA095 6033	107 l	E_p -2.37	SCE	$i_p/ACv^{\frac{1}{2}}$	53.8	-	-	-	-	-	CP(-2.40 V, C=0.25) → benzonitrile which is further reduced at E_{app}, GLC	C, R for $v \geq 3E4$, ece, r	
		-2.32			20F	-		-		-		A	
		-			25.3	-	-	-	-	-		C, r	
	107 m	-1.68F		i_p	1.0F	-						$C, 0$ on first scan, R, r	
		-2.10F			1.3F							$C, 0$ on first scan	
		-2.36			33.7F						4,4'-dicyanobiphenyl, ESR, VR	C, ece	
		-2.30			20F							$A, i_{p,A}/i_{p,C} \downarrow$ as $C \uparrow$ for v=const.	
		-2.09			1.3F							A, R	
		-1.65			2.5F							A, R	
JE043 0365	266 b	$E_{\frac{1}{2}}$ -0.27	SCE	i_ℓ	7.0	i:i	2	αn_a	1.27	Elog	-	$A, \not\models, i_d, i_\ell = kC$ for C= 0.1-2, Tc=1.6, r	EB99 cc43
C0030 4219	65 v	$E_{\frac{1}{2}}$ -1.004F	SCE	-	-	-	-	-	-	-	-	C, r	EC00 ai93 cc44
		-1.144F										C	
		-0.987F										C, r	
		-1.160F										C	
		-0.975F										C, r	
		-1.163F										C	
		-0.977F										C, r	
		-1.204F										C	
		-0.983F										C, r	
		-1.242F										C	
JE043 0365	266 b	$E_{\frac{1}{2}}$ -0.093	SCE	-	-	-	-	$dE_{\frac{1}{2}}/d\log[OH^-]$	136	Elog	-	A, r	
		-0.255		i_ℓ	6.9	PQ i:i	1.20 2	αn_a	1.28		-	$A, \not\models, i_d, i_\ell = kC$ for C= 0.1-2, Tc=1.6, r	
		-0.298		-	-	-	-		136		-	A, r	
JE040 0345	-	$E_{\frac{1}{2}}$ -2.07	Hg/Hg^+ 0.02, $HClO_4$ 0.1	-	-	-	-	-	-	-	-	C, p	
		-2.23											
		-2.23										C	
JE040 0345	-	$E_{\frac{1}{2}}$ -2.08	Hg/Hg^+ 0.02, $HClO_4$ 0.1	-	-	-	-	-	-	-	-	C, p	
		-2.25										C	

EC01 C₇H₅ClO₂

Code No.	Empirical Formula	Name and C.A. Number	Structural Formula	Solvent	Tech.	Medium		μ, M	pH	T, °C	Electrodes	App.	Experimental Parameters
EC01	C₇H₅ClO₂	2-chloro-6-methyl-1,4-benzoquinone C.A. 1123-64-4	Table II	H₂O	PY	BR		-	5.3 6.8	-	DME/SCE	O4AO	C=1.0, m=2.72, t=3.75, h=50
EC02 aj09 cc45	C₇H₅N	benzonitrile C.A. 100-47-0	C₆H₅CN	DMF	VR	Et₄NClO₄	0.1	-	-	-	Pt/SCE	125A F02	0→-2.7→0, C=1.44, A=0.25, v=80.6, Pt Aux
EC03 aj14 cc47	C₇H₅NO₃	3-nitrobenzaldehyde C.A. 99-61-6	3-O₂NC₆H₄CHO	EtOH 10	PY	KOH	1	-	-	-	DME/MSE	2-0	C=1, m=2.73, t=3.55, v=3.3
				MeCN 10	PY	KOH	0.04	-	-	-	DME/MSE	2-0	C=0.1, m=2.73, t=3.55, v=3.3
													C=1
						KOH	1.8						C=0.1
													C=1
EC04 aj15 cc48	C₇H₅NO₃	4-nitrobenzaldehyde C.A. 555-16-8	4-O₂NC₆H₄CHO	EtOH 10	PY	KOH	1	-	-	-	DME/MSE	2-0	C=1, m=2.73, t=3.55, v=3.3
				MeCN 10	PY	KOH	1	-	-	-	DME/MSE	2-0	C=1, m=2.73, t=3.55, v=3.3
				MeCN 20	PY	KOH	1	-	-	-	DME/MSE	2-0	C=1, m=2.73, t=3.55, v=3.3
				MeCN 30	PY	KOH	1	-	-	-	DME/MSE	2-0	C=1, m=2.73, t=3.55, v=3.3
EC05	C₇H₅NO₄	2,3-pyridinedicarboxylic acid C.A. 89-00-9	Table II	H₂O	PY	KCl H₂SO₄ GEL	0.1 ns	-	-1 0 1 2 3 5	25	DME/SCE	OAO	C=1, m=3.57, t=2.8
EC06 CONT	C₇H₅NO₄	2,4-pyridinedicarboxylic acid C.A. 499-80-9	Table II	H₂O	PY	KCl H₂SO₄ GEL	0.1 ns	-	-1 0 1.1	25	DME/SCE	OAO	C=1, m=3.57, t=2.8

TABLE I. Electrochemical Data $C_7H_5NO_4$ (CONT.) EC06

Ref.	C/M	Charact. Potential		Response Const.		Tech.	n	Electrokinetic Data			Products and Identification	Description and Remarks	Code No.
		Value	vs.		Value			Parameter	Value	From			
C0028 2163	-	$E_{\frac{1}{2}}$	0.13	SCE	-	-	-	-	-	-	-	-	C,p	EC01
			0.06		-	-	-	-	-	-	-		C,p	
JA095 6033	-	E_p	-2.30F -2.25	SCE	i_p	87.5F 84.4F	-	-	-	-	-	-	C,R,p A,R	EC02 aJ09 cc45
JE043 0365	266 b	$E_{\frac{1}{2}}$	-0.34	SCE	i_ℓ	6.0F 12	PQ i:i	1.42 2 4	αn_a	0.88 -	Elog	-	A,≠,i_d,i_ℓ=kC for C= 0.1-2,Tc=1.6,r C,-NO_2 redn.	EC03 aJ14 cc47
JE043 0365	266 b	$E_{\frac{1}{2}}$	-0.258	SCE	-	-	-	-	$dE_{\frac{1}{2}}/dlog[OH^-]$	104	sttd	-	A,r	
			-0.211		-	-	-	-		112		-	A,r	
			-0.423		-	-	-	-		104		-	A,r	
			-0.396		-	-	-	-		112		-	A,r	
JE043 0365	266 b	$E_{\frac{1}{2}}$	-0.355	SCE	i_ℓ	6.05	i:i	2	αn_a	0.92	Elog	-	A,≠,i_d,i_ℓ=kC for C= 0.1-2,Tc=1.6,r	EC04 aJ15 cc48
JE043 0365	266 b	$E_{\frac{1}{2}}$	-0.435	SCE	-	-	-	-	-	-	-	-	A,r	
JE043 0365	266 b	$E_{\frac{1}{2}}$	-0.375	SCE	-	-	-	-	-	-	-	-	A,r	
JE043 0365	266 b	$E_{\frac{1}{2}}$	-0.285	SCE	-	-	-	-	-	-	-	-	A,r	
JE047 0499	-	$E_{\frac{1}{2}}$	-	SCE	i_ℓ	-	QE	-	αn_a	-	Elog	-	C,i_a,Pr,p	EC05
			-0.68F -			10.3F -		2 -		1.85 -			C,pK_1=-0.15,pK_2=2.4 C	
			-0.70F -			12.5F -		- -		- -			C,i_a,p C C	
			-0.75F -			14.3F -		- -		- -			C,i_a,p C C	
			- -0.81F -			- 14.3F -		- -		- -			C,i_a,p C C	
			- -0.92F -			- 14F -		- -		- -			C,i_a,p C C	
			- -1.17F -			- 7.7F -		- -		- -			C,i_a,p C C	
JE047 0499	-	$E_{\frac{1}{2}}$	-0.60F -	SCE	i_ℓ	13F -	-	-	αn_a	1.9 -	Elog	-	C,pK_1=0.15,pK_2=1.95,p C C	EC06
			-0.62F -			14.7F -		- -		- -			C,p C C	
			-0.65F -			14.7F -		- -		- -			C,p C C	
														CONT

109

EC06 (CONT.) C₇H₅NO₄

Code No.	Empirical Formula	Name and C.A. Number	Structural Formula	Solvent	Tech.	Medium		μ, M	pH	T, °C	Electrodes	App.	Experimental Parameters
EC06	C₇H₅NO₄	2,4-pyridinedicarboxylic acid	Table II	H₂O	PY	KCl H₂SO₄ GEL	0.1 ns	-	2.0 3.0 4.2 4.8	25	DME/SCE	OAO	C=1, m=3.57, t=2.8
EC07	C₇H₅NO₄	2,5-pyridinedicarboxylic acid C.A. 100-26-5	Table II	H₂O	PY	KCl H₂SO₄ GEL	0.1 ns	-	-2 0 0.6 1.3 2.2 3 4.8	25	DME/SCE	OAO	C=1, m=3.57, t=2.8
EC08	C₇H₅NO₄	2,6-pyridinedicarboxylic acid C.A. 499-83-2	Table II	H₂O	PY	KCl H₂SO₄ GEL	0.1 ns	-	-2 -1 0 1.8 3 5	25	DME/SCE	OAO	C=1, m=3.57, t=2.8
EC09	C₇H₅NO₄	3,4-pyridinedicarboxylic acid C.A. 490-11-9	Table II	H₂O	PY	KCl H₂SO₄ GEL	0.1 ns	-	-1 0 2 2.6 3.7 4.7	25	DME/SCE	OAO	C=1, m=3.57, t=2.8
CONT													

TABLE I. Electrochemical Data $C_7H_5NO_4$ (CONT.) EC09

| Ref. | C/M | Charact. Potential | | Response Const. | | n | Electrokinetic Data | | | Products and | Description and | Code |
		Value	vs.		Value	Tech.	Parameter	Value	From	Identification	Remarks	No.
JE047 0499	-	$E_{\frac{1}{2}}$ -0.72F - -	SCE	i_ℓ	14.7F - -	-	-	-	-	-	-	C,p C C	EC06
		-0.84F - -			14.7F - -	-	-	-	-	-	C,p C C	
		-0.95F - -			14.7F - -	-	-	-	-	-	C,p C C	
		-1.02F - -			13.7F - -	-	-	-	-	-	C,p C C	
JE047 0499	-	$E_{\frac{1}{2}}$ - -0.65 -	SCE	i_ℓ	- 16F -	-	-	αn_a	- 1.55 -	Elog	-	C,i_a,Pr,p C,pK_1=0.6,pK_2=2.35 C	EC07
		- -0.65F -			- 17F -	-	-	-	-	-	-	C,i_a,p C C	
		- -0.65F -			- 17F -	-	-	-	-	-	-	C,i_a,p C C	
		- -0.88F -			- 17F -	-	-	-	-	-	-	C,i_a,p C C	
		- 0.92F -			- 17F -	-	-	-	-	-	-	C,i_a,p C C	
		- -1.02F -			- 17F -	-	-	-	-	-	-	C,i_a,p C C	
		- -1.23F -			- 12.5F -	-	-	-	-	-	-	C,i_a,p C C	
JE047 0499	-	$E_{\frac{1}{2}}$ -0.63F -	SCE	i_ℓ	15F -	-	-	αn_a	1.6 -	Elog	-	C,pK_1=-0.9,pK_2=2.3,p C	EC08
		-0.65F -			16F -	-	-	-	-	-	-	C,p C	
		-0.67F -			16F -	-	-	-	-	-	-	C,p C	
		-0.77F -			16F -	-	-	-	-	-	-	C,p C	
		-0.88 -			16F -	-	-	-	-	-	-	C,p C	
		-1.11F -			13F -	-	-	-	-	-	-	C,p C	
JE047 0499	-	$E_{\frac{1}{2}}$ -0.56F - -	SCE	i_ℓ	11.7F - -	-	-	-	-	-	-	C,p C C	EC09
		-0.61 - -			13F - -	-	-	αn_a	1.9 -	Elog	-	C,pK_1=-0.25,pK_2=2.65,p C C	
		0.70F - -			13F - -	-	-	-	-	-	-	C,p C C	
		0.75F - -			13F - -	-	-	-	-	-	-	C,p C C	
		0.83F - -			13F - -	-	-	-	-	-	-	C,p C C	
		0.92F - -			12F - -	-	-	-	-	-	-	C,p C C	
													CONT

EC09 (CONT.) $C_7H_5NO_4$

Code No.	Empirical Formula	Name and C.A. Number	Structural Formula	Solvent	Tech.	Medium		μ, M	pH	T, °C	Electrodes	App.	Experimental Parameters
EC09	$C_7H_5NO_4$	3,4-pyridinedicarboxylic acid	Table II	H_2O	PY	KCl H_2SO_4 GEL	0.1 ns	-	6	25	DME/SCE	OAO	C=1, m=3.57, t=2.8
EC10	$C_7H_5NO_4$	3,5-pyridinedicarboxylic acid C.A. 499-81-0	Table II	H_2O	PY	KCl H_2SO_4 GEL	0.1 ns	-	-2 -1 0 1 2.4 4.2 5	25	DME/SCE	OAO	C=1, m=3.57, t=2.8
EC11	$C_7H_5NS_2$	benzothiazole-2-thiol C.A. 149-30-4	Table II	MeCN	CP	$NaClO_4$	0.1	-	-	-	Pt/SCE	---	E_{app}=0.84 V
EC12	$C_7H_8N_2O_2$	1-oxo-2-hydroxyimino-1H-2,3-dihydroindolizine C.A. 41634-03-1	Table II	H_2O	PY	BR		-	2.15 3.84 9.25 11.28	18±0.1	DME/SCE	23AO	C=0.0163, EC=1.69 at -1.0 V, h=80, MP Aux
EC13 aj80 cc59 db66	C_7H_8O	benzaldehyde C.A. 100-52-7	C_6H_5CHO	H_2O	PW	BENZ	0.1	-	10.1	25	DME/SCE	---	-
EC14	$C_7H_8O_2$	4-hydroxybenzaldehyde C.A. 123-08-0	4-HOC_6H_4CHO	EtOH 10	PY	KOH	1	-	-	-	DME/MSE	2-O	C=1, m=2.73, t=3.55, v=3.3
EC15 cc60 db69	$C_7H_8O_2$	2-methyl-1,4-benzoquinone C.A. 553-97-9	Table II-3	2-PrOH 40	PY	PHOS		-	7	-	DME/NCE	05AO	C=0.1

TABLE I. Electrochemical Data $C_7H_6O_2$ EC15

Ref.	C/M	Charact. Potential Value	vs.	Response Const. Value		Tech.	n	Electrokinetic Data Parameter	Value	From	Products and Identification	Description and Remarks	Code No.
JE047 0499	-	$E_{\frac{1}{2}}$	-1.03F	SCE	i_ℓ	4F	-	-	-	-	-	-	C,p / C / C	EC09
JE047 0499	-	$E_{\frac{1}{2}}$	-1.04	SCE	i_ℓ	8	-	-	αn_a	1.45	Elog	-	C,pK$_1$=0.05,pK$_2$=2.7, p / C	EC10
			-1.04			10	-	-	-	-	-	-	C,p / C	
			-1.04			11.7F	-	-	-	-	-	-	C,p / C	
			-1.09F			11.8	-	-	-	-	-	-	C,p / C	
			-1.17F			11.8F	-	-	-	-	-	-	C,p / C	
			-1.37F			11.8F	-	-	-	-	-	-	C,p / C	
		-	-	-		10F	-	-	-	-	-	-	C,p / C	
JE038 0245	417 a	-	-	-	-	-	-	-	-	-	-	monosulfide,2,2'-thiobis(benzo-thiazole),NMR,IRS,UVS	A,r	EC11
JE042 0049	391 a	$E_{\frac{1}{2}}$	-0.3F	SCE	i_ℓ	0.19F	QE	3.9 ± 0.1	-	-	-	1,2-dihydro-2-amino-pyrrolizin-1-one, UVS	C,i_d,i_ℓ=kC for C= 0.01-0.5,Tc=1.6 for T=10-30,r	EC12
			-1.05F			0.21F		-				1-hydroxy-1,2-di-hydropyrrolizine, UVS	C	
			-0.4F			0.19F		3.9 ± 0.1	-	-	-	1,2-dihydro-2-amino-pyrrolizin-1-one, UVS	C,i_d,i_ℓ=kC for C= 0.01-0.5,Tc=1.6 for T=10-30,r	
			-1.14F			0.1F		-				1-hydroxy-1,2-di-hydropyrrolizine, UVS	C	
			-0.81F			0.19F		3.9 ± 0.1	-	-	-	1,2-dihydro-2-amino-pyrrolizin-1-one, UVS	C,i_d,i_ℓ=kC for C= 0.01-0.5,Tc=1.6 for T=10-30,r	
			-1.54F			0.09F		-				1,2-dihydropyrroli-zin-1-one,UVS	C	
	391 b		-0.92F			0.11F		3.9 ± 0.1	-	-	-	1,2-dihydro-2-amino-pyrrolizin-1-one, UVS	C,i_k,r	
			-1.16F			0.09F							C;$\Sigma i_{\ell_1} + i_{\ell_2} = k_1 C + k_2$, i_d and $\neq$	
			-1.65F			0.09F						1,2-dihydropyrroli-zine-1-one,UVS	C	
JE048 0146	63 s	-	-	-	-	-	-	-	-	-	-	-	see AJ80,CC59,DB66	EC13 aj80 cc59 db66
JE043 0365	-	-	-	-	-	-	-	-	-	-	-	-	A,o,p	EC14
C0030 4192	-	$E_{\frac{1}{2}}$	-0.086	-	-	-	-	-	-	-	-	-	C,p	EC15 cc60 db69

EC16 $C_7H_6O_2$

Code No.	Empirical Formula	Name and C.A. Number	Structural Formula	Solvent	Tech.	Medium		μ, M	pH	T, °C	Electrodes	App.	Experimental Parameters
EC16 cc61 db70	$C_7H_6O_2$	salicylaldehyde C.A. 90-02-8	Table II-3	EtOH 10	PY	KOH	1	-	-	-	DME/MSE	---	-
EC17	$C_7H_6O_4$	2,5-dihydroxybenzoic acid C.A. 490-79-9	Table II	H_2O	VR	$HClO_4$	0.1	-	-	-	CPE/SCE	2A0	$C\approx 1, v=83.3$
EC18	$C_7H_6O_5$	2,3,6-trihydroxy-benzoic acid C.A. 16534-78-4	Table II	H_2O	VA	$HClO_4$	0.1	-	-	-	CPE/SCE	2A0	$C\approx 1, v=83.3$
EC19	$C_7H_6O_5$	2,4,5-trihydroxy-benzoic acid C.A. 610-90-2	Table II	H_2O	VA	$HClO_4$	0.1	-	-	-	CPE/SCE	2A0	$C\approx 1, v=83.3$
EC20	C_7H_7ClOS	4-chlorophenyl methyl sulfoxide C.A. 934-73-6	$4\text{-}ClC_6H_4SOCH_3$	EtOH	PY	Et_4NBr	0.1	-	-	30.0 ±0.5	DME/SCE	2-0	$C=0.5, t(c)=3$
				EtOH 80 DMF 20	PY	Et_4NBr	0.1	-	-	30.0 ±0.5	DME/SCE	2-0	$C=0.5, t(c)=3$
				EtOH 80 dioxane 20	PY	Et_4NBr	0.1	-	-	30.0 ±0.5	DME/SCE	2-0	$C=0.5, t(c)=3$
				MeCN	PY	Et_4NBr	0.1	-	-	30.0 ±0.5	DME/SCE	2-0	$C=0.5, t(c)=3$
				DMF	PY	Et_4NBr	0.1	-	-	30.0 ±0.5	DME/SCE	2-0	$C=0.6, t(c)=2$
													$t(c)=8$
						PHEN	1E-3						
EC21	C_7H_7NO	benzamide C.A. 55-21-0	$C_6H_5CONH_2$	H_2O	PO	H_2SO_4	1.0	-	-	-	DME/SCE	04A2	$C=2.0$
EC22 ak34	$C_7H_7NO_3$	2-nitroanisole C.A. 91-23-6	$2\text{-}O_2NC_6H_4OCH_3$	DMSO	PY	Bu_4NClO_4	0.1	-	-	-	DME/MP	2-0	$C=1$
EC23 ak35 cc67	$C_7H_7NO_3$	3-nitroanisole C.A. 555-03-3	$3\text{-}O_2NC_6H_4OCH_3$	DMSO	PY	Bu_4NClO_4	0.1	-	-	-	DME/MP	0-0	$C=1$
EC24 ak36 cc68	$C_7H_7NO_3$	4-nitroanisole C.A. 100-17-4	$4\text{-}O_2NC_6H_4OCH_3$	MeOH 10	PY	buffer GEL	7E-3	-	1 7 10.4 11.2	-	DME/SCE	04A0	$C=0.5$
CONT													

TABLE I. Electrochemical Data $C_7H_7NO_3$ (CONT.) EC24

Ref.	C/M	Charact. Potential		Response Const.		n		Electrokinetic Data			Products and Identification	Description and Remarks	Code No.
		Value	vs.		Value	Tech.		Parameter	Value	From			
JE043 0365	-	-	-	-	-	-	-	-	-	-	-	see CC61,DB70	EC16 cc61 db70
JE038 0389	318 b	E_p	0.58F	SCE	i_p	194.4F	QE at 1.0 V	4.2 ± 0.1	-	-	-	-	A,r	EC17
			0.45F			61.1F						CP→2,3,6-trihydroxy-benzoic acid,ESR	C	
			0.28F			61.1F						-	C,R	
			0.34F			33.3F							A,0 on first scan,R	
JE038 0389	318 c	E_p	0.28F	SCE	i_p	80.6F	-	-	-	-	-	-	A,r	EC18
			0.35F			50F							C	
JE038 0389	-	E_p	0.49F	SCE	i_p	83.3F	-	-	-	-	-	-	A,p	EC19
			0.39F			38.9F							C	
JE054 0221	47 c	$E_{\frac{1}{2}}$	-2.085	SCE	-	-	sttd	2	αn_a $k_{s,h}$	0.5 8.9E-19	Kplot	-	C,∓,$\Delta E_{\frac{1}{2}}$=3.6σ,r	EC20
JE054 0221	47 c	$E_{\frac{1}{2}}$	-2.095	SCE	-	-	sttd	2	-	-	-	-	C,∓,$\Delta E_{\frac{1}{2}}$=3.7σ,r	
JE054 0221	47 c	$E_{\frac{1}{2}}$	-2.185	SCE	-	-	sttd	2	-	-	-	-	C,∓,$\Delta E_{\frac{1}{2}}$=3.4σ,r	
JE054 0221	47 d	$E_{\frac{1}{2}}$	-2.275	SCE	-	-	sttd	1	-	-	-	-	C,∓,$\Delta E_{\frac{1}{2}}$=2σ,r	
			-2.475					1					C,∓	
JE054 0221	47 d	$E_{\frac{1}{2}}$	-2.247	SCE	i_ℓ	3.48	sttd	1	-	-	-	-	C,∓,$\Delta E_{\frac{1}{2}}$=3.8σ,r	
			-2.445			-		1					C,∓	
			-2.269 -			2.96 -	-	-					C,r C	
	47 c		-2.263 -			5.92 -	-	-					C,r C	
C0028 0838	382 a	E_i	-0.75 -0.95	SCE	-	-	-	-	-	-	-	-	C,p C	EC21
JE057 0191	-	$E_{\frac{1}{2}}$	-0.92 -1.9	MP	$i_\ell(u)$	1 3	sttd	1 -	-			radical anion,ESR	C,R,p C	EC22 ak34
JE050 0073	-	$E_{\frac{1}{2}}$	-1.02	MP	$i_\ell(u)$	1	Elog!	1	$dlogi_\ell/dlogt$ Elog	0.18 60	-	CP→ radical anion, ESR	C,R,i_d,p	EC23 ak35 cc67
			-2.15			3		-		-			C,∓	
JE036 0223	-	$E_{\frac{1}{2}}$	-0.204F	SCE	-	-	-	-	$dE_{\frac{1}{2}}/dpH$	60	-	-	C,p	EC24 ak36 cc68
			-0.53F		-	-	-	-		60	-	-	C,p	
			-0.72F		-	-	-	-		60	-	-	C,p	
			-0.75F		-	-	-	-		0	-	-	C,p	

CONT

Code No.	Empirical Formula	Name and C.A. Number	Structural Formula	Solvent	Tech.	Medium		μ, M	pH	T, °C	Electrodes	App.	Experimental Parameters
EC24 ak36 cc68	$C_7H_7NO_3$	4-nitroanisole	$4\text{-}O_2NC_6H_4OCH_3$	DMSO	PY	Bu_4NClO_4	0.1	-	-	-	DME/MP	2-O	C=1
EC25	$C_7H_7N_3O_2$	1,2-dioxo-1H,2,3-dihydroindolizine dioxime C.A. 18332-84-8	Table II	H_2O	PY	BR		-	1.84	22.0 ±0.1	DME/SCE	-A-	C=0.099, EC= 1.74, h=80
									2.99				
									4.64				
									6.84				
									8.0				
									9.56				
									10.16				
									11.41				
EC26 ak40	$C_7H_7N_5O$	6-methylpterin C.A. 708-75-8	Table II-2	H_2O	PV	H_2SO_4	1.0	-	-	-	DME/SCE	OAO	C=0.1, Δe=10
						BR KNO_3	0.1		2.0				
									4.0				
									6.0				
									8.0				
									10				
EC27	$C_7H_7N_5O_2$	2-amino-4,6-dihydroxy-7-methyl-pteridine C.A. 492-10-4	Table II	H_2O	PV	BR KNO_3 NaOH	0.1 1.0	-	2.0 -	-	DME/SCE	OAO	C=0.1, Δe= 10 mV

TABLE I. Electrochemical Data $C_7H_7N_5O_2$ EC27

Ref.	C/M	Charact. Potential		Response Const.		n Tech.	Electrokinetic Data			Products and Identification	Description and Remarks	Code No.
		Value	vs.		Value		Parameter	Value	From			
JE057 0191	-	$E_{\frac{1}{2}}$	-1.14 -2.17	MP	$i_\ell(u)$	11F 33F	sttd	1 -	- -	- -	- -	-	C,i_d,p C	EC24 ak36 cc68
JE048 0277	392 a	$E_{\frac{1}{2}}$	-0.45F	SCE	i_ℓ	1.9F	I:I	6	$dE_{\frac{1}{2}}/dpH$	60	-		$C,Mx,E_{\frac{1}{2}}=-0.33-0.06pH,r$	EC25
			-0.71F			0.65F		-		40			$C,E_{\frac{1}{2}}=-0.67-0.04pH$	
			-1.02F			0.95F				40			$C,E_{\frac{1}{2}}=-0.99-0.04pH, i_k+i_d$	
			-0.51F		I i_ℓ	10.15 1.9F	I:I, QE	6		60		CP→1,2-dihydro-2-aminopyrrolizin-1-one, UVS; NH_3	C,W,r	
			-0.75F			0.6F		≈8		40		1-aminopyrrolizine, UVS, $\lambda_{max}=313$	C	
			-1.06F			0.48F		-		40		1,2-dihydro-1-amino-pyrrolizine, UVS	C,i_k+i_d	
			-0.63F -0.86F -1.12F			1.9F 0.58F 0.38F		-		60 68 40	-	-	C,W,r C C,i_k+i_d	
			-0.76F -1.01F -1.26F			1.9F 0.35F 0.2F	-	-		60 68 40	-	-	C,r C C	
			-0.82F -1.07F -			1.75F 0.1F small				60 68 -			C,r C C	
	392 b		-0.95 -1.32 -1.59			1.01F 0.7F 0.18F	-	-		0 0 0	-		C,r $C,i_\ell\uparrow$ as pH $\uparrow$ $C,i_\ell\uparrow$ as pH $\uparrow$	
			0.93F 1.33F 1.65F			0.33F 1.0F 0.3F	-	-	-	-	-	-	C,r C C	
	392 c		-0.93F -1.38F -1.65F			0.1F 0.5F 0.8F	-	-	-	-	-	-	$C,0$ for pH > 11.5, r C C	
C0030 2460	-	E_{su}	-0.205 -0.260	SCE	i_{su}	2.0F 2.5F	-	-	-	-	-	- dihydro derivative, E_{su}	C,i_a,r C	EC26 ak40
			-0.880			-						-	C	
			-0.34F			10.0F	-	-	-	-	-	-	C,i_a	
			-0.41F			7.5F						dihydro derivative, E_{su}	C	
			-1.05F			2.8F						-	C	
			-0.52F			2.8F	-	-	-	-	-	-	C,i_a,r	
			-0.58F			2.7F						dihydro derivative, E_{su}	C	
			-1.08F			0.4F						-	C	
			-0.66F			-						dihydro derivative, E_{su}	C,r	
			-0.79F									dihydro derivative, E_{su}	C,r	
			-0.84F									-	C,i_a	
			-0.850			-	-	-	-	-	-	-	C,i_a,r	
			-0.93F									dihydro derivative, E_{su}	C	
C0030 2460	-	E_{su}	0.3 0.18	SCE	i_{su}	≈10F 7F	-	-	-	-	-	Hg salt	A,R,p C,i_a or i_{cat}	EC27
			-0.2 -0.3			- -	-	-	-	-	-	Hg salt	A,R,p C,i_a or i_{cat}	
	-	$E_{\frac{1}{2}}$	-1.14		$i_\ell(u)$		sttd							

Code No.	Empirical Formula	Name and C.A. Number	Structural Formula	Solvent	Tech.	Medium		μ, M	pH	T, °C	Electrodes	App.	Experimental Parameters
EC28	$C_7H_8N_2O_4S$	N-methyl-4-nitro-benzenesulfonamide C.A. 6319-45-5	$4\text{-}O_2NC_6H_4SO_2\text{-}(NHCH_3)$	DMF	IR	Et_4NClO_4	0.1	-	-	22.5 ±0.5	Pt/SCE	256A FO	E_{app}= -1.05 V, A=0.25, t= 0.01-10, Pt Aux
													E_{app}= -1.25 V, t=0.01
													t=1.2
													t=10
													E_{app}= -1.6 V, t=0.01-10
EC29	C_7H_8O	benzyl alcohol C.A. 100-51-6	$C_6H_5CH_2OH$	MeCN	CP	$LiClO_4$	0.1	-	-	-	Pt Ag/Ag⁺	25A-	A=12.9, E_{app}= 1.9, stainless steel Aux
					CT	Bu_4NBF_4	0.5	-	-	-	RPDE Ag/AgClO₄ 0.01, Bu₄NBF₄ 0.5	2--	C=5, A=0.12, ω=31.8
					IR	Bu_4NBF_4	0.5	-	-	-	PDE Ag/AgClO₄ 0.01, Bu₄NBF₄ 0.5	2-2	C=2, E_{app}= 1.77 V, A= 0.18, t=0.01
													E_{app}=2.19 V
EC30 ak67	C_7H_8OS	methyl phenyl sulfoxide C.A. 1193-82-4	$C_6H_5SOCH_3$	EtOH 80 DMF 20	PY	Et_4NBr	0.1	-	-	30.0 ±0.5	DME/SCE	2-0	C=0.5, t(c)=3
				EtOH 80 dioxan 20	PY	Et_4NBr	0.1	-	-	30.0 ±0.5	DME/SCE	2-0	C=0.5, t(c)=3
				MeCN	PY	Et_4NBr	0.1	-	-	30.0 ±0.5	DME/SCE	2-0	C=0.5, t(c)=3
				DMF	PY	Et_4NBr	0.1	-	-	30.0 ±0.5	DME/SCE	2-0	C=0.5, t(c)=3
EC31 ak70	$C_7H_8O_2$	4-methoxyphenol C.A. 150-76-5	$4\text{-}H_3CC_6H_4OH$	H_2O	EE	H_2SO_4	0.2	-	-	-	CPE/SCE (?)	---	C=1.58, $i_{reverse}/i_{forward}$= -0.414
						H_2SO_4 4-nitroaniline satd	0.2						
					VR	H_2SO_4	0.2	-	-	-	CPE/SCE (?)	---	0.2→0.7→ 0.2 V, C= 1.58, v=83.3
													0.4→1.0→ 0.4 V
						H_2SO_4 4-nitroaniline satd	0.2						
	$C_7H_8N_2O_4S$					Et_4NClO_4	0.1						

TABLE I. Electrochemical Data $C_7H_8O_2$ EC31

Ref.	C/M	Charact. Potential Value	vs.	Response Const.	Value	Tech.	n	Electrokinetic Data Parameter	Value	From	Products and Identification	Description and Remarks	Code No.
JE053 0293	412 b	-	-	-	$it^{\frac{1}{2}}/C$	33	-	1	-	-	-	N-methyl-4-nitro-benzenesulfonamide anion,TLC,MPT	C,i_d,r	EC28
		-	-	-	-	-	-	1.14	-	-	-	-	C,i_k,r	
		-	-	-	-	-	-	1.32	-	-	-	-	C,i_k,r	
		-	-	-	-	-	-	1.7	-	-	-	-	C,i_k,r	
		-	-	-	$it^{\frac{1}{2}}/C$	64	-	2	-	-	-	-	C,i_d,r	
JA094 6812	-	-	-	-	-	-	-	-	-	-	-	benzaldehyde(30%); benzoic acid(40%), NMR,GLC,IRS	A,p	EC29
JE038 0185	401 a	E_p	2.0	Ag/AgClO$_4$ 0.01, Bu$_4$NBF$_4$ 0.5	i_p	1.6E3F	-	-	-	-	-	CP→benzaldehyde, GLC	A,p	
			2.5			3.0E3F							A	
JE038 0185	401 a	-	-	-	logi	1.1E3F	-	-	-	-	-	-	A,logi≠f(t),p	
		-	-	-		7.2E3F	-	-	-	-	-	-	A,logi=k_1(logt) + k_2, p	
JE054 0221	47 c	$E_{\frac{1}{2}}$	-2.17	SCE	-	-	sttd	2	-	-	-	-	$C,\not\models,i_d,r$	EC30 ak67
JE054 0221	47 c	$E_{\frac{1}{2}}$	-2.27	SCE	-	-	sttd	2	-	-	-	-	$C,\not\models,i_d,r$	
JE054 0221	47 d	$E_{\frac{1}{2}}$	-2.315	SCE	-	-	sttd	2	-	-	-	-	$C,\not\models,i_d,r$	
JE054 0221	47 d	$E_{\frac{1}{2}}$	-2.32	SCE	-	-	sttd	2	-	-	-	-	$C,\not\models,i_d,r$	
JE045 0467	45 c	$E_{T/4}$ $E_{0.22}$	0.64F 0.27F	SCE(?)	τ_f τ_r	11.5F 10F	-	-	-	-	-	-	A,p C	EC31 ak70
		$E_{T/4}$ $E_{0.22}$	0.64F 0.46F 0.20F		τ_f τ_r	11.5F 3.5F 5.3F							A,p C C	
JE045 0467	45 c	E_p	0.64	SCE(?)	i_p	246.3F	-	-	-	-	-	4-methoxycarbonium ion	A,i_p↓ on later scans,p	
			0.35F 0.53F			75.6F 41.5F								C,R,quinone redn. A,O on first scan, hydroquinone oxidn.	
			0.77F -			234F -	-						A,p C,O	
			0.75F			-						4-methoxyphenyl carbonium ion	A,p	
			0.43F			-						-	C,N-(4-nitrophenyl)-4-benzoquinone-imine redn.	
			0.47F										A	

Code No.	Empirical Formula	Name and C.A. Number	Structural Formula	Solvent	Tech.	Medium		μ, M	pH	T, °C	Electrodes	App.	Experimental Parameters
EC32 cc76	$C_7H_8O_2$	methyl hydroquinone	Table 11-3	H_2O	QE	$HClO_4$	0.5	-	-	-	C/SCE	---	-
EC33	$C_7H_8O_2Se$	4-methylbenzene-seleninic acid C.A. 20753-52-0	$4-CH_3C_6H_4SeO_2H$	H_2O	VY	$HClO_4$ $NaClO_4$	0.05 0.2	-	-	25	Pt PDS MSE	126A0	C=1, v=5.6, MSE Aux, PE
EC34	$C_7H_9IN_2O$	1-methylpyridinium-3-carbamoyl iodide C.A. 6456-44-6	Table 11	H_2O	PY	HCl	0.5	-	0.58	25.0 ±0.2	DME Ag/AgCl, KCl satd	2A0	C=0.1, m=0.86±0.03, t(c)=0.1, h=35
													t(c)=0.5
						PHOS	0.5		1.86				C=0.10, t(c)=0.5
													C=1
													C=10
						$H_2NCH_2COOH \cdot H^+Cl^-$	0.5		2.4				C=0.1
						FORM	0.5		3.5				
													C=1
						ACET	0.5		4.7				C=0.1
						imidazole-H^+Cl^-	0.5		7.15				
						tris-H^+Cl^-	0.5		8.2				
						BOR	0.5		9.6				
CONT		methyl hydroquinone	Table 11-3	H_2O	QE	$HClO_4$	0.5						

TABLE I. Electrochemical Data $C_7H_9IN_2O$ (CONT.) EC34

Ref.	C/M	Charact. Potential		Response Const.		n	Tech.	Electrokinetic Data			Products and Identification	Description and Remarks	Code No.
		Value	vs.		Value			Parameter	Value	From			
JE038 0389	-	-	-	-	-	-	-	-	-	-		see CC76	EC32 CC76
JE040 0339	404 a	$E_{\frac{1}{2}}$	0.46F	SCE	-	-	-	-	-	-	-	-	$C,\ne,\Delta E_{\frac{1}{2}}=0.10\sigma,p$	EC33
JE039 0447	42 d	$E_{\frac{1}{2}}$	-0.72F -1.04F	NHE	i_ℓ/C	2.6F 16F	-	-	-	-	-	-	C,p C	EC34
			-0.68F			3.2F	-	-	$dE_{\frac{1}{2}}/dlogt$ αn_a Elog $dE_{\frac{1}{2}}/dpH$ Tomeš	30±2 ≈1 50 74 50±5	- Elog -	-	$C,E_{\frac{1}{2}}=-0.633-0.074pH$ (± 0.010 V), $dE_{\frac{1}{2}}/dT=$ 1 and Tc=0.92 for T=0-40,p	
			-1.0F			22±9	-	-	$dE_{\frac{1}{2}}/dlogt$ $dE_{\frac{1}{2}}/dpH$	22±4 45	-	-	$C,E_{\frac{1}{2}}=-0.970-0.045pH$ (± 0.005 V), $E_{\frac{1}{2}}\ne f(T)$ and Tc=3.5 for T= 0-40,H	
			-0.78F		i_ℓ	<0.1F	-	-	$dE_{\frac{1}{2}}/dlogC$	0	plot	-	C,p	
			-1.04F			1F				27.7F			C	
			-0.77F -0.88F -			2.4F 1.7F -	- -	- -		+40F 40F -		-	C,p C C,O	
			-0.73F -0.92F			2.4F 25.6F	-	-		+40F 40F		-	C,p C	
			-0.82F		i_ℓ/C	3.8	-	-	$dE_{\frac{1}{2}}/dlogt$ αn_a Elog $dE_{\frac{1}{2}}/dpH$ Tomeš	30±2 ≈1 50 74 50±5	- Elog -	-	C,p	
			-1.06F			22±9			$dE_{\frac{1}{2}}/dlogt$ $dE_{\frac{1}{2}}/dpH$	22±4 45			C,H	
	42 c		-0.82F			-	-	-	$dE_{\frac{1}{2}}/dlogC$	+28.9	-	-	C,p	
			-0.80F -1.02F			-	-	-		0 0		-	C,p C	
			-0.82F		i_ℓ/C	1.2F	-	-		-	-	-	C,p	
			-1.01F			3.9F			$dE_{\frac{1}{2}}/dpH$	60	sttd		$C,M,E_{\frac{1}{2}}=-0.75-0.06pH$ (± 0.03 V)	
			-0.87F			1.3F	-	-	$dE_{\frac{1}{2}}/dlogC$ $dE_{\frac{1}{2}}/dpH$	+28 0	plot	-	$C,dE_{\frac{1}{2}}/dlogC=0$ for C>3,p	
			-1.2F			3F				65 60	sttd		C,M	
			-1.33F -1.49F			-				-			C C,Mx	
			-0.89F			1.3F	-	-		+28 0	plot	-	$C,dE_{\frac{1}{2}}/dlogC=0$ for C>3,p	
			-1.3F			1.3F				53 60			C	
			-			-				=			O	
			-0.85			1.2F	-	-		+28 0	plot	-	$C,dE_{\frac{1}{2}}/dlogC=0$ for C>0.3,p	
			-1.38F			1.2F				53 60	sttd		C	CONT

Code No.	Empirical Formula	Name and C.A. Number	Structural Formula	Solvent	Tech.	Medium		μ, M	pH	T, °C	Electrodes	App.	Experimental Parameters
EC34	$C_7H_9IN_2O$	1-methylpyridinium-3-carbamoyl iodide	Table II	H_2O	PY	BOR	0.5	-	9.6	25±0.2	DME Ag/AgCl, KCl satd	2AO	C=1, m=0.86±0.03, t(c)=1, h=35
													C=10
					VA	imidazole H^+ Cl^-	0.5	-	7.2	25±0.2	GCE Ag/AgCl, KCl satd	2AO	0.5→-1.3→0.5 V, C=1, d=0.12, v=500
						tris H^+ Cl^-	0.5		8.2				
EC35 ak83	C_7H_9N	N-methylaniline C.A. 100-61-8	$C_6H_5NHCH_3$	H_2O	VY	ACET PHOS		-	0.7	-	Pt PDS SCE	2-0	C=0.5, v=0.92
									2.5				
									4.0				
									6.0				
									7.6				
									8.5				
									10.0				
									12.0				
EC36 ak84	C_7H_9N	2-toluidine C.A. 95-53-4	$2-CH_3C_6H_4NH_2$	H_2O	IL	ACET PHOS		-	0.7	-	CPE/SCE	2-0	C=0.5
									4.0				
									6.0				
									8.0				
									10.0				
									11.5				
					VR	ACET PHOS		-	0.7	-	CPE/SCE	2-0	-0.25→1.2→-0.25, C=0.5, v=5
					VY	ACET PHOS		-	0.7	-	Pt PDS SCE	2-0	C=0.5, v=0.92
									2.5				
									4.5				
									6.0				
									7.6				
									10.0				
									12.0				

TABLE I. Electrochemical Data C$_7$H$_9$N EC36

Ref.	C/M	Charact. Potential		Response Const.		n Tech.	n	Electrokinetic Data			Products and Identification	Description and Remarks	Code No.
		Value	vs.		Value			Parameter	Value	From			
JE039 0447	42 c	$E_{\frac{1}{2}}$	-0.82F	NHE	i_ℓ	1.15F	-	-	$dE_{\frac{1}{2}}/dlogC$	0	-	-	C,$E_{\frac{1}{2}}\neq f(T)$ and Tc= 1.76 for T=0-40, E_0= -0.87±0.02,p	EC34
								$dE_{\frac{1}{2}}/dlogt$	13	plot			
			-1.4F			1.1F	-	-	$dE_{\frac{1}{2}}/dlogt$	29±2	sttd		C,$E_{\frac{1}{2}}\neq f(T)$ and Tc= 1.76 for T=0-40.	
								Elogi$\frac{2}{3}$	57±4	-			
								$dE_{\frac{1}{2}}/dlogC$	53±2	-			
			-0.82F		-	-	-	-	-	-	-	-	C,p	
			-1.46F										C	
JE039 0447	42 c	E_p	-0.25F	NHE	-	-	-	-	-	-	-	-	C,p	
			-0.85± 0.02										C	
			-0.65F		-			-		-			A	
			-0.11± 0.02										A	
			0.19F										A	
			-0.85± 0.02		-	-	-	-	-	-	-	-	C,$E_p\neq f(pH)$ for pH= 6-10,p	
			-0.11± 0.02										A,$E_p\neq f(pH)$ for pH= 6-10	
JE052 0093	-	$E_{\frac{1}{2}}$	0.90F	SCE	-	-	-	-	-	-	-	-	A,p	EC35 ak83
			0.86F		-	-	-	-	-	-	-	-	A,p	
			0.78F		-	-	-	-	-	-	-	-	A,p	
			0.71F		-	-	-	-	-	-	-	-	A,p	
			0.64F		-	-	-	-	-	-	-	-	A,p	
			0.60F		-	-	-	-	-	-	-	-	A,p	
			0.55F		-	-	-	-	-	-	-	-	A,p	
			0.54F		-	-	-	-	-	-	-	-	A,p	
JE052 0093	38 n	$E_{p/2}$	0.87F	SCE	-	-	-	-	α	0.8	$E_{p/2}$-E_p	-	A,p	EC36 ak84
			0.72F		-	-	-	-	-	-		-	A,p	
			0.63		-	-	-	-	-	-		-	A,p	
	38 p		0.60F		-	-	-	-		0.56		-	-	
			0.54F		-	-	-	-	-	-		-	A,p	
			0.54F		-	-	-	-		0.6		-	A,p	
JE052 0093	38 n	E_p	0.85F	SCE	$i_p(u)$	25F	-	-	-	-	-	2-tolidine(20%),VA	A,p	
			0.5F			2F							C,2-tolidine redn.	
			0.25F			6F							C	
			0.36F			3F							A,0 on first scan	
			0.58F			2F							A,0 on first scan	
JE052 0093	38 n	$E_{\frac{1}{2}}$	0.92F	SCE	i_ℓ	9.8F	QE	2.15 ± 0.15	α dlogi/dlogX	1.6 -0.5	Elog	-	A,X=renewal time of diffusion layer,Ic= 1.05,i_d,r	
			0.86F		-	-		-		-			A,p	
			0.76F		i_p	7.6F		2.15 ± 0.15		-			A,r	
			0.69F		-	-		-		-			A,p	
			0.66F		-	-		-		-			A,p	
	38 p		0.60F		i_p	10.4F	QE	2.0 ± 0.1	-	-		-	A,p	
			0.59F		-	-	-	-	-	-		-	A,p	
JE039 0447	42 c	$E_{\frac{1}{2}}$	-0.82F	NHE	i_ℓ	1.15F	-	-	$dE_{\frac{1}{2}}/dlogC$	0	-	-	C,$E_{\frac{1}{2}}\neq f(T)$ and Tc=	

Code No.	Empirical Formula	Name and C.A. Number	Structural Formula	Solvent	Tech.	Medium	μ, M	pH	T, °C	Electrodes	App.	Experimental Parameters
EC37 ak85	C_7H_9N	3-toluidine C.A. 108-44-1	$3\text{-}CH_3C_6H_4NH_2$	H_2O	IL	ACET PHOS	-	0.7	-	CPE/SCE	2-0	C=0.5
								2.5				
								4.0				
								6.0				
								8.0				
								10.0				
								11.5				
								11.5				
					VR	ACET PHOS	-	0.7	-	CPE/SCE	2-0	C=0.5, v=5
					VY	ACET PHOS	-	0.7	-	Pt PDS SCE	2-0	C=0.5, v=0.92
								2.5				
								4.0				
								6.0				
								8.0				
								10.0				
								12.0				
EC38 ak86	C_7H_9N	4-toluidine C.A. 106-49-0	$4\text{-}CH_3C_6H_4NH_2$	H_2O	IL	ACET PHOS	-	0.7	-	CPE/SCE	2-0	C=0.5
								2.5				
								4.0				
								5.0				
								6.5				
								7.3				
								8.0				
								10.0				
								12.0				
					VY	ACET PHOS	-	0.7	-	Pt PDS SCE	2-0	C=0.5
								3.8				
								5.0				
								6.0				
								8.0				
								10.0				
								12.0				
EC39	$C_7H_9NO_2$	N-isopropylmaleimide C.A. 1073-93-4	Table II	EtOH 50	PY	BR	-	7.25	-	DME/SCE Tl	04A0	C=1.0, m=3.3, t=3.5, h=40

TABLE I. Electrochemical Data $C_7H_9NO_2$ EC39

Ref.	C/M	Charact. Potential		Response Const.		n	Tech.	Electrokinetic Data			Products and Identification	Description and Remarks	Code No.
		Value	vs.		Value			Parameter	Value	From			
JE052 0093	38 o	$E_{p/2}$ 0.87F	SCE	-	-	-	-	α	0.82	$E_{p/2}-E_p$	-	A,p	EC37 ak85
		0.82F		-	-	-	-		-			A,p	
		0.75F		-	-	-	-		0.7			A,p	
		0.66F		-	-	-	-		-			A,p	
	38 p	0.63F		-	-	-	-		0.5			A,p	
		0.54F		-	-	-	-		-			A,p	
		0.52F		-	-	-	-		0.56			A,p	
JE052 0093	38 o	E_p 0.81F	SCE	$i_p(u)$	29F	-	-	-	-	-	2,3'-dimethyl-4-aminodiphenylamine, VA	A,p	
		0.31F			7F						-	C	
		0.36F			4.5F							A,O on first scan	
JE052 0093	38 o	$E_{\frac{1}{2}}$ 0.91F	SCE	-	-	QE	2	α dlogi/dlogX	1.4 -0.28	Elog -	-	A,X=renewal time of diffusion layer, Tc=2.56,p	
		0.84F		-	-	-	-		-		-	A,p	
		0.86F		-	-	-	-		-		-	A,p	
		0.65F		-	-	-	2		-		-	A,p	
	38 p	0.62F		-	-	-	2		-		-	A,p	
		0.57F		-	-	-	-		-		-	A,p	
		0.55F		-	-	-	2		-		-	A,p	
JE052 0105	38 s	$E_{p/2}$ 0.89F	SCE	-	-	QE	2	α	0.75	$E_{p/2}-E_p$	-	A,p	EC38 ak86
		0.85F		-	-		-		-		-	A,p	
		0.75F 0.85F		-	-	}	2		0.65 -		-	A,p A	
		0.68F 0.83F		-	-		-		-		-	A,p A	
		0.65F 0.73F		-	-		2		-		-	A,p A	
		0.65F		-	-		-		-		-	A,merging of two waves,p	
	38 p	0.64F		-	-		2		0.5		-	A,p	
		0.58F		-	-		-		-		-	A,p	
		0.54F		-	-		2		0.55		-	A,p	
JE052 0105	38 s	$E_{\frac{1}{2}}$ 0.90F	SCE	-	-	-	-		-		-	A,p	
		0.76F		-	-	-	-		-		-	A,p	
		0.59F		i_ℓ	1.38F		-	dlogi/dlogX	-0.42	-	-	A,X,renewal time of diffusion layer, Tc=1.3,i_ℓ=kC,S,p	
		0.73F			1F		-		-0.40			A,Tc=1.3,i_ℓ=kC,p	
		0.63F		-	-	-	-		-	-		A,p	
	38 p	0.58F		-	-	-	-		-	-		A,p	
		0.51F		-	-	-	-		-	-		A,p	
		0.49F		-	-	-	-		-	-		A,p	
C0026 2749	-	$E_{\frac{1}{2}}$ -0.831 -1.299	SCE	-	-	sttd	1 1	-	-	-	-	C,i=f(t),p C,i=f(t),p	EC39

125

EC40 $C_7H_9NO_2$

Code No.	Empirical Formula	Name and C.A. Number	Structural Formula	Solvent	Tech.	Medium	μ, M	pH	T, °C	Electrodes	App.	Experimental Parameters
EC40	$C_7H_9NO_2$	N-(1-propyl)male-imide C.A. 21746-40-7	Table II	EtOH 50	PY	BR	-	7.25	-	DME/SCE Tl	04A0	C=1.0, m=3.3, t=3.5, h=40
EC41	$C_7H_9N_3O$	2,4-dimethylpyrimidine-5-carboxamide C.A. 53554-30-6	Table II	H_2O	PY	CITR	-	4.5	28.0 ±0.2	DME/SCE	0A0	C=0.1, EC=1.3
EC42	$C_7H_{11}KOS_2$	potassium O-cyclohexyldithiocarbonate C.A. 2720-77-6	Table II	Me_2CO	PY	Et_4NClO_4 0.1	-	-	20.0 ±0.1	DME Ag/AgCl, LiCl 0.1	12A0	C=1, Pt Aux
EC43 cc95	$C_7H_{11}N_3O_2$	5-tert-butyl-6-aza-uracil C.A. 52236-30-3	Table II-3	H_2O	PY	BR	-	2.06 2.30 2.90 3.90 4.30 6.43 8.9 9.51 10.62 11.68	25	DME/SCE Tl$^+$	0A0	C=0.24, h=50
EC44	$C_7H_{12}O$	cycloheptanone C.A. 502-42-1	Table II	EtOH? 75	PY	Et_4NClO_4 0.1	-	-	-	DME/SCE	2-0	C=1
				EtOH 90	PY	Et_4NClO_4 0.1	-	-	20	DME/SCE	2-0	C=0.5
EC45	$C_7H_{12}OS$	2-methyl-4-thiepanone	Table II	H_2O	PY	Et_4NClO_4 0.1	-	-	20	DME/SCE	2-0	C=0.5
EC46	$C_7H_{12}OS$	7-methyl-3-thiepanone	Table II	H_2O	PY	Et_4NClO_4 0.1	-	-	20	DME/SCE	2-0	C=0.5
EC47	$C_7H_{12}OS$	5-thiocanone C.A. 20701-80-8	Table II	H_2O	PY	Et_4NClO_4 0.1	-	-	20	DME/SCE	2-0	C=0.5
				EtOH(?) 75	PY	Et_4NClO_4 0.1	-	-	-	DME/SCE	2-0	C=1

TABLE I. Electrochemical Data $C_7H_{12}OS$ EC47

Ref.	C/M	Charact. Potential		Response Const.		n Tech.	n	Electrokinetic Data			Products and Identification	Description and Remarks	Code No.
		Value	vs.		Value			Parameter	Value	From			
C0026 2749	-	$E_{\frac{1}{2}}$	-0.800 -1.263	SCE	-	-	sttd	1 1	-	-	-	-	$C,i=f(t),p$ $C,i=f(t)$	EC40
JF055 0445	-	$E_{\frac{1}{2}}$	-1.1	SCE	-	-	i:i	2	-	-	-	-	$C,W,i_d,T_c=1.6,E_{\frac{1}{2}}\neq$ $f(pH)$ for $pH=2-10,p$	EC41
JE048 0071	333 a	$E_{\frac{1}{2}}$	-0.357	Ag/AgCl, LiCl 0.1	-	-	-	-	Tomeš	64	-	-	$A,i_\ell=kh^{\frac{1}{2}},i_d,i_\ell=kC$ for $C=0.5-10, E_{\frac{1}{2}}\rightarrow$ more neg. as $C\uparrow,r$	EC42
			0.086							72			$A,i_\ell=kh^{\frac{1}{2}},i_d,i_\ell=kC$ for $C=0.5-10, E_{\frac{1}{2}}\rightarrow$ more pos. as $C\uparrow$ A,M	
			-				-		-					
C0027 0546	-	$E_{\frac{1}{2}}$	-0.69F	SCE	i_ℓ	0.27F	sttd	2	-	-	-	-	C,r	EC43 cc95
			-0.74F			0.28		2	-	-	-	-	C,r	
			-0.84F			0.15_5		2	-	-	-	-	C,r	
			-0.92F			0.06F		2	-	-	-	-	C,r	
			-1.30F			0.28F		-	-	-	-	-	C	
			-1.27F			0.34F		2	-	-	-	-	C,r	
			-1.34F			0.34F		2	-	-	-	-	C,r	
			-1.40F			0.33F		2	-	-	-	-	C,r	
			-1.48F			0.26F		2	-	-	-	-	C,r	
			-1.54F			0.11F		2	-	-	-	-	C,r	
			-1.5F			0.05F		2	-	-	-	-	C,r	
JE035 0363	-	$E_{\frac{1}{2}}$	-2.48	SCE	-	-	-	-	-	-	-	-	C,p	EC44
JE039 0195	-	$E_{\frac{1}{2}}$	-2.48	SCE	-	-	-	-	-	-	-	-	C,p	
JE039 0195	400 ab	$E_{\frac{1}{2}}$	-2.14	SCE	-	-	-	-	-	-	-	-	C,p	EC45
JE039 0195	400 ab	$E_{\frac{1}{2}}$	-1.88	SCE	-	-	-	-	-	-	-	-	C,p	EC46
JE039 0195	400 ab	$E_{\frac{1}{2}}$	-2.15	SCE	-	-	-	-	-	-	-	-	C,p	EC47
JE035 0363	-	$E_{\frac{1}{2}}$	-2.15	SCE	-	-	-	-	-	-	-	-	C,p	
		$E_{\frac{1}{2}}$					sttd						$C,i=f(t),p$	EC40

Code No.	Empirical Formula	Name and C.A. Number	Structural Formula	Solvent	Tech.	Medium		μ, M	pH	T, °C	Electrodes	App.	Experimental Parameters
EC48	$C_7H_{12}O_3S$	5-thiocanone-1,1-dioxide C.A. 36165-02-3	Table II	EtOH(?) 75	PY	Et_4NClO_4	0.1	-	-	-	DME/SCE	2-0	C=1
EC49	$C_7H_{13}BF_4N_2$	2-ethyl-2,3-diazeniumbicyclo[2.2.1]-hept-2-ene tetrafluoroborate C.A. 43008-09-9	Table II	MeCN	VR	Bu_4NClO_4	0.1	-	-	-	DME?/SCE	---	v=190
EC50 db99	$C_7H_{14}N_2$	2,3-dimethyl-2,3-diazabicyclo[2.2.1]-heptane C.A. 14287-89-9	Table II-4	MeCN	VR	Bu_4NClO_4		-	-	-	DME?/SCE	---	-
EC51	C_7H_{16}	heptane C.A. 142-82-5	$CH_3(CH_2)_5CH_3$	CF_3COOH	IL	FSO_3H	2.5	-	-	-	Pt/SCE	25-2	C=30, v=50, Pt Aux
EC52	$C_7H_{16}B_9Co$	(η-cyclopentadienyl)-[7-11-η-undecaborane(11)$^{2-}$]cobalt^{1-} C.A. 51850-04-5	$(\pi$-$C_5H_5)Co(\pi$-1,2-$B_9C_2H_{11})$	MeCN	PY	Bu_4NPF_6	0.3	-	-	25	DME/SCE	2AO	ns
					PVI	Bu_4NPF_6	0.3	-	-	25	DME/SCE	2AO	ns
EC53	$C_8H_4ClNO_2$	4-chlorophthalimide C.A. 7147-90-2	Table II	MeCN 66.7	VY	ACET		-	8	22±2	Pb RDE SCE	-AO	C=40, v=10, ω=35.8
				MeCN 63 H_2SO_4 5	VY	$HClO_4$ $NaClO_4$		-	1	22±2	Pb RDE SCE	-AO	C=40, v=10, ω=15.9
EC54 a147	$C_8H_4O_3$	phthalic anhydride C.A. 85-44-9	Table II-2	DMF	PY	$KClO_4$	0.1	-	-	20±2	DME/SCE	24AO	C=1
						$LiClO_4$	0.1						
						Et_4NClO_4	0.1						
						$NaClO_4$	0.1						
EC55	$C_8H_5NO_2$	isatin C.A. 91-56-5	Table II	DMF	PY	Bu_4NClO_4	0.5	-	-	-	DME/MP	0-0	C=1
						Bu_4NClO_4	>0.5						
CONT													

TABLE I. Electrochemical Data $C_8H_5NO_2$ (CONT.) EC55

Ref.	C/M	Charact. Potential		Response Const.		n Tech.	n	Electrokinetic Data			Products and Identification	Description and Remarks	Code No.
		Value	vs.		Value			Parameter	Value	From			
JE035 0363	-	$E_{\frac{1}{2}}$	-1.84	SCE	-	-	-	-	-	-	-	-	C,p	EC48
JA095 6454	-	$E_{p/2}$ (?)	-0.70	SCE	-	-	-	-	-	-	-	-	C,R at v=1.9E3,p A,0 for v<380	EC49
JA095 6454	-	-	-	-	-	-	-	-	-	-	-	-	see DB99	EC50 db99
JE054 0181	-	$E_{p/2}$	2.32	SCE	-	-	-	-	-	-	-	-	A,≠,p	EC51
JE050 0031	-	$E_{\frac{1}{2}}$	-1.21	SCE	D	53	sttd	1	Elog	≈60	-	$[(\pi-C_5H_5)Co(\pi-1,2-B_9C_2H_{11})]^-$	C,R,r	EC52
			-2.11			53		1		≈60		$[(\pi-C_5H_5)Co(\pi-1,2-B_9C_2H_{11})]^{2-}$	C,R	
JE050 0031	-	-	-	-	-	-	-	-	α $k_{s,h}^{app}$ $k_{s,h}$	0.5 1 2.3	eqn. -	-	C,r	
									α $k_{s,h}^{app}$ $k_{s,h}$	0.4 0.26 8.9			C	
JE057 0351	13 g	$E_{\frac{1}{2}}$	-1.1 -1.375	SCE	-	-	-	-	-	-	-	-	C,p C	EC53
JE057 0351	13 b	$E_{\frac{1}{2}}$	-0.9	SCE	-	-	-	-	dlogi/dE	90	-	-	C,p	
JE042 0253	-	$E_{\frac{1}{2}}$	-1.267	SCE	-	-	-	-	-	-	-	-	C,$E_{\frac{1}{2}}$→more pos. as $[K^+]$ ↑,p C	EC54 a147
			-1.225		-	-	-	-	-	-	-	-	C,$E_{\frac{1}{2}}$→more pos. as $[Li^+]$ ↑,p C	
			-1.33		-	-	QE	0.94	Elog	62	-	-	C,R,$E_{\frac{1}{2}}$≠f$[Et_4N^+]$,p C	
			-1.27		-	-	-	-	$dE_{\frac{1}{2}}/dlog[Na^+]$	22	-	-	C,$E_{\frac{1}{2}}$→more pos. as $[Na^+]$ ↑,p C	
JE045 0397	394 b	$E_{\frac{1}{2}}$	-0.77	MP	$i_\ell(u)$	10F	-	-	-	-	-	CP→anion radical, ESR	C,i_d,i_ℓ ↑ as $[H^+$ donor] ↑,p	EC55
			-1.35			8.5F						-	C,i_ℓ ↓ as $[H^+$ donor] ↓	
			-1.62			11.5F						CP→electroinactive compd.	C	
			-1.35F -1.62F			15F 9F	-	-	-	-	-	- CP→dianion radical, ESR	C,p C	
														CONT
	-	$E_{\frac{1}{2}}$	-1.84											

EC55 (CONT.) $C_8H_5NO_2$

Code No.	Empirical Formula	Name and C.A. Number	Structural Formula	Solvent	Tech.	Medium		μ, M	pH	T, °C	Electrodes	App.	Experimental Parameters
EC55	$C_8H_5NO_2$	isatin	Table II	DMF	PA	Bu_4NClO_4	0.5	-	-	-	DME/SCE	0-0	C=1, v=600
EC56 a168 cd17	$C_8H_5NO_2$	phthalimide C.A. 85-41-6	Table II-2	MeCN 66.7	VY	ACET		-	8	22±2	Pb RDE/ SCE	-A0	C=40, v=10, ω=35.8
				MeCN 63 H_2SO_4 5	VA		ns	-	-	22±2	MPCuD/ SCE	-A0	C=40, A=3.14, ω=300
					VY		ns	-	-	22±2	C RDE SCE	-A0	C=40, v=10, ω=6.37
											Pb RDE SCE		
						$HClO_4$ $NaClO_4$	ns		1				ω=15.9
											RMAuD/ SCE		
											RMCuD/ SCE		A=3.14
											Sn RDE/ SCE		
				DMF	PY	$Ba(ClO_4)_2$	0.05	-	-	20±2	DME/SCE	24A0	C=1, m=2.36, t=5.55
						$KClO_4$	0.1						
						$LiClO_4$	0.1						
						Et_4NClO_4	0.1						
						$NaClO_4$	0.1						
				EE		Et_4NClO_4	0.1	-	-	20±2	XEl/SCE	24A2	C=50, A=0.23, τ_f=0.22, j= 15, PE
													j=84.10
													j=1000
					VA	Et_4NClO_4	0.1	-	-	-	XEl/SCE	24A0	C=1-3, A= 0.23, PE
EC57	$C_8H_5N_3O$	3-diazo-1,3-dihydro- 2H-indol-2-one C.A. 3265-29-0	Table II	H_2O	PY	BR NaCl	2	-	5.22	23.0 ±0.1	DME/SCE	0A0	C=0.11, EC= 1.73(-1.05 V), h=79.8
									7.98				
									8.5				
									9.0				
									9.90				
									12.0				
EC58 a197	$C_8H_6O_2$	isophthalaldehyde C.A. 626-19-7	3-OHCC$_6$H$_4$CHO	EtOH 10	PY	KOH	1	-	-	-	DME/MSE	2-0	C=1, m=2.73, t=3.55, v= 3.3
CONT													

TABLE I. Electrochemical Data $C_8H_6O_2$ (CONT.) EC58

Ref.	C/M	Charact. Potential Value	vs.	Response Const.	Value	Tech.	n	Electrokinetic Data Parameter	Value	From	Products and Identification	Description and Remarks	Code No.
JE045 0397	394 b	E_p -0.89F -1.88F -2.2F -1.8F -0.78F	MP	$i_p(u)$	12.5F 14F 16F 5F 7F	-	-	-	-	-	-	C,p C C A A	EC55
JE057 0351	13 g	$E_{\frac{1}{2}}$ -1.195 -1.43	SCE	-	-	-	-	-	-	-	-	C,p C	EC56 a168 cd17
JE057 0351	13 b	E_p -0.98F -	SCE	i_ℓ	1E5F -	-	-	-	-	-	-	C,≠,$i_p=kv^{\frac{1}{2}}$,r A,O	
JE057 0351	13 b	$E_{\frac{1}{2}}$ -1.1	SCE	-	-	-	-	dlogi/dE	60	-	-	C,p	
		-0.93		-	-	-	-		60	-	-	C,p	
		-0.96		-	-	-	-		60	-	-	C,p	
		-0.81		-	-	-	-		60	-	-	C,p	
		-0.875		i_ℓ i_D	9.3E4F ≈8	QE	2		60	-	CP→hydroxyphthal-imidine,NMR,IRS,UVS,CHN	C,$i_\ell=kw^{\frac{1}{2}}$,p	
		-0.95		-	-	-	-		120	-	-	C,p	
JE052 0229	13 f	$E_{\frac{1}{2}}$ -1.434 -	SCE	-	-	-	-	-	-	-	-	C,p C	
		-1.499 -		-	-	-	-	-	-	-	-	C,p C	
		-1.468 -		-	-	-	-	-	-	-	-	C,p C	
		-1.51 -		-	-	QE	0.62	-	-	-	-	C,p C	
		-1.486 -		-	-	-	-	-	-	-	-	C,p C	
JE052 0229	13 f	-	-	τ_r/τ_f	0.3F	-	-	-	-	-	-	A,XEl=Pt coated with Ag(Hg),p	
		-	-		0.2F	-	-	-	-	-	-	A,p	
		-	-		0.085F	-	-	-	-	-	-	A,p	
JE052 0229	13 f	-	-	-	-	-	-	ΔE_p	60 -	-	-	C,XEl=Pt coated with Ag(Hg),p A	
JE034 0543	-	$E_{\frac{1}{2}}$ -0.54F	SCE	i_ℓ	1.7F	QE	5.8 ±0.2	αn_a	1.9	Elog	CP→3-aminooxindole, UVS	C,≠,Tc=1.6 for T=15-30,Mx,r	EC57
		-0.72F			1.9F	-	-	-	-	-	-	C,2 merging waves,p	
		-0.67F -0.78F				-	-	-	-	-	-	C,p C	
	389 a	0.68r -0.80F									-	C,µ C	
		-0.7F			0.60F	QE	2		1.9		Isatin-3-hydrazone	C,r	
		-0.83F			1.0F		4		0.9		3-aminooxindole	C,$\Sigma i_\ell \neq f(pH)$,$\Sigma i_\ell=kC$ for C=0.01-0.5,$i_\ell=kh^{\frac{1}{2}}$,$i_\ell \uparrow$ on adding isatin-3-hydrazone	
		-0.74F -0.89F		-		-	-	-	-	-	-	C,p C	
JE043 0365	266 b	$E_{\frac{1}{2}}$ -0.26 -0.17	SCE	i_ℓ	12.1	PQ i:i	2.75 4	αn_a	0.74 0.84	Elog	-	A,≠,i_d,$i_\ell=kC$ for C=0.1-2,Tc=1.6,r A	EC58 a197
													CONT

EC58 (CONT.) $C_8H_6O_2$

Code No.	Empirical Formula	Name and C.A. Number	Structural Formula	Solvent	Tech.	Medium		μ, M	pH	T, °C	Electrodes	App.	Experimental Parameters
EC58 a197	$C_8H_6O_2$	isophthalaldehyde	3-OHCC_6H_4CHO	EtOH 10	IL	KOH	1	-	-	-	HMDE MSE	2-0	C=1,v=3.3
				MeCN 10	PY	KOH	1	-	-	-	DME/MSE	2-0	C=1,m=2.73, t=3.55,v=3.3
				MeCN 20	PY	KOH	1	-	-	-	DME/MSE	2-0	C=1,m=2.73, t=3.55,v=3.3
EC59 a198	$C_8H_6O_2$	phthalaldehyde C.A. 643-79-8	2-OHCC_6H_4CHO	EtOH 10	PY	KOH	1	-	-	-	DME/MSE	2-0	C=1,m=2.73, t=3.55,v=3.3
				DMF	PY	KClO$_4$	0.1	-	-	20.0 ±0.2	DME/SCE	24A0	C=1
						LiClO$_4$	0.1						
						Et$_4$NClO$_4$	0.1						
						NaClO$_4$	0.1						
EC60 a199	$C_8H_6O_2$	terephthalaldehyde C.A. 623-27-8	4-OHCC_6H_4CHO	EtOH 10	PY	KOH	1	-	-	-	DME/MSE	2-0	C=1,m=2.73, t=3.55,v=3.3
				MeCN 10	PY	KOH	1	-	-	-	DME/MSE	2-0	C=1,m=2.73, t=3.55,v=3.3
				MeCN 20	PY	KOH	1	-	-	-	DME/MSE	2-0	C=1,m=2.73, t=3.55,v=3.3
				MeCN 30	PY	KOH	1	-	-	-	DME/MSE	2-0	C=1,m=2.73, t=3.55,v=3.3
EC61	$C_8H_6O_3$	2-formylbenzoic acid C.A. 119-67-5	2-OHCC_6H_4COOH	EtOH 10	PY	KOH	1	-	-	-	DME/MSE	2-0	C=1,m=2.73, t=3.55,v=3.3
EC62	$C_8H_6O_3$	4-formylbenzoic acid C.A. 619-66-9	4-OHCC_6H_4COOH	EtOH 10	PY	KOH	1	-	-	-	DME/MSE	2-0	C=1,m=2.73, t=3.55,v=3.3
EC63	$C_8H_7NO_2$	1,3-dihydro-3-hydroxy-2H-indol-2-one C.A. 61-71-2	Table II	DMF	PY	Bu$_4$NClO$_4$	0.5	-	-	-	DME/MP	0-0	C=1
EC64	$C_8H_7NO_2$	phenylglyoxal aldoxime C.A. 532-54-7	C_6H_5COCH:NOH	H$_2$O	PY	BR		-	1.8	23.0 ±0.1	DME/SCE	23A0	C=0.2,EC= 1.875,h=80, MP Aux
									2.9				
									4.7				
									6.0				
CONT													

TABLE I. Electrochemical Data $C_8H_7NO_2$ (CONT.)   EC64

Ref.	C/M	Charact. Potential		Response Const.		n Tech.		Electrokinetic Data			Products and Identification	Description and Remarks	Code No.
		Value	vs.		Value			Parameter	Value	From			
JE043 0365	266 b	E_p	-0.7F	MSE	i_p	2.4F	-	-	-	-	-	-	$A,S,E_p \to$ more pos. on later scans, r	EC58 a197
			-0.63F			1.9F							$A,E_p \to$ more pos. and $i_p \downarrow$ on later scans	
JE043 0365	266 b	$E_{\frac{1}{2}}$	-0.33 -0.24	SCE	-	-	-	-	α	0.44 0.65	Elog	-	A,r A	
JE043 0365	266 b	$E_{\frac{1}{2}}$	-0.265 -0.185	SCE	-	-	-	-	-	-	-	-	A,r A	
JE043 0365	266 b	$E_{\frac{1}{2}}$	-0.245	SCE	i_ℓ	3.9	i:i	2	αn_a	0.94	Elog	-	$A, \not\models, i_d, i_\ell = kC$ for C= 0.1-2, Tc=1.6, r	EC59 a198
JE042 0253	-	$E_{\frac{1}{2}}$	-1.337	SCE	-	-	-	-	-	-	-	-	$C, E_{\frac{1}{2}} \to$ more pos. as $[K^+] \uparrow$, p C	
			-											
			-1.275		-	-	-	-	-	-	-	-	$C, E_{\frac{1}{2}} \to$ more pos. as $[Li^+] \uparrow$, p C	
			-											
			-1.37		-	-	QE		Elog	0.97	-	-	$C, R, E_{\frac{1}{2}} \neq f[Et_4N^+]$, p C	
			-							-				
			-1.329		-	-	-	-	$dE_{\frac{1}{2}}/dlog[Na^+]$	21	-	-	$C, E_{\frac{1}{2}} \to$ more pos. as $[Na^+] \uparrow$, p C	
			-							-				
JE043 0365	266 b	$E_{\frac{1}{2}}$	-0.305 -0.195	SCE	i_ℓ }	12.2	PQ i:i	2.75 4	αn_a	0.74 0.68	Elog	-	$A, \not\models, i_d, i_\ell = kC$ for C= 0.1-2, Tc=1.6, r A	EC60 a199
JE043 0365	266 b	$E_{\frac{1}{2}}$	-0.36 -0.25	SCE	-	-	-	-	α	0.55 0.48	Elog	-	A,r A	
JE043 0365	266 b	$E_{\frac{1}{2}}$	-0.285 -0.195	SCE	-	-	-	-	α	0.43 0.52	-	-	A,r A	
JE043 0365	266 b	$E_{\frac{1}{2}}$	-0.215 -	SCE	-	-	-	-	-	-	-	-	A,r A	
JE043 0365	-	-	-	-	-	-	-	-	-	-	-	-	A,O,p	EC61
JE043 0365	266 b	$E_{\frac{1}{2}}$	-0.24	SCE	i_ℓ	6.12	PQ i:i	1.58 2	αn_a	1.04	Elog	-	$A, \not\models, i_d, i_\ell = kC$ for C= 0.1-2, Tc=1.6, r	EC62
JE045 0397	-	$E_{\frac{1}{2}}$	-2.24	MP	$i_\ell(u)$	15F	-	-	-	-	-	-	C,p	EC63
JE047 0335	414 a	$E_{\frac{1}{2}}$	-0.2F	SCE	-	-	-	-	-	-	-	-	C, i_d, r	EC64
			-0.91F										C, i_d, Mx	
			-0.25F			2.3F	-	-	-	-	-	-	C, i_d, r	
			-0.96F			1.1F							C, i_d	
			-0.35F			-	-	-	-	-	-	-	C, i_d, r	
			-1.03F										C, i_d	
			-0.45F			-	-	-	-	-	-	-	C, i_d, r	
			-1.10F										C, i_d	

CONT

133

EC64 (CONT.) $C_8H_7NO_2$

Code No.	Empirical Formula	Name and C.A. Number	Structural Formula	Solvent	Tech.	Medium	μ, M	pH	T, °C	Electrodes	App.	Experimental Parameters
EC64	$C_8H_7NO_2$	phenylglyoxal aldoxime	$C_6H_5COCH:NOH$	H_2O	PY	BR	—	6.3	23±0.1	DME/SCE	23A0	C=0.2, EC=1.875, h=80 MP Aux
								7.15				
								8.8				
								9.3				
								10.0				
								10.5				
								11.6				
EC65	$C_8H_8N_2O$	phenylglyoxal 2-hydrazone C.A. 20292-75-5	$C_6H_5COCH:NNH_2$	H_2O	PY	BR	—	1.8	23.0±0.1	DME/SCE	23A0	C=0.2, EC=1.875, h=80, MP Aux
								2.9				
								3.6				
								4.8				
								7.15				
								8.5				
								10.5				
								11.6				
	$C_8H_7NO_2$		$C_6H_5COCH:NOH$	H_2O	PY							

TABLE I. Electrochemical Data $C_8H_8N_2O$ EC65

Ref.	C/M	Charact. Potential		Response Const.		Tech.	n	Electrokinetic Data			Products and Identification	Description and Remarks	Code No.
		Value	vs.		Value			Parameter	Value	From			
JE047 0335	414 a	$E_{\frac{1}{2}}$ -0.5F	SCE	i_ℓ	2.3F	QE	4.0 ± 0.2	-	-	-	CP → 3-amino-1-phenylpropanone, UVS, λ_{max}=250	C, i_d, r	EC64
		-1.1F			1.0F		≈2				CP → NH_3, acetophenone, UVS, λ_{max}=247	C, i_d, Mx	
		-1.5F			-		-				-	$C, i_k + i_d, Mx$	
		-0.55F			2.3F							C, r	
		-1.16F			1.2F							C, Mx	
		-1.52F			0.5F							C, Mx	
		-0.65F			-		-				-	C, i_d, r	
		-1.25F										C, i_d	
		-1.55F										$C, i_k + i_d$	
		-0.70F			-		-				-	C, r	
		-1.29F										C	
		-1.55F										C	
	414 b	-0.77F			-		-				-	C, i_d, r	
		-1.36F										C, i_d	
		-1.57F										$C, i_k + i_d$	
		-0.87F			2.1F		-				-	C, i_d, r	
		-1.43F			0.9F							C, i_d	
		-1.6F			0.2F							$C, i_k + i_d$	
		-0.93F			-		-				-	C, i_d, r	
		-1.52F										C, i_d	
		-1.63F										$C, i_k + i_d$	
JE047 0335	390 a	$E_{\frac{1}{2}}$ -0.18F	SCE	-	-	QE	2	-	-	-	CP → phenylgloxal, UVS, PY	C, i_d, r	EC65
		-					-				-	C, redn. of phenylglyoxal	
		-0.92F										C, i_d	
		-0.25F		i_ℓ	1.7F		-				-	C, i_d, r	
		-0.84F			0.5F							C	
		-0.95			0.9F							C, i_d	
		-0.34F			-		-				-	CP → phenylglyoxal, UVS, PY	C, i_d, r	
		-											C, i_d	
	390 b	-0.42F			-		4	-	-	-	CP → 3-amino-1-phenylpropanone, UVS, λ_{max}=250	C, i_d, r	
		-1.05F					≈2				CP → NH_3, acetophenone, UVS, λ_{max}=247	C, i_d	
		-1.37F					-					$C, i_k + i_d$	
		-0.55F			2.3F		-	-	-	-		C, i_d, r	
		-1.15F			1.1F							C	
		-1.51F			0.5F							C	
		-0.67F			2.27F		-	-	-	-		C, i_d, r	
		-1.22F			1.07F							C, Mx, i_d	
		-1.57F			0.5F							$C, Mx, i_k + i_d$	
	390 c	-0.77F			2.27F		-	-	-	-		C, i_d, r	
		-1.4F			0.83F							C, i_d	
		-1.6F			0.3F							$C, i_k + i_d$	
		-0.87F			-		-				-		C, i_d, r	
		-1.5F											C, i_d	
		-1.65F											$C, i_k + i_d$	

EC66 C_8H_8O

Code No.	Empirical Formula	Name and C.A. Number	Structural Formula	Solvent	Tech.	Medium		μ, M	pH	T, °C	Electrodes	App.	Experimental Parameters
EC66 dc28	C_8H_8O	phenylacetaldehyde C.A. 122-78-1	$C_6H_5CH_2CHO$	EtOH 2	PY	LiOH LiCl	0.1 0.4	-	-	25.00 ±0.02	DME Hg/Hg_2Cl_2, LiCl 2	O34A 0	C=0.2, m=2.6, ±0.4, t=3.6± 0.5, h=65±5
				EtOH 15	PY	BARB LiCl		0.5	8	25.00 ±0.02	DME/SCE	O34A 0	C=0.2, m=2.6± 0.4, t=3.6± 0.5, h=65±5
						BOR LiCl			9.8				
						LiOH LiCl			12.3				
									14				
EC67	C_8H_8O	3-tolualdehyde C.A. 620-23-5	$3\text{-}CH_3C_6H_4CHO$	EtOH 10	PY	HCl	0.1	-	1.1	-	DME/SCE	2AO	C=0.03, t=0.25
													C=0.03, t=3.0
													C=0.3, t=0.25
													C=0.3, t=3.0
													C=2.0, t=0.25
													C=3.0, t=3.0
						NaOH	0.1		13				C=0.03, t=3.0
													C=0.2, t=0.17
													C=0.2, t=2.0
													C=0.2, t=0.48
													C=0.3, t=3.0
													C=3.0, t=3.0
													C=10.0, t=3.0
EC68	$C_8H_8O_2$	2-anisaldehyde C.A. 135-02-4	$2\text{-}CH_3OC_6H_4CHO$	EtOH 10	PY	KOH	1	-	-	-	DME/MSE	2-O	C=1, m=2.73, t=3.55, v=3.3
EC69 am70 cd45	$C_8H_8O_2$	2,5-dimethyl-1,4-benzoquinone C.A. 137-18-8	Table II-2	2-PrOH 40	PY	PHOS		-	7	-	DME/NCE	---	-
EC70 am79	$C_8H_8O_3$	vanillin C.A. 121-33-5	Table II-2	EtOH 10	PY	KOH	1	-	-	-	DME/MSE	2-O	C=1, m=2.73, t=3.55, v=3.3
EC71 cd54 dc33	$C_8H_9NO_3$	pyridoxal C.A. 66-72-8	Table II-3	H_2O	PY	NaOH	0.1	-	-	-	DME/SCE	O4AO	C=0.5
EC72 cd63 dc36	$C_8H_{10}NO_6P$	pyridoxal 5-phosphate C.A. 54-47-7	Table II-3	H_2O	PY	NaOH	0.1	-	-	-	DME/SCE	O4AO	C=0.5
EC73	$C_8H_{10}N_2O_3$	ethyl 2-methyl-4-hydroxypyrimidine-5-carboxylate C.A. 53135-24-3	Table II	H_2O	PY	CITR		-	4.5	28.0 ±0.2	DME/SCE	OAO	C=0.1, EC=1.3

TABLE I. Electrochemical Data $C_8H_{10}N_2O_3$ EC73

Ref.	C/M	Charact. Potential		Response Const.		Tech.	n	Electrokinetic Data			Products and Identification	Description and Remarks	Code No.
		Value	vs.		Value			Parameter	Value	From			
JE046 0323	-	$E_{\frac{1}{2}}$	-1.75F	SCE	-	-	-	-	-	-	-	-	$C, \Delta E_{\frac{1}{2}}=0.2\sigma^*, r$	EC66 dc28
JE046 0343	32 e	-	-	-	i_ℓ	0.05	-	-	-	-	-	-	C,r	
		-	-	-	-	0.73	-	-	-	-	-	-	A,r C	
		-	-	-	-	1.4	-	-	-	-	-	-	$A, i_\ell \neq f(pH)$ for pH > 12, r C	
		-	-	-	-	0.7	-	-	-	-	-	-	A,r C	
C0030 4219	65 v	$E_{\frac{1}{2}}$	-0.896F	SCE	-	-	-	-	-	-	-	-	C,r	EC67
			-0.890F		-	-	-	-	-	-	-	-	C,r	
			-0.870F		-	-	-	-	-	-	-	-	C,r	
			-0.870F		-	-	-	-	-	-	-	-	C,r	
			-0.872F		-	-	-	-	-	-	-	-	C,r	
			-0.883F		-	-	-	-	-	-	-	-	C,r	
	65 u		-1.490F		-	-	-	-	-	-	-	-	C,r	
			-1.497F		-	-	-	-	-	-	-	-	C,r	
			-1.462F		-	-	-	-	-	-	-	-	C,r	
			-1.480F		-	-	-	-	-	-	-	-	C,r	
			-1.466F		-	-	-	-	-	-	-	-	C,r	
			-1.441F		-	-	-	-	-	-	-	-	C,r	
			-1.430F		-	-	-	-	-	-	-	-	C,r	
JE043 0365	266 b	$E_{\frac{1}{2}}$	-0.175	SCE	i_ℓ	6.5	i:i	2	αn_a	1.4	Elog	-	$A, \neq, i_d, i_\ell=kC$ for C=0.1-2, Tc=1.6, r	EC68
C0030 4192	-	-	-	-	-	-	-	-	-	-	-	-	see AM70, CD45	EC69 am70 cd45
JE043 0365	-	-	-	-	-	-	-	-	-	-	-	-	A,O,p	EC70 am79
C0027 0486	-	$E_{\frac{1}{2}}$	-1.25	SCE	-	-	-	-	-	-	-	-	C,p	EC71 cd54 dc33
C0027 0486	-	$E_{\frac{1}{2}}$	-1.630	SCE	-	-	-	-	-	-	-	-	C,p	EC72 cd63 dc36
JE055 0445	-	$E_{\frac{1}{2}}$	-1.2	SCE	-	-	i:i	2	-	-	-	-	$C, W, i_d, Tc=1.6, E_{\frac{1}{2}} \neq f(pH)$ for pH=2-10, p	EC73

EC74 $C_8H_{10}N_2O_3S$

Code No.	Empirical Formula	Name and C.A. Number	Structural Formula	Solvent	Tech.	Medium		μ, M	pH	T, °C	Electrodes	App.	Experimental Parameters
EC74	$C_8H_{10}N_2O_3S$	ethyl 2-methylthio-4-hydroxypyrimidine-5-carboxylate C.A. 53554-29-3	Table II	H_2O	PY	CITR		–	4.5	28.0 ±0.2	DME/SCE	OAO	C=0.1, EC=1.3
EC75	$C_8H_{10}N_2O_4S$	N,N-dimethyl-4-nitrobenzenesulfonamide C.A. 17459-03-9	$4\text{-}O_2NC_6H_4SO_2N\text{-}(CH_3)_2$	DMF	VA	Et_4NClO_4	0.1	–	–	22.5 ±0.5	Pt/SCE	256A FO	$0 \to -2.4 \to$ OF, C=1.03, A=0.25, v=100, Pt Aux
EC76	$C_8H_{10}OS$	4-methylphenyl methyl sulfoxide C.A. 934-72-5	$4\text{-}CH_3C_6H_4SOCH_3$	EtOH	PY	Et_4NBr	0.1	–	–	30.0 ±0.5	DME/SCE	2-O	C=0.5, t(c)=3
				EtOH 80 DMF 20	PY	Et_4NBr	0.1	–	–	30.0 ±0.5	DME/SCE	2-O	C=0.5, t(c)=3
				EtOH 80 dioxane 20	PY	Et_4NBr	0.1	–	–	30.0 ±0.5	DME/SCE	2-O	C=0.5, t(c)=3
				MeCN	PY	Et_4NBr	0.1	–	–	30.0 ±0.5	DME/SCE	2-O	C=0.5, t(c)=3
				DMF	PY	Et_4NBr	0.1	–	–	30.0 ±0.5	DME/SCE	2-O	C=0.5, t(c)=3
EC77	$C_8H_{10}O_2$	anisyl alcohol C.A. 105-13-5	$CH_3OC_6H_4CH_2OH$	MeCN	VY	$NaClO_4$	0.5	–	–	–	RPDE Ag/ $AgClO_4$ 0.01	-AO	C=1, v=100, ω=38.2
						$NaClO_4$ PYR	0.5 0.01						
				CH_2Cl_2	CT	Bu_4NBF_4	0.5	–	–	–	RPDE Ag/ $AgClO_4$ 0.01	-AO	C=5, v=300, ω=9.5
EC78 an64	$C_8H_{10}O_2$	1,4-dimethoxybenzene C.A. 150-78-7	$4\text{-}H_3COC_6H_4OCH_3$	$MeNO_2$	VY	$AlCl_3$	1	–	–	–	RPDE Ag/AgCl, Me_4NCl satd/Fc	2AO	C=ns, A=0.01
						Et_4NClO_4	0.1						
EC79	$C_8H_{10}O_2S$	4-methoxyphenyl methyl sulfoxide C.A. 3517-99-5	$4\text{-}CH_3OC_6H_4SOCH_3$	EtOH	PY	Et_4NBr	0.1	–	–	30.0 ±0.5	DME/SCE	2-O	C=0.5, t(c)=3
				EtOH 80 DMF 20	PY	Et_4NBr	0.1	–	–	30.0 ±0.5	DME/SCE	2-O	C=0.5, t(c)=3
				EtOH 80 dioxane 20	PY	Et_4NBr	0.1	–	–	30.0 ±0.5	DME/SCE	2-O	C=0.5, t(c)=3
				MeCN	PY	Et_4NBr	0.1	–	–	30.0 ±0.5	DME/SCE	2-O	C=0.5, t(c)=3
				DMF	PY	Et_4NBr	0.1	–	–	30.0 ±0.5	DME/SCE	2-O	C=0.5, t(c)=3

TABLE I. Electrochemical Data $C_8H_{10}O_2S$ EC79

Ref.	C/M	Charact. Potential		Response Const.		n Tech.	n	Electrokinetic Data			Products and Identification	Description and Remarks	Code No.
		Value	vs.		Value			Parameter	Value	From			
JE055 0445	-	$E_{\frac{1}{2}}$	-1.05	SCE	-	-	i:i	2	-	-	-	-	$C,W,i_d,Tc=1.6,E_{\frac{1}{2}}\neq f(pH)$ for pH=2-10,p	EC74
JE053 0293	-	E_p	-0.92F	SCE	i_p	47F	IR	1	-	-	-	radical anion	C,R,r	EC75
			-1.64F			45F		1				dianion	C,R	
			-1.60F			50F		-				-	A,R	
			-0.84F			47F							A,R	
JE054 0221	47 c	$E_{\frac{1}{2}}$	-2.23	SCE	-	-	sttd	2	αn_a $k_{s,h}$	0.81 5E-30	Kplot	-	$C,\neq,i_d,\Delta E_{\frac{1}{2}}=3.6\sigma,r$	EC76
JE054 0221	47 c	$E_{\frac{1}{2}}$	-2.235	SCE	-	-	sttd	2	-	-	-	-	$C,\neq,i_d,\Delta E_{\frac{1}{2}}=3.7\sigma,r$	
JE054 0221	47 c	$E_{\frac{1}{2}}$	-2.33	SCE	-	-	sttd	2	-	-	-	-	$C,\neq,i_d,\Delta E_{\frac{1}{2}}=3.4\sigma,r$	
JE054 0221	47 d	$E_{\frac{1}{2}}$	-2.35	SCE	-	-	sttd	2	-	-	-	-	$C,\neq,i_d,\Delta E_{\frac{1}{2}}=2.0\sigma,r$	
JE054 0221	47 d	$E_{\frac{1}{2}}$	-2.395	SCE	-	-	sttd	2	-	-	-	-	$C,\neq,i_d,\Delta E_{\frac{1}{2}}=3.8\sigma,r$	
JE034 0505, JA094 6812	401 b	$E_{\frac{1}{2}}$	1.30	Ag/ AgClO₄ 0.01	j_ℓ	5E3F	-	-	-	-	-	CP at 1.3V (in 0.1M LiClO₄) → 4-methoxy-benzaldehyde, NMR, GLC, IRS	$A,\neq,S,r$	EC77
			1.8			4E3F							$A,\neq,M$	
			1.25F 1.73F			250F 180F	-	-	-	-	-	-	A,r A,a.s.,O with [PYR]=0.25	
JE034 0505	401 b	E_p	-	Ag/ AgClO₄ 0.01	j_p	small	-	-	-	-	-	-	A,Pr,r	
			1.83F -			5.6E3F small							A C,related to first anodic peak	
JE039 0385	-	$E_{\frac{1}{2}}$	0.93	Fc	-	-	-	-	-	-	-	-	A,r	EC78 an64
			0.93	-	-	-	-	-	-	-	-	-	A,r	
JE054 0221	47 c	$E_{\frac{1}{2}}$	-2.295	SCE	-	-	sttd	2	αn_a $k_{s,h}$	0.6 2.5E-25	Kplot	-	$C,\neq,i_d,\Delta E_{\frac{1}{2}}=3.6\sigma,r$	EC79
JE054 0221	47 c	$E_{\frac{1}{2}}$	-2.30	SCE	-	-	sttd	2	-	-	-	-	$C,\neq,i_d,\Delta E_{\frac{1}{2}}=3.7\sigma,r$	
JE054 0221	47 c	$E_{\frac{1}{2}}$	-2.355	SCE	-	-	sttd	2	-	-	-	-	$C,\neq,i_d,\Delta E_{\frac{1}{2}}=3.4\sigma,r$	
JE054 0221	47 d	$E_{\frac{1}{2}}$	-2.385	SCE	-	-	sttd	2	-	-	-	-	$C,\neq,i_d,\Delta E_{\frac{1}{2}}=2.0\sigma,r$	
JE054 0221	47 d	$E_{\frac{1}{2}}$	-2.44	SCE	-	-	sttd	2	-	-	-	-	$C,\neq,i_d,\Delta E_{\frac{1}{2}}=3.8\sigma,r$	

Code No.	Empirical Formula	Name and C.A. Number	Structural Formula	Solvent	Tech.	Medium		μ, M	pH	T, °C	Electrodes	App.	Experimental Parameters
EC80 an81 cd71	$C_8H_{11}N$	N,N-dimethylaniline C.A. 121-69-7	$C_6H_5N(CH_3)_2$	H_2O	VY	ACET PHOS		-	0.7 2.4 3.8 5.1 6.2 9.0	-	Pt PDS SCE	2-0	C=0.5, v=0.92
				$AlCl_3$ 50 NaCl 50	VA	Cl^-	0.083	-	-	175	W button E Al		C=2.88, v=500
					VR	ns		-	-	175	W button E Al	2-2	C=2.14, v=5E2
EC81	$C_8H_{11}NO_2$	N-butylmaleimide C.A. 2973-09-3	Table II	EtOH 50	PY	BR		-	2.48 4.24 5.34 6.12 7.25 9.10	-	DME/SCE	04A0	C=1.0, m=3.2, t=3.5, h=40
EC82	$C_8H_{11}NO_3$	norepinephrine C.A. 138-65-8	Table II	H_2O	VA	$HClO_4$	2	-	-	25	CPE/SCE	---	C≈1, v=100
EC83	$C_8H_{11}N_3$	1-phenyl-3,3-dimethyltriazene C.A. 7227-91-0	$C_6H_5N:NN(CH_3)_2$	EtOH	PY	CITR AM		-	2.9 3.9 4.7 5.6 8.2 9.2	-	DME/SCE (?)	04A0	C=0.2, h=60
EC84 an94	$C_8H_{11}N_3O_6$	6-azauridine C.A. 54-25-1	Table II-2	H_2O	PO	H_2SO_4 $HCOONH_4$ NaOH	1.0 1.0 1.0	-	-	-	SME/-	OB2	C=0.5

TABLE I. Electrochemical Data $C_8H_{11}N_3O_6$ EC84

| Ref. | C/M | Charact. Potential | | Response Const. | | n Tech. | n | Electrokinetic Data | | | Products and Identification | Description and Remarks | Code No. |
		Value	vs.		Value			Parameter	Value	From			
JE052 0093	–	$E_{p/2}$	0.82F	SCE	–	–	–	–	–	–	–	–	A,p	EC80 an81 cd71
			0.75F		–	–	–	–	–	–	–	–	A,p	
			0.68		–	–	–	–	–	–	–	–	A,p	
			0.60F		–	–	–	–	–	–	–	–	A,p	
			0.58F		–	–	–	–	–	–	–	–	A,p	
			0.57F		–	–	–	–	–	–	–	–	A, $E_{p/2} \neq f(pH)$ for pH > 7.5, p	
JE049 0281	147 i	E_p	1.24F	Al	$i_p(u)$	30F	–	–	–	–	–	–	A, $E_p \to$ more pos. as $AlCl_3/NaCl \downarrow$, r	
			1.20F			3F							C, radical cation redn., $i_p \uparrow$ as $AlCl_3/NaCl \downarrow$	
			1.0F			12F							C, $i_p \downarrow$ as $AlCl_3/NaCl \downarrow$	
JE038 0476	–	E_p	1.26F	Al	i_p	8.4F	–	–	–	–	–	–	A,p	
			1.16F			0.6F						4,4'-bis(dimethyl-amino)biphenyl	C, radical cation redn.	
			0.98F			2.6F							C,R	
			1.05			0.8F							A,R,O on first scan	
C0026 2749	–	$E_{\frac{1}{2}}$	-0.48F	SCE	i(u)	21F	sttd	2	–	–	–	–	C, i_d, r	EC81
			-0.72F			19F		–	–	–	–	–	C, i_d, r	
			-0.75F -1.30F			16F 4F	}	2	–	–	–	–	C,r C	
			-0.79F -1.35F			11F 8F	}	2	–	–	–	–	C,r C	
			-0.818 -1.290			–		1 1	–	–	–	–	C, i=f(t), r C, i=f(t)	
			-0.80F -1.35F			10F 10F		1 1	–	–	–	–	C, i=f(t), r C, i=f(t)	
JE042 0397	–	E_p	0.55F 0.48F	SCE	i_p	150F 130F	–	–	–	–	–	–	A,p C	EC82
C0034 1413	–	$E_{\frac{1}{2}}$	-0.78F	SCE(?)	i_ℓ	small	–	–	–	–	–	–	C,p	EC83
			-0.89F			–	–	–	–	–	–	–	C,dr,p	
			-0.96F			–	–	–	–	–	–	–	C, i_d, p	
			-1.00F			–	–	–	–	–	–	–	C, i_d, p	
			-1.12F			–	–	–	–	–	–	–	C,Mn,p	
			-1.18F			–	–	–	–	–	–	–	C,Mn,p	
C0029 0182	–	Q	0.67	–	–	–	–	–	–	–	–	–	C,p	EC84 an94
			0.73	–	–	–	–	–	–	–	–	–	C,p	
			0.14	–	–	–	–	–	–	–	–	–	C,p	

141

Code No.	Empirical Formula	Name and C.A. Number	Structural Formula	Solvent	Tech.	Medium	μ, M	pH	T, °C	Electrodes	App.	Experimental Parameters
EC85	$C_8H_{13}N_3O_5$	diethyl mesoxalate semicarbazone C.A. 31909-48-5	$[CH_3CH_2OC(O)]_2$-C:$NNHCONH_2$	H_2O	PY	MB	-	2.0	-	DME/SCE	OAO	C=0.5(?),m=2.62,t=2.8, h=55
								3.3				
								5.0				
								7.0				
								10.0				
EC86	$C_8H_{14}O$	cyclooctanone C.A. 502-49-8	Table 11	EtOH(?) 75	PY	Et_4NClO_4	0.1	-	-	DME/SCE	2-0	C=1
				EtOH 90	PY	Et_4NClO_4	0.1	-	20	DME/SCE	2-0	C=0.5
EC87	$C_8H_{14}O$	2-ethyl-2-hexenal C.A. 645-62-5	$CH_3(CH_2)_2CH$:$C(CH_2CH_3)CHO$	H_2O	PY	BOR KCl	1.0 ±0.2	2	25.00 ±0.05	DME/SCE	OAO	ns
								10				
EC88 dc45	$C_8H_{14}O_4S_2$	glycol dimercaptopropionate C.A. 22504-50-3	$(HSCH_2CH_2COO$-$CH_2)_2$	H_2O	PY	AM Co^{2+}	0.1 1E-3	-	20.0 ±0.1	DME/SCE	OAO	C=0.05-3, m=1.95, t(oc)=3.22, h=60
				EtOH 10	PY	ACET KNO_3	0.1	3.54	30	DME/SCE	OAO	C=1,EC=1.69 (-0.2 V)
						CL KNO_3	0.1	11.6				
EC89 dc49	$C_8H_{16}N_2$	2,3-dimethyl-2,3-diazabicyclo[2.2.2]-octane C.A. 14287-92-4	Table 11-4	MeCN	VR	Bu_4NClO_4	0.1	-	-	DME?/SCE	---	ns
EC90	C_8H_{18}	octane C.A. 111-65-9	$CH_3(CH_2)_6CH_3$	MeCN	VY	Et_4NBF_4	0.1	-	25	RPDE Ag/Ag$^+$ 0.01	2-0	C≈1,A=0.12, v=50,ω=16.3, Pt Aux
				CH_2Cl_2	VY	Bu_4NBF_4	0.2	-	25	RPDE Ag/Ag$^+$ 0.01	2-0	C≈1,A=0.12, v=50,ω=16.3, Pt Aux
				$EtNO_2$	VY	Et_4NBF_4	0.2	-	25	RPDE Ag/Ag$^+$ 0.01	2-0	C≈1,A=0.12, v=50,ω=16.3, Pt Aux
				$MeNO_2$	VY	Et_4NBF_4	0.1	-	25	RPDE Ag/Ag$^+$ 0.01	2-0	C≈1,A=0.12, v=50,ω=16.3, Pt Aux
				CF_3COOH	IL	FSO_3H	2.5	-	-	Pt/SCE	25-2	C=30,v=50, Pt Aux
EC91 ao63 CONT	C_9H_8ClN	6-chloroquinoline C.A. 612-57-7	Table 11-2	N-methylpyrrolidone	PY	Et_4NClO_4	0.1	-	-	DME Hg/Hg$^+$ 0.02, $HClO_4$ 0.1	2AO	C=1-2

TABLE I. Electrochemical Data C_9H_8ClN (CONT.) EC91

Ref.	C/M	Charact.	Potential Value	vs.	Response	Const. Value	n Tech.		Electrokinetic Data Parameter	Value	From	Products and Identification	Description and Remarks	Code No.
C0036 0331	-	$E_{\frac{1}{2}}$	-0.62F	SCE	-	-	i:i	2	$dE_{\frac{1}{2}}/dpH$	56F	-	-	C, i_d, p	EC85
			-0.70F		-	-		2		56F			C, i_d, p	
			-0.81F		-	-		2		56F			C, i_d, p	
			-0.91F		-	-		2		56F			C, i_d, p	
			-1.07F		-	-		2		56F			C, i_d, p	
JE035 0363	-	$E_{\frac{1}{2}}$	-2.43	SCE	-	-	-	-	-	-	-	-	C, p	EC86
JE039 0195	-	$E_{\frac{1}{2}}$	-2.43	SCE	-	-	-	-	-	-	-	-	C, p	
JE049 0433	-	$E_{\frac{1}{2}}$	-1.03c	-	I	0.13	-	-	$dE_{\frac{1}{2}}/dpH$ p	71±6 1.2±0.2	sttd $dE_{\frac{1}{2}}/dpH$	-	$C, E_{\frac{1}{2}} = -0.886-0.071$ (± 0.006 V)pH for pH= 2-10, p	EC87
			-1.60c			0.80	QE	0.78		71±6 1.2±0.2		-	$C, i_\ell = kh^{0.61}, p$	
JE040 0448	-	E_{max}	-1.4	-	-	-	-	-	-	-	-	-	$C, i_{cat}, i_{max} = f[Co^{2+}] = kh^0 = k_1C$ K_2, p	EC88 dc45
JE036 0515	-	$E_{\frac{1}{2}}$	-0.36F	SCE	i_ℓ	0.4F	-	-	-	-	-	-	$A, Pr, 0$ for $pH > 10$, i_a, p	
			-0.25F		I i_ℓ	2.92 5.0F	Elog (!)	2	Elog	32	-	-	$A, R, E_{\frac{1}{2}} \rightarrow$ more neg. as pH ↑ for pH < 8.2, Tc=1.22, i_d	
		-	-	-	-	-	-	-	k α	2.8E-3 0.583	Kplot	-	A, p	
JA095 6454	-	$E_{p/2}$?	0.10	SCE	-	-	-	-	-	-	-	-	A, p	EC89 dc49
JE042 0133	-	$E_{\frac{1}{2}}$	2.70	Fc	-	-	-	-	-	-	-	-	A, p	EC90
JE042 0133	-	$E_{\frac{1}{2}}$	2.54	Fc	-	-	-	-	-	-	-	-	A, p	
JE042 0133	-	$E_{\frac{1}{2}}$	2.59	Fc	-	-	-	-	-	-	-	-	A, p	
JE042 0133	-	$E_{\frac{1}{2}}$	2.67	Fc	-	-	-	-	-	-	-	-	A, p	
JE054 0181	-	$E_{p/2}$	2.21	SCE	-	-	-	-	-	-	-	-	$A, \rightleftharpoons, p$	
JE040 0345	-	$E_{\frac{1}{2}}$	-2.31	Hg/Hg^+ 0.02, $HClO_4$ 0.1	-	-	-	-	-	-	-	-	C, p	EC91 a063
			-2.56										C	CONT

143

Code No.	Empirical Formula	Name and C.A. Number	Structural Formula	Solvent	Tech.	Medium		μ, M	pH	T, °C	Electrodes	App.	Experimental Parameters
EC91 ao63	C_9H_6ClN	6-chloroquinoline	Table II-2		VY	Et_4NClO_4	0.1	-	-	-	MP Hg/Hg^+ 0.02, $HClO_4$ 0.1	29A0	C=1-2
EC92 ao58	$C_9H_6O_4$	ninhydrin C.A. 485-47-2	Table II-2	H_2O	PY	HCl?		-	1.1	-	DME/SCE	04A0	C=1.0
									1.9				
						ACET			3.8				
									5.6				
						PHOS			6.5				
									7.5				
						Pyrophosphate			8.0				
						$C_6H_5O^-$			10.0				
						PHOS			11.1				
									11.8				
									12.4				
						NaOH			12.4				
EC93 ao79	C_9H_7N	quinoline C.A. 91-22-5	Table II-2	N-methyl-pyrrolidone	PY	Et_4NClO_4	0.1	-	-	-	DME Hg/Hg^+ 0.02, $HClO_4$ 0.1	2A0	C=1-2
					VY	Et_4NClO_4	0.1	-	-	-	MP Hg/Hg^+ 0.02, $HClO_4$ 0.1	29A0	C=1-2
EC94 ao81	C_9H_7NO	8-hydroxyquinoline C.A. 148-24-3	Table II-2	N-methyl-pyrrolidone	PY	Et_4NClO_4	0.1	-	-	-	DME Hg/Hg^+ 0.02, $HClO_4$ 0.1	2A0	C=1-2
					VY	Et_4NClO_4	0.1	-	-	-	MP Hg/Hg^+ 0.02, $HClO_4$ 0.1	29A0	C=1-2
EC95	$C_9H_7NO_2$	N-methylisatin C.A. 2058-74-4	Table II	DMF	PY	Bu_4NClO_4	0.5	-	-	-	DME/MP	0-0	C=1
CONT													

TABLE I. Electrochemical Data $C_9H_7NO_2$ (CONT.) EC95

Ref.	C/M	Charact. Potential		Response Const.		n		Electrokinetic Data			Products and Identification	Description and Remarks	Code No.
		Value	vs.		Value	Tech.		Parameter	Value	From			
JE040 0345	-	$E_{\frac{1}{2}}$	-2.30	Hg/Hg^+ 0.02, $HClO_4$ 0.1	-	-	-	-	-	-	-	-	C,p	EC91 ao63
			-2.55										C	
C0030 4024	381 a	$E_{\frac{1}{2}}$	-0.64F -0.80F	SCE	i_ℓ	14.1F 1.7F	sttd	4 6	-	-	-	-	C,r C	EC92 ao58
			-0.66F -0.80F -0.92F			8.0 5.8 3.2		2 2 6	-	-	-	-	C,r C C	
			-0.81F -0.99F			7.6F 8.9F		2 2	-	-	-	-	C,r C	
			-0.88 -1.15F			8.6F 7.4F		2 2	-	-	-	-	C,r C	
			-0.9F -1.21F			7.5F 8.0F		2 2	-	-	-	-	C,r C	
			0.91F 1.25F			5.7F 10.3F		2 2	-	-	-	-	C,r C	
	381 b		-0.2F -0.5F -0.96F -1.35F			0.8F 2.1F 7.1F 12.9F	-	-	-	-	-	-	C,r C,lx C C,lx	
	381 c		-0.48F -1.10F -1.14F			4.0F 5.9F 9.1F	-	-	-	-	-	-	C,lx,r C,lx C,lx	
			-0.44F -1.10F -1.44F			4.0F 7.0F 6.5F	-	-	-	-	-	-	C,lx,r C,lx C,lx	
			-0.45F -1.15F -1.45F			3.0F 7.2F 5.3F	-	-	-	-	-	-	C,lx,r C,lx C,lx	
			-0.45F -1.18F -1.49F			1.6F 5.4F 3.0F	-	-	-	-	-	-	C,lx,r C,lx C,lx	
			-0.46F -1.18F -1.52F			3.0F 3.0F 2.0F	-	-	-	-	-	-	C,lx,r C,lx C,lx	
JE040 0345	-	$E_{\frac{1}{2}}$	-2.55	Hg/Hg^+ 0.02, $HClO_4$ 0.1	-	-	-	-	-	-	-	-	C,p	EC93 ao79
			-										C	
JE040 0345	-	$E_{\frac{1}{2}}$	-2.61	Hg/Hg^+ 0.02, $HClO_4$ 0.1	-	-	-	-	-	-	-	-	C,p	
JE040 0345	-	$E_{\frac{1}{2}}$	-2.26	Hg/Hg^+ 0.02, $HClO_4$ 0.1	-	-	-	-	-	-	-	$CP \to$ monoanion, $\lambda_{max}=417\pm 2$, 355 ± 5 p	C,i_ℓ=kC for C=0.1-3, p	EC94
			-2.93									$CP \to$ dianion, $\lambda_{max} < 340$	C,i_ℓ=kC for C=0.1-3	
JE040 0345	-	$E_{\frac{1}{2}}$	-2.25	Hg/Hg^+ 0.02, $HClO_4$ 0.1	-	-	-	-	-	-	-	-	C,p	
			-2.87										C	
JE045 0397	394 a	$E_{\frac{1}{2}}$	-0.79	MP	$i_\ell(u)$	15F	-	-	-	-	-	$CP \to$ radical anion, unstable, ESR	C,$i_\ell \uparrow$ on adding H^+ donor,p	EC95
			-1.56			15F						$CP \to$ electroinactive diamagnetic cmpd.	C,$i_\ell \downarrow$ on adding H^+ donor	

CONT

Code No.	Empirical Formula	Name and C.A. Number	Structural Formula	Solvent	Tech.	Medium		μ, M	pH	T, °C	Electrodes	App.	Experimental Parameters
EC95	$C_9H_7NO_2$	N-methylisatin	Table 11	DMF	PA	Bu_4NClO_4	0.5	-	-	-	DME/SCE	0-0	C=1,v=600
EC96 cell	$C_9H_7NO_2$	N-methylphthalimide C.A. 550-44-7	Table 11-3	MeCN 66.7	VY	ACET		-	8	22±2	Pb RDE SCE	-A0	C=40,v=10, ω=35.8
				MeCN 63 H_2SO_4 5	VY	$HClO_4$ $NaClO_4$		-	1	22±2	Pb RDE SCE	-A0	C=40,v=10, ω=15.9
EC97	$C_9H_7N_3OS$	4-(2-thiazolylazo)-phenol C.A. 1823-45-6	Table 11	EtOH 50	PY	ACET	0.2	-	5.65	25	DME/SCE	2-0	C=0.2,m=1.817,t(c)=1,v=5
EC98	$C_9H_7N_3O_2S$	4-(2-thiazolylazo)-resorcinol C.A. 2246-46-0	Table 11	EtOH 50	PY	ACET	0.2	-	5.65	25	DME/SCE	2-0	C=0.2,m=1.817,t(c)=1,v=5
						NaOH	0.2	-					
EC99 ap00 dc72	C_9H_8O	cinnamaldehyde C.A. 104-55-2	$C_6H_5CH:CHCHO$	N-methyl-pyrrolidone	PY	Et_4NClO_4	0.1	-	-	-	DME Hg/Hg^+ 0.02, $HClO_4$ 0.1	2A0	C=1-2
					VY	Et_4NClO_4	0.1	-	-	-	MP Hg/Hg^+ 0.02, $HClO_4$ 0.1	29A0	C=1-2
ED00 ap01	C_9H_8O	1-indanone C.A. 83-33-0	Table 11-2	EtOH 2	PY	H_2SO_4 K_2SO_4	0.05 0.2	-	-	-	DME/SCE	04A0	C=0.2,m=2.1, t=3.02,h=80
						ACET			3.6				
									4.8				
						PHOS			6.00				
									6.44				
									7.02				
						BOR			8.36				
						CARB			10.52				
						PHOS			11.58				
						NaOH			13				
ED01 ap05	$C_9H_8O_4$	acetylsalicylic acid C.A. 50-78-2	Table 11-2	H_2O	CP	PHOS KCl	0.8 0.6	-	-	-	RPDE SCE	---	-
ED02	$C_9H_{10}N_2O$	N^2-hydroxycinnamamidine C.A. 55654-09-6	$C_6H_5CH:CHC-(NOH)NH_2$	EtOH 25	PY	BR		-	2.1	-	DME/SCE	04A0	C=0.5
									3.6				
									4.9				
CONT													

TABLE I. Electrochemical Data $C_9H_{10}N_2O$ (CONT.) ED02

Ref.	C/M	Charact. Potential		Response Const.		n		Electrokinetic Data			Products and Identification	Description and Remarks	Code No.
		Value	vs.		Value		Tech.	Parameter	Value	From			
JE045 0397	394 a	E_p	-0.99F -2.0F -0.93F	MP	$i_p(u)$	14F 16F 9F	-	-	-	-	-	-	C,R,p C A,R	EC95
JE057 0351	411 c	$E_{\frac{1}{2}}$	-1.215 -1.45	SCE	-	-	-	-	-	-	-	-	C,p C	EC96 cell
JE057 0351	411 b	$E_{\frac{1}{2}}$	-1.0	SCE	-	-	-	-	dlogi/dE	60	-	-	C,p	
JE050 0113	-	$E_{\frac{1}{2}}$	-0.255	SCE	D	2.35	i:i, QE	2 4	-	-	-	-	$C, i_\ell \neq f(pH)$ for pH= 1-13, $\Delta E_{\frac{1}{2}}$=0.14σ,p	EC97
JE050 0113	-	$E_{\frac{1}{2}}$	-0.303	SCE	D	2.52	i:i, QE	2 4	-	-	-	-	$C, i_\ell \neq f(pH)$ for pH= 1-8, $\Delta E_{\frac{1}{2}}$=0.14σ,p	EC98
	-	-	-	-	-	-	i:i	3.1	-	-	-	-	C,p	
JE040 0345	-	$E_{\frac{1}{2}}$	-1.97 -2.5L	Hg/Hg^+ 0.02, $HClO_4$ 0.1	-	-	-	-	-	-	-	-	C,p C	EC99 ap00 dc72
JE040 0345	-	$E_{\frac{1}{2}}$	-1.98 -2.30	Hg/Hg^+ 0.02, $HClO_4$ 0.1	-	-	-	-	-	-	-	-	C,p C	
JE056 0285	65 x	$E_{\frac{1}{2}}$	-1.11F -1.19F	SCE	i_ℓ	0.5F 1.2F	-	-	-	-	-	-	C,S,r C,M	ED00 ap01
			-1.32			1.10	PQ	2	$dE_{\frac{1}{2}}/dpH$	60	sttd	-	$C, i_d, dE_{\frac{1}{2}}/dpH$=60 for pH=2.5-6,r	
			-1.4			1.2	-	-		60		-	$C, E_{\frac{1}{2}}$ and $i_\ell \neq f(pH)$ for pH=4.8-5.44,r	
			-1.46			1.15	-	-		0		-	C,r	
			-1.49			0.3	-	-		0		-	C,r	
			-1.54			1.3	-	-		56		-	$C, dE_{\frac{1}{2}}/dpH$=56 for pH=7-9,r	
			-1.62			1.3	-	-		56		-	C,r	
			-1.65			0.8	-	-		0		-	$C, E_{\frac{1}{2}} \neq f(pH)$ for pH > 9.8,r	
			-1.65			0.65	i:i	1		0		-	C, I_d, r	
			-1.65			0.60	-	-		0		-	C,r	
JE048 0167	-	-	-	-	-	-	-	-	-	-	-	acetylsalicylalde- hyde,salicylalde- hyde,GLC	see AP05	ED01 ap05
C0026 2438	-	$E_{\frac{1}{2}}$	-0.89F -0.92F - -0.95F -	SCE	-	-	sttd	2 2 - -	-	-	-	-	C,p C, pK'=4.98, p C, i_{cat} C, pK'=4.98, p C, i_{cat}	ED02 CONT

ED02 (CONT.) $C_9H_{10}N_2O$

Code No.	Empirical Formula	Name and C.A. Number	Structural Formula	Solvent	Tech.	Medium	μ, M	pH	T, °C	Electrodes	App.	Experimental Parameters
ED02	$C_9H_{10}N_2O$	N^2-hydroxycinnamamidine	$C_6H_5CH:CHC-(:NOH)NH_2$	EtOH 25	PY	BR		6.3	-	DME/SCE	04A0	C=0.5
								7.2				
ED03	$C_9H_{10}O$	2-phenylpropanal C.A. 93-53-8	Table II	EtOH 2	PY	BARB LiCl	0.5	7.1	25.00 ±0.02	DME/SCE	034A0	C=0.2, m=2.6± 0.4, t=3.6± 0.5, h=65±5
						BOR LiCl		9.0				
						LiOH LiCl		11.7				
								13				
								14				
ED04 ap31	$C_9H_{10}O$	3-phenylpropanal C.A. 104-53-0	$C_6H_5(CH_2)_2CHO$	EtOH 2	PY	H_2SO_4	ns	-	25.00 ±0.02	DME MSE, H_2SO_4 conc.	034A0	C=0.2, m=2.6± 0.4, t=3.6± 0.5, h=65±5
						BARB LiCl	0.5	7.1		DME/SCE		
						BOR LiCl		9.45				
						LiOH LiCl	ns	11.4		DME Hg/Hg_2Cl_2, LiCl 2		
								12.3				
						NaOH	ns	>12.5				
						LiOH LiCl	0.5	14				
				EtOH 5	PY	BARB LiCl	0.5	7.1	2.00 ±0.02	DME/SCE	034A0	C=0.05, m= 2.6±0.4, t= 3.6±0.5, h= 65±5
						BOR LiCl		9.45				
						LiOH LiCl	ns	11.4				
								13.0				
								14.0				
ED05	$C_9H_{11}BrN_2O_6$	5-bromouridine C.A. 957-75-5	Table II	H_2O	PO	H_2SO_4	1.0	-	-	SME/-	0B2	C=0.5
						$HCOONH_4$	1.0					
						NaOH	1.0					

TABLE I. Electrochemical Data $C_9H_{11}BrN_2O_6$ ED05

Ref.	C/M	Charact. Potential Value	vs.	Response Const.	Value	n	Tech.	Electrokinetic Data Parameter	Value	From	Products and Identification	Description and Remarks	Code No.
C0026 2438	-	$E_{\frac{1}{2}}$	-1.04F	SCE	-	-	-	-	-	-	-	-	C,p C,i_{cat}	ED02
		-1.08F	-	-	-	-	-	-	-	-	-	C,p C,i_{cat}	
JE046 0343	32 e	-	-	-	i_ℓ	0.77	-	-	-	-	-	-	C,r	ED03
		-	-	-	1.23	-	-	-	-	-	-	A,$i_\ell \uparrow$ as pH $\uparrow$,r C	
		-	-	-	1.83	-	-	-	-	-	-	A,$i_\ell \uparrow$ as pH $\uparrow$,r C	
		-	-	-	1.48	-	-	-	-	-	-	A,$i_\ell \neq f(pH)$ for pH > 12,r C	
		-	-	-	-	-	-	-	-	-	-	A,r	
		-	-	-	0.53	-	-	-	-	-	-	C	
JE046 0323	32 d	-	-	-	-	-	-	-	-	-	-	C,i_{cat} for pH=0-5, $i_\ell \downarrow$ as pH $\uparrow$, $i_\ell \downarrow$ as h $\uparrow$,p	ED04 ap31
		$E_{\frac{1}{2}}$	-1.6	SCE	-	-	-	-	-	-	-	-	C,Tc>2, $i_\ell = k_1 h^{\frac{1}{2}} + k_2, i_k$,r	
		-1.7	-	i_ℓ	1.71	i:i	1.6F	-	-	-	CP→3-phenylpropan-1-ol,GLC	C,r	
		-1.82₅	-	-	2.18	-	2	-	-	-	-	C,$i_\ell = kh^{\frac{1}{2}}, i_d$,r	
		-1.80₅	-	-	2.1	-	1.93	-	-	-	-	C,r	
	32 b	-	-	-	-	-	2	$dE_{\frac{1}{2}}/dpH$	90	-	-	A,$\neq$, $i_\ell \neq f(pH)$, $E_{\frac{1}{2}} \to$ more neg. as pH $\uparrow$,p	
		-1.84₅	-	-	2.1	-	1.93	-	-	-	-	C,$\Delta E_{\frac{1}{2}} = 0.2\sigma^*$,r	
JE046 0323	32 d	-	-	-	i_ℓ	0.25	-	-	-	-	-	-	C,r	
		-	-	-	0.45	-	-	-	-	-	-	C,r	
		-	-	-	1.55	-	-	-	-	-	-	C,r	
	32 b	-	-	-	1.9	-	-	-	-	-	-	C,r	
		-	-	-	1.60	-	-	-	-	-	-	C,r	
C0029 0182	-	Q	0.91	-	-	-	-	-	-	-	-	-	C,p	ED05
		0.90	-	-	-	-	-	-	-	-	-	C,p	
		0.10	-	-	-	-	-	-	-	-	-	C,R,p	
		0.10	-	-	-	-	-	-	-	-	-	A,R	

Code No.	Empirical Formula	Name and C.A. Number	Structural Formula	Solvent	Tech.	Medium		μ, M	pH	T, °C	Electrodes	App.	Experimental Parameters
ED06 ap48	$C_9H_{11}NO$	4-dimethylamino-benzaldehyde C.A. 100-10-7	4-$(CH_3)_2NC_6H_4$-CHO	EtOH 10	PY	KOH	1	-	-	-	DME/SCE	2-0	C=1, m=2.73, t=3.55, v=3.3
ED07 ap55	$C_9H_{11}NO_4$	3-(3,4-dihydroxyphenyl)alanine C.A. 587-45-1	Table 11-2	H_2O	VA	$HClO_4$	1	-	-	-	GE/SCE	---	C=1, v=100
					VR	PHOS	-	-	6.3	-	GE/SCE	---	C=1, v=100
					VY	$HClO_4$	6	-	-	-	RPE/SCE	---	C=1, v=100
						$HClO_4$	1			25			C=ns, v=ns
ED08	$C_9H_{11}N_3O_2$	2-hydroxyacetophenone semicarbazone	C_6H_5C(:NNHCO-NH_2)CH_2OH	H_2O	PY	MB		-	2.0	-	DME/SCE	OAO	C=0.5(?), m=2.62, t=2.8, h=55
									4.0				
						ACET NH$_2$NHCONH$_2$	0.04 0.08		4.2				
						ACET NH$_2$NHCONH$_2$	0.16 0.02						
						ACET	0.2						
						MB			5.15				
									6.6				
ED09	$C_9H_{12}N_2O_2$	ethyl 2,4-dimethyl-pyrimidine-5-carboxylate C.A. 2226-86-0	Table 11	H_2O	PY	CITR		-	4.5	28.00 ±0.2	DME/SCE	OAO	C=0.1, EC=1.3
ED10	$C_9H_{12}OS$	4-ethylphenyl methyl sulfoxide C.A. 53120-15-3	4-$C_2H_5C_6H_4SOCH_3$	EtOH	PY	Et_4NBr	0.1	-	-	30.0 ±0.5	DME/SCE	2-0	C=0.5, t(c)=3
				EtOH 80 dioxane 20	PY	Et_4NBr	0.1	-	-	30.0 ±0.5	DME/SCE	2-0	C=0.5, t(c)=3
				EtOH 80 DMF 20	PY	Et_4NBr	0.1	-	-	30.0 ±0.5	DME/SCE	2-0	C=0.5, t(c)=3
				MeCN	PY	Et_4NBr	0.1	-	-	30.0 ±0.5	DME/SCE	2-0	C=0.5, t(c)=3
				DMF	PY	Et_4NBr	0.1	-	-	30.0 ±0.5	DME/SCE	2-0	C=0.5, t(c)=3
ED11 ap87 dc86	$C_9H_{13}N$	N,N-dimethyl-4-toluidine C.A. 99-97-8	4-$H_3CC_6H_4N(CH_3)_2$	$AlCl_3$ 50 NaCl 50	VA			-	pCl= 1.16	175	WBE/Al	2-2	C=6.74, v=500

TABLE I. Electrochemical Data $C_9H_{13}N$ ED11

Ref.	C/M	Charact. Potential		Response Const.		n Tech.	n	Electrokinetic Data			Products and Identification	Description and Remarks	Code No.
		Value	vs.		Value			Parameter	Value	From			
JE043 0365	–	–	–	–	–	–	–	–	–	–	–	A,0,p	ED06 ap48
JE049 0287	378 a	E_p	0.57F	SCE	i_p	87.8F	sttd	2	$E_{p/2}-E_p$	30	–	–	A,p	ED07 ap55
			0.57F			52.7F		–		–			C	
JE049 0287	378 b	E_p	−0.21F	SCE	i_p	20F	QE	–	–	–	–	–	A,0 on first scan,p	
			0.2F			40F		4					A	
			0.16F			7.5F		–					C	
			−0.24F			25F							C	
JE049 0287	378 a	$E_{\frac{1}{2}}$	0.65F	SCE	i_ℓ	5F	QE	4	–	–	–	CP → 2-hydroxy-5-alanine-1,4-benzoquinone, λ_{max} = 263, 385	A,p	
		E_0'	0.81	NHE	–	–	Elog (!) QE	2 4	Elog	30	–	–	A,W, i_ℓ=kC for C=0.2-10, i_ℓ=k$\omega^{\frac{1}{2}}$ for ω = 6.7-33, p	
C0036 0331	–	$E_{\frac{1}{2}}$	−0.77F	SCE	–	–	–	–	–	–	–	–	C,p	ED08
			−0.91F		–	–	–	–	–	–	–	–	C,p	
			−0.97		i_ℓ	13.0	sttd	4	–	–	–	–	C,p	
			−0.97			13.0		4	–	–	–	–	C,p	
			−0.97			11.5		4	–	–	–	–	C,p	
			−1.00F			4.0		4	–	–	–	CP → 2-hydroxy-1-phenylpropylamine, urea	C,Mx,p	
			−1.11F			2.5		–	–	–	–	–	C,Mx,p	
JE055 0445	–	$E_{\frac{1}{2}}$	−1.01	SCE	–	–	i:i	2	–	–	–	–	C,W,i_d,Tc=1.6, $E_{\frac{1}{2}}\neq$ f(pH) for pH=2-10,p	ED09
JE054 0221	47 c	$E_{\frac{1}{2}}$	−2.215	SCE	–	–	sttd	2	αn_a $k_{s,h}$	0.79 2E-29	Kplot	–	C,≠,i_d,$\Delta E_{\frac{1}{2}}$=3.6σ,r	ED10
JE054 0221	47 c	$E_{\frac{1}{2}}$	−2.315	SCE	–	–	sttd	2	–	–	–	–	C,≠,i_d,$\Delta E_{\frac{1}{2}}$=3.4σ,r	
JE054 0221	47 c	$E_{\frac{1}{2}}$	−2.23	SCE	–	–	sttd	2	–	–	–	–	C,≠,i_d,$\Delta E_{\frac{1}{2}}$=3.7σ,r	
JE054 0221	47 d	$E_{\frac{1}{2}}$	−2.345	SCE	–	–	sttd	2	–	–	–	–	C,≠,i_d,$\Delta E_{\frac{1}{2}}$=2.0σ,r	
JE054 0221	47 d	$E_{\frac{1}{2}}$	−2.38	SCE	–	–	sttd	2	–	–	–	–	C,≠,i_d,$\Delta E_{\frac{1}{2}}$=3.8σ,r	
JE049 0281	147 e	E_p	1.24F	Al	$i_p(u)$	46F	–	–	$dE_{p/2}/dpCl$	90	–	–	A,E_p→more pos. as $AlCl_3$/NaCl ↑,a.s. as $AlCl_3$/NaCl ↑,r	ED11 ap87 dc86
			1.13F			11F				–			C,$E_p\neq$f[$AlCl_3$/NaCl]	

151

ED12 $C_9H_{13}NO_3$

Code No.	Empirical Formula	Name and C.A. Number	Structural Formula	Solvent	Tech.	Medium		μ, M	pH	T, °C	Electrodes	App.	Experimental Parameters
ED12	$C_9H_{13}NO_3$	1-(4-hydroxy-3-methoxyphenyl)-2-aminoethanol C.A. 97-31-4	Table II	H_2O	VR	$HClO_4$	2	-	-	25	CPE/SCE	---	C=1, v=100
ED13	$C_9H_{14}N_3Na_2O_{14}P_3$	cytidine-5'-triphosphate disodium salt C.A. 36051-68-0	Table II	H_2O	PY	KCl Na_2SO_4 $CuSO_4$ Bu_3PO_4	1E-3 0.25 3E-3 4E-4	-	5.0	25.0 ±0.2	DME/SCE	--0	C=(0.005-1) E-3, m=0.463, t=10.3, h=100
ED14	$C_9H_{14}OS$	2-thiabicyclo-[4.4.0]decan-5-one C.A. 38481-84-4	Table II	EtOH(?) 75	PY	Et_4NClO_4	0.1	-	-	-	DME/SCE	2-0	C=1
ED15	$C_9H_{14}O_3S$	2-thiabicyclo-[4.4.0]decan-5-one 2,2-dioxide C.A. 36165-03-4	Table II	EtOH(?) 75	PY	Et_4NClO_4	0.1	-	-	-	DME/SCE	2-0	C=1
ED16	$C_9H_{18}OS$	2,2,6,6-tetramethyl-4-oxothiane C.A. 22842-41-7	Table II	EtOH(?) 75	PY	Et_4NClO_4	0.1	-	-	-	DME/SCE	2-0	C=1
ED17	$C_9H_{18}OS$	3,3,5,5-tetramethyl-4-oxothiane C.A. 17539-61-6	Table II	H_2O	PY	Et_4NClO_4	0.1	-	-	20	DME/SCE	2-0	C=0.5
ED18	$C_9H_{18}O_3S$	2,2,6,6-tetramethyl-4-oxothiane 1,1-dioxide C.A. 28898-50-2	Table II	EtOH(?) 75	PY	Et_4NClO_4	0.1	-	-	-	DME/SCE	2-0	C=1
ED19	$C_9H_{17}BF_4N_2$	2-(1,1-dimethylethyl)-3-aza-2-azoniabicyclo-[2.2.1]hept-2-ene tetrafluoroborate C.A. 41322-56-9	Table II	MeCN	VR	Bu_4NClO_4	0.1	-	-	-	DME?/SCE	---	v=100
ED20	$C_{10}F_8$	octafluoronaphthalene C.A. 313-72-4	Table II	HF	VA	NaF	1	-	-	0	GCE Pd/H_2, NaF 1	24-0	C=1.3, v=5
											Pt Pd/H_2, NaF 1		Pt Aux
ED21	$C_{10}H_6K_2O_2$	dipotassium 1,4-naphthohydroquinone C.A. 57434-97-6	Table II	DMF	PY	Bu_4NClO_4	ns	-	-	-	DME/SCE	2--	Pt Aux
					VR	Bu_4NClO_4	0.5	-	-	20	PDE Ag/AgCl, naphthoquinone	125A 2	C≤0.5, Pt Aux
ED22 aq50 dd00	$C_{10}H_6O_2$	1,2-naphthoquinone C.A. 524-42-5	Table II-2	DMF	PY	$Ba(ClO_4)_2$	0.05	-	-	-	DME/SCE	26A0	C=1, m=3.38, t=3.05
CONT													

TABLE I. Electrochemical Data $C_{10}H_8O_2$ (CONT.) ED22

Ref.	C/M	Charact. Potential		Response Const.		n Tech.		Electrokinetic Data			Products and Identification	Description and Remarks	Code No.
		Value	vs.		Value			Parameter	Value	From			
JE042 0397	46 b	E_p	0.73F	SCE	i_p	183F	QE	2	-	-	-	CP→methanol,MAS	A,p	ED12
			0.45F			67F		0.91				CP→norepinephrine, VA	C	
			0.52F			42F		-				-	A,0 on first scan	
JE035 0219	-	-	-	-	-	-	-	-	-	-	-	-	C,i_{cat},p	ED13
JE035 0363	-	$E_{\frac{1}{2}}$	-2.0	SCE	-	-	-	-	-	-	-	-	C,p	ED14
JE035 0363	-	$E_{\frac{1}{2}}$	-1.71	SCE	-	-	-	-	-	-	-	-	C,p	ED15
JE035 0363	-	$E_{\frac{1}{2}}$	-2.14	SCE	-	-	-	-	-	-	-	-	C,p	ED16
JE039 0195	400 ab	$E_{\frac{1}{2}}$	-2.14	SCE	-	-	-	-	-	-	-	-	C,p	ED17
JE035 0363	-	$E_{\frac{1}{2}}$	-1.84	SCE	-	-	-	-	-	-	-	-	C,p	ED18
JA095 6454	-	$E_{p/2}$	-0.72	SCE	-	-	QE	1	-	-	-	radical,ESR,reacts with O_2, anerobic decompn. prods.= hydrazine deriv. and 3-tert-butyl-2,3-diazanortricyclene	C,R,p	ED19
JE051 0456	-	E_p	1.24	Pd/H_2, NaF 1	$i_p(u)$	1	-	-	-	-	-	-	A,R,$i_p/v^{\frac{1}{2}}\neq f(v)$,p	ED20
			1.18			1							C,R,$i_p/v^{\frac{1}{2}}=f(v)$	
			1.22			1							A,R,$i_p/v^{\frac{1}{2}}\neq f(v)$,p	
			1.16			1							C,R,$i_p/v^{\frac{1}{2}}\neq f(v)$	
JA092 4139	-	$E_{\frac{1}{2}}$	-0.63	SCE	-	-	-	-	-	-	-	-	A,p	ED21
			-1.32										A, dianion oxidn.	
JA095 6688	-	E	-0.68	Ag/AgCl	-	-	sttd	1	-	-	-	-	A,R,r	
			-1.50					1					A,R	
JE055 0277	-	$E_{\frac{1}{2}}$	-0.28	SCE	i_ℓ	3.6	-	-	Elog	60	-	-	C,R,$E_{\frac{1}{2}}\neq f(t)$ for t= 1.0-5.5,r	ED22 aq50 dd00
			-			-				-			C,≠	
		E_p			i_p									CONT

Code No.	Empirical Formula	Name and C.A. Number	Structural Formula	Solvent	Tech.	Medium		μ, M	pH	T, °C	Electrodes	App.	Experimental Parameters
ED22 ag50 dd00	$C_{10}H_8O_2$	1,2-naphthoquinone	Table 11-2	DMF	PY	$KClO_4$	0.1	-	-	-	DME/SCE	26A0	$C=1, m=3.38, t=3.05$
						$LiClO_4$	0.1						
						Et_4NClO_4	0.1						
						$NaClO_4$	0.1						
						$Sr(ClO_4)_2$	0.05						
ED23 aq51 dd01	$C_{10}H_8O_2$	1,4-naphthoquinone C.A. 130-15-4	Table 11-2	DMF	PY	$Ba(ClO_4)_2$	0.05	-	-	-	DME/SCE	26A0	$C=1, m=3.38, t=3.05$
						$Ca(ClO_4)_2$	0.05						
						$KClO_4$	0.1						
						$LiClO_4$	0.1						
						$Mg(ClO_4)_2$	0.05						
						Bu_4NClO_4	ns	-	-	-	DME/SCE	2--	Pt Aux
						Et_4NClO_4	ns	-	-	-	DME/SCE	26A0	$C=1, m=3.38, t=3.05$
						$NaClO_4$	0.1						
						$Sr(ClO_4)_2$	0.05						
					VR	Bu_4NClO_4	0.5	-	-	20	HMDE or Pt Ag/AgCl	125A2	ns
						Bu_4NClO_4	1.0	-	-	20.0 ±0.3	PDE Ag/AgCl	2A2	$C=0.4-1$, Pt Aux
	$C_{10}H_8O_2$	1,2-naphthoquinone	Table 11-2		PY	$KClO_4$	0.1						

TABLE I. Electrochemical Data $C_{10}H_8O_2$ ED23

Ref.	C/M	Charact. Potential		Response Const.		n Tech.		Electrokinetic Data			Products and Identification	Description and Remarks	Code No.
		Value	vs.		Value			Parameter	Value	From			
JE055 0277	-	$E_{\frac{1}{2}}$	-0.48	SCE	i_ℓ	3.7	-	-	Elog	65	-	-	$C,R,E_{\frac{1}{2}} \neq f(t)$ for t= 1.0-5.5,r	ED22 ag50 dd00
			-			-				-			C,⊬	
			-0.35			3.4	-	-		70	-	-	$C,R,E_{\frac{1}{2}} \neq f(t)$ for t= 1.0-5.5,r	
			-			-				-			C,⊬	
			-0.54			3.7	-	-		60	-	-	$C,R,E_{\frac{1}{2}} \neq f(t)$ for t= 1.0-5.5,r	
			-			-				-			C,⊬	
			-0.45			3.9	-	-		70	-	-	$C,R,E_{\frac{1}{2}} \neq f(t)$ for t= 1.0-5.5,r	
			-			-				-			C,⊬	
			-0.24			3.4	-	-		70	-	-	$C,R,E_{\frac{1}{2}} \neq f(t)$ for t= 1.0-5.5,r	
			-			-				-			C,⊬	
JE055 0277	-	$E_{\frac{1}{2}}$	-0.61	SCE	i_ℓ	4.5	-	-	Elog	60	-	-	$C,R,E_{\frac{1}{2}} \neq f(t)$ for t= 1.0-5.5,r	ED23 aq51 dd01
			-			-				-			C,⊬	
			-0.59			4.2	-	-		58	-	-	$C,R,E_{\frac{1}{2}} \neq f(t)$ for t= 1.0-5.5,r	
			-			-				-			C,⊬	
			-0.65			4.3	-	-		60	-	-	$C,R,E_{\frac{1}{2}} \neq f(t)$ for t= 1.0-5.5,r	
			-			-				-			C,⊬	
			-0.65			4.4	-	-		65	-	-	$C,R,E_{\frac{1}{2}} \neq f(t)$ for t= 1.0-5.5,r	
			-			-				-			C,⊬	
			-0.56			4.8	-	-		60	-	-	$C,R,E_{\frac{1}{2}} \neq f(t)$ for t= 1.0-5.5,r	
			-			-				-			C,⊬	
JA092 4139	-	$E_{\frac{1}{2}}$	-0.65	SCE	-	-	-	-	-	-	-	-	C,p	
			-1.42										C	
JE055 0277	-	$E_{\frac{1}{2}}$	-0.65	SCE	i_ℓ	4.3	-	-	Elog	61	-	-	$C,R,E_{\frac{1}{2}} \neq f(t)$ for t= 1.0-5.5,r	
			-			-				-			C,⊬	
			-0.65			4.3	-	-		60	-	-	$C,R,E_{\frac{1}{2}} \neq f(t)$ for t= 1.0-5.5,r	
			-			-				-			C,⊬	
			-0.6			4.5	-	-		60	-	-	$C,R,E_{\frac{1}{2}} \neq f(t)$ for t= 1.0-5.5,r	
			-			-				-			C,⊬	
JA095 6688	-	$E_{\frac{1}{2}}$	-0.68	Ag/AgCl	-	-	-	-	-	-	-	-	C,r	
			-1.51										C	
JA096 0249	50 a	XE	-0.64	Ag/AgCl	-	-	sttd	1	-	-	-	radical anion	$C,XE=(E_{p,A}+E_{p,C})/2$ → (v-0),R,p	
			-1.47					1					dianion	C,R	

ED24 $C_{10}H_7NO_2$

Code No.	Empirical Formula	Name and C.A. Number	Structural Formula	Solvent	Tech.	Medium		μ, M	pH	T, °C	Electrodes	App.	Experimental Parameters
ED24 aq66 cf18	$C_{10}H_7NO_2$	1-nitronaphthalene C.A. 86-57-7	Table II-2	DMF	PY	$KClO_4$	0.1	-	-	-	DME/SCE	26-0	C=0.5, m=1.61, t=3.6
						Et_4NClO_4	0.1						
						Et_4NClO_4 H_2O	0.1 1						
						Et_4NClO_4 H_2O	0.1 6						
ED25 aq67 cf19	$C_{10}H_7NO_2$	2-nitronaphthalene C.A. 581-89-5	Table II-2	DMF	PY	$KClO_4$	0.1	-	-	-	DME/SCE	26-0	C=0.5, m=1.61, t=3.6
						Et_4NClO_4	0.1						
					VR	$KClO_4$	0.1	-	-	-	HMDE SCE	26-0	C=0.5
ED26	$C_{10}H_7NO_2$	N-phenylmaleimide C.A. 941-69-5	Table II	EtOH 50	PY	BR		-	2.37 4.16 5.28 6.28 7.25	-	DME/SCE	04AO	C=1.0, m=3.2, t=3.5, h=40
ED27 aq80 cf20 dd03	$C_{10}H_8$	naphthalene C.A. 91-20-3	Table II-2	MeCN	VY	$AlCl_3$	1	-	-	-	RPDE Ag/AgCl, Me_4NCl satd	2AO	ns
				$MeNO_2$	VY	$AlCl_3$ or Et_4NClO_4	1 0.1	-	-	-	RPDE Ag/AgCl, Me_4NCl satd	2AO	C=ns, A= 0.01
				HF	VY	KF	0.1	-	-	0	RDGC Cu/CuF_2, KF 0.2	2-0	C=satd, A= 0.07, v=100, ω=1, Pt Aux
ED28 ar03	$C_{10}H_8O_2$	1,4-naphthohydroquinone C.A. 571-60-8	Table II-2	EtOH 50	PY	ACET	0.1	-	5.6	-	DME/SCE	---	ns
ED29	$C_{10}H_9N$	(1-pyridinium)cyclopentadienide C.A. 1962-02-3	Table II	EtOH	PY	$HClO_4$ LiCl	0.001	-	-	25	DME/SCE	OAO	C=1.0
						$HClO_4$ LiCl	0.0005 0.1						
						LiCl	0.1						
						$HClO_4$ $NaClO_4$	0.001 0.1						
CONT													

TABLE I. Electrochemical Data $C_{10}H_9N$ (CONT.) ED29

Ref.	C/M	Charact. Potential		Response Const.		n		Electrokinetic Data			Products and Identification	Description and Remarks	Code No.
		Value	vs.		Value		Tech.	Parameter	Value	From			
JE039 0395	-	$E_{\frac{1}{2}}$	-1.01	SCE	I	1.9	-	-	Elog	56	-	-	$C, \Delta E_{\frac{1}{2}}=0.010\sigma, \Delta E_{\frac{1}{2}} \propto \Delta\beta, r$	ED24 aq66 cf18
			-1.50			4.3				-			C	
			-1.05			1.8	-	-		60	-	-	$C, \Delta E_{\frac{1}{2}}=0.008\sigma, r$	
			-1.72			4.4				90			C	
			-1.03c		-	-	-	-		-			C,p C	
			-0.95c		-	-	-	-		-			C,p C	
JE039 0395	-	$E_{\frac{1}{2}}$	-1.04	SCE	I	2.2	-	-	Elog	83	-	-	$C, \Delta E_{\frac{1}{2}}=0.010\sigma, \Delta E_{\frac{1}{2}} \propto \Delta\beta, r$	ED25 aq67 cf19
			-1.465			4.5				-			C	
			-1.03			2.1	-	-		64	-	-	C,r	
			-1.7			4.3				-			C	
JE039 0395	-	E_p	-1.1F	SCE	i_p	1.2F	-	-	-	-	-	-	C,p	
			-1.55F			4.8F							C	
			-0.95F			0.6F							A	
			-0.5F			0.4F							A	
			-0.6F			small							C,0 on first scan	
			-0.78F			0.6F							C,0 on first scan	
C0026 2749	-	$E_{\frac{1}{2}}$	-0.55F	SCE	i(ℓ)	21F	sttd	2	-	-	-	-	C,dr(iR?),p	ED26
			-0.80F			20F	-	-					C,Mx,dr(iR?),r	
			-1.6F			2F							C,dr(iR?)	
			-0.80F			16F							C,Mx,dr(iR?),p	
			-1.4F			6F							C,dr(iR?)	
			-0.80F			11F	sttd	1					C,Mx,dr(iR?),p	
			-1.40F			9F		1					C,dr(iR?)	
			-0.754			-		1					C,i=f(t),p	
			-					1					C,i=f(t)	
JE039 0385	-	$E_{\frac{1}{2}}$	1.35	Fc	-	-	-	-	-	-	-	-	A,Σ,p	ED27 aq80 cf20 dd03
JE039 0385	-	$E_{\frac{1}{2}}$	1.30	Fc	-	-	-	-	-	-	-	-	A,r	
JE054 0232	-	$E_{\frac{1}{2}}$	0.655	NHE	i_ℓ	5.5F	-	-	Elog	62	-	-	A,p	
JA092 4139	-	$E_{\frac{1}{2}}$	-0.113	SCE	-	-	-	-	-	-	-	-	A,p	ED28 ar03
C0026 0370	395 a	$E_{\frac{1}{2}}$	-0.95F	SCE	i_ℓ	3.5F	-	-	-	-	-	-	C,p	ED29
			-1.3F		Σl	0.5F 1.51c							C	
			-0.97F		i_ℓ	1.5F	sttd	1					C, i_k, p	
			-1.64F			1.4F							C, i_d	
	395 b		-1.34			0.8F		1					C, i_k, p	
			-1.64			2.5F							C, i_d	
	395 a		-			-		-					A,0,p	
														CONT

Code No.	Empirical Formula	Name and C.A. Number	Structural Formula	Solvent	Tech.	Medium		μ, M	pH	T, °C	Electrodes	App.	Experimental Parameters
ED29	$C_{10}H_9N$	(1-pyridinium)cyclo-pentadienide	Table II	EtOH	PY	$HClO_4$ $NaClO_4$	0.0005 0.1	-	-	25	DME/SCE	OAO	C=1.0
						$NaClO_4$	0.1						
ED30	$C_{10}H_9NO_2$	N-ethylphthalimide C.A. 5022-29-7	Table II	MeCN 66.7	VY	ACET		-	8	22±2	Pb RDE/ SCE	-AO	C=40,v=10, ω=35.8
				MeCN 63 H_2SO_4 5	VY	$HClO_4$ $NaClO_4$		-	1	22±2	Pb RDE/ SCE	-AO	C=40,v=10, ω=15.9
ED31	$C_{10}H_{10}ClCoO_4$	bis(η-cyclopenta-dienyl)cobalt(III) perchlorate C.A. 11077-17-1	[(π-C_5H_5)$_2$Co-(III)]$^+$$ClO_4^-$	PYR	PY	$LiClO_4$	0.1	-	-	25.0 ±0.2	DME Ag/Ag$^+$ 0.1	--1	C=0.28-5.1, m=0.718,t= 7.48(-1.0 V), h=55
					PV	$LiClO_4$	0.1	-	-	25.0 ±0.2	DME Ag/Ag$^+$ 0.1	--1	C=1,Δe=10, f=55,m= 0.718,t=7.48 (-1.0 V),h= 55
				α-pico-line	PY	$LiClO_4$	0.1	-	-	25.0 ±0.2	DME Ag/Ag$^+$ 0.1	--1	C=1.03,m= 0.718,t=7.48 (-1.0 V),h= 55
				γ-pico-line	PY	$LiClO_4$	0.1	-	-	25.0 ±0.2	DME Ag/Ag$^+$ 0.1	--1	C=1.06,m= 0.718,t=7.48 (-1.0 V),h= 55
ED32	$C_{10}H_{10}CoF_6P$	bis(η-cyclopenta-dienyl)cobalt(III) hexafluorophosphate C.A. 12427-42-8	[(π-C_5H_5)$_2$Co]$^+$-PF_6^-	MeCN	PY	Bu_4NPF_6	0.3	-	-	25	DME/SCE	2AO	ns
					PVI	Bu_4NPF_6	0.3	-	-	25	DME/SCE	2AO	ns
ED33 ar36 cf30 dd14	$C_{10}H_{10}Fe$	ferrocene C.A. 102-54-5	Table II-2	MG	VY	Bu_4NClO_4	0.1	-	-	-	RPDE Ag/ $AgClO_4$ 0.01, Bu_4NClO_4 0.1	2-0	A=0.01,ω= 11
				pyrrol-idone	VA	$HClO_4$	0.1	-	-	30	PDE Ag/AgCl, LiCl satd	2-0	C=1,v=50
				THF	VY	Bu_4NClO_4	0.1	-	-	-	RPDE Ag/ $AgClO_4$ 0.01, Bu_4NClO_4 0.1	2-0	A=0.01,ω= 11
ED34 ar52	$C_{10}H_{10}O$	1-tetralone C.A. 529-34-0	Table II-2	EtOH 2	PY	H_2SO_4	ns	-	1	-	DME/SCE	O4AO	C=0.2,m=2.1, t=3.02,h=80
						ACET			4.7				
						PHOS			5.65				
CONT									7.0				

TABLE I. Electrochemical Data $C_{10}H_{10}O$ (CONT.) ED34

Ref.	C/M	Charact. Potential		Response Const.		n		Electrokinetic Data			Products and Identification	Description and Remarks	Code No.
		Value	vs.		Value		Tech.	Parameter	Value	From			
C0026 0370	-	$E_{\frac{1}{2}}$	0.01	SCE	i_ℓ	1F	-	-	-	-	-	-	A,p	ED29
	395 b		0.01			3F	llk	1	-	-	-	-	A,i_d,p	
JE057 0351	411 c	$E_{\frac{1}{2}}$	-1.23 -1.45	SCE	-	-	-	-	-	-	-	-	C,p C	ED30
JE057 0351	411 b	$E_{\frac{1}{2}}$	-1.03	SCE	-	-	-	-	dlogi/dE	80	-	-	C,p	
JE034 0521	403 a	$E_{\frac{1}{2}}$	-0.820	Ag/Ag^+ 0.1	I	2.04	Elog (!)	1	Elog	57.5±0.5	-	-	C,W,R,i_d,$i_\ell=kh^{\frac{1}{2}}$,r	ED31
JE034 0521	403 a	E_{su}	-0.82F	Ag/Ag^+ 0.1	i_{su}	5.8F	-	-	$\Delta E_{su/2}$	93±1	sttd	-	C,R,E_{su} and $i_{su}\neq$ f(time),p	
JE034 0521	403 a	$E_{\frac{1}{2}}$	-0.87	Ag/Ag^+ 0.1	I	2.14	-	-	Elog	62.8	-	-	C,W,R,i_d,$i_\ell=kh^{\frac{1}{2}}$,r	
JE034 0521	403 a	$E_{\frac{1}{2}}$	-0.839	Ag/Ag^+ 0.1	I	2.17	-	-	Elog	56.0	-	-	C,W,R,i_d,$i_\ell=kh^{\frac{1}{2}}$,r	
JE050 0031	-	$E_{\frac{1}{2}}$	-0.94 -1.88	SCE	D	36 36	sttd	1 1	Elog	≈59 ≈59	-	$(\pi-C_5H_5)_2Co$ $[(\pi-C_5H_5)_2Co]^-$	C,R,r C,R	ED32
JE050 0031	-	-	-	-	-	-	-	-	α $k_{s,h}^{app}$ $k_{s,h}$ α $k_{s,h}^{app}$ $k_{s,h}$	0.52 0.86 0.46 0.5 0.27 0.86	eqn	-	C,$k_{s,h}$ is Frumkin-corrected,r C	
JE040 0069	-	E_0	-0.144 ±0.005	$Ag/AgClO_4$ 0.01, Bu_4NClO_4 0.1	-	-	-	-	-	-	-	-	A,r	ED33 ar36 cf30 dd14
JE034 0439	-	$E_{p/2}$	0.53F 0.53F	Ag/AgCl LiCl satd	i_p	2.8F 2.5F	-	-	-	-	-	-	A,R,E_0=0.54±0.01 V,r C,R	
JE040 0069	-	E_0	0.103 ±0.005	$Ag/AgClO_4$ 0.01, Bu_4NClO_4 0.1						-	-	-	A,i	
JE056 0285	65 x	$E_{\frac{1}{2}}$	- -1.0 - -1.28 -1.36 -1.41 -1.47	SCE	i_ℓ	- 0.9 - 0.84 0.9 0.3 1.0	PQ	- 1 - - - - -	$dE_{\frac{1}{2}}/dpH$	- 60 - 60 60 - -	-	-	C,i_a,Pr,r C,i_d C,i_a,Pr,r C C,r C,r C	ED34 ar52
														CONT

ED34 (CONT.) $C_{10}H_{10}O$

Code No.	Empirical Formula	Name and C.A. Number	Structural Formula	Solvent	Tech.	Medium		μ, M	pH	T, °C	Electrodes	App.	Experimental Parameters
ED34 ar52	$C_{10}H_{10}O$	1-tetralone	Table II-2	EtOH 2	PY	BOR		-	8.0	-	DME/SCE	04A0	C=0.2,m=2.1, t=3.02,h=80
						CARB			11.0				
						NaOH	ns		14.0				
ED35 cf35	$C_{10}H_{11}FO_3S_4$	2-(methylthio)-4-phenyl-1,3-dithiolan-2-ylium fluorosulfonate C.A. 51348-29-9	Table II-3	MeCN	VA	Et_4NClO_4	0.1	-	-	-	Pt Ag/$AgNO_3$ 0.01, Et_4NClO_4 0.1	2A2	v < 1E3
ED36	$C_{10}H_{12}N_5Na_3O_{10}P_2$	adenosine 5'-diphosphate trisodium salt C.A. 2092-65-1	Table II	H_2O	PY	KCl Na_2SO_4 $CuSO_4$ Bu_3PO_4	1E-3 0.25 3E-3 4E-4	-	5.0	25.0 ±0.2	DME/SCE	--0	C=(0.005-1)E-3,m=0.463, t=10.3,h=100
ED37	$C_{10}H_{12}N_5O_6P$	adenosine cyclo-2',-3'-hydrogen phosphate C.A. 634-01-5	Table II	H_2O	PY	KCl Na_2SO_4 $CuSO_4$ Bu_3PO_4	1E-3 0.25 3E-2 4E-4	-	5.0	25.0 ±0.2	DME/SCE	--0	C=(0.005-1)E-3,m=0.463, t=10.3,h=100
ED38	$C_{10}H_{12}N_5O_6P$	adenosine cyclo-3'-5'-hydrogen phosphate C.A. 60-92-4	Table II	H_2O	PY	KCl Na_2SO_4 $CuSO_4$ Bu_3PO_4	1E-3 0.25 3E-2 4E-4	-	5.0	25.0 ±0.2	DME/SCE	--0	C=(0.005-1)E-3,m=0.463, t=10.3,h=100
ED39 ar88	$C_{10}H_{12}O_2$	duroquinone C.A. 527-17-3	Table II-2	2-PrOH 40	PY	PHOS		-	7	-	DME/SCE	---	-
ED40	$C_{10}H_{13}NO_2$	N-cyclohexylmaleimide C.A. 1631-25-0	Table II	EtOH 50	PY	BR		-	7.25	-	DME/SCE Tl$^+$	04A0	C=1.0,m=3.3, t=3.5,h=40
ED41 as02	$C_{10}H_{13}N_5O_4$	adenosine C.A. 58-61-7	Table II-2	H_2O	PY	KCl Na_2SO_4 $CuSO_4$ Bu_3PO_4	1E-3 0.25 3E-3 4E-4	-	5.0	25.0 ±0.2	DME/SCE	--0	C=(0.25-1)E-3,m=0.463, t=10.3,h=100
					PVI	MB		0.5	3.4	0	DME/SCE	12A0	C=0.1,m=1.0, t(c)=3,f=50, Δe=3.54(rms), Pt Aux
									5.0				
ED42	$C_{10}H_{14}N_2O_5S$	S-methyl-4-thiouridine	Table II	H_2O	PY	BR KCl		0.2	2	25	DME/SCE	0A0	C=0.5,m=2.54, t=3.3,h=60
									3.5				
									5				
									6.5				
CONT													

TABLE I. Electrochemical Data $C_{10}H_{14}N_2O_5S$ (CONT.) ED42

Ref.	C/M	Charact. Potential Value	vs.	Response Const.	Value	n Tech.	Electrokinetic Data Parameter	Value	From	Products and Identification	Description and Remarks	Code No.
JE056 0285	65 x	$E_{\frac{1}{2}}$ -1.52	SCE	i_ℓ	1.14	- -	$dE_{\frac{1}{2}}/dpH$	30	-	-	C,r	ED34 ar52
		-1.6			0.98	- -		0	-	-	$C,E_{\frac{1}{2}} \neq f(pH)$ for $pH \geq 11$, r	
		-1.62			0.7	- -		0	-	-	C,r	
JE049 0105	338 c	E_p -0.42	SCE	-	-	QE 0.99	$E_p - E_{p/2}$	44	-	2,2'-bis(methyl-thio)-4,4'-diphen-yl-2,2'-bis(1,3-di-thiolanyl),MAS,NMR; tetrathioethylene (trace),MAS,NMR	$C,\neq,E_p-E_{p/2}$ and $J_p/Cv^{\frac{1}{2}} \neq f(v),p$	ED35 cf35
		-				-		-		-	A,X	
JE035 0219	-	-	-	-	-	- -	-	-	-	-	C,i_{cat},p	ED36
JE035 0219	-	-	-	-	-	- -	-	-	-	-	C,i_{cat},p	ED37
JE035 0219	-	-	-	-	-	- -	-	-	-	-	C,i_{cat},p	ED38
C0030 4192	-	-	-	-	-	- -	-	-	-	-	see AR88	ED39 ar88
C0026 2749	-	$E_{\frac{1}{2}}$ -0.845	SCE	-	-	sttd 1	-	-	-	-	$C,i=f(t),p$	ED40
		-1.363				1					$C,i=f(t)$	
JE035 0219	-	-	-	-	-	- -	-	-	-	-	C,i_{cat},p	ED41 as02
JA095 8495	113 n	E_{su} -1.33	SCE	i_{su}/C	1.40	- -	-	-	-	-	$C,\neq,r$	
		-1.43			0.80	- -	-	-	-	-	$C,\neq,r$	
JE048 0433	09 e	$E_{\frac{1}{2}}$ -0.79F	SCE	i_ℓ	0.15F	- -	$dE_{\frac{1}{2}}/dpH$	80	sttd	-	$C,E_{\frac{1}{2}}=-0.78-0.08pH$, i_d,Tc=1.2,r	ED42
		-0.89F			0.13F	- -		80		-	C,i_d,Tc=1.2,r	
		-1.21F			0.21F	- -		40			$C,X,E_{\frac{1}{2}}=-1.22-0.04pH$, i_d	
		-1.03F			0.13F	- -		80		-	C,i_d,Tc=1.2,r	
		-1.29F			0.18F	- -		40			C,X,i_d	
		-1.16F			0.28F	- -		40		-	C,r	

CONT

161

ED42 (CONT.) $C_{10}H_{14}N_2O_5S$

Code No.	Empirical Formula	Name and C.A. Number	Structural Formula	Solvent	Tech.	Medium		μ, M	pH	T, °C	Electrodes	App.	Experimental Parameters
ED42	$C_{10}H_{14}N_2O_5S$	S-methyl-4-thio-uridine	Table II	H_2O	PY	BR KCl		0.2	7	25	DME/SCE	OAO	C=0.5,m=2.54, t=3.3,h=60
									8.0				
									9.4				
									12				
					PV	BR KCl		0.2	4.3	25	DME/SCE	OAO	C=1,m=2.8, t=4.8,h=55, Δe=20,f=78
									10.5				
					VA	BR KCl		0.2	5	25	HMDE SCE	OAO	C=ns,v=25
									8				
ED43	$C_{10}H_{14}N_5Na_2O_{13}P_3$	adenosine 5'-tri-phosphate disodium salt C.A. 987-65-5	Table II	H_2O	PY	KCl Na_2SO_4 $CuSO_4$ Bu_3PO_4	1E-3 0.25 3E-3 4E-4	--	5.0	25.0 ±0.2	DME/SCE	--O	C=(0.005-1) E-3,m=0.463, t=10.3,h=100
ED44	$C_{10}H_{14}N_5Na_2O_{14}P_3$	guanosine 5'-tri-phosphate disodium salt C.A. 56001-37-7	Table II	H_2O	PY	KCl Na_2SO_4 $CuSO_4$ Bu_3PO_4	1E-3 0.25 3E-3 4E-4	--	5.0	25.0 ±0.2	DME/SCE	--O	C=(0.005-1) E-3,m=0.463, t=10.3,h=100
ED45 as22	$C_{10}H_{14}N_5O_7P$	5'-adenylic acid C.A. 61-19-8	Table II-2	H_2O	PY	MB		--	3.4	0	DME/SCE	12AO	C=0.1,m=1.7-2.2,t=3-4, Pt Aux
													C=0.26
									5.0				C=0.1
						KCl Na_2SO_4 $CuSO_4$ Bu_3PO_4	1E-3 0.25 3E-3 4E-4	--	5.0	25.0 ±0.2	DME/SCE	--O	C=(0.001-1) E-3,m=0.463, t=10.3,h=100
					PVI	MB		0.5	3.4	0	DME/SCE	12AO	C=0.1,m=1.0, t(c)=3,f=50, Δe=3.54 rms, Pt Aux
													C=0.26
									5.0				C=0.1

TABLE I. Electrochemical Data $C_{10}H_{14}N_5O_7P$ ED45

Ref.	C/M		Charact. Potential		Response Const.		n Tech.	n	Electrokinetic Data			Products and Identification	Description and Remarks	Code No.
			Value	vs.		Value			Parameter	Value	From			
JE048 0433	89 e	$E_{\frac{1}{2}}$	-1.19F	SCE	i_ℓ	0.28F	QE	3	$dE_{\frac{1}{2}}/dpH$	40	sttd	methylthiol, dimer of N-riboside of 6,6'-bis-(3,6-dihydropyrimidone-2, UVS	C,i_d,r	ED42
			-1.39F			0.1F	-	-		40		dimer,UVS,3,6-dihydropyrimidone-2-riboside	C,i_d	
			-1.23F -1.44F			0.25F 0.10F	-	-		40 40		-	C,r C	
			-1.28F -1.44F -1.60F			0.32F 0.16F 0.10F	-	-		40 - 40		-	C,r C C	
			-1.39F -1.58F			0.36F 0.1F	-	-		40 40		-	C,r C	
JE048 0433	89 e	E_{su}	-1.1F	SCE	i_{su}	0.6F	-	-	dE_{su}/dpH	700	-	-	C,p	
			-1.49			0.4F	-	-		0	-	-	$C, E_{su} \neq f(pH)$ for pH= 8-11, p	
JE048 0433	89 e	E_p	-0.9	SCE	-	-	-	-	-	-	-	-	C,i_a,X,p C A	
			-1.1 -0.2											
			-		-	-	-	-	-	-	-	-	C,i_a,X,p C A	
			-1.3 -0.37											
JE035 0219	-	-	-	-	-	-	-	-	-	-	-	-	C,i_{cat},p	ED43
JE035 0219	-	-	-	-	-	-	-	-	-	-	-	-	C,i_{cat},p	ED44
JA095 8495	113 n	$E_{\frac{1}{2}}$	-1.289	SCE	i_ℓ i	1.1 8.2	QE	4.5 to 4.8	Elog	43	sttd	-	$C,i_d,\mp,r$	ED45 as22
			-1.293			2.6 7.2		-		50		-	$C,i_d,\mp,r$	
			-1.409			0.9 6.5		-		45		-	$C,i_d,\mp,r$	
JE035 0219	-	-	-	-	-	-	-	-	-	-	-	-	C,i_{cat},p	
JA095 8495	113 n	E_{su}	-1.35	SCE	i_{su}/C	1.20	-	-	-	-	-	-	$C,\mp,r$	
			-1.33			1.08	-	-		-	-	-	$C,\mp,r$	
			-1.44			0.10	-	-		-	-	-	$C,\mp,r$	

ED46 C₁₀H₁₅Br

Code No.	Empirical Formula	Name and C.A. Number	Structural Formula	Solvent	Tech.	Medium		μ, M	pH	T, °C	Electrodes	App.	Experimental Parameters
ED46 as34 cf59	$C_{10}H_{15}Br$	1-bromoadamantane C.A. 768-90-1	Table II-2	MeCN	IL	Bu_4NBF_4	0.1	-	-	-	Pt Ag/Ag⁺ 0.1	2AO	C=10, v=100
ED47	$C_{10}H_{15}Cl$	1-chloroadamantane C.A. 935-56-8	Table II	MeCN	IL	Bu_4NBF_4	0.1	-	-	-	Pt Ag/Ag⁺ 0.1	2AO	C=10, v=100
ED48	$C_{10}H_{15}F$	1-fluoroadamantane C.A. 768-92-3	Table II	MeCN	IL	Bu_4NBF_4	0.1	-	-	-	Pt Ag/Ag⁺ 0.1	2AO	C=10, v=100
ED49 cf63	$C_{10}H_{16}$	adamantane C.A. 281-23-2	Table II-3	MeCN	IL	Bu_4NBF_4	0.1	-	-	-	Pt Ag/Ag⁺ 0.1	---	C=10, v=100
					QE	$LiClO_4$	0.1	-	-	-	Pt gauze Ag/Ag⁺ 0.1	2AO	1.48 mmol, A=38, E_{app}=2.35, Pt Aux
						Bu_4NBF_4	0.1			0			1.62 mmol, E_{app}=2.15 V
ED50	$C_{10}H_{16}N_2O_3S$	biotin C.A. 58-85-5	Table II	DMF 60	PY	KNO_3	0.05	-	-	-	DME/SCE	-A-	C=0.66, EC=1.772(-1.8 V), m=1.93, t=2.24, h=35
ED51	$C_{10}H_{16}N_2O_8$	ethylenediamine-tetraacetic acid C.A. 60-00-4	$(HOOCCH_2)_2N$- $CH_2CH_2N(CH_2$- $COOH)_2$	H_2O	PY	ACET		-	5.5	25.0 ±0.1	DME/SCE	2AO	C=1, m=2.14, t=4.22, h=50
					PV	ACET	0.2	-	5.5	25.0 ±0.1	DME/SCE	2AO	C=1, m=2.14, t=4.22, h=50, Δe=10, f=35
					VA	ACET	0.2	-	5.5	25.0 ±0.1	HMDE SCE	---	C=0.5, v=135
ED52	$C_{10}H_{16}O$	1-adamantanol C.A. 768-95-6	Table II	MeCN	IL	Bu_4NBF_4	0.1	-	-	-	Pt Ag/Ag⁺ 0.1	2AO	C=10, A=1, v=100, Pt Aux

TABLE I. Electrochemical Data $C_{10}H_{16}O$ ED52

Ref.	C/M	Charact. Potential Value	vs.	Response Const. Value		Tech.	n	Electrokinetic Data Parameter	Value	From	Products and Identification	Description and Remarks	Code No.
JA095 8631	393 b	$E_{p/2}$ 2.54	Ag/Ag^+ 0.1	-	-	-	-	-	-	-	QE at 2.35 V(0.1 $\underline{M}$ $LiClO_4$),n=3.2→ acetamidoadamantane (89%),MPT,IRS,NMR	A,p	ED46 as34 cf59
JA095 8631	393 a	$E_{p/2}$ 2.64	Ag/Ag^+ 0.1	-	-	-	-	-	-	-	QE at 2.50 V(0.1 $\underline{M}$ $LiClO_4$),n=2.1→1-chloro-3-acetamido-adamantane(91%), MPT,IRS,NMR	A,p	ED47
JA095 8631	393 a	$E_{p/2}$ 2.64	Ag/Ag^+ 0.1	-	-	-	-	-	-	-	QE at 2.50 V(0.1 $\underline{M}$ $LiClO_4$),n=2.4→1-fluoro-3-acetamido-adamantane(65%), MPT,IRS,NMR	A,p	ED48
JA095 8631	393 a	$E_{p/2}$ 2.36	Ag/Ag^+ 0.1	-	-	-	-	-	-	-	-	A,⊭,$i_p \propto v^{1/2}$ for v= 50-500,p A	ED49 cf63
		E_p 2.51		-	-	i:i	2	-	-	-	-	A,⊭,$i_p \propto v^{1/2}$ for v= 50-500,p	
		3.10					-					A,⊭	
JA095 8631	393 a	-	-	-	-	-	2.1	-	-	-	acetamidoadamantane, 90%,MPT,IRS,NMR	A,yield=80% in Et_4NBF_4,65% in Me_4NBF_4,p	
						-	2.1				acetamidoadamantane, 60%,MPT,IRS,NMR	A,p	
JE045 0156	-	$E_{1/2}$ -1.62	DME	i_l	0.6 0.52	XEq, PQ	0.985	α k_s,h	0.5 1E-14	Kplot	-	C,⊭,i_d,Tc=2 for T= 26-50,$E_{1/2} \neq f(T)$,i_l= $k/\eta^{1/2}$ for DMF=10-70%,XEq=Koutecký eqn for i_{max},r	ED50
JE049 0027	-	$E_{1/2}$ 0.13F	SCE	i_l	7.3F	sttd	2	Elog	33	-	-	A,$E_{1/2}=f(pH)$,$E_{1/2} \neq f(C)$,r	ED51
JE049 0027	-	E_{su} -0.03F	SCE	D $i_{su}(Y)$	5.01 0.78F	-	-	-	-	-	-	A,a.s.,$i_{su} \neq f(C)$ for $C \geq 1$,i_{su} ↑ as f ↑ for f=35-100,i_a,r	
		0.12F			1.83F							A,$i_{su} \neq f(C$ and $pH)$	
JE049 0027	-	E_p -0.04F	SCE	i_p	0.86F	-	-	-	-	-	-	A,$i_p \neq f(C)$ for C= 0.25-1,$i_p=kv$,E_p→ more neg. as pH ↑, i_a,r	
		0.13F			12.3F							A,$i_p=kC$,QI	
		0.09F			10.5F							C,QI	
JA095 8631	393 c	$E_{p/2}$ 2.12	Ag/Ag^+ 0.1	-	-	-	-	-	-	-	QE at 2.35 V(0.1 $\underline{M}$ $LiClO_4$)n=1.2→ acetamidoadaman-tane(41%),MPT,IRS, NMR(product prob-ably due to nonelec-trochem. reaction)	A,p	ED52

ED53 $C_{10}H_{18}O$ CRC Handbook Series in Organic Electrochemistry

Code No.	Empirical Formula	Name and C.A. Number	Structural Formula	Solvent	Tech.	Medium		μ, M	pH	T, °C	Electrodes	App.	Experimental Parameters
ED53	$C_{10}H_{18}O$	3,7-dimethyl-2,6-octadienal C.A. 5392-40-5	$CH_3C(CH_3):CH-(CH_2)_2C(CH_3):CHCHO$	H_2O	PY	BOR KCl		1.0± 0.2	10	25.00 ±0.05	DME/SCE	OAO	ns
ED54 as65	$C_{10}H_{17}N_3O_6S$	glutathione C.A. 70-18-8	$HOOCCH(NH_2)-(CH_2)_2CONHCH-(CH_2SH)CONHC-H_2COOH$	H_2O	PY	PHOS		–	7.1	–	DME/MSE	OAO	C=1.0, m=3.2, t=2.6
ED55	$C_{10}H_{19}BF_4N_2$	2-(1,1-dimethyl-ethyl)-3-aza-2-azoniabicyclo[2.2.2]oct-2-ene tetrafluoroborate C.A. 43008-10-2	Table II	MeCN	VR	Bu_4NClO_4	0.1	–	–	–	DME?/SCE	---	v=100
ED56 dd39	$C_{10}H_{20}N_2$	2-tert-butyl-3-methyl-2,3-diazanorbornane C.A. 42842-99-9	Table II-4	MeCN	VR	Bu_4NClO_4	0.1	–	–	–	DME?/SCE	---	ns
ED57	$C_{10}H_{22}$	decane C.A. 124-18-5	$CH_3(CH_2)_8CH_3$	CF_3COOH	IL	FSO_3H	2.5	–	–	–	Pt/SCE	25-2	C=30, v=50, Pt Aux
ED58	$C_{10}H_{26}N_2S_2$	thiocholine disulfide C.A. 4468-11-5	$[(CH_3)_3\overset{+}{N}CH_2-CH_2S^-]_2$	H_2O	PY	PHOS PHOS Thymol	2	–	3.6 5.8 7.4 8.8 9.9	–	DME/MSE	OAO	C=2.0, m=3.2, t=2.6
ED59	$C_{11}H_8BrN_3O_2$	5-bromopyridine-2-azo-(2',4'-dihydroxybenzene) C.A. 17091-08-6	Table II	EtOH 50	PY	ACET	0.2	–	5.6	25	DME/SCE	2-0	C=0.2, m=1.82, t(c)=1, v=5
ED60	$C_{11}H_9NO_2$	N-benzylmaleimide C.A. 52641-57-3	Table II	EtOH 50	PY	BR		–	7.25	–	DME/SCE Tl^+	O4AO	C=1.0, m=3.3, t=3.5, h=40
ED61	$C_{11}H_9N_3O$	2-(2'-hydroxyphenyl-azo)pyridine C.A. 10335-29-2	Table II	EtOH 50	PY	ACET	0.2	–	5.65	25	DME/SCE	2-0	C=0.2, m=1.82, t(c)=1, v=5
ED62	$C_{11}H_9N_3O$	2-(4'-hydroxyphenyl-azo)pyridine C.A. 7687-22-1	Table II	EtOH 50	PY	ACET	0.2	–	5.65	25	DME/SCE	2-0	C=0.2, m=1.82, t(c)=1, v=5
ED63	$C_{11}H_9N_3O$	3-(4'-hydroxyphenyl-azo)pyridine C.A. 6759-49-5	Table II	EtOH 50	PY	ACET	0.2	–	5.65	25	DME/SCE	2-0	C=0.2, m=1.82, t(c)=1, v=5
ED64	$C_{11}H_9N_3O$	4-(4'-hydroxyphenyl-azo)pyridine C.A. 20815-66-1	Table II	EtOH 50	PY	ACET	0.2	–	5.65	25	DME/SCE	2-0	C=0.2, m=1.82, t(c)=1, v=5

TABLE I. Electrochemical Data $C_{11}H_9N_3O$ ED64

Ref.	C/M	Charact. Potential		Response Const.		n Tech.		Electrokinetic Data			Products and Identification	Description and Remarks	Code No.
			Value	vs.		Value			Parameter	Value	From			
JE049 0433	-	$E_{\frac{1}{2}}$	-0.147	SCE	I	1.43	QE	0.91	-	-	-	-	C,X for pH<1, i_ℓ = $kh^{0.95}$,p	ED53
C0027 0693	-	$E_{\frac{1}{2}}$	-0.40F	MSE	-	-	-	-	-	-	-	-	A,p	ED54 as65
JA095 6454	-	$E_{p/2}$?	-0.79	SCE	-	-	QE	1	-	-	-	radical, stable in absence of air, ESR	C,R,p	ED55
JA095 6454	-	$E_{p/2}$?	0.17	SCE	-	-	-	-	-	-	-	-	A,p	ED56 dd39
JE054 0181	-	$E_{p/2}$	1.99	SCE	-	-	-	-	-	-	-	-	A,≢,p	ED57
C0027 0693	-	$E_{\frac{1}{2}}$	-0.32F	MSE	-	-	-	-	-	-	-	-	A,Mx,p	ED58
			-0.46F		-	-	-	-	-	-	-	-	A,p	
			-0.57F		-	-	-	-	-	-	-	-	A,p	
			-0.63F		-	-	-	-	-	-	-	-	A,p	
			-0.63F		-	-	-	-	-	-	-	-	A,p	
JE050 0113	-	$E_{\frac{1}{2}}$	-0.286	SCE	D	2.06	i:i QE	2.0 4	-	-	-	-	C,n≠f(pH) for pH= 1-8,$\Delta E_{\frac{1}{2}}$=0.14σ,r	ED59
C0026 2749	-	$E_{\frac{1}{2}}$	-0.791 -1.195	SCE	-	-	sttd	1 1	-	-	-	-	C,i=f(t),p C,i=f(t)	ED60
JE050 0113	-	$E_{\frac{1}{2}}$	-0.19	SCE	D	1.9	i:i QE	2.0 4	-	-	-	-	C,n≠f(pH) for pH= 1-13,r	ED61
JE050 0113	-	$E_{\frac{1}{2}}$	-0.2	SCE	D	1.99	i:i QE	2.0 4	-	-	-	CP → 2-aminopyridine, UVS, 4-aminophenol, UVS	C,n≠f(pH) for pH= 1-13,$\Delta E_{\frac{1}{2}}$=0.14σ,r	ED62
JE050 0113	-	$E_{\frac{1}{2}}$	-0.335	SCE	D	1.94	i:i QE	2.0 4	-	-	-	-	C,$\Delta E_{\frac{1}{2}}$=0.14σ,r	ED63
JE050 0113	-	$E_{\frac{1}{2}}$	-0.135	SCE	D	2.26	i:i QE	2.0 4	-	-	-	-	C,n≠f(pH) for pH= 1-8,$\Delta E_{\frac{1}{2}}$=0.14σ,r	ED64

ED65 $C_{11}H_9N_3O_2$

Code No.	Empirical Formula	Name and C.A. Number	Structural Formula	Solvent	Tech.	Medium		μ, M	pH	T, °C	Electrodes	App.	Experimental Parameters
ED65	$C_{11}H_9N_3O_2$	2,4-dihydroxybenzene-1-azo-2'-pyridine C.A. 1141-59-9	Table II	H_2O	PY	chloroacetate	0.2	–	1.50	25	DME/SCE	2-0	C=0.2, m=1.82, t(c)= v=5
						ACET	0.2		4.2				
									5.06				
									7.8				
									9.2				
						ethylenediamine buffer	0.2		10				
									10.8				
									12				
						NaOH	0.1		13				
				EtOH 50	PY	ACET	0.2	–	5.65	25	DME/SCE	2-0	C=0.2, m=1.82, t(c)=1, v=5
						NaOH	0.2						
ED66	$C_{11}H_9N_3O_2$	2,4-dihydroxybenzene-1-azo-3'-pyridine C.A. 16566-56-6	Table II	EtOH 50	PY	ACET	0.2	–	5.65	25	DME/SCE	2-0	C=0.2, m=1.82, t(c)=1, v=5
ED67	$C_{11}H_9N_3O_2$	2,4-dihydroxybenzene-1-azo-4'-pyridine C.A. 20815-56-9	Table II	EtOH 50	PY	ACET	0.2	–	5.65	25	DME/SCE	2-0	C=0.2, m=1.82, t(c)=1, v=5
ED68	$C_{11}H_{10}N_4O$	4-amino-2-hydroxybenzene-1-azo-2-pyridine C.A. 51790-22-8	Table II	EtOH 50	PY	ACET	0.2	–	5.65	25	DME/SCE	2-0	C=0.2, m=1.82, t(c)=1, v=5
ED69	$C_{11}H_{11}NO_2$	N-propylphthalimide C.A. 5323-50-2	Table II	MeCN 66.7	VY	ACET		–	8	22±2	Pb RDE SCE	-AO	C=40, v=10, ω=35.8
				MeCN 63 H_2SO_4 5	VY	$HClO_4$ $NaClO_4$	ns	–	1	22±2	Pb RDE SCE	-AO	C=40, v=10, ω=15.9
ED70	$C_{11}H_{11}NS$	1,2-dimethylindolizine-3-thial C.A. 25369-27-1	Table II	MeCN	PY	Pr_4NClO_4	0.05	–	–	–	DME Ag/$AgClO_4$ 0.1	2AO	C=0.5, m=2.12, h=0.42, Pt Aux
					VA	Pr_4NClO_4	0.05	–	–	–	HMDE Ag/$AgClO_4$ 0.1	2AO	-0.8→0.2→ -0.8 V, C=ns, v=400, Pt Aux
					VR	Pr_4NClO_4	0.05	–	–	–	HMDE Ag/$AgClO_4$ 0.1	2AO	-1.0→-2.5→ -1.0→-1.5 V, C=ns, v=200, Pt Aux
CONT													

TABLE I. Electrochemical Data $C_{11}H_{11}NS$ (CONT.) ED70

Ref.	C/M	Charact. Potential		Response Const.		n Tech.		Electrokinetic Data			Products and Identification	Description and Remarks	Code No.
		Value	vs.		Value			Parameter	Value	From			
JE050 0113	-	$E_{\frac{1}{2}}$	0.050	SCE	-	-	-	-	$dE_{\frac{1}{2}}/dpH$	90	-	-	C,r	ED65
			-0.18F		-	-	-	-		90	-	-	C,r	
			-0.263		D	5.4	i:i	2		90	-	-	C,r	
			-0.47F		$i_\ell(u)$	2		2		66	-	-	C,r	
			-0.56F			2.18F		-		66	-	-	C,r	
			-0.63F			2.85F		-		66	-	-	C,r	
			-0.65F			3.20	-	-		66	-	-	C,r	
			-0.76F			3.33F	-	-		66	-	-	C,r	
			-0.783			3.33F	-	-		0	-	-	$C,E_{\frac{1}{2}} \neq f(pH)$ for pH ≥ 12,r	
JE050 0113	-	$E_{\frac{1}{2}}$	-0.287	SCE	D	2.3	i:i QE	2 4	-	-	-	-	$C,n \neq f(pH)$ for pH = 1-8, $\Delta E_{\frac{1}{2}}$=0.14σ,r	
			-		-	-	i:i	3.7	-	-	-	-	C,p	
JE050 0113	-	$E_{\frac{1}{2}}$	-0.298	SCE	-	-	i:i QE	2.4 4	-	-	-	-	$C,\Delta E_{\frac{1}{2}}$=0.14σ,r	ED66
JE050 0113	-	$E_{\frac{1}{2}}$	-0.205	SCE	D	1.94	i:i QE	2 4	-	-	-	-	$C,n \neq f(pH)$ for pH= 1-8, $\Delta E_{\frac{1}{2}}$=0.14σ,r	ED67
JE050 0113	-	$E_{\frac{1}{2}}$	-0.346	SCE	D	2.02	i:i QE	2.0 4	-	-	-	-	$C,n \neq f(pH)$ for pH= 1-8,r	ED68
JE057 0351	411 c	$E_{\frac{1}{2}}$	-1.275 -1.5	SCE	-	-	-	-	-	-	-	-	C,p C	ED69
JE057 0351	411 b	$E_{\frac{1}{2}}$	-1.02	SCE	-	-	-	-	dlogi/dE	75	-	-	C,p	
JE056 0427	406 a	$E_{\frac{1}{2}}$	≈-0.1	Ag/ AgClO$_4$ 0.1	$i_\ell(u)$	1	i:i	1	Elog	-	-	-	A,r	ED70
			-1.94			2	QE	2		26.2			C	
			-2.60			2	i.i	2		179			C	
JE056 0427	406 a	E_p	-0.23F	Ag/ AgClO$_4$ 0.1	i_p	2F	-	-	-	-	-	-	A,p	
			0.05F			6.5F							A,QI	
			0.15F			small							A	
			-0.05F			small							C	
			-0.28F			6F							C,QI	
			-0.55F			small							C	
JE056 0427	406 a	E_p	-2.0F	Ag/ AgClO$_4$ 0.1	i_p	3.5F	-	-	-	-	-	-	C,p	
			-1.91F			0.25F							A	
			-1.35			0.59F							A,0 unless scan includes cathodic peak	
			-1.6			1.2F							C,0 on first scan	CONT

ED70 (CONT.) $C_{11}H_{11}NS$ CRC Handbook Series in Organic Electrochemistry

Code No.	Empirical Formula	Name and C.A. Number	Structural Formula	Solvent	Tech.	Medium		μ, M	pH	T, °C	Electrodes	App.	Experimental Parameters
ED70	$C_{11}H_{11}NS$	1,2-dimethylindolizine-3-thial	Table II	MeCN	VR	Pr_4NClO_4	0.05	–	–	–	HMDE Ag/ $AgClO_4$ 0.1	2AO	$-0.2 \to -2.5 \to -0.2 \to -1.5$ V, C=ns, v=200, Pt Aux
ED71	$C_{11}H_{12}O$	6,7,8,9-tetrahydro-5H-benzocyclohepten-5-one C.A. 826-73-3	Table II	EtOH 2	PY	ACET		–	3.6	–	DME/SCE	04AO	C=0.2, m=2.1, t=3.02, h=80
									4.62				
						PHOS			7.02				
						BOR			7.9				
						PHOS			11.6				
						NaOH	ns		12				
									13				
				EtOH	PY	H_2SO_4	ns	–	1	–	DME/SCE	04AO	C=0.2, m=2.1, t=3.02, h=80
ED72 at50 cf93	$C_{11}H_{12}O_2$	ethyl cinnamate C.A. 103-36-6	$C_6H_5CH:CH-COOC_2H_5$	DMF 90	VA	Et_4NClO_4	0.1	–	–	–	HMDE Ag/AgCl	---	–
ED73	$C_{11}H_{15}N$	1-cyanoadamantane C.A. 23074-42-2	Table II	MeCN	IL	Bu_4NBF_4	0.1	–	–	–	Pt Ag/Ag^+ 0.1	2AO	C=10, A=1, v=100, Pt Aux
ED74	$C_{11}H_{16}OS$	4-isobutylphenyl methyl sulfoxide	$4-(CH_3)_2CHCH_2-C_6H_4SOCH_3$	EtOH	PY	Et_4NBr	0.1	–	–	30.0 ±0.5	DME/SCE	2-0	C=0.5, t(c)=3
				EtOH 80 DMF 20	PY	Et_4NBr	0.1	–	–	30.0 ±0.5	DME/SCE	2-0	C=0.5, t(c)=3
				EtOH 80 dioxane 20	PY	Et_4NBr	0.1	–	–	30.0 ±0.5	DME/SCE	2-0	C=0.5, t(c)=3
				MeCN	PY	Et_4NBr	0.1	–	–	30.0 ±0.5	DME/SCE	2-0	C=0.5, t(c)=3
				DMF	PY	Et_4NBr	0.1	–	–	30.0 ±0.5	DME/SCE	2-0	C=0.5, t(c)=3
ED75	$C_{11}H_{16}O_3$	2-(2-methyl-2-methoxypropyl)hydroquinone C.A. 17208-01-4	Table II	EtOH 50	PY	ACET	0.2	–	–	25±0.1	DME/SCE	04AO	ns
ED76 cg16	$C_{11}H_{18}$	1-methyladamantane C.A. 768-91-2	Table II-3	MeCN	IL	Bu_4NBF_4	0.1	–	–	–	Pt Ag/Ag^+ 0.1	2AO	C=10, v=100

TABLE I. Electrochemical Data $C_{11}H_{18}$ ED76

Ref.	C/M	Charact. Potential		Response Const.		n Tech.	n	Electrokinetic Data			Products and Identification	Description and Remarks	Code No.
		Value	vs.		Value			Parameter	Value	From			
JE056 0427	406 a	E_p	−2.0F	Ag/AgClO$_4$ 0.1	i_p	9.1F	−	−	−	−	−	−	C,p	ED70
			−1.78F			1.9F							A	
			−1.37F			1F							A	
			−0.67F			1.3F							A	
			−0.48F			0.9F							A	
			−0.57F			2F							C,0 on first scan	
JE056 0285	65 x	$E_{\frac{1}{2}}$	−	SCE	i_ℓ	−	PQ	−	−	−	−	−	C,i_a,Pr,r	ED71
			−1.18			0.7		1					C,⊧,i_d for pH=2-4	
			−			−		−					C	
			−1.31			1.24		−					C,r	
			−1.39			0.2		−					C,r	
			−1.57			1.4		2					C,i_ℓ≠f(pH) for pH= 7.9-10,i_d,r	
			−1.69			1.24		−					C,r	
			−1.7			1		−					C,r	
			−1.7			0.7	sttd	1					C,⊧,i_d,r	
JE056 0285	65 n	$E_{\frac{1}{2}}$	−	SCE	i_ℓ	−	−	−	−	−	−	−	C,i_a,Pr,r	
			−1.05			0.9							C	
JE042 0189	−	−	−	−	−	−	−	−	−	−	−	−	see AT50,CF93	ED72 at50 cf93
JA095 8631	393 a	$E_{p/2}$	2.68	Ag/Ag$^+$ 0.1	−	−	−	−	−	−	−	QE at 2.50 V (0.1 M LiClO$_4$) n=2.2→1-cyano-3-acetamido-adamantane(41%), MPT,IRS,NMR	A,p	ED73
JE054 0221	47 c	$E_{\frac{1}{2}}$	−2.225	SCE	−	−	sttd	2	αn_a $k_{s,h}$	0.8 2E-29	Kplot	−	C,⊧,i_d,$\Delta E_{\frac{1}{2}}$=3.6σ,r	ED74
JE054 0221	47 c	$E_{\frac{1}{2}}$	−2.235	SCE	−	−	sttd	2	−	−	−	−	C,⊧,i_d,$\Delta E_{\frac{1}{2}}$=3.7σ,r	
JE054 0221	47 c	$E_{\frac{1}{2}}$	−2.315	SCE	−	−	sttd	2	−	−	−	−	C,⊧,i_d,$\Delta E_{\frac{1}{2}}$=3.4σ,r	
JE054 0221	47 d	$E_{\frac{1}{2}}$	−2.345	SCE	−	−	sttd	2	−	−	−	−	C,⊧,i_d,$\Delta E_{\frac{1}{2}}$=2.0σ,r	
JE054 0221	47 d	$E_{\frac{1}{2}}$	−2.385	SCE	−	−	sttd	2	−	−	−	−	C,⊧,i_d,$\Delta E_{\frac{1}{2}}$=3.8σ,r	
C0032 2140	−	$E_{\frac{1}{2}}^\circ$	0.689	NHE	−	−	−	−	−	−	−	−	A,$E_{\frac{1}{2}}$ measured at pH=3.8-5.6 and extrapolated,$\Delta E_{\frac{1}{2}}$= −0.07σ*,r	ED75
JA095 8631	393 a	$E_{p/2}$	2.40	Ag/Ag$^+$ 0.1	−	−	−	−	−	−	−	QE at 2.35 V (0.1 M LiClO$_4$),n=2.3→1-methyl-3-acetamido-adamantane(91%), MPT,IRS,NMR	A,p	ED76 cg16

ED77 $C_{11}H_{18}O$

Code No.	Empirical Formula	Name and C.A. Number	Structural Formula	Solvent	Tech.	Medium		μ, M	pH	T, °C	Electrodes	App.	Experimental Parameters
ED77	$C_{11}H_{18}O$	1-adamantylmethanol C.A. 770-71-8	Table II	MeCN	IL	Bu_4NBF_4	0.1	-	-	-	Pt Ag/Ag$^+$ 0.1	2AO	C=10,A=1,v= 100,Pt Aux
ED78	$C_{11}H_{18}O$	1-methoxyadamantane C.A. 6221-74-5	Table II	MeCN	IL	Bu_4NBF_4	0.1	-	-	-	Pt Ag/Ag$^+$ 0.1	2AO	C=10,A=1,v= 100,Pt Aux
ED79	$C_{12}H_8O_2$	acenaphthenequinone C.A. 82-86-0	Table II	DMF	PY	$Ba(ClO_4)_2$	0.05	-	-	-	DME/SCE	26AO	C=1,m=3.38, t=3.05
						$KClO_4$	0.1						
						$LiClO_4$	0.1						
						Et_4NClO_4	0.1						
						$NaClO_4$	0.1						
						$Sr(ClO_4)_2$	0.05						
ED80 ba15	$C_{12}H_8$	acenaphthylene C.A. 208-96-8	Table II-2	EtOH 50	PY	BR		-	5.6 ±0.1	25.0	DME/SCE	2AO	C=1,m(oc)= 0.245,t(c)= 1,h=60
									7.5				
									8.6				
ED81	$C_{12}H_8BrNO_4S$	4'-bromo-2-nitro-diphenyl sulfone	Table II	EtOH 50	PY	buffer		-	0	-	DME/SCE	04AO	C=0.1,m= 4.0,t=3.5
ED82	$C_{12}H_8BrNO_4S$	4'-bromo-4-nitro-diphenyl sulfone	Table II	EtOH 50	PY	buffer		-	0	-	DME/SCE	04AO	C=0.1,m= 4.0,t=3.5
ED83	$C_{12}H_8ClNO_2S$	2'-chloro-4-nitro-diphenyl sulfide C.A. 33667-99-1	Table II	EtOH 50	PY	buffer		-	0	-	DME/SCE	04AO	C=0.1,m= 4.0,t=3.5

TABLE I. Electrochemical Data $\quad C_{12}H_8ClNO_2S \quad$ ED83

Ref.	C/M	Charact. Potential		Response Const.		n Tech.		Electrokinetic Data			Products and Identification	Description and Remarks	Code No.
		Value	vs.		Value			Parameter	Value	From			
JA095 8631	393 b	$E_{p/2}$	2.15	Ag/Ag^+ 0.1	-	-	-	-	-	-	-	QE at 2.2 V (0.1 M $LiClO_4$), n=2.2 → 1-acetamidoadamantane(37%), MPT, IRS, NMR; 1-adamantyl-carbinyl acetate (20%), MPT, IRS, NMR	A,p	ED77
JA095 8631	393 c	$E_{p/2}$	2.05	Ag/Ag^+ 0.1	-	-	-	-	-	-	-	QE at 2.15 V (0.1 M $LiClO_4$), n=1.0 → 1-acetamidoadamantane(58%), MPT, IRS, NMR (product probably due to non-electrochem. reaction)	A,p	ED78
JE055 0277	-	$E_{\frac{1}{2}}$	-0.69	SCE	i_ℓ	2.1	-	-	Elog	65	-	-	$C,R,E_{\frac{1}{2}} \neq f(t)$ for t= 1.0-5.5, r C,≠	ED79
			-			-				-				
			-0.83			2.6				62		-	$C,R,E_{\frac{1}{2}} \neq f(t)$ for t= 1.0-5.5, r C,≠	
			-			-				-				
			-0.72			2.1				70		-	$C,R,E_{\frac{1}{2}} \neq f(t)$ for t= 1.0-5.5, r C,≠	
			-			-				-				
			-0.81			2.8				65		-	$C,R,E_{\frac{1}{2}} \neq f(t)$ for t= 1.0-5.5, r C,≠	
			-			-				-				
			-0.78			2.5				65		-	$C,R,E_{\frac{1}{2}} \neq f(t)$ for t= 1.0-5.5, r C,≠	
			-			-				-				
			-0.66			2.7				65		-	$C,R,E_{\frac{1}{2}} \neq f(t)$ for t= 1.0-5.5, r C,≠	
			-			-				-				
JE046 0077	215 m	$E_{\frac{1}{2}}$	-1.56F	SCE	i_ℓ	0.83F	QE	2	$dE_{\frac{1}{2}}/dpH$ Elog	30 ≈30	-	-	$C, i_\ell=kh^{\frac{1}{2}}, i_d, Tc=1.5$ for T=15-35, ≠, ece, r	ED80 ba15
			-1.62F			0.76F		2		30 ≈30		-	$C, i_\ell=kh^{\frac{1}{2}}, i_d, Tc=1.5$ for T=15-35, ≠, ece, r	
	215 n		-1.64F			0.8F		2	Elog	≈30		-	$C, i_\ell=kh^{\frac{1}{2}}, i_d, Tc=1.5$ for T=15-35, ≠, $E_{\frac{1}{2}} \neq f(pH)$ for pH ≥ 8.5, ece, r	
C0027 0525	-	$E_{\frac{1}{2}}^\circ$	-0.08	NCE	-	-	-	-	$dE_{\frac{1}{2}}/dpH$	51	sttd	-	$C, E_{\frac{1}{2}}$ measured at pH= 1-5 and extrapolated, p	ED81
C0027 0525	-	$E_{\frac{1}{2}}^\circ$	-0.05	NCE	-	-	-	-	$dE_{\frac{1}{2}}/dpH$	55	sttd	-	$C, E_{\frac{1}{2}}$ measured at pH= 1-5 and extrapolated, p	ED82
C0027 0525	-	$E_{\frac{1}{2}}^\circ$	-0.08	NCE	-	-	-	-	$dE_{\frac{1}{2}}/dpH$	55	sttd	-	$C, E_{\frac{1}{2}}$ measured at pH= 1-10 and extrapolated, p	ED83

Code No.	Empirical Formula	Name and C.A. Number	Structural Formula	Solvent	Tech.	Medium	μ, M	pH	T, °C	Electrodes	App.	Experimental Parameters
ED84	$C_{12}H_8ClNO_4S$	2'-chloro-2-nitro-diphenyl sulfone	Table II	EtOH 50	PY	buffer	-	0	-	DME/SCE	04A0	C=0.1, m=4.0, t=3.5
ED85	$C_{12}H_8ClNO_4S$	2'-chloro-4-nitro-diphenyl sulfone C.A. 42085-93-8	Table II	EtOH 50	PY	buffer	-	0	-	DME/SCE	04A0	C=0.1, m=4.0, t=3.5
ED86	$C_{12}H_8ClNO_4S$	3'-chloro-2-nitro-diphenyl sulfone	Table II	EtOH 50	PY	buffer	-	0	-	DME/SCE	04A0	C=0.1, m=4.0, t=3.5
ED87	$C_{12}H_8ClNO_4S$	3'-chloro-4-nitro-diphenyl sulfone	Table II	EtOH 50	PY	buffer	-	0	-	DME/SCE	04A0	C=0.1, m=4.0, t=3.5
ED88	$C_{12}H_8N_2O_2$	2,2'-pyridil C.A. 492-73-9	Table II	EtOH 50	PY	BARB	-	3.2	-	DME/SCE	04A0	C=0.9, m=3.02, t=3.14, h=40
								5.8				
								7.2				
								8.3				
								9.25				
								10.15				
						BARB H_3BO_3 0.002		10.3				
						BARB H_3BO_3 0.0051		10.3				
						BARB H_3BO_3 0.0102		10.3				
					PO	BARB	-	9.2	-	DME/SCE	04A0	C=0.5, m=3.02, t=3.15, h=40
ED89 dd66	$C_{12}H_8S$	dibenzothiophene C.A. 132-65-0	Table II-4	MeCN	VA	$NaClO_4$ 0.1	-	-	25±0.1	PBE Ag/$AgNO_3$ 0.1	2A0	C=3.2, d=0.16, v=200, MP Aux
					VV	$NaClO_4$ 0.1	-	-	25±0.1	Pt Ag/$AgNO_3$ 0.1	12A0	0.75 → 1.45 V, C=1, Δe=10, f=56, v=10, MP Aux

TABLE I. Electrochemical Data $C_{12}H_8S$ ED89

Ref.	C/M	Charact. Potential		Response Const.		n		Electrokinetic Data			Products and Identification	Description and Remarks	Code No.
			Value	vs.		Value	Tech.		Parameter	Value	From			
C0027 0525	-	$E_{\frac{1}{2}}^{O}$	-0.09	NCE	-	-	-	-	$dE_{\frac{1}{2}}/dpH$	54	sttd	-	$C,E_{\frac{1}{2}}$ measured at pH= 1-5 and extrapolated,p	ED84
C0027 0525	-	$E_{\frac{1}{2}}^{O}$	-0.03	NCE	-	-	-	-	$dE_{\frac{1}{2}}/dpH$	64	sttd	-	$C,E_{\frac{1}{2}}$ measured at pH= 1-10 and extrapolated,p	ED85
C0027 0525	-	$E_{\frac{1}{2}}^{O}$	-0.10	NCE	-	-	-	-	$dE_{\frac{1}{2}}/dpH$	49	sttd	-	$C,E_{\frac{1}{2}}$ measured at pH= 1-5 and extrapolated,p	ED86
C0027 0525	-	$E_{\frac{1}{2}}^{O}$	0.01	NCE	-	-	-	-	$dE_{\frac{1}{2}}/dpH$	69	sttd	-	$C,E_{\frac{1}{2}}$ measured at pH= 1-10 and extrapolated,p	ED87
C0025 3292	-	$E_{\frac{1}{2}}$	-0.18F -0.98F	SCE	i(u)	12F 14F	i:i	2 2	-	-	-	2,2'-pyridoin,PY	C,i_d,p C,i_d	ED88
			-0.43F -1.15F			14F 15F		2 2	-	-	-	2,2'-pyridoin,PY	C,i_d,p C,i_d	
			-0.58F -1.27F			14F 13F		2 2	-	-	-	2,2'-pyridoin,PK	C,i_d,p C,Mx	
			-0.59F -1.29F			14F 12F		2 -	-	-	-	-	C,i_d,p C,Mx	
			-0.60F -1.34F -1.58F			14F 3F 12F		2 - -	-	-	-	-	C,i_d,p C,Mx C	
			-0.60F -1.58F			14F 14F		2 2	-	-	-	-	C,i_d,p C,i_d	
			-0.60F			10F		-	-	-	-	-	C,i_d,H_3BO_3 complex,p	
			-0.60F			4F		-	-	-	-	-	C,i_d,H_3BO_3 complex,p	
			-0.60F			2F		-	-	-	-	-	C,i_d,H_3BO_3 complex,p	
C0025 3292	-	Q	0.3 0.7	-	-	-	-	-	-	-	-	-	C,p C	
JE043 0377	371 c	E_p	1.59	SCE	i_p	173F	QE(at 1.6 V)	1	-	-	-	CP → 2-(dibenzo[b,d]-thiophen-5-ylium)-dibenzo[b,d]thiophene, UVS, IRS	A,E_p → more pos. as v ↑, $i_p=kv^{\frac{1}{2}},r$	ED89
			1.00			155F	QE(at 1.9 V)	1.5	-	-	-		A	
			2.27			133F	-	-	-	-	-		A	
			0.0			46.7F			-	-	-		C,0 on first scan	
JE043 0377	371 c	E_{su}	1.662 ±0.025	SCE	i_{su}	8.2F	-	-	α k	0.34 1E-2.8	-	-	A,α and k from eqn. of Smith et al., JE021/0005	

Code No.	Empirical Formula	Name and C.A. Number	Structural Formula	Solvent	Tech.	Medium		μ, M	pH	T, °C	Electrodes	App.	Experimental Parameters
ED90 ba42 cg75 dd67 dh36 dh37	$C_{12}H_8S_2$	thianthrene C.A. 92-85-3	Table II-2	$AlCl_3$ 63 NaCl 37	VA	ionic melt		-	-	156	Pt Al/Al^{3+}	25-0	C=4.73, A=0.1, v=100
											W Al/Al^{3+}		C=13.5
ED91 ba56 cg86	$C_{12}H_9NO_2$	2-nitrobiphenyl C.A. 86-00-0	2-$O_2NC_6H_4C_6H_5$	DMF	PY	$KClO_4$	0.1	-	-	-	DME/SCE	26-0	C=0.5, m=1.61, t=3.6
						Et_4NClO_4	0.1						
ED92 ba57 cg87	$C_{12}H_9NO_2$	4-nitrobiphenyl C.A. 92-93-3	4-$O_2NC_6H_4C_6H_5$	DMF	PY	$KClO_4$	0.1	-	-	-	DME/SCE	26-0	C=0.5, m=1.61, t=3.6
						Et_4NClO_4	0.1						
ED93 cg88	$C_{12}H_9NO_2S$	2-nitrodiphenyl sulfide	2-$O_2NC_6H_4S$-C_6H_5	EtOH 50	PY	buffer		-	0	-	DME/SCE	04A0	C=0.1, m=4.0, t=3.5
ED94	$C_{12}H_9NO_4S$	2-nitrodiphenyl sulfone C.A. 31515-43-2	2-$O_2NC_6H_4$-$SO_2C_6H_4$	EtOH 50	PY	buffer		-	0	-	DME/SCE	04A0	C=0.1, m=4.0, t=3.5
									1.7				
									3.1				
									7.0				
									8.3				
									9.0				
ED95	$C_{12}H_9NO_4S$	4-nitrodiphenyl sulfone C.A. 1146-39-0	4-$O_2NC_6H_4SO_2$-C_6H_5	EtOH 50	PY	buffer		-	0	-	DME/SCE	04A0	C=0.1, m=4.0, t=3.5
									1.7				
									2.9				
									3.2				
									7.0				
									7.9				
									8.5				
									9.2				
ED96 ba62 dd69	$C_{12}H_9N_3O_2$	4-nitroazobenzene C.A. 2491-52-3	4-$O_2NC_6H_4N$·NC_6H_5	EtOH 50	PY	ACET TX100	0.2 2E-3	-	5.65	25	DME/SCE	2-0	C=0.2, m=1.82, t(c)=1, v=5

TABLE I. Electrochemical Data $C_{12}H_9N_3O_2$ ED96

Ref.	C/M	Charact. Potential		Response Const.		n Tech.	Electrokinetic Data			Products and Identification	Description and Remarks	Code No.
		Value	vs.		Value		Parameter	Value	From			
JE047 0081	–	E_p	0.96	Al/Al^{3+}	i_p	750F	–	–	–	–	–	–	A,r	ED90 ba42 cg75 dd67 dh36 dh37
			1.73			1000F						A	
			1.64			100F						C	
			0.89			850F						C	
			0.66			750F						C	
			0.925			217F	–	–	–	–	–	radical cation, ESR dication	A,R for v=5-2E5,r
			1.67			125F							A,R for v=5-2E5
			1.76			250F							C,R
			1.0			150F							C,R
JE039 0395	–	$E_{\frac{1}{2}}$	-1.175	SCE	I	1.8	–	–	Elog	78	–	–	$C,\Delta E_{\frac{1}{2}}\neq 0.009\sigma, \Delta E_{\frac{1}{2}} \propto \Delta\beta, r$	ED91 ba56 cg86
			-1.50			4.0				–			C	
			-1.19			1.7				65			$C,\Delta E_{\frac{1}{2}}=0.010\sigma, \Delta E_{\frac{1}{2}} \propto \Delta\beta, r$	
			-1.72			4.4				–			C	
JE039 0395	–	$E_{\frac{1}{2}}$	-1.02	SCE	I	1.9	–	–	Elog	78	–	–	$C,\Delta E_{\frac{1}{2}}=0.009\sigma, \Delta E_{\frac{1}{2}} \propto \Delta\beta, r$	ED92 ba57 cg87
			-1.535			4.0				–			C	
			-1.045			1.7				60			$C,\Delta E_{\frac{1}{2}}=0.010\sigma, \Delta E_{\frac{1}{2}} \propto \Delta\beta, r$	
			-1.715			4.3				–			C	
C0027 0525	–	$E_{\frac{1}{2}}^{o}$	-0.12	NCE	–	–	–	–	$dE_{\frac{1}{2}}/dpH$	78	sttd	–	$C, E_{\frac{1}{2}}$ measured at pH= 1-10 and extrapolated, p	ED93 cg88
C0027 0525	–	$E_{\frac{1}{2}}^{o}$	-0.09	NCE	–	–	–	–	$dE_{\frac{1}{2}}/dpH$	67	sttd	–	$C, E_{\frac{1}{2}}$ measured at pH= 1-10 and extrapolated, p	ED94
		$E_{\frac{1}{2}}$	-0.20F		–	–	–	–		67	–	–	C	
			-0.30F		–	–	–	–		67	–	–	C,p	
			-0.57F		–	–	–	–		67	–	–	C,p	
			-0.65F		–	–	–	–		67	–	–	C,p	
			-0.70F		–	–	–	–		67	–	–	C,p	
C0027 0525	–	$E_{\frac{1}{2}}^{o}$	0.02	SCE	–	–	–	–	$dE_{\frac{1}{2}}/dpH$	68	sttd	–	$C, E_{\frac{1}{2}}$ measured at pH= 1-10 and extrapolated, p	ED95
		$E_{\frac{1}{2}}$	-0.08F		–	–	–	–		68	–	–	C,p	
			-0.17F		–	–	–	–		68	–	–	C,p	
			-0.18F		–	–	–	–		68	–	–	C,p	
			-0.46F		–	–	–	–		68	–	–	C,p	
			-0.53F		–	–	–	–		68	–	–	C,p	
			-0.56F		–	–	–	–		68	–	–	C,p	
			-0.61F		–	–	–	–		68	–	–	C,p	
JE052 0115	–	$E_{\frac{1}{2}}$	-0.100	SCE	–	–	i:i	2	$dE_{\frac{1}{2}}/dpH$	56	–	–	$C, E_{\frac{1}{2}}=0.214-0.056$pH for pH=2-6, $\Delta E_{\frac{1}{2}}=0.14\sigma$	ED96 ba62 dd69

ED97 $C_{12}H_9N_3O_3$

Code No.	Empirical Formula	Name and C.A. Number	Structural Formula	Solvent	Tech.	Medium		μ, M	pH	T, °C	Electrodes	App.	Experimental Parameters
ED97	$C_{12}H_9N_3O_3$	4-hydroxy-4'-nitro-azobenzene C.A. 1435-60-5	Table II	EtOH 50	PY	ACET TX100	0.2 2E-3	-	5.65	25	DME/SCE	2-0	C=0.2, m=1.82, t(c)=1, v=5
ED98	$C_{12}H_{10}NO_3Tl$	diphenylthallium-(III) nitrate C.A. 57437-80-6	$(C_6H_5)_2TlNO_3$	DMF	IL	Et_4NClO_4	0.1	-	-	25.0 ±0.1	Pt PDS SCE	2A0	C=1.2, d=0.19, Pt Aux
					VA	Et_4NClO_4	0.1	-	-	25.0 ±0.1	Pt/SCE	2A0	C=1.02, v=200, Pt Aux
ED99 ba88 cg92 dd72	$C_{12}H_{10}N_2$	azobenzene C.A. 103-33-3	$C_6H_5N:NC_6H_5$	H_2O	PY	$HClO_4$ $NaClO_4$	ns	1	-0.2 2.25	25	DME/SCE	2-0	ns
					PA	$HClO_4$	0.1	-	-	25	DME/SCE	2-2	0.17 → -0.13 → 0.17 V, C=0.0145, m=0.232, v=500, delay time=8
					PW	$HClO_4$	4.5	-	-	25	DME/SCE	2-2	C=0.015, v=1E3
													v=4E4
						$HClO_4$ $NaClO_4$	ns	1	-0.8 0.9 2.3				ns
				EtOH 50	PY	ACET TX100	0.2 2E-3	-	5.65	25	DME/SCE	2-0	C=0.2, m=1.82, t(c)=1, v=5
				EtOH 50 w/w	QU	$HClO_4$	0.1	-	-	-	HMDE SCE	27A2	0.2 → -0.2 → 0.2 V, C=0.56, t(E_1)=30, A=0.03, Pt Aux
				N-methyl-pyrrol-idone	PY	Et_4NClO_4	0.1	-	-	-	DME Hg/Hg⁺ 0.02, $HClO_4$ 0.1	2A0	C=1-2
					VY	Et_4NClO_4	0.1	-	-	-	MP Hg/Hg⁺ 0.02, $HClO_4$ 0.1	29A0	C=1-2
EE00 ba92	$C_{12}H_{10}N_2O$	azoxybenzene C.A. 495-48-7	$C_6H_5N(O):NC_6H_5$	MeCN	VA	Et_4NClO_4	0.1	-	-	22.5 ±0.5	PBE/SCE F0	256A	C=3.84, A=0.25, v=100, Pt Aux
				DMF	PY	$KClO_4$	0.0125	-	-	-	DME/SCE	2A0	C=0.5, m=2.74, t=2.55
						Et_4NClO_4	0.1						
CONT						$NaClO_4$	0.075						

TABLE I. Electrochemical Data $C_{12}H_{10}N_2O$ (CONT.) EE00

Ref.	C/M	Charact. Potential		Response Const.		n	Tech.	Electrokinetic Data			Products and Identification	Description and Remarks	Code No.
		Value	vs.		Value			Parameter	Value	From			
JE052 0115	197 e	$E_{\frac{1}{2}}$ -0.245	SCE	-	-	i:i	2	$dE_{\frac{1}{2}}/dpH$	60	sttd	-	$C, E_{\frac{1}{2}}=0.104-0.060pH$ for pH=2-6, $\Delta E_{\frac{1}{2}}=0.14\sigma,p$	ED97
JE036 0117	233 b	$E_{\frac{1}{2}}$ -0.85	SCE	$i_\ell(u)$	1	QE	1	-	-	-	triphenylthallium, thallium, IRS	$C,\neq,p$	ED98
		-1.70			0.5		-				-	$C,\neq$	
JE036 0117	233 b	E_p -1.07F	SCE	i_p	35F	-	-	-	-	-	-	C,p	
		-0.75F			41F							A	
JE042 0415	-	$E_{\frac{1}{2}}$ 0.08F	SCE	-	-	-	-	$dE_{\frac{1}{2}}/dpH$	59	-	-	C,r	ED99 ba88 cg92 dd72
		-0.12F							-			C	
		-0.05F		-	-	-	-		59	-	-	C,r	
		-0.64F							-			C	
JE042 0415	-	E_p 0.02F	SCE	i_p	0.57F	-	-	-	-	-	-	C,r	
		0.03F			0.63F							A	
JE042 0415	-	E_p 0.14F	SCE	-	-	-	-	$dE_p/dlogv$	29.6	-	-	C,r	
		-							-			C	
		0.25F		-	-	-	-		-	-	-	$C, E_p \neq f(v)$ for $v \geq$ 4E4,r	
		-										C	
		0.08F		-	-	-	-		-	-	-	C,r	
		-0.4F										C	
		0.04F		-	-	-	-		-	-	-	C,r	
		-0.84F										C	
		-0.96F		-	-	-	-		-	-	-	C,r	
JE052 0115	-	$E_{\frac{1}{2}}$ -0.295	SCE	-	-	QE	2	$dE_{\frac{1}{2}}/dpH$	86	-	-	$C, E_{\frac{1}{2}}=0.195-0.086pH, n\neq f(pH),p$	
JE034 0283	197 a	-	-	D	3.6	-	2	-	-	-	-	C,r	
JE040 0345	-	$E_{\frac{1}{2}}$ -1.77	Hg/Hg$^+$ 0.02, HClO$_4$ 0.1	-	-	-	-	-	-	-	-	C,p	
		-2.31										C	
JE040 0345	-	$E_{\frac{1}{2}}$ -1.75	Hg/Hg$^+$ 0.02, HClO$_4$ 0.1	-	-	-	-	-	-	-	-	C,p	
		-2.25										C	
JE057 0179	-	E_p -1.43F	SCE	i_p	257.1F	-	-	-	-	-	-	C,R,r	EE00 ba92
		-1.37F			85.7F							A,R	
JE054 0313	-	$E_{\frac{1}{2}}$ -1.4	SCE	-	-	-	-	Elog	74	-	-	C,r	
		-1.65							-			C	
	86 g	-1.4		$i_\ell(u)$	1	-	-		65	-	CP → azobenzene radical anion, VIS, λ_{max}=429	C,r	
		-1.675			1.8				-		azobenzene radical anion	C	
		-1.87			0.8				-		azobenzene radical anion	C	
	-	-1.36		-	-	-	-		70	-	-	C,r	CONT

EE00 (CONT.) $C_{12}H_{10}N_2O$

Code No.	Empirical Formula	Name and C.A. Number	Structural Formula	Solvent	Tech.	Medium		μ, M	pH	T, °C	Electrodes	App.	Experimental Parameters
EE00 ba92	$C_{12}H_{10}N_2O$	azoxybenzene	$C_6H_5N(O){:}NC_6H_5$	DMF	VA	Et_4NClO_4	0.1	-	-	-	HMDE SCE	2AO	C=0.5, d=0.068
						$NaClO_4$	0.0125						
						$NaClO_4$	0.5						
				VR		$KClO_4$	0.0125	-	-	-	HMDE SCE	2AO	-0.9→-1.8→ -0.5→-1.0 V, C=0.5, d=0.068
EE01	$C_{12}H_{10}N_2O$	2-hydroxyazobenzene C.A. 2362-57-4	Table II	EtOH 50	PY	ACET TX100	0.2 2E-3	-	5.65	25	DME/SCE	2-0	C=0.2, m=1.82, t(c)=1, v=5
EE02 ba93 dd73	$C_{12}H_{10}N_2O$	4-hydroxyazobenzene C.A. 1689-82-3	Table II-2	EtOH 50	PY	ACET TX100	0.2 2E-3	-	5.65	25	DME/SCE	2-0	C=0.2, m=1.82, t(c)=1, v=5
EE03	$C_{12}H_{10}N_2O_2$	2,2'-dihydroxyazobenzene C.A. 2050-14-8	Table II	EtOH 50	PY	ACET TX100	0.2 2E-3	-	5.65	25	DME/SCE	2-0	C=0.2, m=1.82, t(c)=1, v=5
EE04	$C_{12}H_{10}N_2O_2$	2,4-dihydroxyazobenzene	Table II	EtOH 50	PY	ACET TX100	0.2 2E-3	-	5.65	25	DME/SCE	2-0	C=0.2, m=1.82, t(c)=1, v=5
EE05	$C_{12}H_{10}N_2O_2$	3,3'-dihydroxyazobenzene C.A. 2050-15-9	Table II	EtOH 50	PY	ACET TX100	0.2 2E-3	-	5.65	25	DME/SCE	2-0	C=0.2, m=1.82, t(c)=1, v=5
EE06	$C_{12}H_{10}N_2O_2$	4,4'-dihydroxyazobenzene	Table II	EtOH 50	PY	ACET TX100	0.2 2E-3	-	5.65	25	DME/SCE	2-0	C=0.2, m=1.82, t(c)=1, v=5
EE07	$C_{12}H_{10}N_2O_2$	2,2'-pyridoin C.A. 1141-06-6	Table II	EtOH 50	PY	BARB		-	3.2	-	DME/SCE	04AO	C=0.9, m=3.02, t=3.14, h=40
									5.8				
									7.2				
CONT													

TABLE I. Electrochemical Data $C_{12}H_{10}N_2O_2$ (CONT.) EE07

Ref.	C/M	Charact. Potential		Response Const.		n Tech.	n	Electrokinetic Data			Products and Identification	Description and Remarks	Code No.
		Value	vs.		Value			Parameter	Value	From			
JE054 0313	86 g	E_p	-1.44 -1.72 -1.9 -1.36F	SCE	i_p	6F 10F 3.3F 3F	-	-	-	-	-	-	C,r C C A	EE00 ba92
	-		-1.45 -1.3 -0.6 -0.77			4F 1F 5F 5.5F	-	-	-	-	-	-	C,r A A C	
	-		-1.36 -0.5F -0.65F			5F 5.5F 7F	-	-	-	-	-	-	C,r A C	
JE054 0313	-	E_p	-1.445 -1.69 -1.32 -0.64	SCE	i_p	2.7F 1.8F 1.3F 2.3F	-	-	-	-	-	CP → azobenzene, PY	C,R,r C A A, E_p → more pos. as $[K^+]$ ↑	
			-0.89			4.3F							C, E_p → more pos. as $[K^+]$ ↑	
JE052 0115	-	$E_{\frac{1}{2}}$	-0.335	SCE	-	-	i:i	2	$dE_{\frac{1}{2}}/dpH$	68	sttd	-	C, $E_{\frac{1}{2}}$=0.05-0.068pH for pH=2-6, $\Delta E_{\frac{1}{2}}$= 0.14σ,p	EE01
JE052 0115	197 e	$E_{\frac{1}{2}}$	-0.35	SCE	-	-	i:i	2	$dE_{\frac{1}{2}}/dpH$	77	sttd	-	C, $E_{\frac{1}{2}}$=0.085-0.077pH for pH=2-6, $\Delta E_{\frac{1}{2}}$= 0.14σ,p	EE02 ba93 dd73
JE052 0115	-	$E_{\frac{1}{2}}$	-0.353	SCE	-	-	i:i	2.4	$dE_{\frac{1}{2}}/dpH$	78	sttd	-	C, $E_{\frac{1}{2}}$=0.09-0.078pH for pH=2-6, $\Delta E_{\frac{1}{2}}$= 0.14σ,p	EE03
JE052 0115	-	$E_{\frac{1}{2}}$	-0.398	SCE	-	-	i:i	4	-	-	-	-	C,p	EE04
JE052 0115	-	$E_{\frac{1}{2}}$	-0.292	SCE	-	-	i:i, QE	2	$dE_{\frac{1}{2}}/dpH$	66	sttd	-	C, $E_{\frac{1}{2}}$=0.085-0.066pH for pH=2-6, $\Delta E_{\frac{1}{2}}$= 0.14σ,p	EE05
JE052 0115	197 e	$E_{\frac{1}{2}}$	-0.427	SCE	-	-	i:i	2.9	$dE_{\frac{1}{2}}/dpH$	103	sttd	-	C, $\Delta E_{\frac{1}{2}}$=0.14σ,p	EE06
C0025 3292	-	$E_{\frac{1}{2}}$	0.02F -0.95F	SCE	i(u)	15.4F 22.5F	sttd i:i	2 2	-	-	-	-	A, i_d,p C,dr, i_d	EE07
			-0.06F -1.14F			15.4F 20F	sttd	2 -	-	-	-	-	A, i_d,p C, i_d	
			-0.14F -1.25F			15.4F 19F		2 -	-	-	-	2,2'-pyridil, PK	A, i_d,p C, i_d	
														CONT

Code No.	Empirical Formula	Name and C.A. Number	Structural Formula	Solvent	Tech.	Medium		μ, M	pH	T, °C	Electrodes	App.	Experimental Parameters
EE07	$C_{12}H_{10}N_2O_2$	2,2'-pyridoin	Table II	EtOH 50	PY	BARB		-	8.3	-	DME/SCE	04A0	C=0.9,m=3.02,t=3.14,h=40
									9.25				
									10.15				
EE08	$C_{12}H_{10}N_2O_2S$	4'-amino-2-nitro-diphenyl sulfide C.A. 1144-81-6	Table II	EtOH 50	PY	buffer		-	0	-	DME/SCE	04A0	C=0.1,m=4.0,t=3.5
EE09	$C_{12}H_{10}N_2O_3$	2,2',3-trihydroxy-azobenzene	Table II	EtOH 50	PY	ACET TX100	0.2 2E-3	-	5.65	25	DME/SCE	2-0	C=0.2,m=1.82,t(c)=1,v=5
EE10	$C_{12}H_{10}N_2O_4$	2,2'-pyridoin bis-N-oxide	Table II	EtOH 50	PY	BARB		-	3.12	-	DME/SCE	04A0	C=0.56,m=3.02,t=3.15,h=40
						ACET			3.42				
						BARB			5.9				
									7.0				
									8.0				
									8.95				
									9.7				
									10.3				
EE11	$C_{12}H_{10}N_2O_4S$	N-(4-nitrophenyl)-benzenesulfonamide C.A. 1829-81-8	$4\text{-}O_2NC_6H_4NHSO_2\text{-}C_6H_5$	MeOH 10	PY	buffer GEL	7E-3	-	1.2	-	DME/SCE	04A0	C=0.5
									3.0				
									5.0				
									7.0				
									7.4				
									8				
									11				
									11.6				

TABLE I. Electrochemical Data $C_{12}H_{10}N_2O_4S$ EE11

Ref.	C/M	Charact. Potential		Response Const.		n Tech.	n	Electrokinetic Data			Products and Identification	Description and Remarks	Code No.
		Value	vs.		Value			Parameter	Value	From			
C0025 3292	-	$E_{\frac{1}{2}}$	-0.19F	SCE	i(u)	15.4F	sttd	2	-	-	-	-	A, i_d, p	EE07
			-1.28			18F		-					C, i_d	
			-0.22F			15.4F		2	-	-	-	-	A, i_d, p	
			-1.31F			8F		-					C, i_k	
			-1.55F			8F							C	
			-1.7F			7F							C, i_{cat}	
			-1.32F			1.0F		-	-	-	-	-	C, i_k, p	
			-1.55F			16F							C, i_d	
C0027 0525	-	$E_{\frac{1}{2}}^{\circ}$	-0.04	SCE	-	-	-	-	$dE_{\frac{1}{2}}/dpH$	69	sttd	-	$C, E_{\frac{1}{2}}$ measured at pH= 1-10 and extrapolated, p	EE08
JE052 0115	-	$E_{\frac{1}{2}}$	-0.372	SCE	-	-	i;i	4	$dE_{\frac{1}{2}}/dpH$	72	sttd	-	$C, E_{\frac{1}{2}}=0.035-0.072$pH for pH=2-6, $\Delta E_{\frac{1}{2}}=0.14\sigma, p$	EE09
C0025 3292	-	$E_{\frac{1}{2}}$	-0.90F	SCE	i(u)	24	i;i	6	-	-	-	-	C,p	EE10
			0.14F			9F		2	-	-	-	-	A, i_d, p	
			-			-		6					C, i_d	
			0.00F			8.2F		2	-	-	-	-	A, i_d, p	
			-1.10F			22F		-					C	
			-0.05F			8.2F		2	-	-	-	-	A, i_d, p	
			-1.23F			21F		-					C	
			-0.10F			8.2F		2	-	-	-	-	A, i_d, p	
			-1.26F			20F		-					C, Mx	
			-0.14F			8.2F		2	-	-	-	-	A, i_d, p	
			-1.28F			17.5F		-					C, Mx	
			-1.59F			5F							C	
			-1.7			6F							C, i_{cat}	
			-1.29F			14F		-	-	-	-	-	C, p	
			-1.59			7F		2					C	
			-1.30F			12F		-	-	-	-	-	C, i_d, p	
						7F		2					C, i_d	
JE036 0223	-	$E_{\frac{1}{2}}$	-0.16F	SCE	-			-	$dE_{\frac{1}{2}}/dpH$	55	-	-	C, r	EE11
			-0.26F		-	-	-	-		55	-	-	C, r	
			-0.68F		-	-	-	-		55	-	-	C, r	
			-0.47F		-	-	-	-		55	-	-	C, r	
			-0.51F		-	-	-	-		80	-	-	C, r	
			-0.57F		-	-	-	-		80	-	-	C, r	
			-0.8F		-	-	-	-		80	-	-	C, r	
			-0.83F		-	-	-	-		0	-	-	C, r	
			-0.19F		i(u)	15.4F	sttd	2						

Code No.	Empirical Formula	Name and C.A. Number	Structural Formula	Solvent	Tech.	Medium		μ, M	pH	T, °C	Electrodes	App.	Experimental Parameters
EE12	$C_{12}H_{10}N_4O_2$	1,2-diisonicotinoyl-hydrazine C.A. 4329-75-3	Table II	H_2O	PY	HCl	6	-	-	-	DME/SCE	0-0	C=0.20,m= 2.45,t(oc)= 5.7,h=48.5
						HCl	2.5						
						HCl KCl	ns		0.40				
									0.70				
									0.90				
									1.50				
						GLYC			2.10				
						CITR			2.60				
									3.05				
						ACET			4.25				
									4.75				
						succinate			5.05				
						PHOS			6.15				
									6.60				
									7.15				
						BOR			8.25				
									9.00				
									9.7				
						PHOS			11.05				
									11.80				
						KOH	0.1		-				
EE13 bb10	$C_{12}H_{10}OS$	diphenyl sulfoxide C.A. 945-51-7	$(C_6H_5)_2SO$	MeCN	IL	$NaClO_4$	0.1	-	-	25.0 ±0.1	PBE Ag/ $AgNO_3$ 0.1	2A02	C=0.6,v=79
													v=2.9E4
													v=8.1E4
					VR	$NaClO_4$	0.1	-	-	25.0 ±0.1	PBE Ag/ $AgNO_3$ 0.1	2A02	C=5,v=200
CONT													

TABLE I. Electrochemical Data $C_{12}H_{10}OS$ (CONT.) EE13

Ref.	C/M	Charact. Potential		Response Const		Tech.	n	Electrokinetic Data			Products and Identification	Description and Remarks	Code No.
		Value	vs.		Value			Parameter	Value	From			
AS017 1077	387 d	$E_{\frac{1}{2}}$	-0.535	SCE	i_ℓ	2.55	QE	6	-	-	-	pyridine-4-aldehyde, (can be further reduced, consuming 4e)	C,Mx,r	EE12
			-			-		-					C,O	
			-0.55			2.55	-	-	-	-	-	-	C,Mx,r	
			-			-							C,O	
	-		-0.585			2.60	-	-	-	-	-	-	C,Mx,r	
			-			-							C,O	
			-0.59			2.64	-	-	-	-	-	-	C,Mx,r	
			-			-							C,O	
			-0.605			2.76	-	-	-	-	-	-	C,Mx,r	
			-			-							C,O	
			-0.615 -0.69		Σi_ℓ	3.12	-	-	-	-	-	-	C,Mx,r C	
			-0.63 -0.71			3.32	-	-	-	-	-	-	C,Mx,r C	
			-0.66 -0.73			3.35	-	-	-	-	-	-	C,Mx,r C	
			-0.68 -0.76			3.40	-	-	-	-	-	-	C,Mx,r C	
			-0.76 -0.87			3.52	-	-	-	-	-	-	C,Mx,r C	
			-0.80 -0.91			4.00	-	-	-	-	-	-	C,Mx,r C	
			-0.82 -0.94			4.05	-	-	-	-	-	-	C,Mx,r C	
			-0.91 -1.03			4.00	-	-	-	-	-	-	C,Mx,r C	
			-0.94 -1.07			3.72	-	-	-	-	-	-	C,Mx,r C	
			-0.98 -1.11			3.44	-	-	-	-	-	-	C,Mx,r C	
			-1.07 -1.18			2.64	-	-	-	-	-	-	C,Mx,r C	
			1.10 1.21			2.64	-	-	-	-	-	-	C,Mx,r C	
			-1.23 -			2.44 -	-	-	-	-	-	-	C,r C,O	
			-1.34 -			1.60 -	-	-	-	-	-	-	C,r C,O	
			-1.38			1.40	-	-	-	-	-	-	C,r	
			-1.46			1.35	-	-	-	-	-	-	C,r	
JE055 0109	407 a	E_p	2.17	SCE	i_p	21.7F	-	-	-	-	-	-	A,r	EE13 bb10
			-	-		234.8F	-	-	-	-	-	-	A,r	
			-			330.4F	-	-	-	-	-	-	A,QI for $v > 3.5E4$,r	
JE055 0109	407 a	E_p	2.17	SCE	i_p	278.3F	QE	1.1 ± 0.1	-	-	-	CP → diphenylsulfone (50%), IRS	A,i_d,i_p=kC,r	
			0.91			56.5F		-					C	
			-0.9			43.5F							C,H	
			0.12			26.1F							C	
			1.50			-							A,O on first scan	
														CONT

EE13 (CONT.) $C_{12}H_{10}OS$

Code No.	Empirical Formula	Name and C.A. Number	Structural Formula	Solvent	Tech.	Medium		μ, M	pH	T, °C	Electrodes	App.	Experimental Parameters
EE13 bb10	$C_{12}H_{10}OS$	diphenyl sulfoxide	$(C_6H_5)_2SO$	MeCN	VR	$NaClO_4$ C_6H_6	0.1 0.1	-	-	25.0 ±0.1	PBE Ag/ $AgNO_3$ 0.1	2A02	C=4, v=200
EE14	$C_{12}H_{10}O_2$	4,4'-dihydroxy-biphenyl C.A. 92-88-6	$(4-HOC_6H_4)_2$	HF	VA	NaF	1	-	-	0	GCE Pd/H_2, NaF 1	24-0	C=ns, v=5
EE15 bb12	$C_{12}H_{10}O_2S$	diphenyl sulfone C.A. 127-63-9	$(C_6H_5)_2SO_2$	MG 97	PY	Bu_4NClO_4	0.3	-	-	25.0 ±0.5	DME/SCE	24A0	C=1, t=5
				MG	PY	Bu_4NClO_4	0.3	-	-	25.0 ±0.5	DME/SCE	24A0	C=1, t=5
						Bu_4NClO_4 PHEN	0.3 0.03						
				DMF 97	PY	Bu_4NClO_4	0.3	-	-	25.0 ±0.5	DME/SCE	24A0	C=1, t=5
				DMF	PY	Bu_4NClO_4	0.3	-	-	25.0 ±0.5	DME/SCE	24A0	C=1, t=5
						Bu_4NClO_4 PHEN	0.3 0.06						
				DMSO 97	PY	Bu_4NClO_4	0.3	-	-	25.0 ±0.5	DME/SCE	24A0	C=1, t=5
				DMSO	PY	Bu_4NClO_4	0.3	-	-	25.0 ±0.5	DME/SCE	24A0	C=1, t=5
						Bu_4NClO_4 PHEN	0.3 0.04						
					IR	Bu_4NClO_4	0.3	-	-	25.0 ±0.5	HMDE SCE	24A0 2	C=0.8, A=0.032, t=0.001-0.05, E_{app} = -2.65 V
						Bu_4NClO_4 PHEN	0.3 0.04						
					VA	Bu_4NClO_4	0.3	-	-	25.0 ±0.5	HMDE SCE	24A0 2	-1.5→-2.6→-1.5 V, C=1, A=0.032, v=54
													v=126
													-1.5→-2.6→-0.5→-1.5 V, v=270
													-1.5→-2.7→-1.5 V, v=9E3
													v=1.6E4
CONT		diphenyl sulfoxide	$(C_6H_5)_2SO$	MeCN	VR	$NaClO_4$	0.1						

TABLE I. Electrochemical Data $C_{12}H_{10}O_2S$ (CONT.) EE15

Ref.	C/M	Charact. Potential		Response Const.		Tech.	n	Electrokinetic Data			Products and Identification	Description and Remarks	Code No.
		Value	vs.		Value			Parameter	Value	From			
JE055 0109	407 b	E_p	2.17	SCE	i_p	217.4F	QE	1.0 ± 0.05	-	-	-	CP → diphenylsulfone (50%), IRS	A,r	EE13 bb10
			-0.09			78.3F		-				H_2	C,H	
			0.34			21.7F							A, H_2 oxidn.	
JE051 0456	-	E_p	0.49	Pd/H_2, NaF 1	$i_p(u)$	1	-	-	-	-	-	-	A, $i_p/v^{\frac{1}{2}} \neq f(v)$, R,p	EE14
			0.7			1							A, $i_p/v^{\frac{1}{2}} \neq f(v)$, R	
			0.64			1							C, $i_p/v^{\frac{1}{2}} \neq f(v)$, R	
			0.43			1							C, $i_p/v^{\frac{1}{2}} \neq f(v)$, R	
JE051 0075	43 c	$E_{\frac{1}{2}}$	-2.03 -	SCE	$i_\ell(u)$	1.00 0.02	-	-	-	-	-	-	C,p C	EE15 bb12
JE051 0075	43 c	$E_{\frac{1}{2}}$	-2.08 -2.39			1.00 0.52	-	-	-	-	-	-	C,p C	
			-2.03			-	-	-	-	-	-	-	C,p	
JE051 0075	43 c	$E_{\frac{1}{2}}$	-2.03 -	SCE	$i_\ell(u)$	1.00 0.03	-	-	-	-	-	-	C,p C	
JE051 0075	43 c	$E_{\frac{1}{2}}$	-2.05 -2.38	SCE	$i_\ell(u)$	1.00 0.23	-	-	-	-	-	-	C,p C	
			-2.03 -			1.00 0.02	-	-	-	-	-	-	C,p C	
JE051 0075	43 c	$E_{\frac{1}{2}}$	-2.01 -2.35	SCE	$i_\ell(u)$	1.00 0.13	-	-	-	-	-	-	C,p C	
JE051 0075	43 c	$E_{\frac{1}{2}}$	-2.03	SCE	$i_\ell(u)$	1.00	QE	2.0 ± 0.2	-	-	-	CP → phenylsulfonyl anion, λ_{max}=324; benzene; GLC	C,ece,r	
			-2.36			0.23		2.0 ± 0.2				CP → phenylsulfonyl cation, λ_{max}=324; benzene, GLC	C	
			-2.01 -2.35			1.00 0.17	-	-	-	-	-	-	C,p C	
JE051 0075	43 c	-	-	-	$it^{\frac{1}{2}}$	4.81± 0.07	-	-	-	-	-	-	C,Av(10),r	
			-			4.88± 0.09	-	-	-	-	-	-	C,Av(10),r	
JE051 0075	43 c	E_p	-2.08 -2.52 -	SCE	$J_p/Cv^{\frac{1}{2}}$	18.28 4.02 -	-	-	-	-	-	-	C,r C A,O	
			-2.09 -2.52 -			15.7 4.87 4.87	-	-	-	-	-	-	C,r C A	
			-2.11 -2.52 -2.05 -0.68			14.82 7.56 8.15 -	-	-	-	-	-	-	C,r C A A,O if limit of scan is less neg. than -2.1 V	
			-2.12 -2.65 -2.06			14.53 14.53 14.53	-	-	-	-	-	-	C,r C A	
			-2.12 -2.06			14.53 14.53	-	-	-	-	-	-	C,R,r A,R	
		E_p				217.4F								CONT

Code No.	Empirical Formula	Name and C.A. Number	Structural Formula	Solvent	Tech.	Medium	μ, M	pH	T, °C	Electrodes	App.	Experimental Parameters
EE15 bb12	$C_{12}H_{10}O_2S$	diphenyl sulfone	$(C_6H_5)_2SO_2$	DMSO	VA	Bu_4NClO_4 0.3 PHEN 0.04	-	-	25.0 ±0.5	HMDE SCE	24AO2	C=1, A=0.032, v=54
												v=270
												v=540
												v=1.62E3
EE16 bb14 dd77	$C_{12}H_{10}S$	diphenyl sulfide C.A. 139-66-2	$(C_6H_5)_2S$	MeCN	VA	$NaClO_4$ 0.1	-	-	25.0 ±0.1	PBE Ag/$AgNO_3$ 0.1	2AO	C=6, v=200, d=0.16, MP Aux
EE17 bb15	$C_{12}H_{10}S_2$	diphenyl disulfide C.A. 882-33-7	$(C_6H_5S)_2$	DMSO	PY	$KClO_4$ 0.1	-	-	25.0 ±0.1	DME Hg/Hg_2Cl_2, LiCl 1	256AO	C=0.05, EC=2.06, h=48.4, Pt Aux
												C=0.96
				MeCN	VA	Bu_4NBF_4 0.1	-	-	25.0 ±0.1	PBE Ag/Ag^+ 0.1	25AO	0→2.2→ -1.0→0 V, C=6.4, d=0.6, v=200, MP Aux
					VR	$NaClO_4$ 1	-	-	25.0 ±0.1	PBE Ag/Ag^+ 0.1	25AO	0→2→-1.2→ 0 V, C=4.24, d=0.6, v=200, MP Aux
EE18 bb20 ch10	$C_{12}H_{11}NO$	4-(phenylamino)-phenol C.A. 122-37-2	4-$HOC_6H_4NHC_6H_5$	MeCN	VA	$HClO_4$ 0.03 Et_4NClO_4 0.1	-	-	-	GCE Ag/Ag^+ 0.01 Pt Ag/Ag^+ 0.01	--O	C=2, v=33.3
					VY	Et_4NClO_4 0.1	-	-	-	RDGC Ag/Ag^+ 0.01	--O	C=2, ω=10
						$HClO_4$ 0.03 Et_4NClO_4 0.1						
						Et_4NClO_4 0.1				RPE Ag/Ag^+ 0.01		
						$HClO_4$ 0.03 Et_4NClO_4 0.1						

TABLE I. Electrochemical Data $C_{12}H_{11}NO$ EE18

Ref.	C/M	Charact. Potential		Response Const.		n		Electrokinetic Data			Products and Identification	Description and Remarks	Code No.
		Value	vs.		Value	Tech.		Parameter	Value	From			
JE051 0075	43 c	E_p	-2.07 -2.47 -	SCE	$J_p/Cv^{\frac{1}{2}}$	25.2 3.78 -	-	-	-	-	-	-	C,r C A,O	EE15 bb12
			-2.09 -2.49 -			16.6 4.65 6.31	-	-	-	-	-	-	C,r C A	
			-2.09 -2.50 -2.03			15.02 7.06 8.1	-	-	-	-	-	-	C,r C A	
			-2.08 -2.50 -2.01			14.7 9.87 10.58	-	-	-	-	-	-	C,r C A	
JE036 0389	409 a	E_p	1.48F	SCE	i_p	314.3F	QE	0.97	-	-	-	$CP \to (C_6H_5)_2S^+-$ $C_6H_4SC_6H_5$ ClO_4^-, CHA, IRS, λ_{max}=300	A,≠,r	EE16 bb14 dd77
			1.71F			108.6F		1.50	-	-	-	$CP \to (C_6H_5)_2S^+-$ $C_6H_4S^+(C_6H_5)-$ $C_6H_4SC_6H_4S^+(C_6H_5)_2$ 3 ClO_4^-, CHA, IRS, λ_{max}=320	A,≠	
			1.95F			74.3F		1.98	-	-	-	$(C_6H_5)_2S^+C_6H_4S^+-$ $(C_6H_5)C_6H_4S(:O)-$ $C_6H_4S^+(C_6H_5)_2 3ClO_4^-$ (unstable)	A,≠	
			0.07F			62.9F	-					-	C,H	
JE056 0373	-	$E_{\frac{1}{2}}$	-0.71	SCE	i_ℓ	0.24	QE	2	-	-	-	$CP \to$ sodium thio-phenolate, PY	$C, i_d, i_\ell/C \neq f(C)$ for $C=0.05-0.5, r$	EE17 bb15
			-0.78			4.35	-	-				-	C, i_d, r	
JE042 0057	410 ab	E_p	1.81F	SCE	i_p	120F	QE	3.29 ± 0.04	-	-	-	-	A,r	
			-0.08F			13.3F		-					C,H	
JE042 0057	410 ab	E_p	1.5F	SCE	i_p	40.8F	QE	1.07 ± 0.05	-	-	-	-	$A, i_p \uparrow$ on later scans due to oxidn. of Cl^-, r	
			1.81F			42.5F		-					A	
			0.91F			small							C, redn. of Cl_3^-	
			-0.41F			14.2F							C, redn. of ClO_2	
JE034 001A	105 d	E_p	0.63F	Ag/Ag^+ 0.01	i_p	8.9F	-	-	-	-	-	-	A,p	EE18 bb20 ch10
			0.58F			6.4F							C	
			0.7F			0.92F	-	-				-	A, E_p and $I_p(Ap)$, rounded peak, p	
			0.53F			2.7F							C	
JE034 001A	105 d	$E_{\frac{1}{2}}$	0.34F	Ag/Ag^+ 0.01	i_ℓ	12.8F	-	-	-	-	-	-	A,p	
			0.59F			4.3F							A	
			0.6F			20.3F	-	-				-	A,r	
			0.35F			3.5F	-	-	-	-	-	-	A,p	
			0.59F			4.72F	-	-					A,p	

Code No.	Empirical Formula	Name and C.A. Number	Structural Formula	Solvent	Tech.	Medium		μ, M	pH	T, °C	Electrodes	App.	Experimental Parameters
EE19	$C_{12}H_{11}NO_2$	N-(phenethyl)maleimide C.A. 6943-90-4	Table II	EtOH 50	PY	BR		-	7.25	-	DME/SCE Tl$^+$	04A0	C=1.0, m=3.3, t=3.5, h=40
EE20	$C_{12}H_{11}N_3O$	1-isonicotinoyl-2-phenylhydrazine C.A. 58481-06-4	Table II	H_2O	PY	HCl	6	-	-	-	DME/SCE	0-0	C=0.20, m=2.45, t(oc)=3.7, h=48.5
						HCl	0.25						
						HCl KCl			0.40				
									0.90				
									1.50				
						GLYC			2.10				
						CITR			3.05				
						ACET			4.25				
						succinate			5.05				
						PHOS			6.15				
									7.15				
						BOR			8.25				
									9.00				
									9.70				
						PHOS			11.05				
									11.80				
						KOH	0.1		-				
EE21	$C_{12}H_{12}N_2O_3S$	4-aminodiphenyl-amine-2-sulfonic acid C.A. 91-30-5	Table II	H_2O	VA	ACET PHOS		-	2.5	25.0 ±0.1	CPE/MSE	2-0	-0.35→1.25→-0.35 V, C=0.3, A=0.038, v=5

TABLE I. Electrochemical Data $C_{12}H_{12}N_2O_3S$ EE21

Ref.	C/M	Charact. Potential		Response Const.		Tech.	n	Electrokinetic Data			Products and Identification	Description and Remarks	Code No.
		Value	vs.		Value			Parameter	Value	From			
C0026 2749	-	$E_{\frac{1}{2}}$	-0.809 -1.200	SCE	-	-	sttd	1 1	-	-	-	-	C,i=f(t),p C,i=f(t)	EE19
AS017 1077	387 e	$E_{\frac{1}{2}}$	-0.45 -0.58	SCE	i_ℓ	1.00 1.02	-	-	$dE_{\frac{1}{2}}/dpH$	72 -	sttd	-	C,r C	EE20
			-0.46			1.05	QE	2.5		72		QE (0.4 M HCl) → iso-nicotinic amide,PY; aniline, colori-metry	C,r	
			-0.595			1.15	-	-		-			C	
			-0.495 -0.605			1.10 1.15	-	-		72 -		-	C,r C	
			-0.535 -0.645			1.10 1.20	-	-		72 -		-	C,r C	
			-0.57 -0.675			1.05 1.40	-	-		72 -		-	C,r C	
			-0.61 -0.71			1.05 1.60	-	-		72 -		-	C,r C	
			-0.665 -0.755			1.10 1.80	-	-		72 -		-	C,r C	
		-	-0.74 -0.86			1.10 2.10	-	-		72 -		-	C,r C	
			-0.805 -0.935			1.05 2.15	-	-		72 -		-	C,r C	
			-0.89 -1.025			1.10 2.15	-	-		72 -		-	C,r C	
			-0.965 -1.105			1.10 1.70	-	-		72 -		-	C,r C	
			0.0 -1.04 -1.17			1.9 1.05 1.05	sttd	4 - -		72 -		-	A,r C C	
			-0.03 -1.11 -1.21			1.9 1.05 1.00	-	-		72 -		-	A,r C C	
			-0.06 -1.15 -1.23			1.19 1.00 1.00	-	-				-	A,r C C	
			-0.14 - -1.275			1.8 - 1.85	-	-		-		-	A,r C,0 C	
			-0.17 -1.315			1.7 1.75	QE	2.5 -		-		N_2? -	A,r C	
			-0.24			1.9		2.5		-		isonicotinic acid, QE (0.2 M KOH),n=4 → aniline(90%), colorimetry	A,r	
			-1.44			1.90	-					-	C	
JE043 0387	38 g	E_p	0.45F 0.95F 0.35F	SCE	$i_p(u)$	10F small 9F	-	-		-		-	A,r A C	EE21
C0026 2749			-0.809				sttd						C,i=f(t),p	

191

EE22 $C_{12}H_{12}N_2O_4$

Code No.	Empirical Formula	Name and C.A. Number	Structural Formula	Solvent	Tech.	Medium		μ, M	pH	T, °C	Electrodes	App.	Experimental Parameters
EE22	$C_{12}H_{12}N_2O_4$	1,2-bis(2-pyridyl)-1,2-dihydroxyethane N,N'-dioxide	Table II	EtOH 50	PY	BARB		–	3.12	–	DME/SCE	O4AO	C=0.5(?),m=3.02,t=3.15,h=40
									5.9				
									7.0				
									8.0				
									8.95				
EE23	$C_{12}H_{12}N_4O$	4-hydroxyaminodiazo-aminobenzene C.A. 34529-57-2	Table II	MeOH 50	PY	UB GEL	0.01	–	7	21±1	DME/SCE	0-0	C=ns,m=4.9,t(oc)=1.9
									8.5				
									10				
									12				
EE24	$C_{12}H_{16}O_4S_4$	2-ethylthio-4-phenyl-1,3-dithiolanium methyl sulfate C.A. 51348-31-3	Table II	MeCN	VA	Et_4NClO_4	0.1	–	–	–	Pt Ag/$AgNO_3$ 0.01 Et_4NClO_4 0.1	2A2	v < 1E3
EE25	$C_{12}H_{18}O$	1-acetyladamantane C.A. 18220-83-2	Table II	MeCN	IL	Bu_4NBF_4	0.1	–	–	–	Pt Ag/Ag^+ 0.1	2AO	C=10,A=1,v=100,Pt Aux
EE26	$C_{12}H_{18}O_2$	methyl 1-adamantanecarboxylate C.A. 711-01-3	Table II	MeCN	IL	Bu_4NBF_4	0.1	–	–	–	Pt Ag/Ag^+ 0.1	2AO	C=10,A=1,v=100,Pt Aux
EE27	$C_{12}H_{18}O_3$	2-(2-ethoxy-2-methyl-1-propyl)-1,4-benzohydroquinone C.A. 17075-79-5	Table II	EtOH 50	PY	ACET	0.2	–	3.8-5.6	25±0.1	DME/SCE	O4AO	ns
EE28	$C_{12}H_{19}NO$	N-(1-adamantyl)-acetamide C.A. 19026-73-4	Table II	MeCN	IL	Bu_4NBF_4	0.1	–	–	–	Pt Ag/Ag^+ 0.1	2AO	C=10,A=1,v=100,Pt Aux
EE29	$C_{12}H_{20}N_2O_8$	1,4-diaminobutan-N,N,N',N'-tetra-acetic acid C.A. 1798-13-6	[(HOOCCH_2)$_2$N-(CH_2)$_2$-]$_2$	H_2O	PY	ACET		0.1	3.6	–	DME/SCE	O4AO	ns
						BR			8.15				
						BOR			9.3				

TABLE I. Electrochemical Data $C_{12}H_{20}N_2O_8$ EE29

Ref.	C/M	Charact. Potential		Response Const.		n	Electrokinetic Data			Products and Identification	Description and Remarks	Code No.
		Value	vs.	Value	Tech.		Parameter	Value	From			
C0025 3292	-	$E_{\frac{1}{2}}$ -1.06F	SCE	-	-	sttd	4	-	-	-	2,2'-bispyridylgly-col	C, i_d, redn. of two $N \to O$ groups, p	EE22
		-1.22F		-	-		4	-	-	-	2,2'-bispyridylgly-col	C,p	
		-1.32F		-	-		-	-	-	-	2,2'-bispyridylgly-col	C,p	
		-1.37F		-	-		-	-	-	-	2,2'-bispyridylgly-col	C,p	
		-1.37F		-	-		-	-	-	-	2,2'-bispyridylgly-col	C,p	
JE035 0369	386 a	$E_{\frac{1}{2}}$ -1.03F	SCE	-	-	-	-	$dE_{\frac{1}{2}}/dpH$	62	sttd	-	$C, E_{\frac{1}{2}}= -0.60-0.062$pH for pH=7-12, p	EE23
		-1.13c -1.58		-	-	-	-		62 0		-	C,p C	
		-1.26F		-	-	-	-		62		-	C,p	
		-1.58F		-	-	-	-		0		-	C	
		-1.34c -1.58		-	-	-	-		62 0		-	C,p C	
JE049 0105	338 c	E_p -0.46	SCE	-	-	QE	0.82	$E_p-E_{p/2}$	45	-	2,2'-bis(ethylthio)-4,4'-diphenyl-2,2'-bi(1,3-dithiolanyl), MAS,NMR; tetrathio-ethylene(trace), MAS,NMR	$C, *, E_p-E_{p/2}$ and $J_p/Cv^{\frac{1}{2}} \neq f(v)$, p	EE24
		-				-					-	A,X	
JA095 8631	393 b	$E_{p/2}$ 1.96	Ag/Ag$^+$ 0.1	-	-	-	-	-	-	-	QE at 2.35 V (in 0.1 M LiClO$_4$) → 1-acetamidoadamantane (44%), MPT, IRS, NMR	A,p	EE25
JA095 8631	393 a	$E_{p/2}$ 2.56	Ag/Ag$^+$ 0.1	-	-	-	-	-	-	-	QE at 2.45 V (in 0.1 M LiClO$_4$) → 1-carbomethoxy-3-acetamidoadamantane (64%), MPT, IRS, NMR	A,p	EE26
C0032 2140	-	$E_{\frac{1}{2}}^o$ 0.690	NHE	-	-	-	-	-	-	-	-	$A, E_{\frac{1}{2}}$ measured at pH= 3.8-5.6 and extrapolated, $\Delta E_{\frac{1}{2}}=0.07\sigma^*$, p	EE27
JA095 8631	-	$E_{p/2}$ 1.90	Ag/Ag$^+$ 0.1	-	-	-	-	-	-	-	-	A,p	EE28
C0027 1997	-	$E_{\frac{1}{2}}$ 0.305	SCE	-	-	-	-	-	-	-	mercury complex	A,R,r	EE29
		0.060		-	-	-	-	-	-	-	-	A,R,r	
		0.060		-	-	-	-	-	-	-	-	A,R,r	

Code No.	Empirical Formula	Name and C.A. Number	Structural Formula	Solvent	Tech.	Medium		μ, M	pH	T, °C	Electrodes	App.	Experimental Parameters
EE30	$C_{12}H_{27}O_4P$	tributyl phosphate C.A. 126-73-8	$[CH_3(CH_2)_3O]_3$-$P(O)$	H_2O	PV	$HClO_4$ $NaClO_4$	0.01 1	-	-	-	DME/SCE	2-0	C=0.2,t=3, Δe=10,f=32, Pt Aux
EE31	$C_{13}H_7BrO$	1-bromo-9-fluorenone C.A. 36804-63-4	Table II	DMF	VA	Pr_4NClO_4	0.1	-	-	20	PBE/SCE	26A0	C=0.1,v=3.68 v=72
EE32	$C_{13}H_7BrO$	3-bromo-9-fluorenone C.A. 2041-19-2	Table II	DMF	VA	Pr_4NClO_4	0.1	-	-	20	PBE/SCE	26A0	C=0.1,v=4.3 v=40
EE33	$C_{13}H_7ClO_2$	2-chloro-1-acenaph-thylenecarboxylic acid C.A. 13152-82-4	Table II	EtOH 50	PY	BR		-	3 5.6 7 8.8 12.8	25.0 ±0.1	DME/SCE	2A0	C=0.1,m= 2.057,t= 3.9,h=50
EE34	$C_{13}H_8O_2$	2-hydroxy-1-ace-naphthylenecarbox-aldehyde C.A. 30013-72-0	Table II	EtOH 50	PY	BR		-	2.8 6 8.1 11.3	25.0 ±0.1	DME/SCE	2A0	C=1.5,m(oc)= 245,t(c)=1, h=60
					VA	BR		-	5	25.0 ±0.1	HMDE SCE	2A0	C=1.5,v=500
EE35 bc87	$C_{13}H_9N$	acridine C.A. 260-94-6	Table II-2	MeCN	IL	Et_4NClO_4		-	-	-	Pt/SCE	2-0	A=0.22
					VR	Et_4NClO_4		-	-	-	Pt/SCE	2-0	0→-1.8→ 1.8→-0.8 V, A=0.22

TABLE I. Electrochemical Data $C_{13}H_9N$ EE35

Ref.	C/M	Charact. Potential		Response Const.		n Tech.	n	Electrokinetic Data			Products and Identification	Description and Remarks	Code No.
		Value	vs.		Value			Parameter	Value	From			
JE033 0061	–	E_{su}	≈0	SCE	i_{su}	0.15F	–	–	–	–	–	–	C,i_a,p	EE30
			≈–1.3			0.12F							C,i_a	
JE056 0443	–	E_p	–1.22	SCE	–	–	–	–	–	–	–	radical anion	$A,E_p=k_1+k_2\log v,p$ A C,0	EE31
			–											
			–1.228		–	–	–	–	–	–	–	–	$A,E_p \ne f(v),p$ C	
			–1.16											
JE056 0443	–	E_p	–1.214	SCE	i_p	5.6F	–	–	–	–	–	radical anion	$A,E_p=k_1+k_2\log v,p$ A C,0	EE32
			–1.33F			5.6F								
			–			–								
			–1.224			16F	–	–	–	–	–	–	$A,E_p \ne f(v),p$ C	
			–1.16F			4.4F								
JE055 0417	399 a	$E_{\frac{1}{2}}$	–0.94F	SCE	i_ℓ	0.38F	QE	2	Elog	59	–	–	$C,i_d,R,i_\ell=kC$ for $C=$ 0.03–0.1, Tc=1.6±0.2,r	EE33
			–1.16F			0.39F				59			$C,i_d,R,i_\ell=kC$ for $C=$ 0.03–0.1, Tc=1.6±0.2	
	399 b		–1.11F			0.38F		2		59	–	–	$C,i_d,R,i_\ell=kC$ for $C=$ 0.03–0.1, Tc=1.6±0.2,r	
			–1.25F			0.39F				59			$C,i_d,R,i_\ell=kC$ for $C=$ 0.03–0.1, Tc=1.6±0.2	
	399 c		–1.28F			0.74F		2	–	–	–	–	C,r	
			–1.45F			0.78F		2	–	–	–	–	C,r	
	399 d		–1.57F			0.78F		2	–	–	–	–	C,r	
JE046 0077	380 a	$E_{\frac{1}{2}}$	–0.96F	SCE	i_ℓ	1.4F	QE	1	$dE_{\frac{1}{2}}/dpH$	66	–	–	$C,\ne,p$	EE34
			–			–		1		–			C	
			–1.25F			1.4F	–	–		66	–	–	$C,\ne,p$	
			–1.48F			1F				0		CP at –1.5 V → 2-hydroxy-1-acenaphthene carbaldehyde, IRS	$C,E_{\frac{1}{2}} \ne f(pH)$ for pH= 4.5–6,$\ne$,Tc=2 for T= 15–35	
	380 b		–1.5F			1.4F	–	–	–	–	–	–	$C,\ne,p$	
			–1.63F			1.1F							$C,\ne$	
	380 c		–1.71F			2.4F	QE	2		30	–	–	$C,\ne,p$	
JE046 0077	380 a	E_p	–1.25F	SCE	i_p	11.1F	–	–	–	–	–	–	C,p	
			–1.5F			small							C,i_k	
JE049 0111	396 a	$E_{p/2}$	1.58	SCE	–	–	–	–	–	–	–	–	A,p	EE35 bc87
JE049 0111	396 a	E_p	–1.75F	SCE	i_p	100F	QE	–	–	–	–	–	C,p A	
			–1.6F			small								
			1.68F			75F		1.02				CP → 9-acridylacridinium perchlorate, CHN	A	
			–0.38F			17.6F		–				–	C,0 on first scan	

EE36 $C_{13}H_9N$

Code No.	Empirical Formula	Name and C.A. Number	Structural Formula	Solvent	Tech.	Medium		μ, M	pH	T, °C	Electrodes	App.	Experimental Parameters
EE36	$C_{13}H_9N$	benzo[c]isoquinoline	Table II	MeCN	IL	Et_4NClO_4	ns	-	-	-	Pt/SCE	2-0	A=0.22
EE37 bc88	$C_{13}H_9N$	benzo[f]quinoline C.A. 85-02-9	Table II-2	MeCN	IL	Et_4NClO_4	ns	-	-	-	Pt/SCE	2-0	A=0.22
EE38 bc89	$C_{13}H_9N$	benzo[h]quinoline C.A. 230-27-3	Table II-2	MeCN	IL	Et_4NClO_4	ns	-	-	-	Pt/SCE	2-0	A=0.22
EE39	$C_{13}H_9N_3OS$	1-(2-thiazolylazo)-2-naphthol C.A. 1147-56-4	Table II	EtOH 50	PY	ACET	0.2	-	5.65	25	DME/SCE	2-0	C=0.2, m=1.82, t(c)=1, v=5
EE40	$C_{13}H_{10}N_2O$	1-(2-pyridyl)-3-(3-pyridyl)-2-propen-1-one C.A. 13344-54-2	Table II	EtOH 50	PY	BR GEL	0.02	-	2.6	22±1	DME/SCE	OAO	C=0.5, m=1.15, t=5.2, h=64
									6.9				
									10.0				
									11.0				
									12.6				
				dioxane 75	PY	Bu_4NI GEL	0.175 0.02	-	-	22±1	DME/SCE	OAO	C=0.5, m=1.15, t=5.2, h=64
EE41	$C_{13}H_{10}N_2O$	1-(3-pyridyl)-3-(2-pyridyl)-2-propen-1-one C.A. 13309-06-3	Table II	EtOH 50	PY	BR GEL	0.02	-	2.6	22±1	DME/SCE	OAO	C=0.5, m=1.15, t=5.2, h=64
									6.9				
									8.0				
									10.0				
									11.0				
									12.6				
				dioxane 75	PY	Bu_4NI GEL	0.175 0.02	-	-	22±1	DME/SCE	OAO	C=0.5, m=1.15, t=5.2, h=64

TABLE I. Electrochemical Data $C_{13}H_{10}N_2O$ EE41

Ref.	C/M	Charact. Potential		Response Const.		Tech.	n	Electrokinetic Data			Products and Identification	Description and Remarks	Code No.
		Value	vs.		Value			Parameter	Value	From			
JE049 0111	-	$E_{p/2}$ 1.80	SCE	-	-	-	-	-	-	-	-	A,p	EE36
JE049 0111	-	$E_{p/2}$ 1.69	SCE	-	-	-	-	-	-	-	-	A,p	EE37 bc88
JE049 0111	-	$E_{p/2}$ 1.72	SCE	-	-	-	-	-	-	-	-	A,p	EE38 bc89
JE050 0113	-	$E_{\frac{1}{2}}$ -0.3	SCE	D	1.75	i:i QE	2 4	-	-	-	-	C,n≠f(pH) for pH= 1-8, n=3.2 for pH= 13, $\Delta E_{\frac{1}{2}}=0.16\sigma$,p	EE39
JE039 0419	-	$E_{\frac{1}{2}}$ -0.3 -0.56 -1.23	SCE	-	-	-	-	-	-	-	-	C,i_d,r C,i_d C	EE40
			-0.64 -0.78 -1.48 -1.65	-	-	-	-	-	-	-	-	C,r C C C,i_{cat},H	
			-0.8 -1.03 -1.75 -1.85	-	-	-	-	-	-	-	-	C,r C C C,i_{cat},H	
			-0.9 -1.12 -	-	-	-	-	-	-	-	-	C,r C C,O	
			-0.9 -1.12 -1.23	-	-	-	-	-	-	-	-	C,r C C	
JE039 0419	-	$E_{\frac{1}{2}}$ -0.92 -1.33 -2.0	SCE	-	-	-	-	-	-	-	-	C,$\Delta E_{\frac{1}{2}}=0.72\Sigma\sigma^*$,r C C	
JE039 0419	-	$E_{\frac{1}{2}}$ -0.25 -0.65	SCE	-	-	-	-	-	-	-	-	C,i_d,r C,i_d	EE41
			-0.53 -0.62 -1.08	-	-	-	-	-	-	-	-	C,r C C	
			-0.63 -0.71 -1.18 -1.75	-	-	-	-	-	-	-	-	C,r C C C	
			-0.78 -0.86 -1.35 -1.80	-	-	-	-	-	-	-	-	C,r C C C	
			-0.83 -0.93 -1.45	-	-	-	-	-	-	-	-	C,r C C	
			-0.90 -1.0 -1.15	-	-	-	-	-	-	-	-	C,r C C	
JE039 0419	-	$E_{\frac{1}{2}}$ -0.93 -1.47 -2.03	SCE	-	-	-	-	-	-	-	-	C,$\Delta E_{\frac{1}{2}}=0.72\Sigma\sigma^*$,r C C	
	-	$E_{p/2}$										A,p	

Code No.	Empirical Formula	Name and C.A. Number	Structural Formula	Solvent	Tech.	Medium		μ, M	pH	T, °C	Electrodes	App.	Experimental Parameters
EE42	$C_{13}H_{10}N_2O$	1-(3-pyridyl)-3-(3-pyridyl)-2-propen-1-one C.A. 13309-07-4	Table II	EtOH 50	PY	BR GEL	0.02	-	2.6	22±1	DME/SCE	OAO	C=0.5, m=1.15, t=5.2, h=64
									3.6				
									6.05				
									8.0				
									10.0				
									12.6				
				dioxane 75	PY	Bu_4NI GEL	0.175 0.02	-	-	22±1	DME/SCE	OAO	C=0.5, m=1.15, t=5.2, h=64
EE43	$C_{13}H_{10}N_2O$	1-(3-pyridyl)-3-(4-pyridyl)-2-propen-1-one C.A. 13328-57-9	Table II	EtOH 50	PY	BR GEL	0.02	-	2.6	22±1	DME/SCE	OAO	C=0.5, m=1.15, t=5.2, h=64
									6.05				
									10.0				
									11.0				
									12.6				
				dioxane 75	PY	Bu_4NI GEL	0.175 0.02	-	-	22±1	DME/SCE	OAO	C=0.5, m=1.15, t=5.2, h=64
EE44 bd15 ch75 de02	$C_{13}H_{10}O$	benzophenone C.A. 119-61-9	Table II-2	H_2O	PW	BENZ	0.01	-	10.1	-	DME/SCE	---	-
EE45	$C_{13}H_{11}NO_2S$	2-methyl-2'-nitro-diphenyl sulfide C.A. 6640-54-6	Table II	EtOH 50	PY	buffer		-	-	-	DME/SCE	O4AO	C=0.1, m=4.0, t=3.5
EE46 ch90	$C_{13}H_{11}NO_2S$	2-methyl-4'-nitro-diphenyl sulfide	Table II-3	EtOH 50	PY	buffer		-	-	-	DME/SCE	O4AO	C=0.1, m=4.0, t=3.5
EE47	$C_{13}H_{11}NO_2S$	3-methyl-2'-nitro-diphenyl sulfide	Table II	EtOH 50	PY	buffer		-	-	-	DME/SCE	O4AO	C=0.1, m=4.0, t=3.5

TABLE I. Electrochemical Data $C_{13}H_{11}NO_2S$ EE47

Ref.	C/M	Charact. Potential		Response Const.		n Tech.	Electrokinetic Data			Products and Identification	Description and Remarks	Code No.
		Value	vs.		Value		Parameter	Value	From			
JE039 0419	-	$E_{\frac{1}{2}}$	-0.41	SCE	-	-	-	-	-	-	-	-	c, i_d, r c, i_d	EE42
			-0.48 -0.87 -1.5		-	-	-	-	-	-	-	-	c, r c c, i_{cat}, H	
			-0.67 -0.74 -1.05 -1.67		-	-	-	-	-	-	-	-	c, r c c c, i_{cat}, H	
			-0.78 -0.87 -1.2 -1.8		-	-	-	-	-	-	-	-	c, r c c c, i_{cat}, H	
			-0.9 -1.02 -1.4		-	-	-	-	-	-	-	-	c, r c c	
			-0.95 -1.12 -1.3 -1.48		-	-	-	-	-	-	-	-	c, r c c c	
JE039 0419	-	$E_{\frac{1}{2}}$	-0.96 -1.33 -2.02	SCE	-	-	-	-	-	-	-	-	$c, \Delta E_{\frac{1}{2}} = 0.72 \Sigma \sigma^*, r$ c c	
JE039 0419	-	$E_{\frac{1}{2}}$	-0.23 -0.83	SCE	i_ℓ	1.2F 0.3F	-	-	-	-	-	-	c, i_d, r c, i_d	EE43
			-0.48 -0.58 -1.09			0.5F 0.5F 0.95F	-	-	-	-	-	-	c, r c c	
			-0.71 -0.84 -1.36			0.45F 0.23F 0.95F	-	-	-	-	-	-	c, r c c	
			-0.83 -1.42			-	-	-	-	-	-	-	c, r c	
			-1.06 -1.45			0.6F 0.7F	-	-	-	-	-	-	c, r c	
JE039 0419	-	$E_{\frac{1}{2}}$	-0.82 -1.45 -2.02	SCE	-	-	-	-	-	-	-	-	$c, \Delta E_{\frac{1}{2}} = 0.72 \Sigma \sigma^*, r$ c c	
JE048 0146	65 t	-	-	-	-	-	-	-	-	-	-	-	see BD15, CH75, DE02	EE44 bd15 ch75 de02
C0027 0525	-	$E_{\frac{1}{2}}^\circ$	-0.10	SCE	-	-	-	-	$dE_{\frac{1}{2}}/dpH$	53	sttd	-	$c, E_{\frac{1}{2}}$ measured at pH= 1-5 and extrapolated, p	EE45
C0027 0525	-	$E_{\frac{1}{2}}^\circ$	-0.06	SCE	-	-	-	-	$dE_{\frac{1}{2}}/dpH$	65	sttd	-	$c, E_{\frac{1}{2}}$ measured at pH= 1-10 and extrapolated, p	EE46 ch90
C0027 0525	-	$E_{\frac{1}{2}}^\circ$	-0.08	SCE	-	-	-	-	$dE_{\frac{1}{2}}/dpH$	62	sttd	-	$c, E_{\frac{1}{2}}$ measured at pH= 1-5 and extrapolated, p	EE47
	-	$E_{\frac{1}{2}}$			-	-	-	-	-	-	-	-		

Code No.	Empirical Formula	Name and C.A. Number	Structural Formula	Solvent	Tech.	Medium	μ, M	pH	T, °C	Electrodes	App.	Experimental Parameters
EE48	$C_{13}H_{11}NO_2S$	3-methyl-4'-nitro-diphenyl sulfide	Table II	EtOH 50	PY	buffer	-	-	-	DME/SCE	O4AO	C=0.1,m=4.0, t=3.5
								3.1				
								7.0				
								7.5				
								8.3				
								8.5				
								9.2				
EE49	$C_{13}H_{11}NO_2S$	4-methyl-2'-nitro-diphenyl sulfide C.A. 20912-17-8	Table II	EtOH 50	PY	buffer	-	-	-	DME/SCE	O4AO	C=0.1,m=4.0, t=3.5
EE50	$C_{13}H_{11}NO_3S$	4-methoxy-2'-nitro-diphenyl sulfide C.A. 3169-69-5	Table II	EtOH 50	PY	buffer	-	-	-	DME/SCE	O4AO	C=0.1,m=4.0, t=3.5
EE51	$C_{13}H_{11}NO_4S$	2-methyl-2'-nitro-diphenyl sulfone	Table II	EtOH 50	PY	buffer	-	-	-	DME/SCE	O4AO	C=0.1,m=4.0, t=3.5
EE52	$C_{13}H_{11}NO_4S$	3-methyl-2'-nitro-diphenyl sulfone	Table II	EtOH 50	PY	buffer	-	-	-	DME/SCE	O4AO	C=0.1,m=4.0, t=3.5
EE53	$C_{13}H_{11}NO_4S$	3-methyl-4'-nitro-diphenyl sulfone	Table II	EtOH 50	PY	buffer	-	-	-	DME/SCE	O4AO	C=0.1,m=4.0, t=3.5
EE54	$C_{13}H_{11}NO_5S$	4-methoxy-4'-nitro-diphenyl sulfone	Table II	EtOH 50	PY	buffer	-	-	-	DME/SCE	O4AO	C=0.1,m=4.0, t=3.5
EE55	$C_{13}H_{12}S_2$	bis(phenylthio)-methane C.A. 3561-67-9	$(C_6H_5S)_2CH_2$	MeCN	IL	Et_4NClO_4 0.1	-	-	-	Pt/SCE	21A2	C=1.2,A=0.221,v=48
EE56	$C_{13}H_{13}BrN_4O$	2-(5-bromo-2-pyridylazo)-5-(dimethylamino)phenol C.A. 50783-82-9	Table II	EtOH 50	PY	ACET 0.2	-	5.65	25	DME/SCE	2-0	C=0.2,m=1.82,t(c)=1,v=5
EE57 bd83 ch98	$C_{13}H_{14}N_2O$	4-amino-4'-methoxy-diphenylamine C.A. 101-64-4	Table II-2	MeCN	VA	$LiClO_4$ 0.25	-	-	25.0 ±0.2	PDE Ag/Ag+ 0.01	2-0	C=10,v=ns, Pt Aux
						$LiClO_4$ 0.25 $HClO_4$(anhydrous) 0.01						
						$LiClO_4$ 0.25 diphenylguanidine 0.012						
CONT												

TABLE I. Electrochemical Data $C_{13}H_{14}N_2O$ (CONT.) EE57

Ref.	C/M	Charact. Potential		Response Const.		Tech.	n	Electrokinetic Data			Products and Identification	Description and Remarks	Code No.
		Value	vs.		Value			Parameter	Value	From			
C0027 0525	-	$E_{\frac{1}{2}}^{o}$	-0.04	SCE	-	-	-	-	$dE_{\frac{1}{2}}/dpH$	68	sttd	-	$C,E_{\frac{1}{2}}^{o}$ extrapolated,p	EE48
			-0.25F		-	-	-	-		68		-	C,p	
			-0.49F		-	-	-	-		68		-	C,p	
			-0.54F		-	-	-	-		68		-	C,p	
			-0.60F		-	-	-	-		68		-	C,p	
			-0.63F		-	-	-	-		68		-	C,p	
			-0.67F		-	-	-	-		68		-	C,p	
C0027 0525	-	$E_{\frac{1}{2}}^{o}$	-0.10	SCE	-	-	-	-	$dE_{\frac{1}{2}}/dpH$	51	sttd	-	$C,E_{\frac{1}{2}}^{o}$ measured at pH= 1-5 and extrapolated,p	EE49
C0027 0525	-	$E_{\frac{1}{2}}^{o}$	-0.08	SCE	-	-	-	-	$dE_{\frac{1}{2}}/dpH$	54	sttd	-	$C,E_{\frac{1}{2}}^{o}$ measured at pH= 1-5 and extrapolated,p	EE50
C0027 0525	-	$E_{\frac{1}{2}}^{o}$	-0.07	SCE	-	-	-	-	$dE_{\frac{1}{2}}/dpH$	49	sttd	-	$C,E_{\frac{1}{2}}^{o}$ measured at pH= 1-5 and extrapolated,p	EE51
C0027 0525	-	$E_{\frac{1}{2}}^{o}$	-0.08	SCE	-	-	-	-	$dE_{\frac{1}{2}}/dpH$	55	sttd	-	$C,E_{\frac{1}{2}}$ measured at pH= 1-5 and extrapolated,p	EE52
C0027 0525	-	$E_{\frac{1}{2}}^{o}$	0.01	SCE	-	-	-	-	$dE_{\frac{1}{2}}/dpH$	65	sttd	-	$C,E_{\frac{1}{2}}$ measured at pH= 1-10 and extrapolated,p	EE53
C0027 0525	-	$E_{\frac{1}{2}}^{o}$	0.01	SCE	-	-	-	-	$dE_{\frac{1}{2}}/dpH$	68	sttd	-	$C,E_{\frac{1}{2}}$ measured at pH= 1-10 and extrapolated,p	EE54
JE056 0459	405 a	E_p	1.46	SCE	$J_p/Cv^{\frac{1}{2}}$	79.1	QE	2.1	-	-	-	CP→ diphenyl disulfide(66%),MPT, NMR,IRS,MAS; formaldehyde,MPT	$A,\not\models,p$	EE55
			1.57			-		-					$A,\not\models$	
JE050 0113	-	$E_{\frac{1}{2}}$	-0.335	SCE	D	2.37	i:i QE	2.0 4	-	-	-	-	$C, n \neq f(pH)$ for pH= 1-8,p	EE56
JE038 0127	196 d	E_p	-0.01F	Ag/Ag$^+$ 0.01	i_p	10.7F	-	-	-	-	-	-	$A, i_p/v^{\frac{1}{2}} \downarrow$ as v ↑,r	EE57 bd83 ch98
			0.46F			10F							A	
			0.39F			2.2F							C	
			-0.05F			4.8F							C	
			0.45F			13.1F	-	-	-	-	-	-	$A, i_p/v^{\frac{1}{2}} \downarrow$ as v ↑,r	
			0.36F			5.8F							C	
	197 c		0.05F -			10F -	-	-	-	-	-	-	A,r C,0	
														CONT

EE57 (CONT.) $C_{13}H_{14}N_2O$

Code No.	Empirical Formula	Name and C.A. Number	Structural Formula	Solvent	Tech.	Medium		μ, M	pH	T, °C	Electrodes	App.	Experimental Parameters
EE57 bd83 ch98	$C_{13}H_{14}N_2O$	4-amino-4'-methoxy-diphenylamine	Table 11-2	MeCN	VY	$LiClO_4$	0.25	-	-	25.0 ±0.2	RPDE Ag/Ag+ 0.01	2-0	C=10, Pt Aux
						$LiClO_4$ $HClO_4$(anhydrous)	0.25 0.005						
						$LiClO_4$ $HClO_4$(anhydrous)	0.25 0.01						
						$LiClO_4$ diphenyl-guanidine	0.25 0.005						
						$LiClO_4$ diphenyl-guanidine	0.25 0.01						
						$LiClO_4$ diphenyl-guanidine	0.25 0.02						
EE58	$C_{13}H_{14}N_2O$	1,4-dihydro-N-(benzyl)-3-pyridine-carboxamide C.A. 21104-13-2	Table 11	H_2O	PY	TRIS		-	-	-	DME/NHE	2-0	C=0.5
					VY	TRIS		-	8.3	25.0 ±0.1	RDGC NHE	2-0	C=0.5
											RGDE NHE		
											RPDE NHE		
EE59	$C_{13}H_{15}N_5O_3S$	2-[(5-nitro-2-thiazolyl)azo]-5-(diethylamino)phenol C.A. 16247-81-7	Table 11	EtOH 50	PY	ACET	0.2	-	5.65	25	DME/SCE	2-0	C=0.2, m=1.82, t(c)=1, v=5
EE60	$C_{13}H_{16}N_4OS$	5-(diethylamino)-2-(2-thiazolylazo)-phenol C.A. 10558-42-6	Table 11	EtOH 50	PY	ACET	0.2	-	5.65	25	DME/SCE	2-0	C=0.2, m=1.82, t(c)=1, v=5
EE61	$C_{13}H_{20}O_2$	1-adamantylcarbinyl acetate C.A. 778-10-0	Table 11	MeCN	IL	Bu_4NBF_4	0.1	-	-	-	Pt Ag/Ag+ 0.1	2A0	C=10, v=100
EE62 be54 ci35	$C_{14}H_8O_2$	anthraquinone C.A. 84-65-1	Table 11-2	DMF	VR	Bu_4NClO_4	1.0	-	-	20.0 ±0.3	PDE Ag/AgCl	0A2	C=0.4-1.0, Pt Aux

TABLE I. Electrochemical Data $C_{14}H_8O_2$ EE62

Ref.	C/M	Charact. Potential		Response Const.		Tech.	n	Electrokinetic Data			Products and Identification	Description and Remarks	Code No.
		Value	vs.		Value			Parameter	Value	From			
JE038 0127	196 d	$E_{\frac{1}{2}}$	-0.02	Ag/Ag^+ 0.01	i_ℓ	24.6F	QE	0.96	-	-	-	CP → N-[4-methoxy-phenyl]quinonedi-imine, λ_{max}=520	$A, E_{\frac{1}{2}} \neq f(\omega), i_\ell = k_1\omega^{\frac{1}{2}} + k_2, r$	EE57 bd83 ch98
			0.50			21.5F		2.13				CP → N-[4-methoxy-phenyl]quinonedi-imine, λ_{max}=520	$A, E_{\frac{1}{2}} \neq f(\omega), i_\ell = k_1\omega^{\frac{1}{2}} + k_2$	
			-0.02F 0.30F			83.3F 341.7F	-	-	-	-	-	-	A,r A	
			- 0.30F			- 400F	-	-	-	-	-	-	A,O,r A	
	196 c		-0.15F 0.00F 0.52F			8.0F 15.2F 8.8F	-	-	-	-	-	-	A,Pr,r A A	
			-0.15F 0.00F 0.52F			18.8F 25.6F 6.9F	-	-	-	-	-	-	A,Pr,r A A	
			-0.5F			51.3F	-	-	-	-	-	-	A,r	
JE047 0543	-	$E_{\frac{1}{2}}$	-1.09± 0.03	NHE	-	-	-	-	-	-	-	-	$C, E_{\frac{1}{2}} \neq f(pH)$ for pH= 8-10,p	EE58
JE047 0543	-	$E_{\frac{1}{2}}$	0.48± 0.04		-	-	-	-	-	-	-	-	$A, \neq, p$	
			0.60± 0.025		-	-	-	-	-	-	-	-	$A, \neq, p$	
			0.66± 0.02		i_ℓ	4F	i:i QE	2 1.55	-	-	-	monocation,UVS	$A, E_{\frac{1}{2}} \neq f(pH)$ for pH= 7-13, $E_{\frac{1}{2}} \neq f(C)$ for C=0.05-0.5, i_ℓ= $f(pH), i_\ell = kC, \neq, p$	
JE050 0113	-	$E_{\frac{1}{2}}$	-0.27	SCE	D	1.92	i:i QE	2 4	-	-	-	-	$C, n \neq f(pH)$ for pH= 1-8,p	EE59
JE050 0113	-	$E_{\frac{1}{2}}$	-0.36	SCE	D	2.15	i:i QE	2 4	-	-	-	-	$C, n \neq f(pH)$ for pH= 1-8,p	EE60
JA095 8631	393 a	$E_{p/2}$	2.36	Ag/Ag^+ 0.1	-	-	-	-	-	-	-	CP at E_{app}=2.45 V (in 0.1 $\underline{M}$ LiClO$_4$) → 7-acetoxymethyl-3-acetamidoadamantane (52%), MPT, IRS, NMR	A,p	EE61
JA096 0249	50 a	XE	-0.93	Ag/AgCl	-	-	sttd	1	-	-	-	radical anion	$C, \lim[(E_{p,A} + E_{p,C})/2]$ as $v \to 0, p$	EE62 be54 ci35
			-1.63					1				dianion	C,R	

Code No.	Empirical Formula	Name and C.A. Number	Structural Formula	Solvent	Tech.	Medium		μ, M	pH	T, °C	Electrodes	App.	Experimental Parameters
EE63 ci44 de31	$C_{14}H_9NO_2$	9-nitroanthracene C.A. 602-60-8	Table 11-3	DMF	PY	Et_4NClO_4	0.1	-	-	-	DME/SCE	26-0	C=0.5, m=1.61, t=3.6
						$KClO_4$	0.1						
					VR	$KClO_4$	0.1	-	-	-	HMDE SCE	26-0	C=0.5
EE64	$C_{14}H_9NO_2$	9-nitrophenanthrene C.A. 954-46-1	Table 11	DMF	PY	$KClO_4$	0.1	-	-	-	DME/SCE	26-0	C=0.5, m=1.61, t=3.6
						Et_4NClO_4	0.1						
EE65	$C_{14}H_9NO_2$	N-phenylphthalimide C.A. 520-03-6	Table 11	MeCN 66.7	VY	ACET		-	8	22±2	Pb RDE SCE	-A0	C=40, v=10, ω=35.8
				MeCN 63 H_2SO_4 5	VY	$HClO_4$ $NaClO_4$	ns	-	1	22±2	Pb RDE SCE	-A0	C=40, v=10, ω=15.9
EE66 be91	$C_{14}H_{10}N_2O_4$	4,4'-dinitrostilbene C.A. 2501-02-2	Table 11-2	DMF	IL	Et_4NClO_4	0.1	-	-	20.0 ±0.5	HMDE SCE	2-2	v=2.5E3
EE67	$C_{14}H_{12}O$	diphenylacetaldehyde C.A. 947-91-1	$(C_6H_5)_2CHCHO$	EtOH 2	PY	BARB LiCl		0.5	7.1	25.0 ±0.02	DME/SCE	O34A0	C=0.2, m=2.6, t=3.6, h=65±5
						BOR LiCl			9.0				
									10.1				
						LiOH LiCl	ns		11.45				
									13.5				
EE68	$C_{14}H_{14}$	bibenzyl C.A. 103-29-7	$(C_6H_5CH_2)_2$	H_2O	PO	H_2SO_4	1.0	-	-	-	DME/SCE	O4A2	C=0.4
EE69	$C_{14}H_{14}N_2O_3$	3-hydroxy-1-(2-hydroxyphenyl)-3-(2-pyridyl)-1-propanone oxime C.A. 51210-86-7	Table 11	DMF 20	PY	BR		-	2.3	-	DME/SCE	OA0	C=0.25, m=3.45, t=2.72, h=40
									3.95				
									6.8				
									7.8				

TABLE I. Electrochemical Data $C_{14}H_{14}N_2O_3$ EE69

Ref.	C/M	Charact. Potential		Response Const.		n	Tech.	Electrokinetic Data			Products and Identification	Description and Remarks	Code No.
		Value	vs.		Value			Parameter	Value	From			
JE039 0395	-	$E_{\frac{1}{2}}$	-0.995 -1.575	SCE	I	1.6 1.2	-	-	Elog	48 -	-	-	C,r C	EE63 ci44 de31
			-0.94			1.9	-	-		53	-	-	C,$\Delta E_{\frac{1}{2}} \neq \rho\sigma$ or k(LUMO), r	
			-1.43			1.6				-			C	
JE039 0395	-	E_p	-0.96F -0.15F -0.35F	SCE	i_p	1.1F 0.3F 0.1F	-	-	-	-	-	-	C,p A C,0 on first scan	
JE039 0395	-	$E_{\frac{1}{2}}$	-0.98	SCE	I	1.8	-	-	Elog	72	-	-	C,$\Delta E_{\frac{1}{2}}=0.009\sigma$,$\Delta E_{\frac{1}{2}} \propto \Delta\beta$,r	EE64
			-1.46			2.5				-			C	
			-1.02			1.5	-	-		61	-	-	C,$\Delta E_{\frac{1}{2}}=0.010\sigma$,$\Delta E_{\frac{1}{2}} \propto \Delta\beta$,r	
			-1.685			2.9				-			C	
JE057 0351	411 c	$E_{\frac{1}{2}}$	-1.25 -1.45	SCE	-	-	-	-	-	-	-	-	C,p C	EE65
JE057 0351	411 b	$E_{\frac{1}{2}}$	-1.025	SCE	-	-	-	-	dlogi/dE	80	-	-	C,p	
JE047 0215	-	E_p	-0.91F	SCE	i_p	30F	-	-	-	-	-	-	C,p	EE66 be91
JE046 0343	32 e	$E_{\frac{1}{2}}$	-1.505	SCE	i_ℓ	0.45	-	-	-	-	-	-	C,r	EE67
			-1.58			0.68	-	-	-	-	-	-	A,r C	
			-1.64			0.95	-	-	-	-	-	-	A,r C	
			-1.70			0.7	-	-	-	-	-	-	A,r C	
			-			0.05	-	-	-	-	-	-	A,$i_\ell \neq f(pH)$ for pH > 12,r C	
C0028 0838	65 w	E_i	0.02	SCE	-	-	-	-	-	-	-	-	C,i_{cat},P	EE68
			-1.18										C,i_{cat}	
			-1.18										A,i_{cat}	
			-0.05										A,i_{cat}	
			0.02										A,i_{cat}	
JE048 0297	109 d	$E_{\frac{1}{2}}$	-0.94	SCE	i_ℓ	2.64	i:i	4	$dE_{\frac{1}{2}}/dpH$	65±5	sttd	-	C,W,i_d,i_ℓ=kC and $kh^{\frac{1}{2}}$,r	EE69
			-1.06			2.5	-	-		65±5		-	C,r	
			-1.3			1.2	-	-		65±5		-	C,r	
			-1.35			0.4	-	-		65±5		-	C,r	
		$E_{\frac{1}{2}}$	-0.995		I	1.6	-	-	Elog	48	-			

EE70 $C_{14}H_{14}N_2O_3$ CRC Handbook Series in Organic Electrochemistry

Code No.	Empirical Formula	Name and C.A. Number	Structural Formula	Solvent	Tech.	Medium	μ, M	pH	T, °C	Electrodes	App.	Experimental Parameters
EE70	$C_{14}H_{14}N_2O_3$	3-hydroxy-1-(2-hydroxyphenyl)-3-(4-pyridyl)-1-propanone oxime C.A. 51210-87-8	Table II	DMF 20	PY	BR	-	2.3 3.95 6.8 7.80	-	DME/SCE	OAO	C=0.25,m=3.45,t=2.72, h=40
EE71	$C_{14}H_{15}ClN_2O_4S$	5-chloro-N-cyclohexyl-4-sulfamyl phthalimide C.A. 3822-99-9	Table II	MeCN 66.7	VY	ACET	-	4.5 8	22±2	Pb RDE SCE	-AO	C=40,v=10, ω=15.9 ω=35.8
				MeCN 63 H_2SO_4 5	VY	$HClO_4$ $NaClO_4$ ns	-	1	22±2	Pb RDE SCE	-AO	C=40,v=10, ω=15.9
EE72	$C_{14}H_{16}$	1-methyl-7-isopropylazulene	Table II	EtOH	PY	LiCl H_2SO_4 LiCl	0.1 1.0 0.1	-	-	25	DME/MP	OAO	C=1.0
EE73 bg19 cj07	$C_{14}H_{16}N_2$	o-tolidine	Table II-2	H_2O	VY	ACET PHOS	-	0.7	-	Pt PDS SCE	2-O	C=0.1,v= 0.92
EE74 cj16	$C_{14}H_{20}O_2$	2,6-di(tert-butyl)-1,4-benzoquinone C.A. 719-22-2	Table II-3	2-PrOH 40	PY	PHOS	-	7	-	DME/NCE	O5AO	C=0.1
EE75	$C_{14}H_{24}N_2O_4S$	tetraethylammonium 4-nitrobenzene-sulfinate	$(C_2H_5)_4N^+$ $4-O_2NC_6H_4SOO^-$	DMF	VA	Et_4NClO_4	0.1	-	-	22.5 ±0.5	PBE/SCE	256A FO	C=0.153,v= 100,Pt Aux
EE76 (see also EB44)	$C_{14}H_{25}N_3O_4S$	tetraethylammonium 4-nitrobenzene-sulfonamidate	$(C_2H_5)_4N^+$ $4-O_2NC_6H_4SO_2NH^-$	DMF	VA	Et_4NClO_4	0.1	-	-	22.5 ±0.5	Pt/SCE	256A FO	C=0.63,A= 0.25,v=100, Pt Aux
EE77	$C_{15}H_{11}N_3O$	1-(2-pyridylazo)-2-naphthol C.A. 85-85-8	Table II	EtOH 50	PY	ACET NaOH	0.2 0.2	-	5.65	25	DME/SCE	2-O	C=0.2,m= 1.82,t(c)= 1,v=5
EE78	$C_{15}H_{11}N_3O$	1-(3-pyridylazo)-2-naphthol C.A. 1533-65-9	Table II	EtOH 50	PY	ACET	0.2	-	5.65	25	DME/SCE	2-O	C=0.2,m= 1.82,t(c)= 1,v=5
EE79	$C_{15}H_{11}N_3O$	1-(4-pyridylazo)-2-naphthol C.A. 16219-95-7	Table II	EtOH 50	PY	ACET	0.2	-	5.65	25	DME/SCE	2-O	C=0.2,m= 1.82,t(c)= 1,v=5

TABLE I. Electrochemical Data $C_{15}H_{11}N_3O$ EE79

Ref.	C/M	Charact. Potential		Response Const.		n		Electrokinetic Data			Products and Identification	Description and Remarks	Code No.
		Value	vs.		Value	Tech.		Parameter	Value	From			
JE048 0297	109 d	$E_{\frac{1}{2}}$	-0.92	SCE	i_ℓ	2.52	i:i	4	$dE_{\frac{1}{2}}/dpH$	65±5	sttd	-	$C,W,i_d,i_\ell=kC$ and $kh^{\frac{1}{2}},r$	EE70
			-1.05			2.60	-	-		65±5		-	C,r	
			-1.32			1.84	-	-		65±5		-	C,r	
			-1.45			1.54	-	-		-		-	C,2 merging waves,r	
JE057 0351	411 c	$E_{\frac{1}{2}}$	-0.97F	SCE	-	-	-	-	-	-	-	-	C,p	EE71
			-1.02		-	-	-	-	-	-	-	-	C,p	
			-1.25		-	-	-	-	-	-	-	-	C	
JE057 0351	411 b	$E_{\frac{1}{2}}$	-0.81	SCE	-	-	-	-	dlogi/dE	70	-	-	C,p	
C0026 0370	-	$E_{\frac{1}{2}}$	-0.55	NCE	i_ℓ	4F	i:i	1	-	-	-	-	C,p	EE72
			-1.78			7.0	llk	2	-	-	-	-	C,p	
JE052 0105	-	$E_{p/2}$	0.43F	SCE	i_ℓ	1.2F	-	-	dlogi/dlogX	-0.5	-	-	A,X=renewal time of diffusion layer,Tc=1.4,$i_\ell=kC$,S,$E_{\frac{1}{2}}(Ap)$,p	EE73 bg19 cJ07
			0.5F			0.78F				-0.46			A,Tc=1.4,$i_\ell=kC$, $E_{\frac{1}{2}}(Ap)$,p	
C0030 4192	-	$E_{\frac{1}{2}}$	-0.320	NCE	-	-	-	-	-	-	-	-	C,p	EE74 cJ16
JE053 0293	-	E_p	-1.15F	SCE	i_p	21.4F	-	-	-	-	-	dianion	C,R,r	EE75
			-1.07F			15.7F							A,R	
JE053 0293	412 c	E_p	-1.15F	SCE	i_p	27.1F	-	-	-	-	-	dianion	C,R,r	EE76
			-1.08F			14.3F							A,R	
JE050 0113	-	$E_{\frac{1}{2}}$	-0.305	SCE	D	2.15	i:i QE	2 4	-	-	-	-	C,i_ℓ,n≠f(pH) for pH=1-8,$\Delta E_{\frac{1}{2}}=0.16\sigma$,p	EE77
		"	"	"	"	"	i:i	3.1	-	-	-	-	C,p	
JE050 0113	-	$E_{\frac{1}{2}}$	-0.406	SCE	-	-	i:i QE	2.9 4	-	-	-	-	C,$\Delta E_{\frac{1}{2}}=0.16\sigma$,p	EE78
JE050 0113	-	$E_{\frac{1}{2}}$	-0.232	SCE	D	2.39	i:i QE	2.0 4	-	-	-	-	C,$i_\ell \neq f(pH)$ for pH=1-8,$\Delta E_{\frac{1}{2}}=0.16\sigma$,p	EE79

Code No.	Empirical Formula	Name and C.A. Number	Structural Formula	Solvent	Tech.	Medium		μ, M	pH	T, °C	Electrodes	App.	Experimental Parameters
EE80	$C_{15}H_{11}N_3O$	4-(2-pyridylazo)-1-naphthol C.A. 35140-51-3	Table II	EtOH 50	PY	ACET NaOH	0.2 0.1	-	5.65	25	DME/SCE	2-0	C=0.2, m=1.82, t(c)=1, v=5
EE81	$C_{15}H_{12}N_4$	4-(2-pyridylazo)-1-naphthylamine	Table II	EtOH 50	PY	ACET	0.2	-	5.65	25	DME/SCE	2-0	C=0.2, m=1.82, t(c)=1, v=5
EE82	$C_{15}H_{17}BrN_4O$	2-(5-bromo-2-pyridylazo)-5-(diethylamino)phenol C.A. 14337-53-2	Table II	EtOH 50	PY	ACET NaOH	0.2 0.2	-	5.65	25	DME/SCE	2-0	C=0.2, m=1.82, t(c)=1, v=5
EE83	$C_{15}H_{17}ClN_4O$	2-(5-chloro-2-pyridylazo)-5-ethylamino C.A. 26015-51-0	Table II	EtOH 50	PY	ACET	0.2	-	5.65	25	DME/SCE	2-0	C=0.2, m=1.82, t(c)=1, v=5
EE84	$C_{15}H_{17}I$	2-(5-iodo-2-pyridylazo)-5-ethylamino C.A. 14493-15-3	Table II	EtOH 50	PY	ACET	0.2	-	5.65	25	DME/SCE	2-0	C=0.2, m=1.82, t(c)=1, v=5
EE85	$C_{15}H_{18}N_4O$	5-(diethylamino)-2-(2-pyridylazo)phenol C.A. 14337-52-1	Table II	EtOH 50	PY	ACET	0.2	-	5.65	25	DME/SCE	2-0	C=0.2, m=1.82, t(c)=1, v=5
EE86	$C_{16}H_9NO_2$	1-nitropyrene C.A. 5522-43-0	Table II	DMF	PY	$KClO_4$ Et_4NClO_4	0.1 0.1	-	-	-	DME/SCE	26-0	C=0.5, m=1.61, t=3.6
EE87	$C_{16}H_{10}N_2$	dibenzo[f,h]quinoxaline C.A. 217-68-5	Table II	MeCN	VA	Et_4NClO_4	ns	-	-	-	Pt/SCE	2-0	A=0.22
EE88	$C_{16}H_{10}O_2$	dibenzoylacetylene C.A. 1087-09-8	$C_6H_5COC{:}CCOC_6H_5$	DMF	VR	Bu_4NClO_4	1.0	-	-	20.0 ±0.3	PDE Ag/AgCl	2A2	C=0.4-1.0, Pt Aux
EE89	$C_{16}H_{11}N_3O_3$	2-hydroxynaphthalene-1-azo-(2'-nitrobenzene) C.A. 6410-09-9	Table II	EtOH 80	PY	ACET TX100	0.2 2E-3	-	4.48	25	DME/SCE	2-0	C=0.2, m=1.82, t(c)=1, v=5
EE90	$C_{16}H_{11}N_3O_3$	2-hydroxynapthalene-1-azo-(4'-nitrobenzene) C.A. 6410-10-2	Table II	EtOH 50 EtOH 80	PY PY	NaOH TX100 ACET TX100	0.2 2E-3 0.2 2E-3	- -	- 4.48	25 25	DME/SCE DME/SCE	2-0 2-0	C=0.2, m=1.82, t(c)=1, v=5 C=0.2, m=1.82, t(c)=1, v=5

TABLE I. Electrochemical Data $C_{16}H_{11}N_3O_3$ EE90

| Ref. | C/M | Charact. Potential | | Response | Const. | Tech. | n | Electrokinetic Data | | | Products and Identification | Description and Remarks | Code No. |
		Value	vs.		Value			Parameter	Value	From			
JE050 0113	-	$E_{\frac{1}{2}}$	-0.278	SCE	D	2.02	i:i QE	2 4	-	-	-	-	$C, i_l \neq f(pH)$ for pH= 1-8,p	EE80
	-	-	-	-	-	-	i:i	2.7	-	-	-	-	C,p	
JE050 0113	-	$E_{\frac{1}{2}}$	-0.375	SCE	D	1.95	i:i	2.0	-	-	-	-	$C, i_l \neq f(pH)$ for pH= 1-8,p	EE81
JE050 0113	-	$E_{\frac{1}{2}}$	-0.33	SCE	D	2.11	i:i QE	2.0 4	-	-	-	-	$C, i_l \neq f(pH)$ for pH= 1-8, $\Delta E_{\frac{1}{2}}=0.14\sigma$, p	EE82
	-	-	-	-	-	-	i:i	2.95	-	-	-	-	C,p	
JE050 0113	-	$E_{\frac{1}{2}}$	-0.328	SCE	D	2.11	i:i QE	2.0 4	-	-	-	-	$C, i_l \neq f(pH)$ for pH= 1-8, $\Delta E_{\frac{1}{2}}=0.14\sigma$, p	EE83
JE050 0113	-	$E_{\frac{1}{2}}$	-0.333	SCE	D	1.93	i:i QE	2 4	-	-	-	-	$C, i_l \neq f(pH)$ for pH= 1-8, $\Delta E_{\frac{1}{2}}=0.14\sigma$, p	EE84
JE050 0113	-	$E_{\frac{1}{2}}$	-0.323	SCE	D	1.95	i:i QE	2.0 4	-	-	-	-	$C, i_l \neq f(pH)$ for pH= 1-8, $\Delta E_{\frac{1}{2}}=0.14\sigma$, p	EE85
JE039 0395	-	$E_{\frac{1}{2}}$	-0.925	SCE	I	1.8	-	-	Elog	64	-	-	$C, \Delta E_{\frac{1}{2}}=0.009\sigma, \Delta E_{\frac{1}{2}} \propto \Delta\beta, r$ C	EE86
			-1.375			4.2				-			C	
			-0.985			1.5				51	-		$C, \Delta E_{\frac{1}{2}}=0.01\sigma, \Delta E_{\frac{1}{2}} \propto \Delta\beta, r$ C	
			-1.605			2.6				-				
JE049 0111	-	$E_{p/2}$	1.85 -	SCE	$j_p/Cv^{\frac{1}{2}}$	9.28 10.3	-	-	-	-	-	-	A,p C, dibenzo[f,h]- quinoxaline redn.	EE87
JA096 0249	50 a	XE	-0.92	Ag/AgCl	-	-	sttd	1	-	-	-	radical anion	C,R for v=(1-10)E3, $XE=lim[(E_{p,A}+E_{p,C})/2]$ as v→0,p	EE88
			-1.58					1				dianion	C,R for v=(1-10)E3	
JE052 0115	-	$E_{\frac{1}{2}}$	-0.265	SCE	-	-	i:i	2	$dE_{\frac{1}{2}}/dpH$	57	-	-	$C, E_{\frac{1}{2}}=-0.013-0.057pH$ for pH=2-6, $\Delta E_{\frac{1}{2}}= 0.14\sigma$, p	EE89
JE052 0115	-	-	-	-	-	-	i:i	2.4	-	-	-	-	$C, \Delta E_{\frac{1}{2}}=0.14\sigma$, p	EE90
JE052 0115	-	$E_{\frac{1}{2}}$	-0.298	SCE	-	-	i:i	2	$dE_{\frac{1}{2}}/dpH$	54	sttd	-	$C, E_{\frac{1}{2}}=-0.056-0.054pH$ for pH=2-6, p	

Code No.	Empirical Formula	Name and C.A. Number	Structural Formula	Solvent	Tech.	Medium		μ, M	pH	T, °C	Electrodes	App.	Experimental Parameters
EE91	$C_{16}H_{12}N_2O$	2-hydroxynaphthalene-1-azobenzene C.A. 842-07-9	Table II	EtOH 50	PY	ACET TX100	0.2 2E-3	-	5.65	25	DME/SCE	2-O	C=0.2,m= 1.82,t(c)= 1,v=5
EE92	$C_{16}H_{12}N_2O$	4-hydroxynaphthalene-1-azobenzene C.A. 3651-02-3	Table II	EtOH 50	PY	ACET TX100	0.2 2E-3	-	5.65	25	DME/SCE	2-O	C=0.2,m= 1.82,t(c)= 1,v=5
EE93	$C_{16}H_{12}N_2O_2$	1-(2-hydroxyphenyl-azo)-2-naphthol C.A. 4866-98-2	Table II	EtOH 50	PY	ACET TX100	0.2 2E-3	-	5.65	25	DME/SCE	2-O	C=0.2,m= 1.82,t(c)= 1,v=5
EE94 bi39 ck14	$C_{16}H_{12}O_2$	trans-dibenzoyl-ethene C.A. 959-28-4	Table II-2	DMF	VR	Bu_4NClO_4	1.0	-	-	20.0 ±0.3	PDE Ag/AgCl	2A2	C=0.4-1.0, Pt Aux
EE95	$C_{16}H_{18}O_2$	2,3-diphenyl-2,3-butanediol C.A. 1636-34-6	Table II	H_2O	PO	H_2SO_4 H_2SO_4	8.0 1.0	-	-	-	DME/SCE	O4A2	C=1.0
EE96	$C_{16}H_{34}$	hexadecane C.A. 544-76-3	$CH_3(CH_2)_{14}CH_3$	CF_3COOH	IL	FSO_3H	2.5	-	-	-	Pt/SCE	25-2	C=30,v=50, Pt Aux
EE97 bJ73	$C_{17}H_{11}N$	benz[c]acridine C.A. 225-51-4		MeCN	IL	Et_4NClO_4	ns	-	-	-	Pt/SCE	2-O	A=0.22
EE98	$C_{18}H_{10}O_2$	1,4-dibenzoylbutadi-yne C.A. 21675-25-2	$C_6H_5COC\vdots CC\vdots CCO-C_6H_5$	DMF	VR	Bu_4NClO_4	1.0	-	-	20.0 ±0.3	PDE Ag/AgCl	2A2	C=0.4-1.0, Pt Aux
EE99	$C_{18}H_{11}NO_2$	7-nitrobenz[a]-anthracene C.A. 20268-51-3	Table II	DMF	PY	$KClO_4$ Et_4NClO_4	0.1 0.1	-	-	-	DME/SCE	26-O	C=0.5,m= 1.61,t=3.6
EFOO c104	$C_{18}H_{11}NO_2$	6-nitrochrysene C.A. 7496-02-8	Table II-3	DMF	PY	$KClO_4$ Et_4NClO_4	0.1 0.1	-	-	-	DME/SCE	26-O	C=0.5,m= 1.61,t=3.6

TABLE I. Electrochemical Data $C_{18}H_{11}NO_2$ EF00

Ref.	C/M	Charact. Potential		Response Const.		n Tech.		Electrokinetic Data			Products and Identification	Description and Remarks	Code No.
		Value	vs.		Value			Parameter	Value	From			
JE052 0115	-	$E_{\frac{1}{2}}$	-0.515	SCE	-	-	i:i	4	$dE_{\frac{1}{2}}/dpH$	90	sttd	-	$C, E_{\frac{1}{2}} = -0.006-0.09$pH for pH=2-6, $\Delta E_{\frac{1}{2}} = 0.14\sigma, p$	EE91
JE052 0115	197 e	$E_{\frac{1}{2}}$	-0.371	SCE	-	-	-	-	-	-	-	-	$C, \Delta E_{\frac{1}{2}} = 0.14\sigma, p$	EE92
JE052 0115	-	$E_{\frac{1}{2}}$	-0.385	SCE	-	-	i:i	4	$dE_{\frac{1}{2}}/dpH$	64	sttd	-	$C, E_{\frac{1}{2}} = -0.033-0.064$pH for pH=2-6, $\Delta E_{\frac{1}{2}} = 0.14\sigma, p$	EE93
JA096 0249	50 a	XE	-0.92	Ag/AgCl	-	-	sttd	1	-	-	-	radical anion	$C, XE= \lim [(E_{p,A} + E_{p,C})/2]$ as $v \to 0, R, p$	EE94 bi39 ck14
			-1.58					1				dianion	C,R	
C0028 0838	-	E_i	0.0	SCE	-	-	-	-	-	-	-	-	C, i_{cat}, X, p	EE95
			0.02										A, i_{cat}	
			-0.30										A, i_{cat}	
			0.0	SCE	-	-	-	-	-	-	-	-	C, i_{cat}, p	
			-0.55										C, i_{cat}	
			0.0										A, i_{cat}	
			-0.45										A, i_{cat}	
JE054 0181	-	$E_{p/2}$	1.95	SCE	-	-	-	-	-	-	-	-	$A, \neq, p$	EE96
JE049 0111	-	$E_{p/2}$	1.73	SCE	-	-	-	-	-	-	-	-	A,p	EE97 bJ73
JA096 0249	50 a	XE	-0.83	Ag/AgCl	-	-	sttd	1	-	-	-	radical anion	$C, XE= \lim [(E_{p,A} + E_{p,C})/2]$ as $v \to 0, R$ for $v=(1-10)E3, p$	EE98
			-1.33					1				dianion	C,R for v=(1-10)E3	
JE039 0395	-	$E_{\frac{1}{2}}$	-1.025	SCE	I	1.9	-	-	Clog	54	-	-	$C, \Delta E_{\frac{1}{2}} \neq \rho\sigma, E_{\frac{1}{2}} \not\propto \Delta\beta, r$	EE99
			-1.375			1.5				-			C	
			-1.045			1.7	-	-		55	-	-	C,r	
			-1.507			2.4				-			C	
JE039 0395	-	$E_{\frac{1}{2}}$	-0.945	SCE	I	1.8	-	-	Elog	77	-	-	$C, \Delta E_{\frac{1}{2}} = 0.009\sigma, \Delta E_{\frac{1}{2}} \propto \Delta\beta, r$	EF00 c104
			-1.39			3.8				-			C	
			-0.985			1.5	-	-		60	-	-	$C, \Delta E_{\frac{1}{2}} = 0.01\sigma, \Delta E_{\frac{1}{2}} \propto \Delta\beta, r$	
			-1.58			2.8				-			C	

Code No.	Empirical Formula	Name and C.A. Number	Structural Formula	Solvent	Tech.	Medium	μ, M	pH	T, °C	Electrodes	App.	Experimental Parameters
EF01 bk91	$C_{18}H_{14}N_4$	4-(phenylazo)azobenzene C.A. 1161-45-1	Table 11-2	EtOH 80	PY	ACET	0.2	3.70	25	DME/SCE	2-0	C=0.1, m=1.82, t(c)=1, v=5
								5.42				
						en	0.2	9.53				
EF02 bl12 cl22	$C_{18}H_{15}N$	triphenylamine C.A. 603-34-9	$(C_6H_5)_3N$	$AlCl_3$ 50 NaCl 50	VA	ionic melt	-	-	175	W button Al	2-2	0.5→1.5→0.5 V, C=5.53, v=500
EF03 df97	$C_{18}H_{24}N_5O_{13}P$	cytidylyl (3'→5') uridine C.A. 2382-64-1	Table 11-4	H_2O	PV	MB	0.5	2.5	25	DME/SCE	12A2	C=0.05, m=1, t(c)=3, f=50, Δe=3.54(rms), Pt Aux
								5				
						MB, CARB		6				
								9.2				
EF04 c137 df98	$C_{18}H_{25}N_6O_{12}P$	cytidylyl (3'→5') cytidine C.A. 2536-99-4	Table 11-3	H_2O	PY	MB	0.5	2.5	25	DME/SCE	12A0	C=0.05, m=1.7-2.2, t=3-4, Pt Aux
								5.5				
								6.0				
								8				
								10.6				
						MB	0.13	5.0	0.5			C=0.025
												C=0.05
												C=0.09
												C=0.18
												C=0.36
CONT		4-(phenylazo)azobenzene		EtOH 80	PY	ACET						

TABLE I. Electrochemical Data $C_{18}H_{25}N_6O_{12}P$ (CONT.) EF04

Ref.	C/M	Charact. Potential		Response Const.		Tech.	n	Electrokinetic Data			Products and Identification	Description and Remarks	Code No.
		Value	vs.		Value			Parameter	Value	From			
JE052 0115	197 f	$E_{\frac{1}{2}}$	-0.100	SCE	-	-	i:i	1.9	$dE_{\frac{1}{2}}/dpH$	64	sttd	-	$C, E_{\frac{1}{2}}=0.128-0.064pH$ for pH=3-8,p	EF01 bk91
			-0.250					5.1		79			$C, E_{\frac{1}{2}}=0.046-0.079pH$ for pH=3-8	
			-0.230		-	-	i:i QE	1.9 1.96		64			C,p	
			-0.402				i:i QE	4.0 5.85		79			C	
	197 g		-0.468		-	-	i:i	2	-	-	-		C,p	
			-0.642					2					C	
JE038 0476	-	E_p	1.14F	Al	i_p	6.5F	-	-	-	-	-	-	A, E_p and $i_p/v^{\frac{1}{2}} \neq$ f(v) for v=100 1E5,p	EF02 b112 c122
			1.05F			4.4F							C	
JA095 8495	-	E_{su}	-1.11c	SCE	-	-	-	-	dE_{su}/dpH	+7	sttd	-	$C, E_{su}=-1.13+0.007pH,p$	EF03
			-1.10c	-	-	-	-	-		+7		-	C,r	
			1.07c	-	-	-	-	-		+38			$C, E_{\frac{1}{2}}=-1.30+0.038pH, p$	
			-0.95c	-	-	-	-	-		+38		-	C,r	
JA095 8495	89 d	$E_{\frac{1}{2}}$	-1.143c	SCE	-	-	-	-	$dE_{\frac{1}{2}}/dpH$	41	sttd	-	$C, Pr, \neq, E_{\frac{1}{2}}=-1.046-0.041pH, r$	EF04 c137 df98
			-1.225c				sttd	2		56			$C, i_d, E_{\frac{1}{2}}=-1.08-0.056pH$ for pH=2.5-6	
			-1.266c	SCE	-	-	-	-		41		-	$C, Pr, \neq, r$	
			-1.37c				sttd	2		104			$C, E_{\frac{1}{2}}=-0.800-0.104pH$ for pH=4.6-8.1	
			-1.393c	-	-	-	-	-		56			C, i_d	
			-1.421c	-	-	-	-	-		56		-	C, i_d, no $Pr, \neq, r$	
			-1.632c	-	-	-	-	-		104		-	$C, i_d, \neq, r$	
			-1.766	-	-	-	-	-		51		-	$C, i_d, \neq, E_{\frac{1}{2}}=-1.225\pm 0.051pH$ for pH=8.1-10.6, r	
			-		I	5.57F	-	-		-	-	-	$C, Pr, \neq, r$	
			-			4.0F	-	-		-	-	-	$C, Pr, \neq, r$	
			-1.384F			1.54F							C, i_d	
			-			2.28F	-	-		-		-	$C, Pr, \neq, r$	
			-1.394F			3.29F							C, i_d	
			-			1.43F	-	-		-		-	$C, Pr, \neq, r$	
			-1.420F			4.14F							C, i_d	
			-			0.86F	-	-		-		-	$C, Pr, \neq, r$	
			-1.446F			4.57F							C, i_d	CONT

213

EF04 (CONT.) $C_{18}H_{25}N_6O_{12}P$

Code No.	Empirical Formula	Name and C.A. Number	Structural Formula	Solvent	Tech.	Medium	μ, M	pH	T, °C	Electrodes	App.	Experimental Parameters
EF04 c137 df98	$C_{18}H_{25}N_6O_{12}P$	cytidylyl (3'→5') cytidine	Table 11-3	H_2O	IL	Cl⁻,MB	0.05	2.0	25.0 ±0.5	HMDE SCE	12A2	C=0.05, A=0.022, v=300, Pt Aux
								5.0				
								5.5				
								5.9				
								6.1				
								8.0				
								9.0				
								10.0				
					PV	Cl⁻,MB	0.5	2.0	25	DME/SCE	12A2	C=0.05, m=1, t(c)=3, f=50, Δe=3.54(rms), Pt Aux
								4.2				
								4.5				
								7.0				
								9				
								10				
EF05	$C_{19}H_{13}N$	9-phenylacridine C.A. 602-56-2	Table 11	MeCN	IL	Et_4NClO_4	ns	-	-	Pt/SCE	2-0	A=0.22
EF06 bm09 c159	$C_{19}H_{14}O$	4-phenylbenzophenone C.A. 2128-93-0	$4\text{-}C_6H_5C_6H_4C(O)\text{-}C_6H_5$	H_2O	PW	BENZ	0.01	-	10.1	-	DME/SCE	2-2	C=1, t=60-100, MP Aux
						BENZ	0.1					
EF07	$C_{19}H_{20}N_4O$	5-(diethylamino)-2-(2-quinolylazo)-phenol C.A. 42485-44-9	Table 11	EtOH 50	PY	ACET	0.2	-	5.65	25	DME/SCE	2-0	C=0.2, m=1.82, t(c)=1, v=5
EF08	$C_{19}H_{21}N$	1,2,4,5,7,8-hexamethylacridine C.A. 40505-10-0	Table 11	MeCN	IL	Et_4NClO_4	ns	-	-	Pt/SCE	2-0	A=0.22
		cytidylyl (3'→5') cytidine		H_2O	IL	Cl⁻,MB	0.05					

TABLE I. Electrochemical Data $\quad C_{19}H_{21}N$ EF08

Ref.	C/M	Charact. Potential		Response Const.		n Tech.		Electrokinetic Data			Products and Identification	Description and Remarks	Code No.
		Value	vs.		Value			Parameter	Value	From			
JA095 8495	89 d	E_p	-1.22c	SCE	-	-	-	-	dE_p/dpH	42	sttd	-	C,≠,E_p= -1.14- 0.042pH for pH=2-5.3,r	EF04 c137 df98
			-1.35c		-	-	-	-		42		-	C,≠,r	
			-1.38c		-	-	-	-		210		-	C,≠,E_p= -0.23- 0.210pH for pH= 5.3-6,r	
			-1.47c		-	-	-	-		210		-	C,≠,r	
			-1.52c		-	-	-	-		101		-	C,≠,E_p= -0.90- 0.101pH for pH=6-8,r	
			-1.71c		-	-	-	-		101		-	C,≠,r	
			-1.76c		-	-	-	-		76		-	C,≠,E_p= -1.08- 0.076pH for pH= 9-10,r	
			-1.84c		-	-	-	-		76		-	C,≠,r	
JA095 8495	89 d	E_{su}	-1.12c		-	-	-	-	dE_{su}/dpH	19		-	C,E_{su}= -1.08- 0.019pH for pH=2-4.5,r	
			-1.29c		-	-	-	-		41		-	C,E_{su}= -1.21- 0.041pH for pH=2.0-4.2	
			-1.16c		-	-	-	-		19		-	C,r	
			-1.38c		-	-	-	-		41		-	C	
			-1.18c		-	-	-	-		0		-	C,E_{su}=-1.18 for pH= 4.5-5.7,r	
			-1.51c		-	-	-	-		108		-	C,E_{su}= -1.02- 0.108pH for pH=4.2-8	
			-1.062c		-	-	-	-		+94		-	C,E_{su}= -1.72 + 0.094pH for pH=5.7-8,r	
			-1.776c							108			C	
			-0.939c		-	-	-	-		+19		-	C,E_{su}= -1.11 + 0.019pH for pH=9-10,r	
			-0.920c		-	-	-	-		+19		-	C,r	
JE049 0111	-	$E_{p/2}$	1.55	SCE	-	-	-	-	-	-	-	-	A,p	EF05
JE048 0146	65 t	-	-	-	-	-	-	-	$dE_p/dlogv$	31	-	-	C,E_p≠f(C),p	EF06 bm09 c159
		-	-	-	-	-	-	-		41	-	-	C,p	
JE050 0113	-	$E_{\frac{1}{2}}$	-0.318	SCE	D	1.83	i:i QE	2.0 4	-	-	-	-	C,i_l≠f(pH) for pH= 1-8,p	EF07
JE049 0111	-	$E_{p/2}$	1.20	SCE	-	-	-	-	-	-	-	-	A,p	EF08

Code No.	Empirical Formula	Name and C.A. Number	Structural Formula	Solvent	Tech.	Medium	μ, M	pH	T, °C	Electrodes	App.	Experimental Parameters
EF09	$C_{19}H_{24}N_7O_{12}P$	uridylyl (3'→5') adenosine C.A. 3256-24-4	Table II	H_2O	PY	MB	0.5	2.0	25	DME/SCE	12A0	C=0.057,m= 1.7-2.2,t= 3-4,Pt Aux
								4.6				
							-	5.0	0			C=0.078
							0.13	5.0	0.5			C=0.15
												C=0.325
							0.5	5.5	25			C=0.057
					IL	MB	0.05	2.0	25.0 ±0.5	HMDE SCE	12A2	C=0.05,A= 0.022,v= 300,Pt Aux
								4.7				
					PV	MB	0.5	2.0	25	DME/SCE	12A2	C=0.05,m=1, t(c)=3,f=50, Δe=3.54(rms), Pt Aux
								5.0				
					PVI	MB	0.5	5.0	0	DME/SCE	12A0	C=0.078,m= 1.0,t(c)=3, f=50,Δe= 3.54(rms), Pt Aux
EF10	$C_{19}H_{24}N_7O_{12}P$	uridylyl (5'→3') adenosine C.A. 3051-84-1	Table II	H_2O	PY	MB	0.5	2.0	25	DME/SCE	12A0	C=0.05,m= 1.7-2.2,t= 3-4,Pt Aux
							-	3.4	0			C=0.094
								5.0				C=0.070
							0.13	5.0	0.5			C=0.02
												C=0.14
												C=0.46
							0.5	5.5	25			C=0.05
					IL	MB	0.05	2.5	25.0 ±0.5	HMDE SCE	12A2	C=0.05,A= 0.022,v=300, Pt Aux
								4.7				
CONT		uridylyl (3'→5') adenosine	Table II	H_2O	PY	MB						

TABLE I. Electrochemical Data $C_{19}H_{24}N_7O_{12}P$ (CONT.) EF10

Ref.	C/M	Charact. Potential		Response Const.		n Tech.	n	Electrokinetic Data			Products and Identification	Description and Remarks	Code No.
		Value	vs.		Value	Tech.		Parameter	Value	From			
JA095 8495	113 n	$E_{\frac{1}{2}}$	-1.140c	-	i_ℓ	1.41F	i:i	4.2	$dE_{\frac{1}{2}}/dpH$	80	sttd	-	$C,i_d,\not\models,E_{\frac{1}{2}}=-0.986-0.080pH$ for pH=2.0-5.5,r	EF09
			-1.348F			1.09F	sttd	4.2	$dE_{\frac{1}{2}}/dpH$ Elog	80 50		-	$C,i_d,\not\models,i_\ell=kC$ and $E_{\frac{1}{2}}=f(C)$ for C=0.05-0.2,r	
			-1.362			6.0		4.2		44		-	$C,i_d,\not\models,E_{\frac{1}{2}}\neq f(C)$ for C<0.14 or C>0.46,r	
			-1.369F			5.17F		4.2	-	-	-	-	$C,i_d,\not\models,r$	
			-1.380F			4.5F		4.2	-	-	-	-	$C,i_d,\not\models,r$	
			-1.42c			0.4F	-	-	$dE_{\frac{1}{2}}/dpH$	80	sttd	-	$C,i_d,\not\models,r$	
JA095 8495	113 n	E_p	-1.21c	SCE	-	-	-	-	dE_p/dpH	80	sttd	-	$C,\not\models,i_p/v^{\frac{1}{2}}=k$ for v=90-300, $E_p=-1.05-0.080pH$ for pH=2.0-4.7,r	
			-1.43c		-	-	-	-		80		-	$C,\not\models,i_p \downarrow$ as pH $\uparrow$, 0 for pH>5-6,r	
JA095 8495	113 n	E_{su}	-1.14c	SCE	-	-	-	-	dE_{su}/dpH	0	sttd	-	$C,E_{su}\neq f(pH)$ for pH=2-5,r	
			-1.26c							66		-	$C,E_{su}=-1.13-0.066pH$ for pH=2-5.0	
			-1.14c -1.46c							0 66			C,r C	
JA095 8495	113 n	E_{su}	-1.40	SCE	i_{su}/C	0.82	-	-	-	-	-	-	$C,\not\models,r$	
JA095 8495	113 n	$E_{\frac{1}{2}}$	-1.145c	SCE	i_ℓ	-	l:l	3.6	$dE_{\frac{1}{2}}/dpH$ Elog	80 50	sttd	-	$C,i_d,\not\models,i_\ell=kC$ and $E_{\frac{1}{2}}=f(C)$ for C=0.05-0.2, $E_{\frac{1}{2}}=-0.985-0.080pH$ for pH=2-5.5,r	EF10
			-1.273		i_ℓ I	0.74 5.6	l:l QE	3.6 ≈7	Elog	56	sttd	-	$C,i_d,\not\models,i_{cat},H,D=5.3,r$	
			-1.366			0.48 5.0		-		52		-	$C,i_d,\not\models,E_{\frac{1}{2}}\neq f(C)$ for C<0.14 or C>0.46,D=4.2,r	
			-1.374F			6.7F	-	-	-	-	-	-	$C,i_d,\not\models,r$	
			-1.374F			4.7F	-	-	-	-	-	-	$C,i_d,\not\models,r$	
			-1.386F			3.0F	-	-	-	-	-	-	$C,i_d,\not\models,r$	
			-1.425c			-	-	-	$dE_{\frac{1}{2}}/dpH$	80	sttd	-	$C,i_d,\not\models,r$	
JA095 8495	113 n	E_p	-1.25c	SCE	-	-	-	-	dE_p/dpH	75	sttd	-	$C,\not\models,i_p/v^{\frac{1}{2}}=k$ and $E_p \rightarrow$ more neg. for v=90-300, $E_p=-1.06-0.075pH$ for pH=2.5-4.7,r	
			-1.41c			-	-	-		75		-	$C,\not\models,i_p \downarrow$ as pH $\uparrow$, 0 for pH>5-6,r	
														CONT

EF10 (CONT.) $C_{19}H_{24}N_7O_{12}P$

Code No.	Empirical Formula	Name and C.A. Number	Structural Formula	Solvent	Tech.	Medium	μ, M	pH	T, °C	Electrodes	App.	Experimental Parameters
EF10	$C_{19}H_{24}N_7O_{12}P$	uridylyl (5'→3') adenosine	Table 11	H_2O	PV	BR	0.5	2.0	25	DME/SCE	12A2	C=0.05,m=1, t(c)=3,f=50, Δe=3.54(rms), Pt Aux
								5.5				
								6.5				
					PVI	MB	0.5	3.4	0	DME/SCE	12A0	C=0.094,m= 1.0,t(c)=3, f=50,Δe=3.54 (rms),Pt Aux
								5.0				C=0.070
EF11 c177	$C_{19}H_{25}N_8O_{11}P$	adenylyl (3'→5') cytidine C.A. 4833-63-0	Table 11-3	H_2O	PY	MB	0.5	2	25	DME/SCE	12A0	C=0.05,m= 7.7-2.2,t= 3-4,Pt Aux
							–	3.4	0			C=0.10
								5.0				C=0.078
							0.13	5.0	0.5			C=0.31
							0.5	5.5	25			C=0.05
					IL	MB	0.05	2.0	25.0 ±0.5	HMDE SCE	12A2	C=0.05,A= 0.022,v=300, Pt Aux
								6.0				
					PV	MB	0.5	2.0	25	DME/SCE	12A2	C=0.05,m=1, t(c)=3,f=50, Δe=3.54(rms), Pt Aux
								2.5				
								5.5				
					PVI	MB	0.5	3.4	0	DME/SCE	12A0	C=0.10,m=1.0, t(c)=3,f=50, Δe=3.54(rms), Pt Aux
								5.0				C=0.078

TABLE I. Electrochemical Data $C_{19}H_{25}N_8O_{11}P$ EF11

Ref.	C/M	Charact. Potential		Response Const.		n Tech.	n	Electrokinetic Data			Products and Identification	Description and Remarks	Code No.
		Value	vs.		Value			Parameter	Value	From			
JA095 8495	113 n	E_{su} -1.16c	SCE	-	-	-	-	dE_{su}/dpH	0	sttd	-	C,E_{su}=1.16 for pH= 2-6.5,r	EF10
		-1.25c							61			C,E_{su}=-1.15-0.061pH for pH=2-5.5	
		-1.16c		-	-	-	-		0			C,r	
		-1.48c							61			C	
		-1.16c		-	-	-	-	-	-	-	-	C,r	
		-										C,0	
JA095 8495	113 n	E_{su} -1.33	SCE	i_{su}/C	0.71	-	-	-	-	-	-	C,⇌,r	
		-1.40			0.72	-	-					C,⇌,r	
JA095 8495	89 d 113 n	$E_{\frac{1}{2}}$ -1.128c	SCE	-	-	1:1	6.2	$dE_{\frac{1}{2}}/dpH$	59	sttd	-	C,Pr,⇌,$E_{\frac{1}{2}}$=-1.010- 0.059pH for pH=2.0-5.5,r	EF11 c177
		-1.183					-		64			C,i_d,$E_{\frac{1}{2}}$=-1.055- 0.064pH for pH=2.0-5.5	
		-1.238		i_ℓ {	0.60 4.3	-	-	Elog	27		-	C,i_d,⇌,Tc=0.5 for T= 8-40,r	
		-1.297		{	0.78 5.6				31			C,i_d	
		-1.320			5.1	-	-		36		-	C,Pr,⇌,$E_{\frac{1}{2}}$≠f(C),I ↓ as C ↑,r	
		-1.394			4.4	QE	11		39			C,i_d,$E_{\frac{1}{2}}$→more neg. and I ↑ as C ↑	
		-		-	-	-	-		-		-	C,Pr,⇌,r	
		-1.421F										C,i_d	
		-1.335c		-	-	sttd	6.2	$dE_{\frac{1}{2}}/dpH$	59		-	C,Pr,⇌,r	
		-1.407c					-		64			C,i_d	
JA095 8495	89 d 113 n	E_p -1.21c	SCE	-	-	-	-	dE_p/dpH	61	sttd	-	C,⇌,$i_p/v^{\frac{1}{2}}$=k for v= 70-350,E_p=-1.09- 0.061pH for pH=2-6, r	
		-1.44c		-	-	-	-		61		-	C,⇌,r	
JA095 8495	89 d 113 n	E_{su} -	SCE	-	-	-	-		-	-	-	C?,r	
		1.26c						dE_{su}/dpH	60	sttd		C,E_{su}=-1.14-0.060pH for pH=2.0-5.8	
		-1.14c		-	-	-	-		22		-	C,E_{su}=-1.08-0.022pH for pH=2.5-5.5,r	
		-1.29c							60			C	
		-1.20c		-	-	-	-		22		-	C,r	
		-1.47c							60			C	
JA095 8495	89 d 113 n	E_{su} -1.27	SCE	i_{su}/C	0.15	-	-	-	-	-	-	C,⇌,r	
		-1.32			1.4							C	
		-1.35			0.45	-	-	-	-	-	-	C,⇌,r	
		-1.42			1.3							C	

219

EF12 $C_{19}H_{25}N_8O_{11}P$

Code No.	Empirical Formula	Name and C.A. Number	Structural Formula	Solvent	Tech.	Medium	μ, M	pH	T, °C	Electrodes	App.	Experimental Parameters
EF12 c178	$C_{19}H_{25}N_8O_{11}P$	cytidylyl (3'→5') adenosine C.A. 2382-66-3	Table 11-3	H_2O	PY	MB	0.5	2.0	25	DME/SCE	12A0	C=0.061, m=1.7-2.2, t=3-4, Pt Aux
								3.7				
								4.5				
							-	5.0	0			C=0.084
							0.13	5.0	0.5			C=0.05
												C=0.09
												C=0.16
								5.5				C=0.061
					IL	MB	0.05	2.0	25.0 ±0.5	HMDE SCE	12A2	C=0.05, A=0.022, v=300, Pt Aux
								5.5				
					PV	MB	0.5	2.0	25	DME/SCE	12A2	C=0.05, m=1, t(c)=3, f=50, Δe=3.54(rms), Pt Aux
								2.5				
								4.5				
								6				
								8				
					PVI	MB	0.5	5.0	-	DME/SCE	12A0	C=0.084, m=1.0, t(c)=3, f=50, Δe=3.54(rms), Pt Aux
EF12	$C_{19}H_{25}N_8O_{11}P$	cytidylyl (3'→5') adenosine C.A.	Table 11-3		PY	MB						

TABLE I. Electrochemical Data $C_{19}H_{25}N_8O_{11}P$ EF12

Ref.	C/M	Charact. Potential		Response	Const.	Tech.		Electrokinetic Data			Products and Identification	Description and Remarks	Code No.
		Value	vs.		Value			Parameter	Value	From			
JA095 8495	89 d 113 n	$E_{\frac{1}{2}}$	-1.131c	SCE	i_ℓ	1.25F	-	-	$dE_{\frac{1}{2}}/dpH$	58	sttd	-	C,Pr,≠,$E_{\frac{1}{2}}$= -1.015- 0.058pH for pH=2.0- 5.5,r	EF12 c178
			-1.197c			2.13F	1:1	6.2		61			C,i_d,$E_{\frac{1}{2}}$= -1.075- 0.061pH for pH=2.0- 5.5	
			-1.230c			1.18F	-	-	$dE_{\frac{1}{2}}/dpH$	58	sttd	-	C,Pr,≠,Tc=0.5 for T= 8-40,r	
			-1.300c			1.93F				61			C,i_d	
			-1.276c			0.88F	-	-	$dE_{\frac{1}{2}}/dpH$	58		-	C,Pr,≠,r	
			-1.350c			1.75F				61			C,i_d	
			-1.320		i_ℓ I	0.53 4.5	-	-	Elog	41		-	C,Pr,≠,$E_{\frac{1}{2}}$≠f(C),I↓ as C↑,r	
			-1.409			0.60 5.2				49			C,i_d,$E_{\frac{1}{2}}$→ more neg. and I↑ as C↑	
			-		I	-	-	-		-		-	C,Pr,≠,r	
			-1.396F			2.86F							C,i_d	
			-			-	-	-		-	-	-	C,Pr,≠,r	
			-1.409F			5.14F							C,i_d	
			-			-	-	-		-	-	-	C,Pr,≠,r	
			-1.424F			6.43F							C,i_d	
			-1.334c		i_ℓ	0.53F	-	-	$dE_{\frac{1}{2}}/dpH$	58	sttd	-	C,Pr,≠,r	
			-1.410c			1.39F				61			C	
JA095 8495	89 d 113 n	E_p	-1.23c	SCE	-	-	-	-	dE_p/dpH	54	sttd	-	C,≠,$i_p/v^{\frac{1}{2}}$ = k for v= 70-350,E_p = -1.12- 0.054pH for pH=2.0- 5.5,r	
			-1.42c		-	-	-	-		54		-	C,≠,r	
JA095 8495	89 d 113 n	E_{su}	-	SCE	-	-	-	-	-	-	-	-	C?,r	
			-1.26c						dE_{su}/dpH	61	sttd		C,E_{su}= -1.14-0.061pH for pH=2.0-4.5	
			-1.09c		-	-	-	-		40		-	C,E_{su}= -0.99-0.040pH for pH=2.5-4.5,r	
			-1.29c							61			C	
			-1.17c		-	-	-	-		40		-	C,r	
			-1.41c							61			C	
			-1.16c		-	-	-	-		+89		-	C,E_{su}= -1.70 + 0.089 pH for pH=0-8,r	
			-0.99c		-	-	-	-		+89		-	C,r	
JA095 8495	89 d 113 n	E_{su}	-1.36	SCE	i_{su}/C	0.13	-	-	-	-	-	-	C,≠,r	
			-1.44			1.1							C	

EF13 $C_{19}H_{25}N_8O_{12}P$

Code No.	Empirical Formula	Name and C.A. Number	Structural Formula	Solvent	Tech.	Medium		μ, M	pH	T, °C	Electrodes	App.	Experimental Parameters
EF13 dg05	$C_{19}H_{25}N_8O_{12}P$	cytidylyl (3'→5') guanosine C.A. 2382-65-2	Table II-4	H_2O	PV	MB		0.5	2.0	25	DME/SCE	12A2	C=0.05, m=1, t(c)=3, f=50, Δe=3.54(rms), Pt Aux
									3.5				
						CARB,MB			4.0				
									9.6				
EF14 bm86	$C_{20}H_{12}N_2$	dibenzo[a,c]phenazine	Table II-2	MeCN	VA	Et_4NClO_4	ns	-	-	-	Pt/SCE	2-0	A=0.22
EF15	$C_{20}H_{12}O_2$	5,14-dimethylcyclo-octadeca-2,4,11,13-tetraen-6,8,15,17-tetryne-1,10-dione C.A. 51606-61-2	Table II	DMF	VR	Bu_4NClO_4	1.0	-	-	20.0 ±0.3	PDE Ag/AgCl	2A2	C=0.4-1, Pt Aux
EF16	$C_{20}H_{12}O_2$	10,15-dimethylcyclo-octadeca-7,9,15,17-tetraen-2,4,11,13-tetryne-1,6-dione C.A. 51606-60-1	Table II	DMF	VR	Bu_4NClO_4	1.0	-	-	20.0 ±0.3	PDE Ag/AgCl	2A2	C=0.4-1.0, Pt Aux
EF17 cm07	$C_{20}H_{25}N_{10}O_{10}P$	adenylyl (3'→5') adenosine C.A. 2391-46-0	Table II-3	H_2O	PY	MB		0.5	1	25	DME/SCE	12A0	C=0.05, m=1.7-2.2, t=3-4, Pt Aux
									2.5				C=0.051
								-	3.4	0			C=0.084
								0.1	3.5	0.5			C=0.017
													C=0.175
													C=0.34
								0.5	3.8	25			C=0.051
									4.5				
								-		0			C=0.094
								0.5	5.5	25			C=0.051
					IL	MB		0.05	2.5	25.0 ±0.5	HMDE SCE	12A2	C=0.05, A=0.022, v=300, Pt Aux
CONT													

TABLE I. Electrochemical Data $C_{20}H_{25}N_{10}O_{10}P$ (CONT.) EF17

Ref.	C/M	Charact.	Potential	Response Const.		Tech.	n	Electrokinetic Data			Products and Identification	Description and Remarks	Code No.
		Value	vs.		Value			Parameter	Value	From			
JA095 8495	-	E_{su}	-1.14c	SCE	-	-	-	-	dE_{su}/dpH	15	sttd	-	$C, E_{su} = -1.11-0.015$pH for pH=2.0-3.7, r	EF13 dg05
			-1.16c		-	-	-	-		15		-	C, r	
			-1.17c		-	-	-	-		+18		-	$C, E_{su} = -1.24+0.018$pH for pH=3.7-9.6	
			-1.07c		-	-	-	-		+18		-	C, r	
JE049 0111	-	$E_{p/2}$	1.82	SCE	$J_p/Cv^{\frac{1}{2}}$	7.02	-	-	-	-	-	-	A, p	EF14 bm86
			-			6.6							C, dibenzo[a,c]phen-azine redn.	
JA096 0249	50 a	XE	-0.68	Ag/AgCl	-	-	sttd	1	-	-	-	radical anion	$C, A, XE = \lim[(E_{p,A} + E_{p,C})/2]$ as $v \to 0$, R, P	EF15
			-0.93					1				dianion	C, A, R	
JE036 0389	50 a	XE	-0.66	Ag/AgCl	-	-	sttd	1	-	-	-	radical anion	$C, R, XE = \lim[(E_{p,A} + E_{p,C})/2]$ as $v \to 0$, p	EF16
			-0.94					1				dianion	C, R	
JA095 8495	113 o	$E_{\frac{1}{2}}$	-1.047c	SCE	-	-	1:1	6	$dE_{\frac{1}{2}}/dpH$	127	sttd	-	$C, i_d, \ne, E_{\frac{1}{2}}$ and $I \ne f(\mu)$ for $\mu = 0.25-0.75$, $i_\ell = f(pH)$, $E_{\frac{1}{2}} = -0.920-0.127$pH for pH=1-2, r	EF17 cm07
			-1.155c		-	1.51F	sttd	6		70		-	$C, i_d, \ne, E_{\frac{1}{2}} = -0.980-0.070$pH for pH=2.5-5.5, r	
			-1.256		i_ℓ I	1.4 12.0	1:1 QE	6 15-16	Elog	28		-	$C, i_d, \ne, Tc=0, i_{cat}, H, D=6.1, r$	
			-1.243F		I	12.3F	-	-	-	-	-	-	$C, i_d, \ne, r$	
			-1.257F			11.4F	1:1	6.7	-	-	-	-	$C, i_d, \ne, r$	
			-1.285F			11.2F	-	-	-	-	-	-	$C, i_d, \ne, r$	
			-1.246c			1.27F	sttd	6	$dE_{\frac{1}{2}}/dpH$ Elog	70 55	sttd	-	$C, i_d, E_{\frac{1}{2}} = -1.221$ V in ACET(pH=4.0), $\ne$, r	
			-1.277c			0.79F	-	-		76 18		-	$C, Pr, S, \ne, i_\ell = k_1 C + k_2$ for C=0.04-0.2, $E_{\frac{1}{2}} = -0.935-0.076$pH for pH=4.0-4.5, p	
			-1.295c			1.36F			$dE_{\frac{1}{2}}/dpH$	70		-	C, i_d, p	
			1.307		i_ℓ I	1.4 11.1	1:1	6.7	Elog	46		-	$C, i_d, \ne, r$	
			-1.365c		i_ℓ	0.85F	-	-	$dE_{\frac{1}{2}}/dpH$	70		-	$C, \ne, r$	
JA095 8495	113 o	E_p	-1.35c	SCE	-	-	-	-	dE_p/dpH	56	sttd	-	$C, \ne, i_p/ACv^{\frac{1}{2}} = v^x, i_d$ and $i_a, i_p \downarrow$ as pH $\uparrow$, $E_p = -1.21-0.056$pH for pH=2.5-5.5, 0 for pH>5-6, r	
	-	E_{su}	-1.14c											CONT

EF17 (CONT.) $C_{20}H_{25}N_{10}O_{10}P$

Code No.	Empirical Formula	Name and C.A. Number	Structural Formula	Solvent	Tech.	Medium	μ, M	pH	T, °C	Electrodes	App.	Experimental Parameters
EF17 cm07	$C_{20}H_{25}N_{10}O_{10}P$	adenylyl (3'→5') adenosine	Table II-3	H_2O	IL	MB	0.05	5.5	25.0 ±0.5	HMDE SCE	12A2	C=0.05, A=0.022, v=300, Pt Aux
					PV	MB	0.5	2.5	25	DME/SCE	12A0	C=0.05, m=1.0, t(c)=3, f=50, Δe=3.54(rms), Pt Aux
								5.5				
					PVI	MB	0.5	3.4	0	DME/SCE	12A0	C=0.084, m=1.0, t(c)=3, f=50, Δe=3.54(rms), Pt Aux
							0.1	3.5				C=0.0936, m=1, t(c)=3, f=300, Pt Aux
												f=3200
												f=6300
									25			f=300
												f=3200
												f=6300
									40			f=300
												f=3200
												f=6300
EF18	$C_{20}H_{25}N_{10}O_{11}P$	adenylyl (3'→5') guanosine C.A. 3352-23-6	Table II	H_2O	PY	MB	0.5	1	25	DME/SCE	12A0	C=0.05, m=1.7-2.2, t=3-4, Pt Aux
								2.5				
								4.5				
					IL	MB	0.05	2.5	25.0 ±0.5	HMDE SCE	12A2	C=0.05, A=0.022, v=300, Pt Aux
CONT		adenylyl (3'→5')	Table II-3	H_2O	IL							

224

TABLE I. Electrochemical Data $C_{20}H_{25}N_{10}O_{11}P$ (CONT.) EF18

Ref.	C/M	Charact. Potential		Response Const.		n Tech.	Electrokinetic Data			Products and Identification	Description and Remarks	Code No.
		Value	vs.		Value		Parameter	Value	From			
JA095 8495	113 o	E_p -1.52c	SCE	-	-	-	-	dE_p/dpH	56	sttd	-	C, E_p is 0.05-0.25 V more neg. in Cl⁻ or ACET,$\neq$,$J_p/Cv^{\frac{1}{2}}$ = f(v), $i_a + i_d$, r	EF17 cm07
JA095 8495	113 o	E_{su} -1.16c	SCE	-	-	-	-	dE_{su}/dpH	30	sttd	-	C, E_{su} = -1.08-0.030pH for pH=2.5-5.5, r	
		-1.31c							60			C, E_{su} = -1.16-0.060pH for pH=2.5-5.5	
		-1.25c		-	-	-	-		30		-	C, r	
		-1.49c							60			C	
JA095 8495	113 o	E_{su} -1.31	SCE	i_{su}	0.20	sttd	6	-	-	-	-	C,$\neq$,E_{su}=k for C→0.4, $E_{\frac{1}{2}} - E_{su}$ ↓ as C→0.2, $E_{\frac{1}{2}} - E_{su}$ = 28 for C>0.2, $\Delta i_{su}/C$ ↓ as C ↑, r	
		-		-		-						C, appears at f>50, E_{su} = -1.4	
		-		XI	0.0122	-	-	-	-	-	-	C, XI=ratio of in-phase current to that expected for a reversible diffusion-controlled process; r	
		-			0.0029	-	-	-	-	-	-	C, r	
		-			0.0030	-	-	-	-	-	-	C, r	
		-			0.0128	-	-	-	-	-	-	C, r	
		-			0.0041	-	-	-	-	-	-	C, r	
		-			0.0033	-	-	-	-	-	-	C, r	
		-			0.0137	-	-	-	-	-	-	C, r	
		-			0.0050	-	-	-	-	-	-	C, r	
		-			0.0042	-	-	-	-	-	-	C, r	
JA095 8495	113 n	$E_{\frac{1}{2}}$ -1.045c	SCE	-	-	-	-	$dE_{\frac{1}{2}}/dpH$	100	sttd	-	C,i_d,$\neq$,$E_{\frac{1}{2}}$ = -0.945-0.100pH for pH=1-2.5, r	EF18
		1.193c		-	-	-	-		-		-	C,i_d,$\neq$,r	
		-1.306c		-	-	-	-		57		-	C,i_a,$E_{\frac{1}{2}}$ = -1.050-0.037pH for pH=2.5-4.5, $E_{\frac{1}{2}}$ = -1.20 V in ACET(pH=4.0),$\neq$,i_ℓ and $E_{\frac{1}{2}}$=kC for C=0.05-0.2, r	
JA095 8495	113 n	E_p -1.31c	SCE	-	-	-	-	dE_p/dpH	65	sttd	-	C,$\neq$,E_p = -1.15-0.065pH for pH=2.5-4.0, r	
		-0.1							-			A, due to redn. prod. from guanine	
		E_p											CONT

EF18 (CONT.) $C_{20}H_{25}N_{10}O_{11}P$

Code No.	Empirical Formula	Name and C.A. Number	Structural Formula	Solvent	Tech.	Medium	μ, M	pH	T, °C	Electrodes	App.	Experimental Parameters
EF18	$C_{20}H_{25}N_{10}O_{11}P$	adenylyl (3'→5') guanosine	Table II	H_2O	IL	MB	0.05	4.0	25.0 ±0.5	HMDE SCE	12A2	C=0.05,A= 0.022,v=300, Pt Aux
					PV	MB	0.5	2.0	25	DME/SCE	12A2	C=0.05,m=1.0, t(c)=3,f=50, Δe=3.54(rms), Pt Aux
								4.5				
EF19	$C_{20}H_{25}N_{10}O_{11}P$	guanylyl (3'→5') adenosine C.A. 6554-00-3	Table II		PY	MB	0.5	2.0	25	DME/SCE	12A0	C=0.052,m= 1.7-2.2,t= 3-4,Pt Aux
								2.5				
								3.8				
					IL	MB	0.05	2.0	25.0 ±0.5	HMDE SCE	12A2	C=0.05,A= 0.022,v=300, Pt Aux
								3.7				
					PV	MB	0.5	2.0	25	DME/SCE	12A2	C=0.05,m=1, t(c)=3,f=50, Δe=3.54(rms), Pt Aux
								3.9				
								5.5				
EF20	$C_{20}H_{42}$	eicosane C.A. 112-95-8	$CH_3(CH_2)_{18}CH_3$	CF_3COOH	IL	FSO_3H	2.5	-	-	Pt/SCE	25-2	C=30,v=50, Pt Aux
EF21	$C_{20}H_{58}B_{18}CoN$	bis[7-11-η-7,8-dicarbaundecaborane-(11)]cobaltate(1-) tetrabutylammonium salt C.A. 11078-84-5	$[(\pi-1,2-B_9C_2-H_{11})_2Co]^-$ $(C_4H_9)_4N^+$	MeCN	PY	Bu_4NPF_6	0.3	-	25	DME/SCE	2A0	ns
					PVI	Bu_4NPF_6	0.3	-	25	DME/SCE	2A0	ns

TABLE I. Electrochemical Data $C_{20}H_{58}B_{18}CoN$ EF21

Ref.	C/M	Charact. Potential		Response Const.		n Tech.		Electrokinetic Data			Products and Identification	Description and Remarks	Code No.
		Value	vs.		Value			Parameter	Value	From			
JA095 8495	113 n	E_p	-1.41c	SCE	-	-	-	-	dE_p/dpH	65	sttd	-	$C,\not{F},i_p \downarrow$ as pH $\uparrow$,0 for pH $>$5-6,p	EF18
			-0.3							-			A,due to redn. prod. from guanine	
JA095 8495	113 n	E_{su}	-1.244c	SCE	-	-	-	-	dE_{su}/dpH	47	sttd	-	$C,M,E_{su}=-1.15-0.047$pH for pH=2.0-4.5,r	
			-1.362c		-	-	-	-		47		-	C,Mn,r	
JA095 8495	113 n	$E_{\frac{1}{2}}$	-1.153c	SCE	i_ℓ	1.41F	-	-	$dE_{\frac{1}{2}}/dpH$	59	sttd	-	$C,i_a,\not{F},E_{\frac{1}{2}}=-1.035-0.059$pH for pH=2.0-4.0,r	EF19
			-1.183c			1.36F	-	-		59		-	$C,i_a,\not{F},r$	
			-1.260c			1.24F	-	-		59		-	$C,i_a,\not{F},i_\ell=kC$ and $E_{\frac{1}{2}}=k(C)$ for C= 0.05-0.2,r	
JA095 8495	113 n	E_p	-1.25c	SCE	-	-	-	-	dE_p/dpH	62	sttd	-	$C,\not{F},E_p=-1.13-0.062$pH for pH=2.0-3.7,$i_p/v^{\frac{1}{2}}$(at v=350)=(0.2-0.5), $i_p/v^{\frac{1}{2}}$(at v=70),r	
			0.06							-			A,due to redn. prod. from guanine,present only after redn. at final current rise	
			-1.36c		-	-	-	-		62		-	$C,\not{F},i_p \downarrow$ as pH $\uparrow$,0 for pH $>$5-6,r	
			-0.2							-			A,due to redn. prod. from guanine	
JA095 8495	113 n	E_{su}	-1.196c	SCE	-	-	-	-	dE_{su}/dpH	68	sttd	-	$C,Mn,E_{su}=-1.06-0.068$pH for pH=2.0-4.0,r	
			-1.321c		i_{su}	1.3F	-	-		68		-	C,Mn,r	
			-1.94c			1.5F	-	-		67		-	$C,Mn,E_{su}=-1.57-0.067$pH for pH=4.0-5.5,r	
JE054 0181	-	$E_{p/2}$	1.88	SCE	-	-	-	-	-	-	-	-	$A,\not{F},p$	EF20
JE050 0031	-	$E_{\frac{1}{2}}$	-1.36	SCE	-	-	-	-	Elog	≈59	-	$[(\pi-1,2-B_9C_2H_{11})_2-Co]^{2-}$	C,R,r	EF21
			-2.24							≈59		$[(\pi-1,2-B_9C_2H_{11})_2-Co]^{3-}$	C,R	
JE050 0031	-	-	-	-	-	-	-	-	α $k_{s,h}^{app}$ $k_{s,h}$	0.48 0.82 12	sttd	-	C,r	
										0.48 0.37 250			C	
		E_p												

EF22 $C_{23}H_{18}Cl_2N_3Na$ CRC Handbook Series in Organic Electrochemistry

Code No.	Empirical Formula	Name and C.A. Number	Structural Formula	Solvent	Tech.	Medium	μ, M	pH	T, °C	Electrodes	App.	Experimental Parameters
EF22	$C_{23}H_{18}Cl_2N_3Na$	4,5-bis(4-chlorophenyl)-2-(4-dimethylaminopheny)limidazole sodium salt C.A. 31909-36-1	Table II	MeCN	VY	Bu_4NClO_4 0.1	-	-	-	RPDE Ag/AgCl	2A0	C=1.9, A=0.012, v=15.3
EF23	$C_{23}H_{18}N_3Na$	2-(4-dimethylaminophenyl)-4,5-(2,2'-biphenylene)imidazole sodium	Table II	MeCN	VY	Bu_4NClO_4 0.1	-	-	-	RPDE Ag/AgCl	2A0	C=1.9, A=0.012, v=15.3
EF24	$C_{23}H_{20}N_3Na$	2-(4-dimethylaminophenyl)-4,5-diphenylimidazole sodium C.A. 31909-31-6	Table II	MeCN	VR	Bu_4NClO_4 0.1	-	-	-	PDE Ag/AgCl	2A0	C=1.9, A=0.012, v=100
					VY	Bu_4NClO_4 0.1	-	-	-	RPDE Ag/AgCl cobaltocinium	2A0	C=1.9, A=0.012, v=15.3
EF25	$C_{26}H_{20}O_2$	dicyclohexeno[d,j]-cyclooctadeca-2,4,-10,12-tetraene-6,8,-15,17-tetrayne-1,14-dione	Table II	DMF	VR	Bu_4NClO_4 1.0	-	-	20.0 ±0.3	PDE Ag/AgCl	2A2	C=0.4-1.0, Pt Aux
EF26	$C_{30}H_{37}N_{15}O_{16}P_2$	adenylyl (3'→5') adenylyl (3'→5') adenosine C.A. 917-44-2	Table II	H_2O	PY	MB	0.5	2.5	25	DME/SCE	12A0	C=0.05, m=1.7-2.2, t=3-4, Pt Aux
							-	3.4	0			C=0.091
							0.1	3.5	0.5			C=0.044
												C=0.14
												C=0.41
							0.5	4.5	25			C=0.05
								6.4				
EF27	$C_{40}H_{49}N_{20}O_{22}P_3$	adenylyl (3'→5') adenylyl (3'→5') adenylyl (3'→5') adenosine C.A. 4042-12-0	Table II	H_2O	PY	MB	-	2.5	25	DME/SCE	12A0	C=0.047, m=1.7-2.2, t=3-4, Pt Aux
								3.4	0			C=0.087
							0.1	3.5	0.5			C=0.01
												C=0.20
							-	3.7	25			C=0.047
								4.6				
								5.4				
CONT								6.4				

TABLE I. Electrochemical Data $C_{40}H_{49}N_{20}O_{22}P_3$ (CONT.) EF27

Ref.	C/M	Charact. Potential		Response Const.		n	Tech.	Electrokinetic Data			Products and Identification	Description and Remarks	Code No.
		Value	vs.		Value			Parameter	Value	From			
C0036 0575	397 a	$E_{\frac{1}{2}}$ 0.083	Ag/AgCl	$i_d(u)$	1.00	1	Elog	αn_a	0.071	Elog	-	A,R,i_d,r	EF22
		0.522			1.18	1			0.070			A,R,i_d	
C0036 0575	397 a	$E_{\frac{1}{2}}$ 0.190	Ag/AgCl	$i_\ell(u)$	1.00	1	Elog	αn_a	0.068	Elog	-	A,R,i_d,r	EF23
		0.584			1.15	1			0.066			A,R,i_d	
C0036 0575	397 a	E_p -0.021	Ag/AgCl	-	-	-	-	-	-	-	-	C,R,r	EF24
		0.446										C,R	
		0.054										A,R	
		0.521										A,R	
C0036 0575	397 a	$E_{\frac{1}{2}}$ 0.020	Ag/AgCl	$i_\ell(u)$	1.00	1	Elog	αn_a	0.070	Elog	radical,CP,color	$A,R,i_d=k\omega^{\frac{1}{2}},i_d=k'C,r$	
		0.503			1.12	1			0.068		cation,CP,color	$A,R,i_d=k\omega^{\frac{1}{2}}$	
JA096 0249	50 a	XE -0.70	Ag/AgCl	-	-	sttd	1	-	-	-	radical anion	$C,R,XE=\lim[(E_{p,A}+E_{p,C})/2]$ as $v \to 0,p$	EF25
		-1.04					1				dianion	C,R	
JA095 8495	113 p	$E_{\frac{1}{2}}$ -1.193c	SCE	-	-	1:1	12	$dE_{\frac{1}{2}}/dpH$	53	sttd	-	$C,i_d,\neq;E_{\frac{1}{2}} \to$ more neg., $I \downarrow$, and Elog $\downarrow$ as $C \uparrow;E_{\frac{1}{2}}=-1.060-0.053pH$ for pH=2.5-4.5,r	EF26
		-1.267		i_ℓ }	2.17 17.3	sttd	2	Elog	18	sttd	-	$C,i_d,\neq,r$	
		-1.255F		-	-	-	-	-	-	-	-	$C,i_d,\neq,r$	
		-1.283F		-	-	-	-	-	-	-	-	$C,i_d,\neq,r$	
		-1.300F		-	-	-	-	-	-	-	-	$C,i_d,\neq,r$	
		-1.283c		-	-	-	-	$dE_{\frac{1}{2}}/dpH$	55	sttd	-	$C,Pr,\neq,E_{\frac{1}{2}}=-1.035-0.055pH$ for pH=4.5-6.4,r	
		-1.298c		-	-	-	-		53			C,i_d	
		-1.387c		-	-	-	-		55		-	$C,Pr,\neq,r$	
JA095 8495	113 p	-	-	i_ℓ	4.3F	1:1	12.7	-	-	-	-	$C,i_d,\neq;E_{\frac{1}{2}} \to$ more neg., $I \downarrow$, and Elog $\downarrow$ as $C \uparrow,r$	EF27
		$E_{\frac{1}{2}}$ -1.271	SCE	i_ℓ }	2.09 17.3	sttd	12.7	Elog	14	sttd	-	$C,i_d,\neq,r$	
		-1.241F		-	-	-	-	-	-	-	-	$C,i_d,\neq,r$	
		-1.305F		-	-	-	-	-	-	-	-	$C,i_d,\neq,r$	
		-	-		3.82F	sttd	12.7				-	$C,i_d,\neq,r$	
		-	-		1.64F	-	-	Elog	18	sttd	-	$C,Pr,\neq,r$	
					2.28F				27			C,i_d	
		-	-		0.88F	-	-	-	-	-	-	$C,Pr,\neq,r$	
		-	-		0.18F	-	-	-	-	-	-	$C,Pr,\neq,r$	CONT

EF27 (CONT.) $C_{40}H_{49}N_{20}O_{22}P_3$

Code No.	Empirical Formula	Name and C.A. Number	Structural Formula	Solvent	Tech.	Medium	μ, M	pH	T, °C	Electrodes	App.	Experimental Parameters
EF27	$C_{40}H_{49}N_{20}O_{22}P_3$	adenylyl (3'→5') adenylyl (3'→5') adenylyl (3'→5') adenosine	Table II	H_2O	IL	MB	0.05	2.5	25.0 ±0.5	HMDE SCE	12A2	C=0.05, A=0.022, v=300, Pt Aux
								6.5				
					PV	MB	0.5	2.5	25	DME/SCE	12A2	C=0.05, m=1.0, t(c)=3, f=50, Δe=3.54(rms), Pt Aux
								4.5				
								6.5				
EF28	$C_{60}H_{73}N_{30}O_{34}P_5$	adenylyl (3'→5') adenylyl (3'→5') adenylyl (3'→5') adenylyl (3'→5') adenylyl (3'→5') adenosine C.A. 2068-77-1	Table II	H_2O	PY	MB	0.5	2.5	25	DME/SCE	12A0	C=0.043, m=1.7-2.2, t=3-4, Pt Aux
								3				
							—	3.4	0			C=0.078
							0.1	3.5	0.5			C=0.01
												C=0.13
							0.5	3.7	25			C=0.043
								4.4				
								5.5				
					IL	MB	0.05	2.5	25.0 ±0.5	HMDE SCE	12A2	C=0.05, A=0.022, v=300, Pt Aux
								3.0				
								4.7				
					PV	MB	0.5	2.5	25	DME/SCE	12A2	C=0.05, m=1.0, t(c)=3, f=50, Δe=3.54(rms), Pt Aux
								4.0				
								4.5				
								5.5				
					IL	MB						

TABLE I. Electrochemical Data $\quad C_{80}H_{73}N_{30}O_{34}P_5 \quad$ EF28

Ref.	C/M	Charact. Potential		Response Const.		n Tech.		Electrokinetic Data			Products and Identification	Description and Remarks	Code No.
		Value	vs.		Value			Parameter	Value	From			
JA095 8495	113 p	E_p	-1.29c	SCE	-	-	-	-	dE_p/dpH	49	sttd	-	$C,\rightleftarrows,J_p/Cv^{\frac{1}{2}} \uparrow$ as $v \uparrow$, $i_a, E_p = -1.17-0.049pH$ for pH=2.5-6.5, r	EF27
			-1.49c		-	-	-	-		49		-	$C,\rightleftarrows,J_p/Cv^{\frac{1}{2}} \uparrow$ as $v \uparrow$, $i_a, i_p \downarrow$ as pH $\uparrow$, 0 for pH>5-6, r	
JA095 8495	113 p	E_{su}	-	SCE	-	-	-	-	-	-	-	-	C,0,r	
			-1.30c						dE_{su}/dpH	60	sttd		$C, E_{su} = -1.15-0.060pH$ for pH=2.5-4.5	
			-1.33		-	-	-	-		0			$C, dE_{su}/dpH=0$ for pH= 4.5-6.5	
			-1.42c							60			C	
			-1.33		-	-	-	-		0			C,r	
			-							-			C,0	
JA095 8495	113 p	$E_{\frac{1}{2}}$	-1.228c	SCE	i_ℓ	5.89	sttd	13	$dE_{\frac{1}{2}}/dpH$	43	sttd	-	$C, i_d, \rightleftarrows, E_{\frac{1}{2}} = -1.120-0.043pH$ for pH=2.5-4.0, r	EF28
			-1.249c			5.7	-	-		43		-	$C, i_d, \rightleftarrows, r$	
			-1.277		i_ℓ I	1.95 18.1	-	-	Elog	19		-	$C, i_d, \rightleftarrows, r$	
			-1.232F		-	-	-	-		-	-	-	$C, i_d, \rightleftarrows, r$	
			-1.302F		-	-	-	-		-	-	-	$C, i_d, \rightleftarrows, r$	
			-1.279c		i_ℓ	5.3F	-	-	$dE_{\frac{1}{2}}/dpH$	43	sttd	-	$C, i_d, \rightleftarrows, r$	
			-1.287c			2.0F	-	-	$dE_{\frac{1}{2}}/dpH$ Elog	39 18		-	$C, Pr, \rightleftarrows, E_{\frac{1}{2}} = -1.115-0.039pH$ for pH=4.0-5.5, r	
			-			-			$dE_{\frac{1}{2}}/dpH$	27			C	
			-1.330c			1.1F	-	-		39	sttd		$C, Pr, \rightleftarrows, r$	
JA095 8495	113 p	E_p	-1.30c	SCE	-	-	1:1	13	dE_p/dpH	42	sttd	-	$C, E_p = -1.20-0.042pH$ for pH=2.5-4.7, E_p on first scan= -1.37 V, $\rightleftarrows$, r	
			-1.33c		-	-	sttd	13		42		-	C, E_p on first scan= -1.37 V, $\rightleftarrows$, $i_p \downarrow$ as pH $\uparrow$, 0 for pH>5-6, r	
			-1.40c		-	-		13		42		-	$C, \rightleftarrows; E_{\frac{1}{2}} \rightarrow$ more neg., I $\downarrow$, and Elog $\downarrow$ as C $\uparrow$; r	
JA095 8495	113 p	E_{su}	-	SCE	-	-	-	-	-	-	-	-	C,0,r	
			-1.32c						dE_{su}/dpH	52	sttd		$C, E_{su} = -1.19-0.052pH$ for pH=2.5-4.0	
			-		-	-	-	-	-	-	-	-	C,0,r	
			-1.40c						dE_{su}/dpH	52	sttd		C	
			1.34		-	-	-	-		0		-	$C, E_{su} \neq f(pH)$ for pH= 4.5-5.5, r	
			-		-	-	-	-		-			C,0	
			1.34		-	-	-	-		0			C,r	

231

TABLE III. Courses and Mechanisms of Half-Reactions

IV(123-3) + H⁺ ⇌ HX–⟨C₆H₄⟩–NH₂OH⁺ (VII)   123-6

VII(123-6) —k₇→ HX⁺=⟨C₆H₄⟩=NH (VIII) + H₂O   123-7

VIII(123-7) + H⁺ + 2e → VI(123-5)   123-8

VIII(123-7) ⇌ V(123-4) + H⁺   123-9

IV(123-3) ⇌ ⁻X–⟨C₆H₄⟩–NHOH (IX) + H⁺   123-10

IX(123-10) → V(123-4) + OH⁻   123-11

2 II(123-1) + 2H⁺ → I(123-1) + III(123-2) + H₂O   123-12

III(123-2) + 2 II(123-1) + 2H⁺ → IV(123-3) + 2 I(123-1)   123-13

V(123-4) + 2 II(123-2) + 2H⁺ → VI(123-5) + 2 I(123-1)   123-14

I(123-1) + II(123-1) → HX–⟨C₆H₄⟩–ṄO₂H (X) + ⁻X–⟨C₆H₄⟩–NO₂ (XI)   123-15

X(123-15) + e → III(123-2) + OH⁻   123-16

III(123-2) + e → HX–⟨C₆H₄⟩–NO⁻ (XII)   123-17

XII(123-17) ⇌ ⁻X–⟨C₆H₄⟩–ṄOH (XIII)   123-18

XIII(123-18) + 2 I(123-1) + e —?→ IV(123-3) + 2 XI(123-15)   123-19

$$XI(123\text{-}15) + e \rightarrow {}^-X\text{-}C_6H_4\text{-}NO_2^{\cdot-} \quad (XIV) \qquad 123\text{-}20$$

$$XIV(123\text{-}20) + I(123\text{-}1) \rightarrow II(123\text{-}1) + XI(123\text{-}15) \qquad 123\text{-}21$$

HX = HO, H_2N, or $(CH_3)_2N$

Code No.	Literature Reference	Conditions or Technique	$k_4(s^{-1})$	k_7, s^{-1}
AG49	AC035 1859	a,b,c,e, plane electrode, ER technique	1.2±0.8	
		a,b,c,e, spherical electrode, ER technique	1.0±0.7	
		a,b,c,e, plane electrode, AE technique	0.65±0.08	
		a,b,c,e, spherical electrode, QE technique	0.59±0.07	
		e, spherical electrode	1.28±0.08 or 1.38±0.1	
	JP070 0396	2 to 3 m$\underline{M}$, EE technique	0.5 ± 0.1$_2$	
	JP073 1406	QR technique IR technique	0.33 0.36	
	AC037 0190	VA technique	0.6±0.1	
	AC038 0542	pH 1.3	0.6±0.1	
		2.0	0.45±0.1	
		3.1	0.4±0.1	
		4.1	0.35±0.1	
		4.9	0.4±0.1	
		5.3	0.45±0.1	
		6.0	0.9±0.2	

TABLE III. Courses and Mechanisms of Half-Reactions

(cont.)

Code No.	Literature Reference	Conditions or Technique	$k_4(s^{-1})$	k_7, s^{-1}
AG52	JP070 0396	2 to 3mM,[c] EE technique	0.14±0.07	
	JA083 0784	[d]	0.20±0.04	
AG54	JP070 0396	2 to 3mM,[c] EE technique	0.8±0.5	
	JP073 1406	[c,] QR technique [c,] IR technique	0.33±0.05 0.32±0.05	
	—			0.5±1.2
AN46	JE014 0119	pH = 2.3		0.9

[a] pH = 4.8,   [b] 0.005% gelatin,   [c] 20% EtOH,   [d] 50% EtOH,   [e] 1×10^{-3} M

124

(aqueous or protic solvent)
(to replace old 2, 78, 124, 125, 130, and 189)

Mechanism:		Proposed for Compound:
124a:	124-1, -3 to -7, and -14 to -16, followed by 103d and 197c	AG47
124b:	124-14	AG41
124c:	124-8, -9, -6, -7, and -14	AG53
124d:	scheme 124c takes place twice (pH > 2)	AG00
124e:	124-1, -3 to -7, and -14, followed by 103e (the whole scheme takes place twice) (pH < 2)	AG00
124f:	124-17 and -14	AK32
124g:	124c, followed by 103f (pH < 4)	DA83
124h:	124-8, -10, -11 (strongly basic)	AF68
124i:	124-1, -2, -6, -7, and -14	AS13
124j:	124i, followed by 103g	AC38, AT19, AT40

124k: 124-7, -14, and -15 (basic)   EB28

124l: 124-8   EB13

124m: 124-16, followed by 197c   BA92

$ArNO_2H^+ \rightleftarrows ArNO_2 + H^+$ 124-1

$ArNO_2H^+ + H^+ + 2e \rightarrow ArN(OH)_2$ 124-2

$ArNO_2H^+ + e \rightarrow ArNO_2H \cdot$ 124-3

$ArN(OH)_2^+ \rightleftarrows ArNO_2H \cdot + H^+$ 124-4

$ArN(OH)_2^+ + e \rightarrow ArN(OH)_2$ 124-5

$ArN(OH)_2 + H^+ \rightarrow ArNOH^+ + H_2O$ 124-6

$ArNOH^+ \rightleftarrows ArNO + H^+$ 124-7

$ArNO_2 + e \rightleftarrows ArNO_2^-$ 124-8

$ArNO_2^- + 2H^+ + e \rightarrow ArN(OH)_2$ 124-9

$ArNO_2^- + e \rightarrow ArNO_2^{2-}$ 124-10

$ArNO_2^{2-} + 3H^+ + 2e \rightarrow ArNHOH + OH^-$ 124-11

$2ArNO_2^- + 2H^+ \rightarrow ArNO + ArNO_2 + H_2O$ 124-12

$ArNO + 2ArNO_2^- + 2H^+ \rightarrow ArNHOH + 2ArNO_2$ 124-13

$ArNO + 2H^+ + 2e \rightleftarrows ArNHOH$ 124-14

$ArNOH^+ + ArNHOH \rightarrow ArN(O):NAr + H_3O^+$ 124-15

$ArN(O):NAr + 4H^+ + 4e \rightarrow ArNHNHAr + H_2O$ 124-16

$ArNO_2 + H_2O + 2e \xrightarrow{slow} ArNO + 2OH^-$ 124-17

Ar = C_6H_5, $3-O_2NC_6H_4$, $4-CH_3C_6H_4$, $4-ClC_6H_4$, [5-furyl-2-CH$_2$SCN], $3-HOC_6H_4$, $4-[(EtO)_2P(S)O]C_6H_4$, $[4-C_6H_4CHOHC(CH_2OH)HNHC(O)CHCl_2]H^+$, [2-methylfuryl], or [1,5-dimethyl-4-nitro-2-(4-C_6H_4)-pyrazol-3(2H)-one]

TABLE III. Courses and Mechanisms of Half-Reactions

Code No.	Conditions	$\Delta E_H = \frac{\Delta(\Delta H^\circ)}{nF}$, mV	$\Delta \epsilon_S = \frac{\Delta(\Delta S^\circ)}{nF}$, mvK^{-1}	k_{15}, dm^{3}mol^{-1}s^{-1}
EB13	[H$_2$O] = 0 (v/v%)	140	-1.00	
	= 2	65	-1.00	
	= 5	25	-1.00	
	= 10	5	-0.96	
EB28				100

125
(to replace old 123, 125, and 127)

Mechanism: Proposed for Compound:
125a: 125-1 to -5 (50% EtOH, or DMSO)   AG52, A067

125b: 125-1, -10, -11, -4, and -12   AG52
       (H$_2$O, basic)

125c: 125-3 to -5 (H$_2$O, 4 < pH < 7)   AQ30, AQ68, AQ69, AQ78

125d: 125-6 to -9 (H$_2$O, pH > 7)   AQ30, AQ68, AQ69, AQ78

$$\text{o-XH-C}_6\text{H}_4\text{-NO}_2 \text{ (I)} + e^- \rightleftarrows [\text{o-XH-C}_6\text{H}_4\text{-NO}_2]^{-\cdot} \text{ (II)} \qquad 125\text{-}1$$

$$\text{II}(125\text{-}1) + 2\text{H}^+ + e^- \rightarrow \text{o-XH-C}_6\text{H}_4\text{-NO} \text{ (III)} + \text{H}_2\text{O} \qquad 125\text{-}2$$

$$\text{III}(125\text{-}2) + 2\text{H}^+ + 2e^- \rightarrow \text{o-XH-C}_6\text{H}_4\text{-NHOH} \text{ (IV)} \qquad 125\text{-}3$$

IV(125-3) $\xrightarrow{k_4}$ [2-iminocyclohexa-2,4-dien-1-ylidene with X substituent] (V) + H_2O 125-4

V(125-4) + $2H^+$ + 2e → [benzene ring with XH and NH_2] (VI)   125-5

III(125-2) ⇌ [benzene ring with X^- and NO] (VII) + H^+ 125-6

VII(125-6) + $2H^+$ + 2e → [benzene ring with X^- and NHOH] (VIII)   125-7

VIII(125-7) ⇌ V(125-4) + OH^- 125-8

V(125-4) + H^+ + 2e → [benzene ring with X^- and NH_2]   125-9

2 II(125-1) + $2H^+$ → III(125-2) + I(125-1) + H_2O 125-10

III(125-2) + 2 II(125-1) + $2H^+$ → IV(125-3) + 2 I(125-1)   125-11

V(125-4) + 2 II(125-1) + $2H^+$ → VI(125-5) + 2 I(125-1)   125-12

126

Mechanism:
126b: 126-1, -4, and -6

Proposed for Compound:
product of 103f(DA83)

TABLE III. Courses and Mechanisms of Half-Reactions

130
(non-aqueous)
(to replace old 2, 78, 123, 130)

Mechanism:		Proposed for Compound:
130a:	130-1, -4, -9, -10, and -5	AP49, AR98, AS01, AS16
130b:	130-1 and -2	AG47, DB63
130c:	130-1 to -3, followed by 86b (no proton donor)	AI77, AI78
130d:	130-1 to -3, followed by 86c (proton donor added)	AI77, AI78
130e:	130-1 and -6	AF58, AF59, AF67, AF68, AF83, AG00, AG92, AI86, AJ17, AK32, AK36, AM31, DB62
130f:	130e, followed by halogen reduction	AF68, AF82
130g:	130-1 to -3, followed by 86d	BA56
130h:	130-1 and -13	AG00
130i:	130-1 to -3, followed by 86e	AG53, AJ18, AJ19, AL71, AL73
130j:	130e, followed by 146b	BF22
130k:	130-7 and -8 (acid added)	AK32
130l:	130-1 and -12	EB29

$$ArNO_2 + e \rightleftarrows ArNO_2^- \qquad \text{130-1}$$

$$ArNO_2^- + e \rightleftarrows ArNO_2^{2-} \qquad \text{130-2}$$

$$ArNO_2^{2-} + 2HSo \rightarrow ArNO + 2So^- + H_2O \qquad \text{130-3}$$

$$ArNO_2^- + HSo \rightarrow ArNO_2H^\cdot + So^- \qquad \text{130-4}$$

$$ArNO_2H^- \rightarrow ArNO + OH^- \qquad \text{130-5}$$

$$ArNO_2^- + 4HSo + 3e \rightarrow ArNHOH + H_2O + 4So^- \qquad \text{130-6}$$

$$ArNO_2 + HA \rightleftarrows ArNO_2H^! + A^- \qquad \text{130-7}$$

$$ArNO_2H^+ + 4e + 2HA \rightarrow ArNHOH + 2A^- + OH^- \qquad \text{130-8}$$

$$ArNO_2^- + ArNO_2H^\cdot \rightarrow ArNO_2 + ArNO_2H^- \qquad \text{130-9}$$

$$ArNO_2H^\cdot + e \rightarrow ArNO_2H^- \qquad \text{130-10}$$

$$ArNO_2H^\cdot + 2HSo + 3e \rightarrow ArNHOH + 2So^- + OH^- \qquad \text{130-11}$$

$$ArNO_2^- + M^+ \rightleftarrows (ArNO_2^- \, M^+) \qquad \text{130-12}$$

$$4ArNO_2^- + 3HB^+ \xrightarrow{k_{13}} ArNHOH + 3ArNO_2 + 3B + OH^- \qquad 130\text{-}13$$

Code No.	k_{13}, s^{-1}	M^+	K_{12}	φ^{eff} $(pK_{12} = pK_{12}^o + m\varphi^{eff})$
AG00	336			
		Et_4N^+	200	0.26
		Cs^+	2500	0.42
		K^+	6200	0.49
		Li^+	6.03E6	0.77
		Na^+	1.41E5	0.61
		Rb^+		0.46

137

Mechanism:
137c: 137-13 to -16 (at foot of first wave)

Proposed for Compound: EA17

$R^1R^2C(X)CHO + H_2O \rightleftarrows R^1R^2C(X)CH(OH)_2$ 137-1

$R^1R^2C(X)CHO + 2e \rightarrow R^1R^2\bar{C}CHO + X^-$ 137-2

$R^1R^2CHCHO \rightleftarrows R^1R^2\bar{C}CHO + H^+$ 137-3

$R^1R^2C(XH^+)CHO \rightleftarrows R^1R^2C(X)CHO + H^+$ 137-4

$R^1R^2C(XH^+)CHO + 2e \rightarrow R^1R^2\bar{C}CHO + XH$ 137-5

$R^1R^2C(XH^+)CH(:OH^+) \rightleftarrows R^1R^2C(XH^+)CHO + H^+$ 137-6

$R^1R^2C(XH^+)CH(:OH^+) + 2e \rightarrow R^1R^2CHCHO + XH$ 137-7

$R^1R^2C(XH^+)CHO + H_2O \rightleftarrows R^1R^2C(XH^+)CH(OH)_2$ 137-8

$R^1R^2C(XH^+)CH(OH)_2 \rightleftarrows R^1R^2C(X)H(OH)_2 + H^+$ 137-9

$R^1R^2\bar{C}CHO \longleftrightarrow R^1R^2C:CH(O^-)$ 137-10

$R^1R^2C:CHOH \rightleftarrows R^1R^2C:CH(O^-) + H^+$ 137-11

TABLE III. Courses and Mechanisms of Half-Reactions

$R^1R^2C{:}CHOH + 2H^+ + 2e \rightarrow R^1R^2CHCH_2OH$ 137-12

$R^1R^2C(X)CHO + H_2O + e \rightleftarrows R^1R^2C(X)\dot{C}HOH_{ads} + OH^- \ (?)$ 137-13

$R^1R^2C(X)\dot{C}HOH_{ads} + e \rightarrow R^1R^2C(X)\overline{C}HOH$ 137-14

$R^1R^2C(X)\overline{C}HOH \rightarrow R^1R^2C{:}CHOH + X^-$ 137-15

$R^1R^2C{:}CHOH \rightleftarrows R^1R^2CHCHO$ 137-16

$R^1 = R^2 = H$

$X = OH$

141

Mechanism: Proposed for Compound:
141k: 141-5, -6, -17, and -18   EB80

$RS^- + Hg \rightleftarrows RSHg + e$ 141-1

$RSH \rightleftarrows RS^- + H^+$ K_2 141-2

$2RSHg \rightarrow Hg(SR)_2 + Hg$ 141-3

$Hg(SR)_2 \rightleftarrows RSSR + Hg$ 141-4

$RSSR + Hg \rightleftarrows (RS)_2Hg_{ads}$ 141-5

$(RS)_2Hg_{ads} + 2e + 2H^+ \rightleftarrows 2RSH + Hg$ 141-6

$R^1SSR^2 + e \rightarrow (R^1SSR^2)^{\cdot -}$ 141-7

$(R^1SSR^2)^{\cdot -} + e \rightleftarrows R^1S^- + R^2S^-$ (protonation possible)   141-8

$2Hg + RSSR \rightarrow Hg_2(SR)_2$ 141-9

$Hg_2(SR)_2 + 2H^+ + 2e \rightarrow 2Hg + 2RSH$ 141-10

$R^1SSR^2 + e \rightarrow R^1S^- + R^2S\cdot$ 141-11

$R^2S\cdot + e \rightarrow R^2S^-$ 141-12

$R^1SSR^2 + 2H^+ + e \rightarrow R^1SH + R^2SH^{+\cdot}$ 141-13

$R^2SH^{+\cdot} + e \rightarrow R^2SH$ 141-14

$2RS^- + Hg \rightleftarrows (RS)_2Hg + 2e$ 141-15

$$RSH + Hg \rightleftharpoons RSHg + H^+ + e \qquad 141\text{-}16$$

$$(RS)_2Hg_{ads} + Hg \rightarrow 2RSHg \qquad 141\text{-}17$$

$$RSHg + H^+ + e \rightleftharpoons Hg + RSH \qquad 141\text{-}18$$

$$R = HOOCCH(NH_2)CH_2$$

146

Mechanism:
146b: 146-1, -2, -4, and -5 (Vol. I)

Proposed for Compound:
product of 130j (BF22)

147

Mechanism:
147e: 147-1 (fused salt)

Proposed for Compound:
ED11

147i: 147-1, -18, -19, and -4 (fused salt)

EC80

$$C_6H_5NR^1R^2 \text{ (I)} \rightleftharpoons [C_6H_5NR^1R^2]^{+} \text{ (II)} + e \qquad 147\text{-}1$$

$$II(147\text{-}1) + I(147\text{-}1) \xrightarrow{k_2} [R^1R^2NC_6H_4C_6H_4NR^1R^2]^{+} \text{ (III)} \quad (+H_2?) \qquad 147\text{-}2$$

$$2\,II(147\text{-}1) \xrightarrow{k_3} R^1R^2N\text{–}C_6H_4\text{–}C_6H_4\text{–}NR^1R^2 \text{ (IV)} + 2H^+ \qquad 147\text{-}3$$

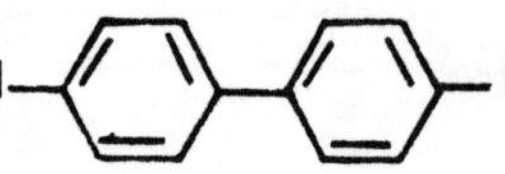

$$IV(147\text{-}3) \rightleftharpoons R^1R^2\overset{+}{N}=C_6H_4=C_6H_4=\overset{+}{N}R^1R^2 \text{ (V)} + 2e \qquad 147\text{-}4$$

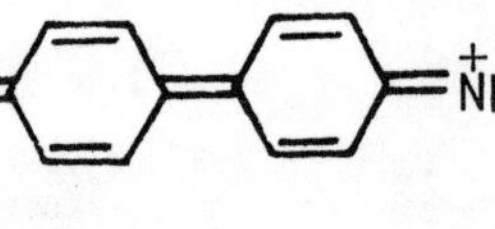

TABLE III. Courses and Mechanisms of Half-Reactions

$HR^1R^2\overset{+}{N}$—⬡—⬡—$\overset{+}{N}R^1R^2H$ $\underset{}{\overset{k_5}{\rightleftarrows}}$ V(147-4) + 2H$^+$ + 2e   147-5

III(147-2) + I(147-1) → coupled product   147-6

In 147-7 and -8, R^2 = CH$_3$

II(147-1) + I(147-1) → $C_6H_5NR^1CH_2$—⬡—NR^1CH_3 (VI) + 2H$^+$ + e   147-7

VI(147-7) + H$^+$ $\rightleftarrows$ $C_6H_5\overset{+}{N}H(R^1)CH_2$—⬡—$NR^1CH_3$ (VII)   147-8

VII(147-8) → $C_6H_5NHR^1$ + $^+CH_2$—⬡—NR^1R^2 (VIII)   147-9

VIII(147-9) + I(147-1) → R^1R^2N—⬡—CH_2—⬡—NR^1R^2 (IX) + H$^+$ 147-10

IX(147-10) + II(147-1) → $(R^1R^2NC_6H_4)_3CH$ (X) + 2H$^+$ + e   147-11

X(147-11) $\rightleftarrows$ $(R^1R^2NC_6H_4)_2C$=⬡=$\overset{+}{N}R^1R^2$ + H$^+$ + e   147-12

In 147-13 and -14, $R^1 = R^2 = CH_3$

II(147-1) ⇌ [3,5-dihydro-N,N-dimethylaniline radical structure] (XI) + H^+ 147-13

2 XI(147-13) → IV(147-3)   147-14

2 II(147-1) + IV(147-3) → 2 I(147-1) + V(147-4)   147-15

IV(147-3) → R^1R^2N—⟨⟩—⟨⟩—$\overset{+\cdot}{N}R^1R^2$ (XII) + e   147-16

XII(147-16) → V(147-4) + e   147-17

2 II(147-1) ⇌ $R^1R^2\overset{+}{N}$=⟨⟩—(H)(H)—⟨⟩=$\overset{+}{N}R^1R^2$ (XIII)   147-18

XIII(147-18) $\xrightarrow{slow}$ IV(147-3) + $2H^+$ 147-19

$R^1 = CH_3$ or C_6H_5

$R^2 = CH_3$ or substituted C_6H_5

TABLE III. Courses and Mechanisms of Half-Reactions

156

Mechanism:
156i: 156-18 to 22, -3, -23, and -24, followed by 65c (protic medium or acidic solution)

Proposed for Compound:
EA58, EA88, EB71

156j: 156-1, -2, and -25, followed by 65k? (aprotic medium or basic solution)

EA58, EA88, EB71

$R^1CH{:}CHCOR^2 + e \rightarrow R^1\dot{C}HCH{:}C(O^-)R^2$ 156-1

$R^1\dot{C}HCH{:}C(O^-)R^2 + H^+ \rightleftarrows R^1\dot{C}HCH{:}C(OH)R^2$ 156-2

$2R^1\dot{C}HCH{:}C(OH)R^2 \rightarrow$ [cyclic structure with R^1, H, CCH$_2$CHR^2, CCH$_2$COR^2] (racemic)   156-3

$2R^1\dot{C}HCH{:}C(OH)R^2 \rightarrow R^1CH_2CH_2COR^2 + R^1CH{:}CHCOR^2$ 156-4

$R^1\dot{C}HCH{:}C(O^-)R^2 + CO_2 \rightarrow {}^-O_2CCHR^1\dot{C}HCOR^2$ 156-5

${}^-O_2CCHR^1\dot{C}HCOR^2 + e \rightarrow {}^-O_2CCHR^1CH{:}C(O^-)R^2$ 156-6

${}^-O_2CCHR^1CH{:}C(O^-)R^2 + CO_2 \rightarrow {}^-O_2CCHR^1CH(CO_2^-)COR^2$ 156-7

$CO_2 + e \rightarrow CO_2^{\cdot -}$ 156-8

$CO_2^{\cdot -} + R^1CH{:}CHCOR^2 \rightarrow {}^-O_2CCHR^1\dot{C}HCOR^2$ 156-9

$2R^1\dot{C}HCH.C(O^-)R^2 \rightarrow$
$\begin{array}{c} H \\ | \\ R^1-C-CH=C(O^-)R^2 \\ | \\ R^1-C-CH=C(O^-)R^2 \\ | \\ H \end{array}$ (I)   156-10

$$\text{I}(156\text{-}10) + 2CO_2 \rightarrow \begin{array}{c} HCO_2^- \\ R^1-C-CH-COR^2 \\ R^1-C-CH-COR^2 \\ HCO_2^- \end{array} \quad (\text{II}) \qquad 156\text{-}11$$

$$2^-O_2CCHR^1\dot{C}HCOR^2 \rightarrow \text{II}(156\text{-}11) \qquad 156\text{-}12$$

$$R^1\dot{C}HCH:C(O^-)R^2 + R^1CH:CHCOR^2 \rightarrow \begin{array}{c} H \\ R^1-C-\dot{C}HCOR^2 \\ R^1-C-\underline{C}HCOR^2 \\ H \end{array} \quad (\text{III}) \qquad 156\text{-}13$$

$$\text{III}(156\text{-}13) + CO_2 \rightarrow \begin{array}{c} H \\ R^1-C-\dot{C}HCOR^2 \\ R^1-C-CHCOR^2 \\ HCO_2^- \end{array} \quad (\text{IV}) \qquad 156\text{-}14$$

$$\text{IV}(156\text{-}14) + e \rightarrow \begin{array}{c} H \\ R^1-C-\overline{C}HCOR^2 \\ R^1-C-CHCOR^2 \\ HCO_2^- \end{array} \quad (\text{V}) \qquad 156\text{-}15$$

$$\text{V}(156\text{-}15) + CO_2 \rightarrow \text{II}(156\text{-}11) \qquad 156\text{-}16$$

$$\text{III}(156\text{-}13) + R^1\dot{C}HCH:C(O^-)R^2 \rightleftarrows \text{I}(156\text{-}10) + RCH:CHCOR^2 \qquad 156\text{-}17$$

$$R^1\overbrace{CH:CHCOR^2}^{H^+} \rightleftarrows R^1CH:CHCOR^2 + H^+ \qquad 156\text{-}18$$

$$R^1\overbrace{CH:CHCOR^2}^{H^+} + e \rightarrow R^1CH_2\dot{C}HCOR^2 \qquad 156\text{-}19$$

TABLE III. Courses and Mechanisms of Half-Reactions

$$R^1CH_2\dot{C}HCOR^2 + e \rightarrow R^1CH_2\overbrace{CH\dot{C}-R^2}^{-} \quad \text{156-20}$$
$$|_O$$

$$R^1CH_2CH_2COR^2 \rightleftarrows R^1CH_2\overbrace{CH\dot{C}R^2}^{-} + H^+ \quad \text{156-21}$$
$$|_O$$

$$R^1\dot{C}HCH{:}C(OH)R^2 \rightleftarrows R^1CH_2\dot{C}HCOR^2 \quad \text{156-22}$$

$$2R^1\dot{C}HCH{:}C(OH)R^2 + Hg \rightarrow \begin{array}{c} R^1 \\ | \\ H-CCH_2COR^2 \\ | \\ Hg \\ | \\ H-CCH_2COR^2 \\ | \\ R^1 \end{array} \quad \text{156-23}$$

$$n\ R^1\overbrace{CH{:}CHCOR^2}^{H^+} \rightarrow \text{polymeric products} \quad \text{156-24}$$

$$R^1\dot{C}HCH{:}C(OH)R^2 + HSo + e \rightarrow R^1CH_2CH_2COR^2 + So^- \quad \text{156-25}$$

$R^1 = C_6H_5$, $R^2 = H$, $R^1CH{:}CHCOR^2 = H_2C{:}CHCOCH_3$, $H_2C = C(CH_3)COCH_3$, or

$(H_3C)_2C{:}CHCOCH_3$

196

Mechanism:	Proposed for Compound:
196c: 196-9 and -10 (basic)	EE57
196d: 196-2, -11 to -15, -4, and -5 (neutral and acidic)	EE57

$$\text{[p-(NR}^1\text{H}_2^+\text{)(NR}^2\text{H}_2^+\text{)C}_6\text{H}_4] \underset{K_1}{\rightleftarrows} \text{[p-(NR}^1\text{H)(NR}^2\text{H}_2^+\text{)C}_6\text{H}_4]\ (I) + H^+} \qquad 196\text{-}1$$

$$I(196\text{-}1) \underset{K_2}{\rightleftarrows} \text{[p-(NR}^1\text{H)(NR}^2\text{H)C}_6\text{H}_4]\ (II) + H^+ \qquad 196\text{-}2$$

$$II(196\text{-}2) \rightleftarrows \text{[p-(=NR}^1\text{H}^+\text{)(=NR}^2\text{H}^+\text{)C}_6\text{H}_4]\ (III) + 2e \qquad 196\text{-}3$$

$$III(196\text{-}3) \underset{K_4}{\rightleftarrows} \text{[p-(=NR}^1\text{)(=NR}^2\text{H}^+\text{)C}_6\text{H}_4]\ (IV) + H^+ \qquad 196\text{-}4$$

$$IV(196\text{-}4) \underset{K_5}{\rightleftarrows} \text{[p-(=NR}^1\text{)(=NR}^2\text{)C}_6\text{H}_4]\ (V) + H^+ \qquad 196\text{-}5$$

TABLE III. Courses and Mechanisms of Half-Reactions

V(196-5) + H$_2$O $\rightleftarrows$ [quinone monoimine structure with =O and =NR^2] (VI) + NR^{1}H$_2$ 196-6

VI(196-6) + H$_2$O $\rightleftarrows$ [p-benzoquinone structure] (VII) + NR^{2}H$_2$ 196-7

R^{1}R^{2}NC$_6$H$_4$NR^{3}R^4 $\rightleftarrows$ (R^{1}R^{2}NC$_6$H$_4$NR^{3}R^4)$^{+\cdot}$ + e   196-8

II(196-2) + B $\rightarrow$ [radical cation with NR^1 and NR^{2}H]$^{\cdot}$ (VIII) + BH$^+$ + e   196-9

VIII(196-9) + B $\rightarrow$ V(196-5) + BH$^+$ + e   196-10

2 II(196-2) $\rightleftarrows$ VIII(196-9) + I(196-1) + e   196-11

VIII(196-9) $\rightleftarrows$ IV(196-4) + e   196-12

I(196-1) $\rightleftarrows$ [dication with NR^{1}H and NR^{2}H$_2$]$^{2+\cdot}$ (IX) + e   196-13

IX(196-13) + B → [structure: benzene ring with NR^{1}H at top and NR^{2}H at bottom, bracketed with + and • charges] (X) + BH$^+$ 196-14

X(196-14) → III(196-3) + e   196-15

R^1 = H, R^2 = 4-CH$_3$OC$_6$H$_4$

197
(aqueous or protic solvent)
(to replace old 197)

Mechanism:		Proposed for Compound:
197a:	197-1 to -7	BA47, BA88, BD53, ED99
197b:	197-1 and -2	BC96, BD08
197c:	197-3 to -6 (pH < 5)	BB55, product of 124 (AG47, BA92)
197d:	197-3, -4, -8, and -2 (5 < pH < 7)	BB55
197e:	197-1 to -3 and -9	ED97, EE02, EE06, EE92
197f:	197-1 and -2 take place twice, followed by -3 and -9 (pH < 9)	EF01
197g:	197b takes place twice (pH > 9)	EF01

$Ar^1 \overset{+}{N}H : \overset{+}{N}HAr^2 \rightleftarrows Ar^1N:NAr^2 + 2H^+$ 197-1

$Ar^1 \overset{+}{N}H : \overset{+}{N}HAr^2 + 2e \rightleftarrows Ar^1NHNHAr^2$ 197-2

$Ar^1 \overset{+}{N}H_2NHAr^2 \rightleftarrows Ar^1NHNHAr^2 + H^+$ 197-3

$Ar^1NH_2{}^+NH_2{}^+Ar^2 \rightleftarrows Ar^1 \overset{+}{N}H_2NHAr^2 + H^+$ 197-4

$Ar^1 \overset{+}{N}H_2 \overset{+}{N}H_2Ar^2 + e \rightarrow Ar^1 \overset{+}{N}H_2 \overset{\cdot}{N}H_2Ar^2$ 197-5

TABLE III. Courses and Mechanisms of Half-Reactions

$$Ar^1\overset{\cdot}{N}H_2\overset{+}{N}H_2Ar^2 + e \rightarrow Ar^1NH_2 + Ar^2NH_2 \qquad \text{197-6}$$

$$Ar^1\overset{+}{N}H_2\overset{+}{N}H_2Ar^2 \xrightarrow{k_7} H_2NAr^1Ar^2NH_2 + 2H^+ \qquad \text{197-7}$$

$$ArNH_2{}^+NHAr^2 + e \rightarrow Ar^1NHNHAr^2 + H\cdot \qquad \text{197-8}$$

In 197-9, Ar^2 =

$Ar^1\overset{+}{N}H_2NH$—⟨⟩—X → products   197-9

Ar^1 = C_6H_5, $4\text{-}HOC_6H_4$, $4\text{-}O_2NC_6H_4$, $4\text{-}ClC_6H_4$, $4\text{-}NCC_6H_4$, $4\text{-}HOOCC_6H_4$, or $4\text{-}CH_3C_6H_4$

Ar^2 = C_6H_5, $4\text{-}HOC_6H_4$, —⟨naphthyl⟩—OH, or $4\text{-}C_6H_4NHNHC_6H_5$

Code No.	k_7, s^{-1}
ED99	4.27 ± 0.5

198
(aprotic solvent)

Mechanism: Proposed for Compound:
198i: 198-1, -2, and -21   product of 86f (EB28)

198j: 198-1 and -2   product of 86j (EE00)

198k: 198-1 to -3, and -8   product of 86h (BA92)

$$Ar^1N{:}NAr^2 + e \rightleftarrows [Ar^1N{:}NAr^2]^{\overline{\cdot}} \qquad \text{198-1}$$

$[Ar^1N:NAr^2]^{\cdot -} + e \rightleftarrows [Ar^1N:NAr^2]^{2-}$ 198-2

$[Ar^1N:NAr^2]^{2-} + HSo \rightarrow Ar^1NH\bar{N}Ar^2 + So^-$ 198-3

$Ar^1NH\bar{N}Ar^2 \rightarrow Ar^1N:NAr^2 + 2e + H^+$ 198-4

$[Ar^1N:NAr^2]^{2-} + O_2 \rightarrow Ar^1N:NAr^2 + (O_2^{-}?)$ 198-5

$[Ar^1N:NAr^2]^{\cdot -} + HQ \rightarrow Ar^1NH\dot{N}Ar^2 + Q^-$ 198-6

$Ar^1NH\dot{N}Ar^2 + e \rightarrow Ar^1NH\bar{N}Ar^2$ 198-7

$Ar^1NH\bar{N}Ar^2 + HSo \rightarrow Ar^1NHNHAr^2 + So^-$ 198-8

$Ar^1N(H^+):NAr^2 \text{ (I)} \rightleftarrows Ar^1N:NAr^2 + H^+$ 198-9

$Ar^1N:NAr^2 + HSo \rightleftarrows Ar^1N(H^+):NAr^2 + So^-$ 198-10

$Ar^1N(H^+):NAr^2 + e \xrightarrow{slow} Ar^1NH\dot{N}Ar^2$ 198-11

$Ar^1NH\dot{N}Ar^2 + e + HA \xrightarrow{fast} Ar^1NHNHAr^2 + A^-$ 198-12

In 198-13 through -16, $Ar^2 = 4\text{-}C_6H_4X$

$Ar^1N(H^+):NC_6H_4XH^+ \rightleftarrows Ar^1N(H^+):NC_6H_4X + H^+$ 198-13

$Ar^1N:NC_6H_4XH^+ \rightleftarrows Ar^1N:NC_6H_4X + H^+$ 198-14

$Ar^1N(H^+):NC_6H_4XH^+ + e \xrightarrow{slow} Ar^1NH\dot{N}C_6H_4XH^+$ 198-15

$Ar^1NH\dot{N}C_6H_4XH^+ + H^+ + e \xrightarrow{fast} Ar^1NHNHC_6H_4XH^+$ 198-16

In 198-17 through -20, $Ar^2 = 4\text{-}C_6H_4YH$

$Ar^1N:NC_6H_4YH \rightleftarrows Ar^1N:NC_6H_4Y^- + H^+$ 198-17

$Ar^1N(H^+):NC_6H_4Y^- \rightleftarrows Ar^1N:NC_6H_4Y^- + H^+$ 198-18

$Ar^1N(H^+):NC_6H_4Y^- + e \xrightarrow{slow} Ar^1NH\dot{N}C_6H_4Y^-$ 198-19

$Ar^1NH\dot{N}C_6H_4Y^- + H^+ + e \xrightarrow{fast} Ar^1NHNHC_6H_4Y^-$ 198-20

$2[Ar^1N:NAr^2]^{\cdot -} \rightarrow Ar^1N:NAr^2 + [Ar^1N:NAr^2]^{2-}$ 198-21

$Ar^1 = Ar^2 = C_6H_5$

HQ = hydroquinone, $ArX = 4\text{-}CH_3OC_6H_4, 4\text{-}(CH_3)_2NC_6H_4,$ or $4\text{-}(CH_3)_3\overset{+}{N}C_6H_4$

$ArYH = 4\text{-}HO_3SC_6H_4$

215

Mechanism:	Proposed for Compound:
215m: 215-1, -11, -4, and -5 (low pH)	ED80
215n: 215-1, -2, -4, and -5 (high pH)	ED80

TABLE III. Courses and Mechanisms of Half-Reactions

233

Mechanism:
233a: 233-1 to -5 (Hg electrode)
233b: 233-6 to -8 (Pt electrode)

Proposed for Compound:

ED98

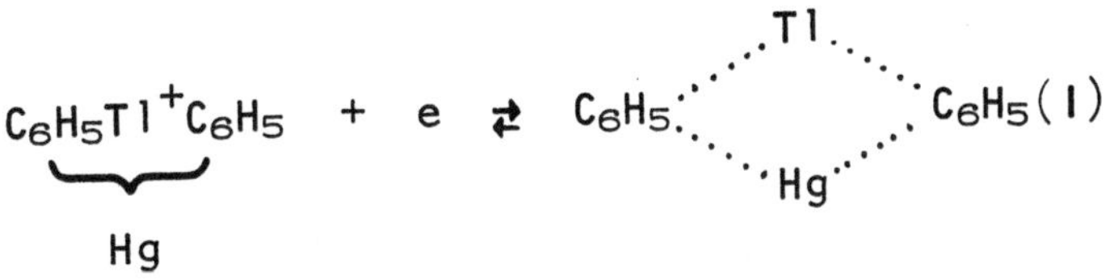

$$C_6H_5Tl^+C_6H_5 \text{ (Hg)} + e \rightleftarrows C_6H_5\cdots\overset{Tl}{\underset{Hg}{\diamond}}\cdots C_6H_5 \quad (I)$$
233-1

$I(233-1) \rightarrow (C_6H_5)_2Hg + Tl$ 233-2

$I(233-1) + H^+ + e \rightarrow Hg\cdots C_6H_5\cdots Tl + C_6H_6$ 233-3

$2Hg\cdots C_6H_5\cdots Tl \rightarrow (C_6H_5)_2Hg + 2Tl + Hg$ 233-4

$Hg\cdots C_6H_5\cdots Tl + H^+ + e \rightleftarrows C_6H_6 + Tl + Hg$ 233-5

$[Ar_2Tl(III)]^+ + e \rightarrow [Ar_2Tl(III)]^\cdot$ 233-6

$3[Ar_2Tl(III)]^\cdot \rightarrow 2[Ar_3Tl(III)] + Tl(0)$ 233-7

$4[Ar_3Tl(III)] + 3e \rightarrow 3[Ar_4Tl(III)]^- + Tl(0)$ 233-8

266

Mechanism:
266a: 266-1 to -3

Proposed for Compound:
CC35, CC36, CC43, CC44,
CC47, CC48, CC59, CD42,
CD43, CD44

266b: 266-4 to -7 (Hg electrode)

EB99, EC00, EC03, EC04,
EC58, EC59, EC60, EC62,
EC68

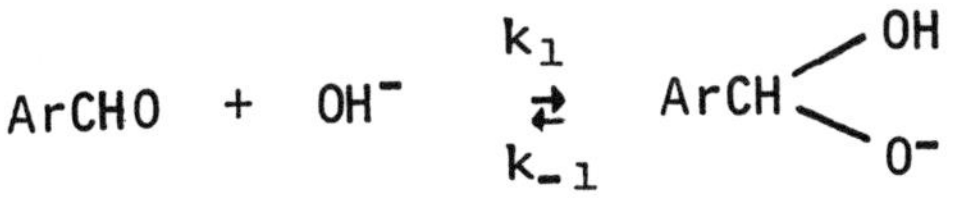

$$ArCHO + OH^- \underset{k_{-1}}{\overset{k_1}{\rightleftarrows}} ArCH\!\!\begin{array}{c}OH\\O^-\end{array}$$
266-1

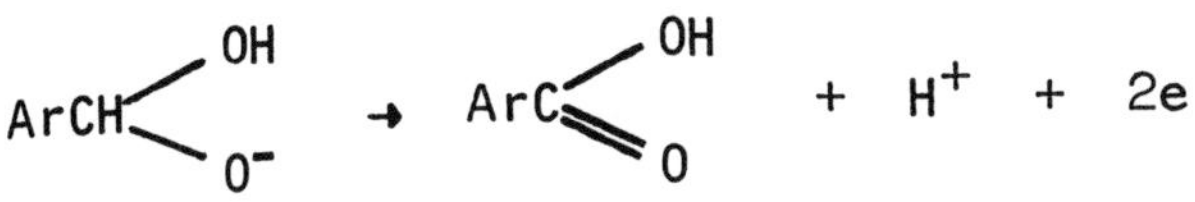

$$ArCH\!\!\begin{array}{c}OH\\O^-\end{array} \rightarrow ArC\!\!\begin{array}{c}OH\\\parallel\\O\end{array} + H^+ + 2e$$
266-2

$$ArCOOH \rightleftarrows ArCOO^- + H^+ \qquad 266\text{-}3$$

$$Hg + 2OH^- \rightleftarrows Hg(OH)_2 + 2e \qquad 266\text{-}4$$

$$Hg(OH)_2 + OH^- \rightarrow \underset{\underset{OH}{|}}{Hg}-O^- + H_2O \qquad 266\text{-}5$$

$$\underset{\underset{OH}{|}}{Hg}-O^- + ArCHO \rightarrow Ar-\underset{\underset{H}{|}}{\overset{\overset{O^-}{|}}{C}}-O-Hg-OH \qquad 266\text{-}6$$

$$Ar-\underset{\underset{H}{|}}{\overset{\overset{O^-}{|}}{C}}-O-Hg-OH \rightarrow ArC\overset{O}{\underset{O^-}{{\diagdown}\!\!\!\diagup}} + Hg + H_2O \qquad 266\text{-}7$$

Ar = C_6H_5, 3-ClC_6H_4, 4-ClC_6H_4, 3-$O_2NC_6H_4$, 4-$O_2NC_6H_4$, 2-C_6H_4CHO, 3-C_6H_4CHO, 4-C_6H_4CHO, 4-C_6H_4COOH, or 2-$C_6H_4OCH_3$

Compound	NaOH			PHOS or CARB buffers		
	pK'	$\log k_1$	$\log k_{-1}$	pK'	$\log k_1$	$\log k_{-1}$
CC35	11.08	5.47	5.28	10.85	5.93	5.74
CC36	10.65	5.61	4.70	10.86	5.19	4.28
CC43	11.24	5.25	5.17	11.45	4.84	4.76
CC44	11.58	5.10	5.54	11.72	4.82	5.26
CC47	10.67	5.56	4.63	11.08	4.73	3.80
CC48	10.48	5.65	4.44	10.80	5.01	3.80
CC59	11.93	4.86	5.76			
CD42	12.23	4.75	6.14			
CD43	11.70	5.03	5.64			
CD44	12.57	4.64	6.60			

TABLE III. Courses and Mechanisms of Half-Reactions

Compound	pK from plot of $E_{\frac{1}{2}}$ vs. pH	$\log k_1$ from pK and pK'	$\log k_{-1}$ from pK and $\log k_1$
CC36	14.82	7.34	8.16
CC43	14.87	6.21	8.16
CC44	14.08	5.74	6.82
CC59	15.30	5.26	6.56
CD42	15.37	4.73	6.07
CD43	15.24	5.66	6.90
CD44	15.55	4.23	5.87

267

Mechanism:
267c: 267-5 to -8 (pH=8.5)

Proposed for Compound:
EA70

$RCHOHCOCH_2OH + 2e \rightarrow RCHOHCO\bar{C}H_2 + OH^-$	267-1
$RCHOHCO\bar{C}H_2 + H_2O \rightarrow RCHOHCOCH_3 + OH^-$	267-2
$RCHOHCOCH_3 + 2e \rightarrow R\bar{C}HCOCH_3 + OH^-$	267-3
$R\bar{C}HCOCH_3 + H_2O \rightarrow RCH_2COCH_3 + OH^-$	267-4
$RCHOHCOCH_2OH + 2H^+ + 2e \rightarrow RCHOHCOCH_3 + H_2O$	267-5
$RCHOHCOCH_3 + 2H^+ + 2e \rightarrow RCH_2COCH_3 + H_2O$	267-6
$RCHOHCOCH_2OH + 2H^+ + 2e \rightarrow RCHOHCHOHCH_2OH$	267-7
$RCHOHCOCH_3 + 2H^+ + 2e \rightarrow RCHOHCHOHCH_3$	267-8

283

Mechanism:
283b: 283-4 to -7, followed by 65d

Proposed for Compound:
product of 414 (EC64),
product of 390 (EC65)

283d: 283-8, followed by 65d

product of 414 (EC64),
product of 390 (EC65)

$$\text{ArC}(:\text{OH}^+)\text{CHR} \rightleftarrows \text{ArCOCHR} + \text{H}^+ \qquad 283\text{-}1$$
$$\quad\quad\quad\; | \quad\quad\quad\quad\quad\; |$$
$$\quad\quad\quad\, X \quad\quad\quad\quad\quad X$$

$$\text{ArCOCHR} + \text{B} \rightleftarrows \text{ArCO}\overline{\text{C}}\text{R} + \text{BH}^+ \qquad 283\text{-}2$$
$$\;\;\;\; | \quad\quad\quad\quad\quad\quad\quad |$$
$$\;\;\;\, X \quad\quad\quad\quad\quad\quad\, X$$

$$\text{ArC}(:\text{OH}^+)\text{CHR} + 2e \rightarrow \text{ArCOCH}_2\text{R} + X^- \qquad 283\text{-}3$$
$$\quad\quad\quad\; |$$
$$\quad\quad\quad\, X$$

$$\text{ArCOCHR} + 2e \rightarrow \text{ArCO}\overline{\text{C}}\text{HR} + X^- \qquad 283\text{-}4$$
$$\;\;\;\; |$$
$$\;\;\;\, X$$

$$\text{ArCOCH}_2\text{R} \rightleftarrows \text{ArCO}\overline{\text{C}}\text{HR} + \text{H}^+ \qquad 283\text{-}5$$

$$\text{ArC}(\text{O}^-):\text{CHR} \leftrightarrow \text{Ar}\overline{\text{C}}\text{OCHR} \qquad 283\text{-}6$$

$$\text{ArC}(\text{OH}):\text{CHR} \rightleftarrows \text{ArC}(\text{O}^-):\text{CHR} + \text{H}^+ \qquad 283\text{-}7$$

In 283-8, X = NH_2

$$\text{ArCOCHR} + 2\text{H}^+ + 2e \rightarrow \text{ArCOCH}_2\text{R} + \text{NH}_3 \qquad 283\text{-}8$$
$$\;\;\;\; |$$
$$\;\;\;\, \text{NH}_2$$

Ar = C_6H_5

R = H

X = NH_3^+(283b) or NH_2(283d)

290

Mechanism:	Proposed for Compound:
290c: 290-1 to -6 (Vol. III)	CB48, CB49, CB50, CB52, CB53, CB54, CB55, CB56, CB57, CB59, CB60, CB61

295

Mechanism:
295c: 295-5 and -6

Proposed for Compound:
EA76, EB79

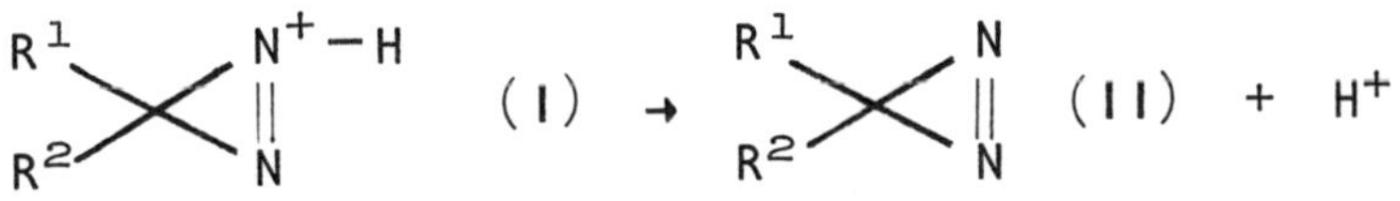

$$R^1R^2C(N^+H)=N \;(I) \rightarrow R^1R^2C(N)=N \;(II) + H^+ \qquad 295\text{-}1$$

$$I(295\text{-}1) + 3H^+ + 4e \rightarrow R^1R^2C(NH_2)(NH_2) \;(III) \qquad 295\text{-}2$$

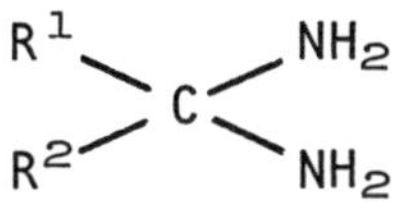

$$III(295\text{-}2) + H_2O \rightarrow R^1R^2CO + 2NH_3 \qquad 295\text{-}3$$

$$II(295\text{-}1) + 2H^+ + 2e \rightarrow R^1R^2C(N-H)(N-H) \;(IV) \qquad 295\text{-}4$$

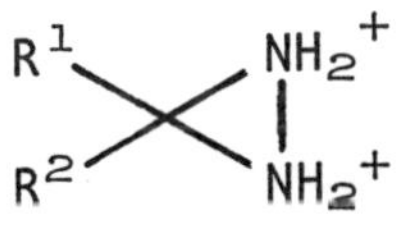

$$R^1R^2C(NH_2^+)(NH_2^+) \;(V) \rightleftarrows IV(295\text{-}4) + 2H^+ \qquad 295\text{-}5$$

$$V(295\text{-}5) + 2H^+ + 2e \rightarrow R^1R^2C(NH_3^+)(NH_3^+) \qquad 295\text{-}6$$

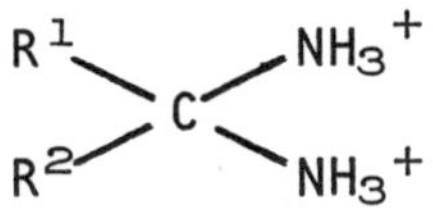

$R^1 = CH_3$, $R_2 - C_2H_5$ or R^1R^2C — cyclohexyl

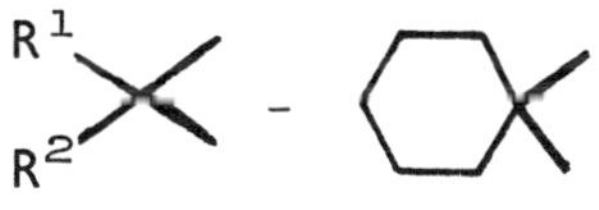

318

Mechanism:
318b: 318-1 and -3 to -8 (acidic)

318c: 318-8 (acidic)

Proposed for Compound:
EC17

EC18

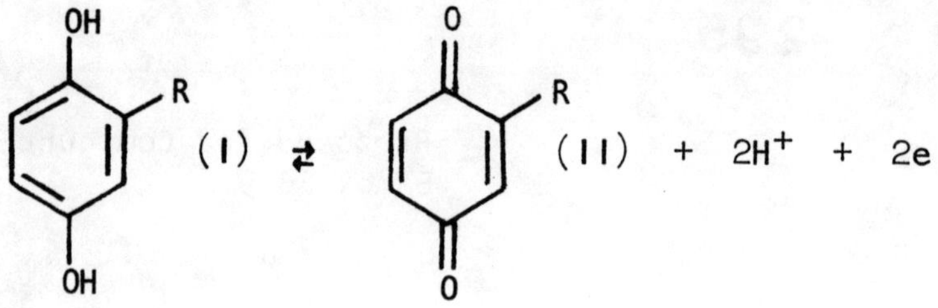

I(318-1) ⇌ II + 2H⁺ + 2e   318-1

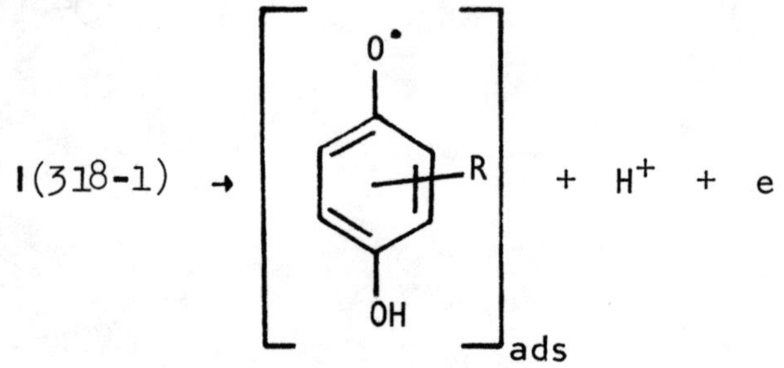

I(318-1) → [semiquinone radical]$_{ads}$ + H⁺ + e   318-2

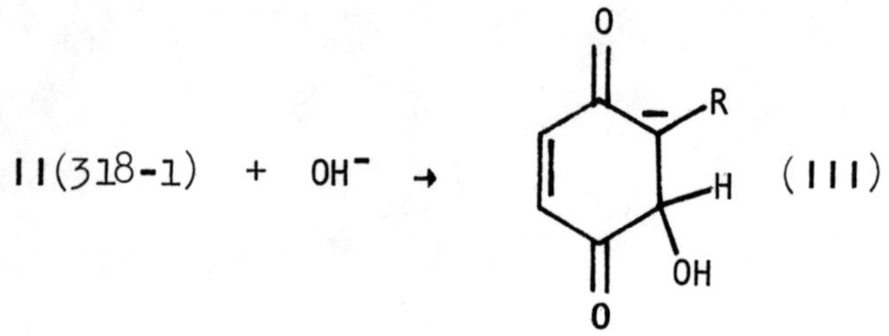

II(318-1) + OH⁻ → (III)   318-3

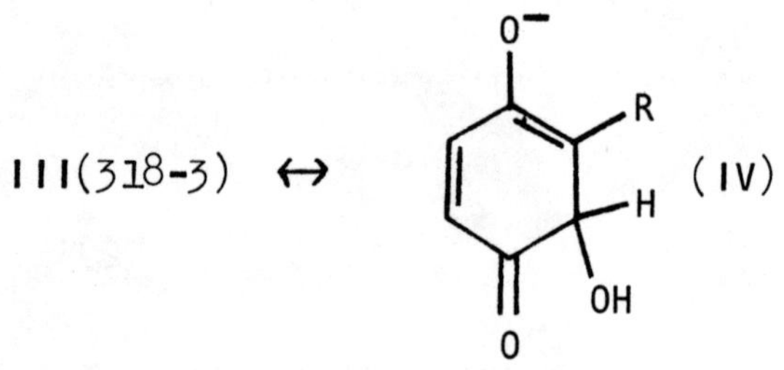

III(318-3) ↔ (IV)   318-4

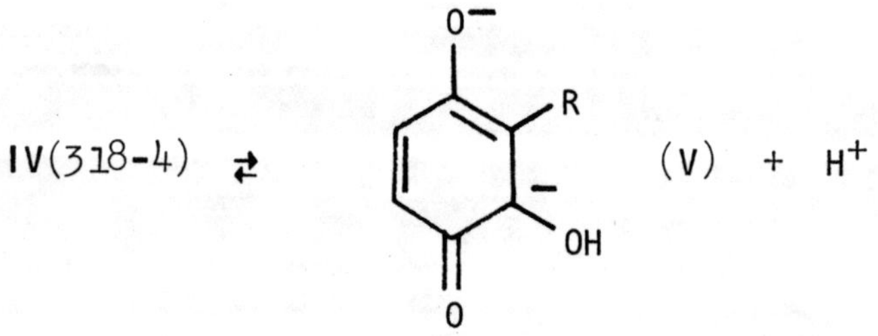

IV(318-4) ⇌ (V) + H⁺   318-5

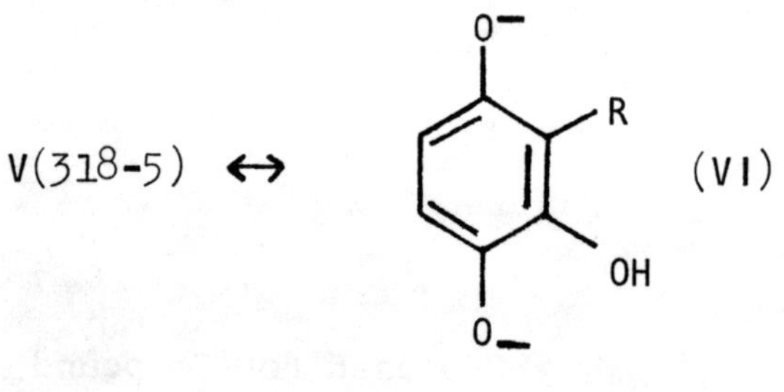

V(318-5) ↔ (VI)   318-6

TABLE III. Courses and Mechanisms of Half-Reactions

VI(318-6) + 2H$^+$ ⇌ [benzene ring with OH, R, OH, OH substituents] (VII) 318-7

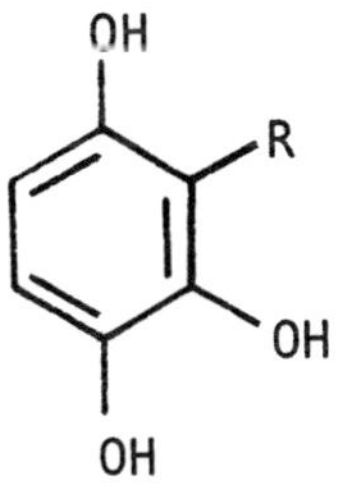

VII(318-7) ⇌ [quinone ring with R, OH substituents] + 2H$^+$ + 2e 318-8

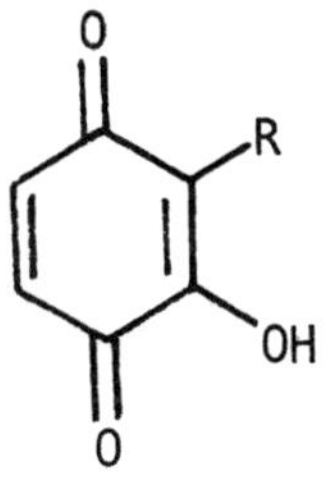

R = COOH

324

Mechanism:
324a: 324-4 to -6 (pH > pK, H_2O)

Proposed for Compound:
CA07, CA16, CA47, CA48, CC19

324b: 324-1 to -3 as one step (pH < pK, H_2O)

CA07, CA16, CA47, CA48, CC19

324c: 324-6 (protic solvent)

CA37, CA38

324d: 324-1 to -3 (protic solvent)

CA16, CA47, CA81, CB31

324e: 324-7 to -11 (unbuffered)

CA12, CA84, CC23

$R^1R^2CHNO_2H^+$ ⇌ $R^1R^2CHNO_2$ + H^+ 324-1

$R^1R^2CHNO_2H^+$ + e → $R^1R^2CHNO_2H\cdot$ 324-2

$R^1R^2CHNO_2H\cdot$ + 3H^+ + 3e → $R^1R^2CHNHOH$ + H_2O 324-3

$R^1R^2CHNO_2$ + OH^- → $R^1R^2\bar{C}NO_2$ + H_2O 324-4

$R^1R^2\bar{C}NO_2$ ↔ $\begin{matrix}R^1\\R^2\end{matrix}C=\overset{+}{N}\begin{matrix}O^-\\O^-\end{matrix}$ 324-5

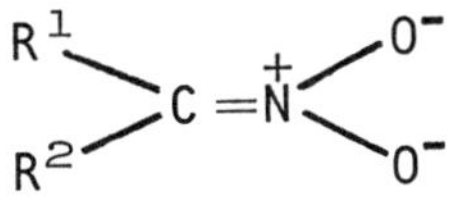

311

$R^1R^2\bar{C}NO_2 \rightarrow$ product $+$ ne   324-6

$R^1R^2CHNH_2OH^+ \rightleftarrows R^1R^2CHNHOH + H^+$ 324-7

$R^1R^2CHNH_2OH^+ + e \rightarrow R^1R^2C\dot{H}NH_2OH$ 324-8

$R^1R^2C\dot{H}NH_2OH + e \rightarrow R^1R^2CHNH_2 + OH^-$ 324-9

$R^1R^2C\dot{H}NH_2OH + H^+ + e \rightarrow R^1R^2CHNHOH + H_2$ 324-10

$R^1R^2CHNH_3^+ \rightleftarrows R^1R^2CHNH_2 + H^+$ 324-11

$R^1R^2CH = CH_3$, CH_3CH_2, $CH_3CH_2CH_2$, $(CH_3)_2CH$, $(CH_3)_3C$, or cyclo-C_6H_{11}

326

Mechanism:	Proposed for Compound:
326a: 326-1 to -4 (DMF, MeCN)	CA16, CA47, CA81, CB31
326b: 326-1, -5, and -6 (basic or aprotic solvent with H_2O or phenol)	CA16, CA47, CA81, CB31

$RNO_2 + e \rightleftarrows RNO_2^{\cdot -}$ 326-1

$RNO_2^{\cdot -} + e \rightarrow RNO_2^{2-}$ 326-2

$RNO_2^{2-} + 2HSo \rightarrow RNO + H_2O + 2So^-$ 326-3

$RNO + 2HSo + 2e \rightarrow RNHOH + 2So^-$ 326-4

$RNO_2^{\cdot -} + HSo \rightarrow RNO_2H\cdot + So^-$ 326-5

$RNO_2H\cdot + 3HSo + 3e \rightarrow RNHOH + H_2O + 3So^-$ 326-6

$R = CH_3(CH_2)_n$; $1 \leq n \leq 4$

HSo = solvent or added proton donor (<u>e.g.</u>, H_2O or phenol)

333

Mechanism:
333a: 333-1 and -2

Proposed for Compound:
EA12, EA28, EA62, EA94, EC42

$$L = R-O-C\begin{matrix}S\\S^-\end{matrix}$$

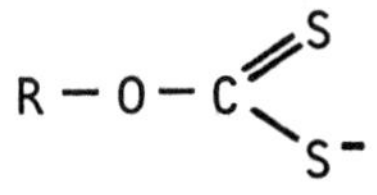

where R = CH_3, C_2H_5, $(CH_3)_2CH$, $CH_3CH_2CH_2CH_2$, or cyclohexyl

337

Mechanism:
337c: 337-4 to -8

Proposed for Compound:
EA25

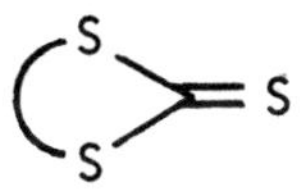

$$\left(\begin{matrix}S\\S\end{matrix}\right\rangle=S \ (I) + e \rightleftarrows \left(\begin{matrix}S\\S\end{matrix}\right\rangle-S^- \ (II) \qquad 337\text{-}1$$

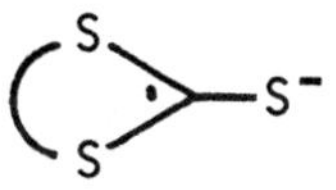

$II(337\text{-}1) \rightarrow product(s)$ \hfill 337-2

products + e → reduced products \hfill 337-3

$2\ I(337\text{-}1) \rightarrow$ (III) + e (possibly in more steps) \hfill 337-4

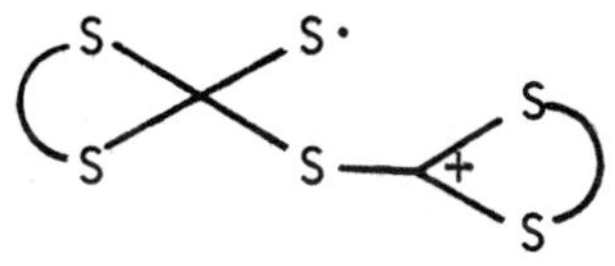

$III(337\text{-}4) \xrightarrow{slow}$ (IV) + S \hfill 337-5

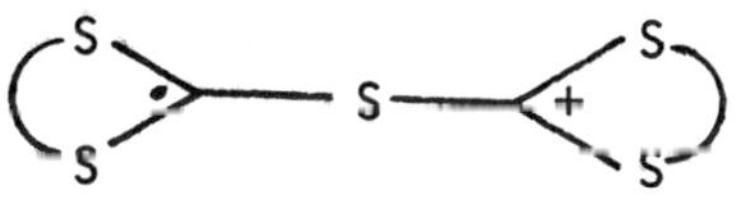

$IV(337\text{-}5) \xrightarrow{fast}$ (V) + e \hfill 337-6

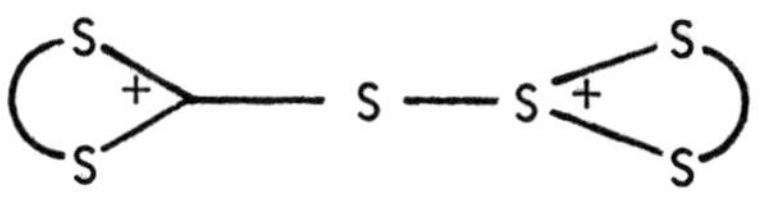

V(337-6) + H₂O → [structure: bicyclic with S, S, OH, S, S⁺ ring] (VI) + H⁺ 337-7

VI(337-7) → [structure: S,S-ring=O] + I(337-1) + H⁺ 337-8

338

Mechanism:	Proposed for Compound:
338b: 338-1 and -2	EA26, EA60, EA93
338c: 338-1, -2, and -4	EA92, EB76, ED35, EE24
338d: 338-5 and -6	EB66

[structure: S,S-ring-C⁺]—S—R + e ⇌ [structure: S,S-ring-C•]—S—R (I) 338-1

2 I(338-1) $\xrightarrow{fast}$ [structure: dimer with two S,S-rings connected with SR, RS groups] (II) 338-2

II(338-2) → products + ne 338-3

II(338-2) → [structure: S,S-ring=ring-S,S] (III) + RSSR 338-4

III(338-4) ⇌ [structure: S,S-ring•—⁺ring-S,S] (IV) + e 338-5

314

TABLE III. Courses and Mechanisms of Half-Reactions

IV(338-5) ⇌ [structure with S-S bridges and two + charges] + e   338-6

371

Mechanism: Proposed for Compound:
371c: 371-12 to -14; or 371-12, -15, and -16; or   ED89
       371-12, -17, and -18 (MeCN, Pt electrode)

(I) + M ⇌ [ ]M·⁺ (II) + e   371-1

II(371-1) + H$_2$O ⇌ [ ]M· (III) + H⁺   371-2

III(371-2) → (IV) + H⁺ + M + e   371-3

2II(371-1) $\xrightarrow{slow}$ [ ]²⁺ (V) + I(371-1) + 2M   371-4

$V(371-4) + H_2O \rightleftarrows IV(371-3) + 2H^+$ 　　　371-5

$II(371-1) + H_2O \rightarrow$ [structure with OH$^+$ on S, X bridge] $(VI) + M + H^+ + e$ 　　　371-6

$VI(371-6) \rightleftarrows IV(371-3) + H^+$ 　　　371-7

$II(371-1) \rightarrow V(371-4) + M + e$ 　　　371-8

$Cl^- + Pt \rightarrow ClPt + e$ 　　　371-9

$ClPt + I(371-1) \rightleftarrows$ [phenothiazine-type structure]$Cl^{\ddagger}$ $(VII) + Pt + e$ 　　　371-10

$VII(371-10) + H_2O \rightarrow VI(371-6) + H^+ + Cl^-$ 　　　371-11

$I(371-1) \rightarrow$ [structure with $^+S^{\bullet}$] $(VIII) + e$ 　　　371-12

$VIII(371-12) \xrightleftharpoons{slow}$ [structure with radical on ring] $(IX) + H^+$ 　　　371-13

TABLE III. Courses and Mechanisms of Half-Reactions

VIII(371-12) + IX(371-13) $\xrightarrow{fast}$ (X) 371-14

VIII(371-12) → (XI) + H$^+$ + e 371-15

I(371-1) + XI(371-15) → X(371-14) 371-16

VIII(371-12) + I(371-1) → (XII) 371-17

XII(371-17) → X(371-14) + H$^+$ + e 371-18

X = single bond in dibenzothiophene

X = no bond and H on each aromatic carbon in diphenylsulfide

372

Mechanism: Proposed for Compound:
372c: 372-1 and -2 as one step, and -12 to -20 EA31

$R^1R^2C:CH_2$ → $(R^1R^2C:CH_2)_{ads}$ 372-1

$(R^1R^2C:CH_2)_{ads} \rightarrow R^1R^2CCH_2^{+\cdot} + e$ 372-2

$R^1R^2CCH_2^{+\cdot} \rightarrow R^1R^2CCH_2^{2+} + e$ 372-3

$$R^1R^2CCH_2^{2+} + 2R^3OH \rightarrow R^1-\underset{OR^3}{\overset{R^2}{C}}-CH_2OR^3 + 2H^+$$ 372-4

$R^1R^2CCH_2^{2+} + R^1R^2C:CH_2 \rightarrow (R^1R^2\overset{+}{C}CH_2)_2$ 372-5

$$(R^1R^2\overset{+}{C}CH_2)_2 + 2R^3OH \rightarrow (R^1-\underset{OR^3}{\overset{R^2}{C}}-CH_2)_2 + 2H^+$$ 372-6

$(R^1R^2\overset{+}{C}CH_2)_2 \rightarrow (R^1R^2C:CH)_2 + 2H^+$ 372-7

$R^1R^2CCH_2^{+\cdot} + R^3OH \rightarrow R^1R^2\dot{C}CH_2OR^3 + H^+$ 372-8

$R^1R^2\dot{C}CH_2OR^3 \rightarrow R^1R^2\overset{+}{C}CH_2OR^3 + e$ 372-9

$$R^1R^2\overset{+}{C}CH_2OR^3 + R^3OH \rightarrow R^1-\underset{OR^3}{\overset{R^2}{C}}CH_2OR^3 + H^+$$ 372-10

$R^1R^2CCH_2^{+\cdot} + R^1R^2C:CH_2 \rightarrow R^1R^2\dot{C}CH_2CH_2\overset{+}{C}R^1R^2$ 372-11

$R^1R^2CCH_2^{+\cdot} + BF_4^- \rightarrow R^1R^2\underset{F}{\dot{C}}CH_2 + BF_3$ 372-12

$R^1R^2\underset{F}{C}-\dot{C}H_2 + HO\overset{O}{\overset{\|}{C}}-CH_3 \rightarrow R^1R^2\underset{F}{C}-CH_2O\overset{O}{\overset{\|}{C}}CH_3 + H^+ + e$ 372-13

In 372-14 through -20, $R^1 = CH_3$ and $R^2 = H$

$R^1R^2CCH_2^{+\cdot} \rightarrow \overbrace{CH_2CHCH_2}^{+} + H^+ + e$ 372-14

TABLE III. Courses and Mechanisms of Half-Reactions

$\overset{+}{\overbrace{CH_2CH_2CH_2}} + HO\overset{O}{\overset{\|}{C}}CH_3 \rightarrow CH_2{:}CHCH_2O\overset{O}{\overset{\|}{C}}CH_3 + H^+$ 372-15

$\overset{+}{\overbrace{CH_2CH_2CH_2}} + BF_4^- \rightarrow CH_2{:}CHCH_2F + BF_3$ 372-16

$\overset{+}{\overbrace{CH_2CH_2CH_2}} + xCH_3CH{:}CH_2 \rightarrow {+}CH{-}CH_2{+}_x$
                                                                   $\hspace{2em}|$
                                                                   $\hspace{1.5em}CH_3$ 372-17

$CH_3\overset{+}{C}HCH_3 \rightleftarrows CH_3CH{:}CH_2 + H^+$ 372-18

$CH_3\overset{+}{C}HCH_3 + HO\overset{O}{\overset{\|}{C}}CH_3 \rightarrow (CH_3)_2CHO\overset{O}{\overset{\|}{C}}CH_3 + H^+$ 372-19

$CH_3\overset{+}{C}HCH_3 + xCH_3CH{:}CH_2 \rightarrow {+}CH{-}CH_2{+}_x$
                                                                   $\hspace{2em}|$
                                                                   $\hspace{1.5em}CH_3$ 372-20

$R^1 = CH_3$

$R^2 = H$

377

Mechanism:	Proposed for Compound:
377a: 377-1 to -6	EA51, EA54
377b: 377-1 and -2	EA06
377c: 377-1, -2, -4, -5, and -7 (at DME)	EA53, EA59
377d: 377-1, -2, and -8 to -10 (QE at MP)	EA53
377e: 377-1, -2, -9, -11, and -12	EA09

$CXYZCOOR + e \rightarrow CXYZCOOR^{\overline{\cdot}}$ 377-1

$$CXYZCOOR^{\bar{\cdot}} \rightarrow {}^{\cdot}CYZCOOR + X^{-} \qquad 377\text{-}2$$

$${}^{\cdot}CYZCOOR \rightarrow \text{unreducible products} \qquad 377\text{-}3$$

$$CXYZCOOR^{\bar{\cdot}} + HA \rightleftarrows CXYZ\overset{\cdot}{C}\!-\!OR + A^{-} \qquad 377\text{-}4$$
$$\phantom{CXYZCOOR^{\bar{\cdot}} + HA \rightleftarrows CXYZ}\underset{H}{\overset{O}{|}}$$

$$CXYZ\overset{\cdot}{C}\!-\!OR + e \rightarrow CXYZ\overset{-}{C}\!-\!OR \qquad 377\text{-}5$$
$$\underset{H}{\overset{O}{|}} \underset{H}{\overset{O}{|}}$$

$$CXYZ\overset{-}{C}\!-\!OR + HA \rightarrow CXYZ\overset{H}{\overset{|}{C}}\!-\!OR + A^{-} \qquad 377\text{-}6$$
$$\underset{H}{\overset{O}{|}} \underset{H}{\overset{O}{|}}$$

$$CXYZ\overset{-}{C}\!-\!OR \xrightarrow{\text{fast}} CXYHCOOR + Z^{-} \qquad 377\text{-}7$$
$$\underset{OH}{|}$$

In 377-8 through -12, X = Y = Cl

$$2CXCl_2COOR^{\bar{\cdot}} \rightarrow 2Cl^{-} + CXCl_2COOR + {:}CZCOOR \qquad 377\text{-}8$$

$$2\overset{\cdot}{C}ClZCOOR + Hg \rightarrow Hg(CClZCOOR)_2 \qquad 377\text{-}9$$

$$Hg(CClZCOOR)_2 + 2e \rightarrow Hg + 2Cl^{-} + 2\,{:}CZCOOR \qquad 377\text{-}10$$

$${}^{\cdot}CClZCOOR + Hg \rightarrow {}^{\cdot}Hg(CClZCOOR) \qquad 377\text{-}11$$

$${}^{\cdot}Hg(CClZCOOR) + e \rightarrow Hg + {:}CClZCOOR \qquad 377\text{-}12$$

X = H, Cl, or F

Y = H, Cl, or F

Z = Y = Cl or F

R = H, C_2H_5, or C_6H_5

HA = 3,4-xylenol

TABLE III. Courses and Mechanisms of Half-Reactions

378

Mechanism:
378a: 378-1 to -8 (pH < 6.3)
378b: 378-1, and -9 to -16 (pH > 6.3)

Proposed for Compound:
ED07
ED07

(I) ⇌ (II) + H⁺	378-1
I(378-1) ⇌ (III) + 2H⁺ + 2e	378-2
III(378-2) + OH⁻ ⇌ (IV)	378-3
IV(378-3) ↔ (V)	378-4
V(378-4) ⇌ (VI) + H⁺	378-5

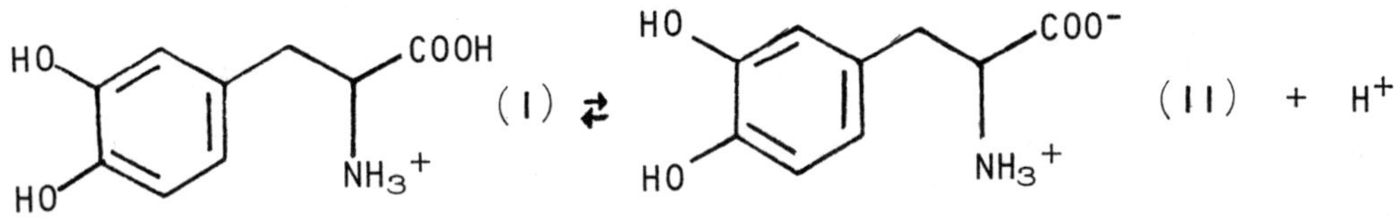

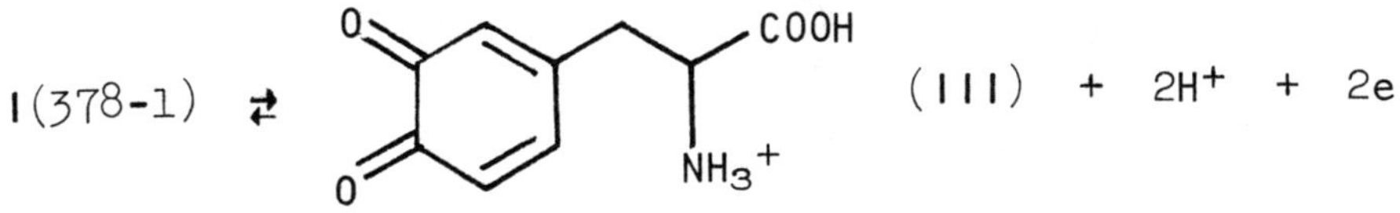

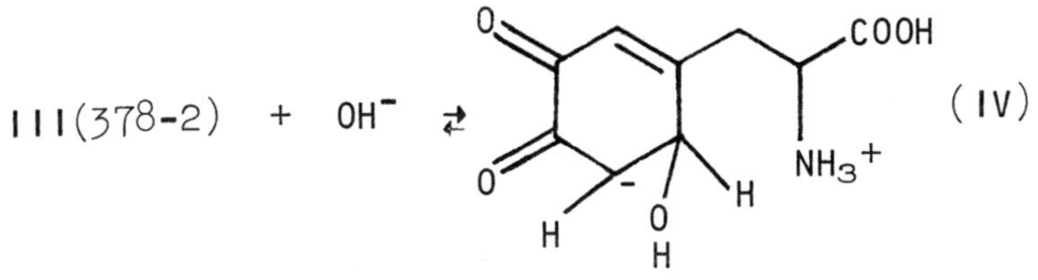

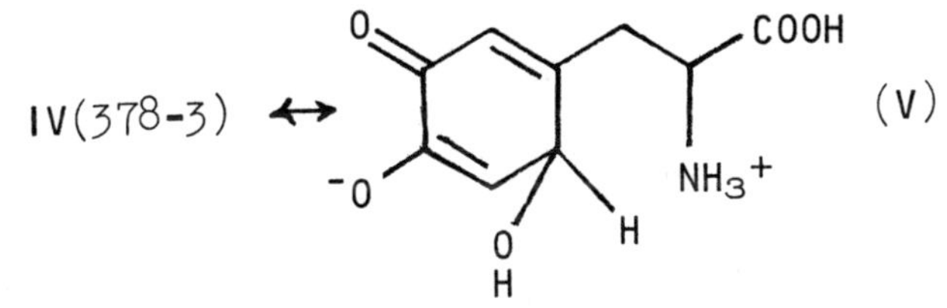

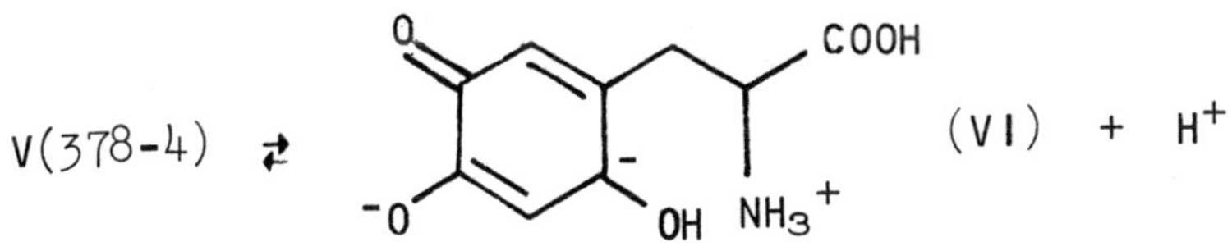

VI(378-5) ↔ [structure: benzene ring with $-O^-$, $-O^-$, $-O_H$, $-OH$, and $-CH_2CH(NH_3^+)COOH$ substituents] (VII)   378-6

VII(378-6) + 2H$^+$ ⇌ [structure: benzene ring with HO, HO, OH, O_H, and $-CH_2CH(NH_3^+)COOH$ substituents] (VIII)   378-7

VIII(378-7) + III(378-2) → [structure: quinone ring with =O, =O, HO, and $-CH_2CH(NH_3^+)COOH$ substituents] + I(378-1)   378-8

II(378-1) ⇌ [structure: ortho-quinone with $-CH_2CH(NH_3^+)COO^-$ substituent] (IX) + 2H$^+$ + 2e   378-9

IX(378-9) ⇌ [structure: ortho-quinone with $-CH_2CH(NH_2)COO^-$ substituent] (X) + H$^+$ 378-10

X(378-10) → [bicyclic structure with two C=O, NH$_2^+$, COO$^-$, and H, H$^-$] (XI)   378-11

TABLE III. Courses and Mechanisms of Half-Reactions

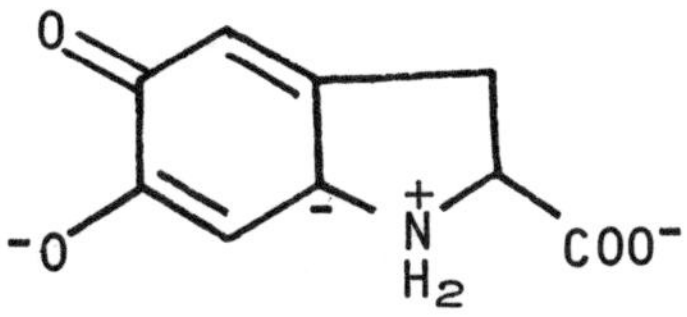

XI(378-11) ↔ [structure] (XII) 378-12

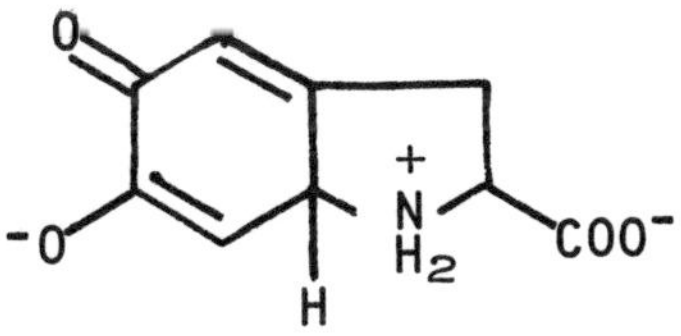

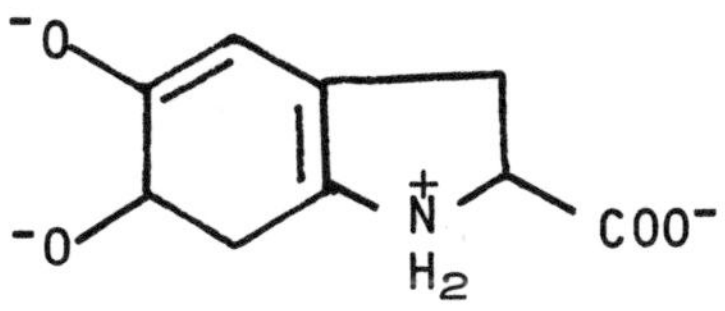

XII(378-12) ⇌ [structure] (XIII) + H⁺ 378-13

XIII(378-13) ↔ [structure] (XIV) 378-14

XIV(378-14) + 2H⁺ ⇌ [structure] (XV) 378-15

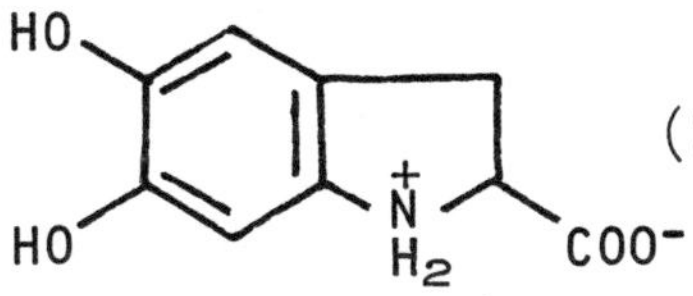

XV(378-15) + IX(378-9) → [structure] + II(378-1) 378-16

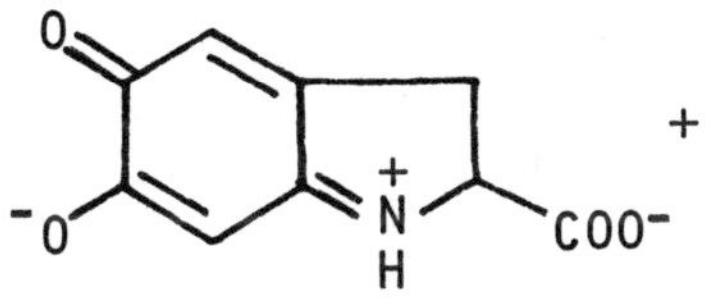

379

Mechanism:
379a: 379-1, -3, and -4 (basic)
379b: 379-1 and -2 (acidic)
379c: 379-1, and -3 to -5 (basic)

Proposed for Compound:
EA35, EB84, EB85, EB86

phenylglycine

$$R-\underset{NH_2}{\overset{H}{C}}-COOH \rightleftarrows R-\underset{NH_2}{\overset{H}{C}}-COO^- + H^+ \qquad 379\text{-}1$$

$$R-\underset{NH_2}{\overset{H}{C}}-COOH \rightarrow R-\underset{H}{C}=N-H + 2H^+ + 2e + CO_2 \qquad 379\text{-}2$$

$$R-\underset{NH_2}{\overset{H}{C}}-COO^- \rightarrow R-\underset{H}{C}=N-H + H^+ + 2e + CO_2 \qquad 379\text{-}3$$

$$R-\underset{H}{C}=N-H \rightarrow R-CN + 2H^+ + 2e \qquad 379\text{-}4$$

$$R-\underset{H}{C}=N-H + H_2O \rightleftarrows R-\underset{H}{C}=O + NH_3 \qquad 379\text{-}5$$

$R = CH_3$, $CH_3CH_2\underset{CH_3}{\overset{|}{C}H}$, $(CH_3)_2CHCH_2$, or $CH_3(CH_2)_3$

TABLE III. Courses and Mechanisms of Half-Reactions

380

Mechanism:
380a: 380-1, and -3 to -6 (pH < 7)
380b: 380-1, -2, and -3 to -5 as one step (7 < pH < 9)
380c: 380-1, -2, -7, and -8 (pH > 9)

Proposed for Compound:
EE34
EE34
EF34

(I) ↔ (II)   380-1

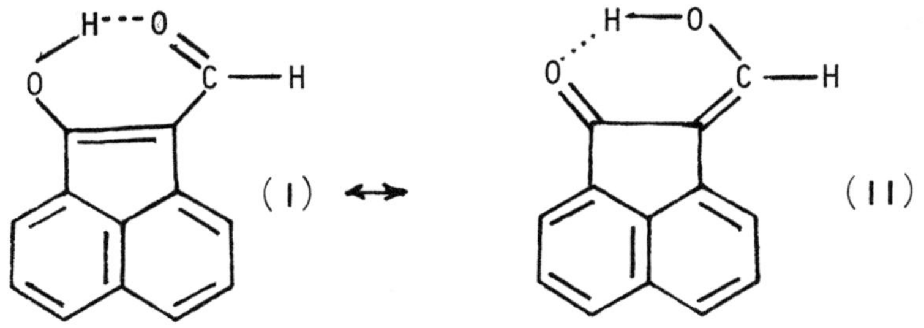

I(380-1) or II(380-1) ⇌ (III) + H$^+$ 380-2

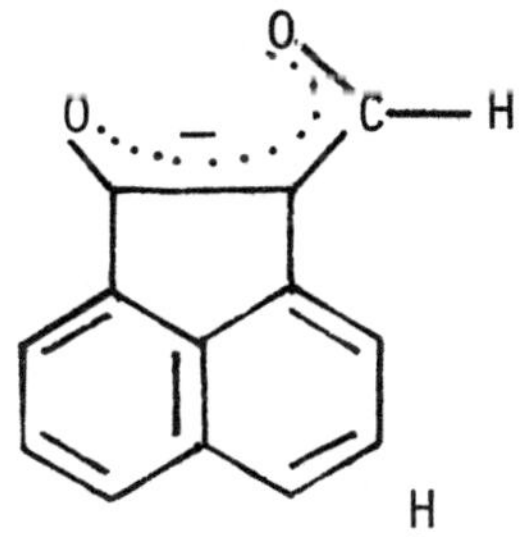

I(380-1) + H$^+$ + e → 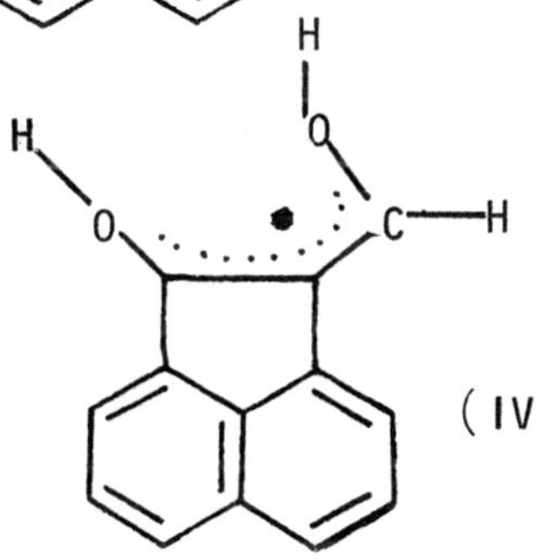 (IV)   380-3

IV(380-3) $\xrightarrow{\text{base catalysis}}$ (V)   380-4

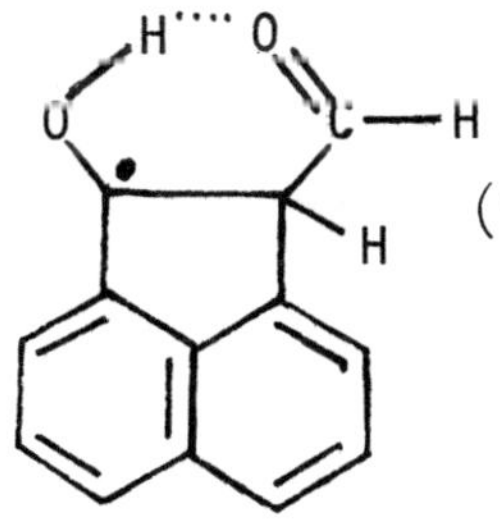

V(380-4) + e → (VI)   380-5

VI(380-5) + H⁺ → (VII)   380-6

III(380-2) + H⁺ + 2e → (VIII)   380-7

VII(380-6) ⇌ VIII(380-7) + 2H⁺   380-8

381

Mechanism:	Proposed for Compound:
381a: 381-1 to -6 (pH < 4.5)	EC92
381b: 381-7 to -15 (8 < pH < 10)	EC92
381c: 381-7, -8, -11 to -14, and -16 to -32 (pH > 10)	EC92

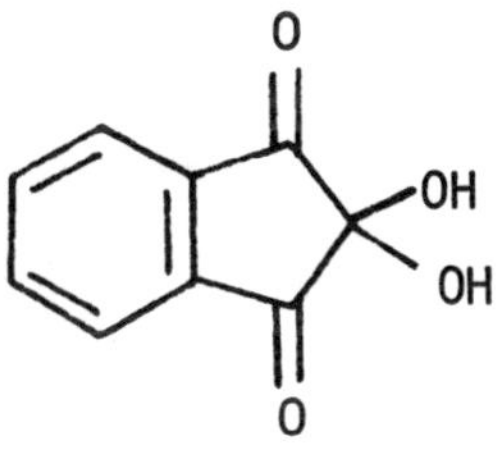

TABLE III. Courses and Mechanisms of Half-Reactions

(I) + 2H⁺ + 2e → 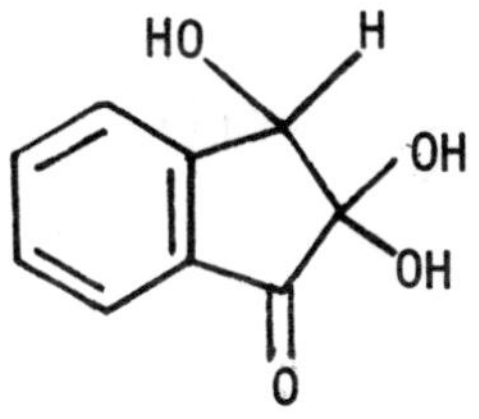 (II) 381-1

II(381-1) + 2H⁺ + 2e → 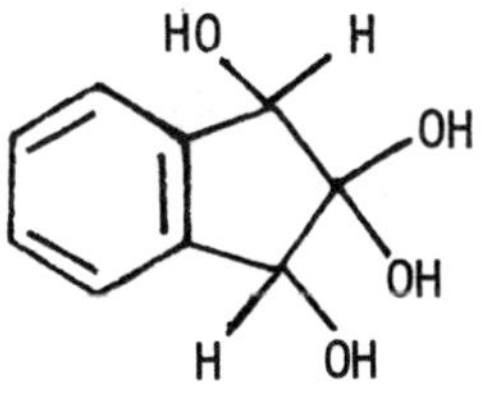 (III) 381-2

III(381-2) ⇌ (IV) + H₂O 381-3

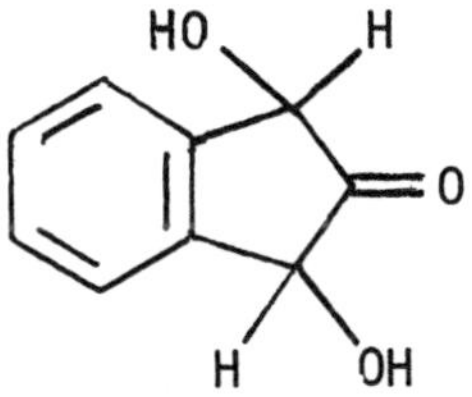

IV(381-3) ⇌ (V) 381-4

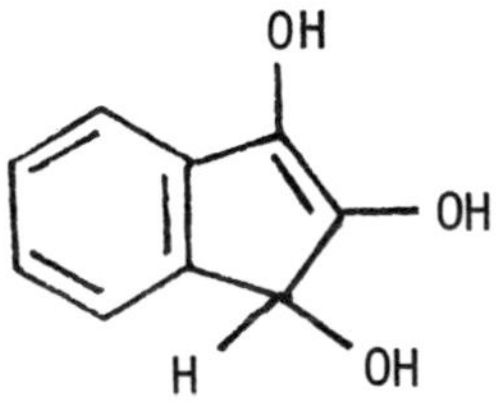

V(381-4) ⇌ (VI) 381-5

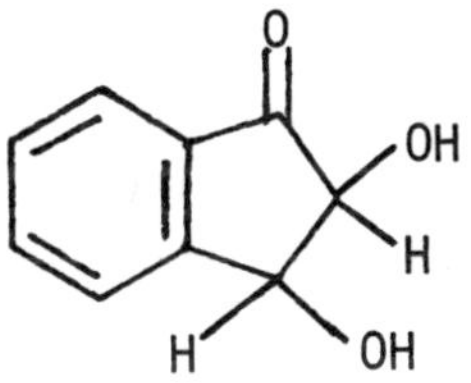

VI(381-5) + 2H⁺ + 2e → 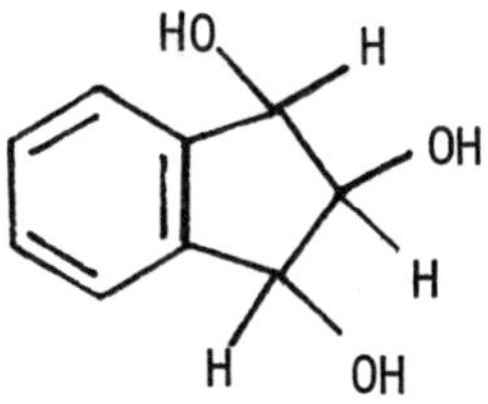 (VII) 381-6

I(381-1) ⇌ [2-hydroxy-2-oxido-1,3-indandione] (VIII) + H⁺   381-7

VIII(381-7) + OH⁻ → [benzene-1,2-bis(carboxylate) with CO—COO⁻ and COO⁻ groups] (IX) + 2H⁺ + 2e   381-8

VIII(381-7) ⇌ [indane-1,2,3-trione] (X) + OH⁻   381-9

X(381-9) + H⁺ + 2e ⇌ [3-hydroxy-2-oxido-1-indanone] (XI)   381-10

XI(381-10) ↔ [1-hydroxy-3-oxido-2-indanone] (XII)   381-11

VIII(381-7) + 2H⁺ + 2e → [2,3-dihydroxy-2-oxido-1-indanone] (XIII)   381-12

TABLE III. Courses and Mechanisms of Half-Reactions

XIII(381-12) → [structure: 3-hydroxy-3H-indane-1,2-dione with HO and H on C3, =O on C2, =O on C1] (XIV) + OH⁻   381-13

XIV(381-13) → XI(381-10) + H⁺   381-14

XI(381-10) or XII(381-11) + 5H⁺ + 4e → VII(381-6)   381-15

VIII(381-7) + OH⁻ → [structure: indanetrione hydrate dianion with HO, O⁻ on C3; OH, O⁻ on C2; =O on C1] (XV)   381-16

XV(381-16) ⇌ [structure: benzene ring with ortho substituents —C(OH)=O and —C(=O)—C(OH)(O⁻)—] (XVI)   381-17

XVI(381-17) ↔ [structure: benzene ring with ortho substituents —C(OH)=O and —C(O⁻)=C(O⁻)—OH] (XVII)   381-18

XVII(381-18) ⇌ [structure: benzene with -C(=O)O⁻ and -CH(O⁻)-C(=O)OH substituents ortho] (XVIII) 381-19

XVIII(381-19) ⇌ [structure: benzene with -C(=O)O⁻ and -CH(OH)-C(=O)O⁻ substituents ortho] (XIX) 381-20

IX(381-8) + 2H⁺ + 2e → XIX(381-20) 381-21

XIII(381-12) → [structure: benzene with -C(=OH⁺)H and -C(=O)-C(OH)-O⁻ substituents ortho] (XX) 381-22

XX(381-22) ⇌ [structure: benzene with -C(=O)H and -C(=O)-CH(OH)-O⁻ substituents ortho] (XXI) 381-23

TABLE III. Courses and Mechanisms of Half-Reactions

XXI(381-23) → [2-formylphenyl-C(=O)-C(=O)-O⁻] (XXII) + 2H⁺ + 2e   381-24

XX(381-22) ↔ [2-(C(OH⁺)=H)phenyl-C(O⁻)=C(OH)-O⁻] (XXIII)   381-25

XXIII(381-25) ↔ [2-(C(OH⁺)=H)phenyl-C(O⁻)(−)-C(OH)=O] (XXIV)   381-26

XXIV(381-26) ⇌ [2-(C(=O)H)phenyl-CH(O⁻)-C(OH)=O] (XXV)   381-27

XXV(381-27) → [2-(C(=O)H)phenyl-CH(OH)-C(O⁻)=O] (XXVI)   381-28

XI(381-10) ⇌ [indanone-enolate structure] (XXVII) + H⁺ 381-29

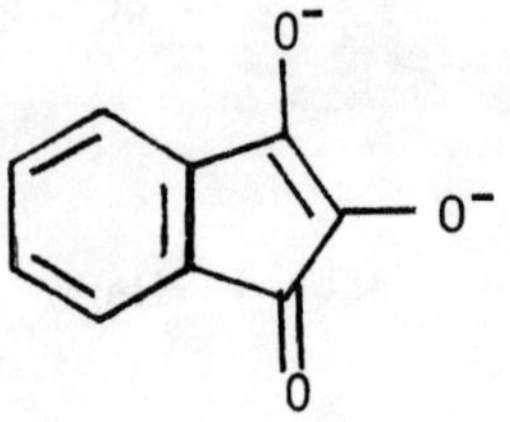

XII(381-11) ⇌ [indandione bis-enolate structure] (XXVIII) + H⁺ 381-30

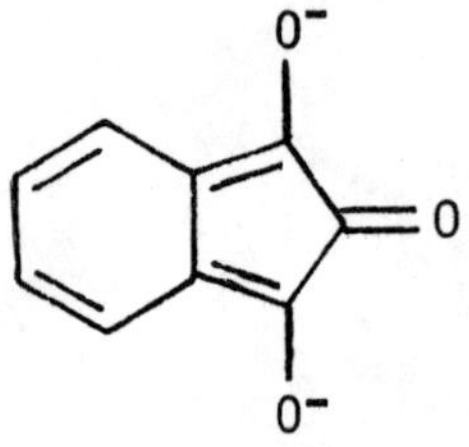

XXVII(381-29) ↔ XXVIII(381-30) 381-31

XXVII(381-29) or XXVIII(381-30) + H₂O → [hydrated structure] 381-32

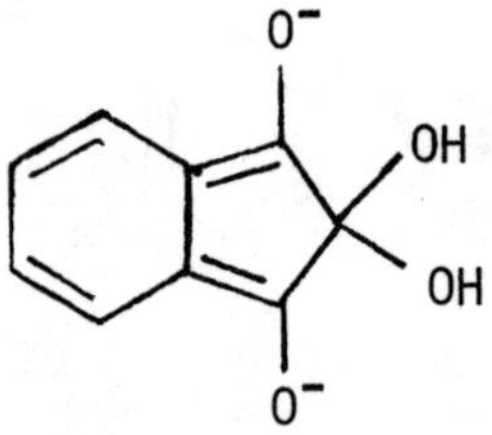

382

Mechanism:
382a: 382-1 to -5, followed by 65c

Proposed for Compound: EC21

$ArC(:OH^+)NH_2 \rightleftarrows ArC(:O)NH_2 + H^+$ 382-1

$ArC(:OH^+)NH_2 + H^+ + 2e \rightarrow ArCH(OH)NH_2$ 382-2

$ArCH(OH)NH_2 \rightleftarrows ArCH:NH + H_2O$ 382-3

$ArCH:NH + 2H^+ + 2e \rightarrow ArCH_2NH_2$ 382-4

$ArCH:NH + H_2O \rightleftarrows ArCHO + NH_3$ 382-5

Ar = C_6H_5

TABLE III. Courses and Mechanisms of Half-Reactions

383

Mechanism:
383a: 383-1 and -2

Proposed for Compound:
EB52

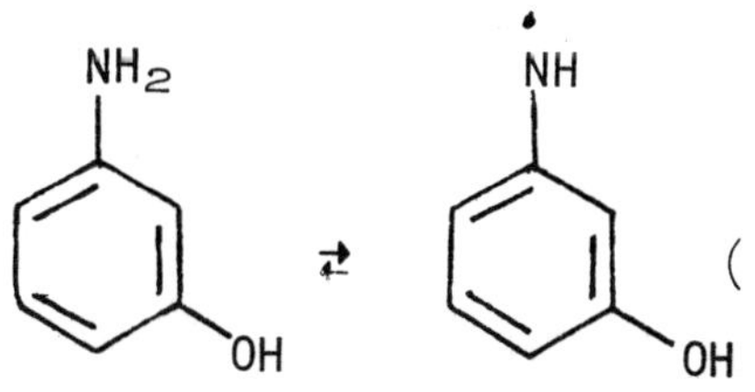

3-aminophenol ⇌ 3-aminophenol radical (I) + H$^+$ + e   383-1

I(383-1) $\xrightarrow{\text{fast}}$ products   383-2

384

Mechanism:
384a: 384-1 to -5

Proposed for Compound:
AB08

$R_2NNH_2 \rightarrow R_2NNH_2^{+\cdot} + e$ 384-1

$R_2NNH_2^{+\cdot} \rightleftarrows R_2NNH^{\cdot} + H^+$ 384-2

$2R_2NNH^{\cdot} \rightarrow R_2NNHNHNR_2$ 384-3

$R_2NNHNHNR_2 \rightarrow 2R_2NH + N_2$ 384-4

$R_2NH_2^+ \rightleftarrows R_2NH + H^+$ 384-5

R = CH$_3$

385

Mechanism:
385a: 385-1 to -6 (acidic)

385b: 385-7 and -8 (basic)

Proposed for Compound:
EA04

EA04

$RNHNH_3^+ \rightleftarrows RNHNH_2 + H^+$ 385-1

$RNHNH_3^+ \rightleftarrows RNHNH\cdot + 2H^+ + e$ 385-2

$RNHNH\cdot \rightleftarrows RNHNH\cdot_{ads}$ 385-3

$RNHNH\cdot_{ads} \rightleftarrows RNHN_{ads} + H^+ + e$ 385-4

$RNHN_{ads} \xrightarrow{slow} RN:NH$ 385-5

$RN:NH + H_2O \xrightarrow{fast} ROH + 2H^+ + N_2 + 2e$ 385-6

$RNHNH_2 + OH^- \rightarrow RNHNH^+ + H_2O + 2e$ 385-7

$RNHNH^+ + 3OH^- \rightarrow ROH + 2H_2O + N_2 + 2e$ 385-8

$R = CH_3$

386

Mechanism: Proposed for Compound:
386a: 386-1 to -12 (neutral and basic)   EE23

$Ar^1NHN:\overset{+}{N}HAr^2 \rightleftarrows Ar^1NHN:NAr^2 + H^+$ 386-1

$Ar^1NHN:\overset{+}{N}HAr^2 + e \rightarrow Ar^1NH\dot{N}NHAr^2$ 386-2

$Ar^1NH\overset{+}{\dot{N}}HNHAr^2 \rightleftarrows Ar^1NH\dot{N}NHAr^2 + H^+$ 386-3

$Ar^1NH\overset{+}{\dot{N}}HNHAr^2 + e \rightarrow Ar^1NHNHNHAr^2$ 386-4

$Ar^1NHNHNHAr^2 + 2H^+ + 2e \rightarrow Ar^1NH_2 + Ar^2NHNH_2$ 386-5

$Ar^1NH\dot{N}NHAr^2 + e \rightarrow Ar^1NH\overline{N}NHAr^2$ 386-6

$Ar^1NH\overline{N}NHAr^2 \rightarrow Ar^1NH^- + Ar^2\overset{+}{N}HN^-$ 386-7

$Ar^1NH^- + H_2O \rightarrow Ar^1NH_2 + OH^-$ 386-8

$Ar^2\overset{+}{N}HN^- \rightarrow Ar^2N:NH$ 386-9

$Ar^2N:NH + 2H_2O + 2e \rightarrow Ar^2NHNH_2 + 2OH^-$ 386-10

$Ar^1NHN:NAr^2 + e \rightarrow [Ar^1NHNNAr^2]^{\cdot-}$ 386-11

TABLE III. Courses and Mechanisms of Half-Reactions

$$[Ar^1NHNNAr^2]^- + H_2O \rightarrow Ar^1NH\dot{N}NHAr^2 + OH^- \qquad 386\text{-}12$$

$Ar^1 = C_6H_5$

$Ar^2 = 4\text{-}C_6H_4NHOH$

387

Mechanism:		Proposed for Compound:
387a:	387-1 and -3 to -5, followed by 32d (strong acid, macro)	EB38, EB59
387b:	387-2, -6, and -8 to -12, followed by 32b ($9 < pH < 11.5$, PHOS, macro)	EB59
387c:	387-2 and -7 to -12, followed by 32b ($pH > 11.5$, KOH, macro)	EB59
387d:	387-13 and -14 (acid, macro)	EE12
387e:	387-1, -3, and -4 (macro)	EE20

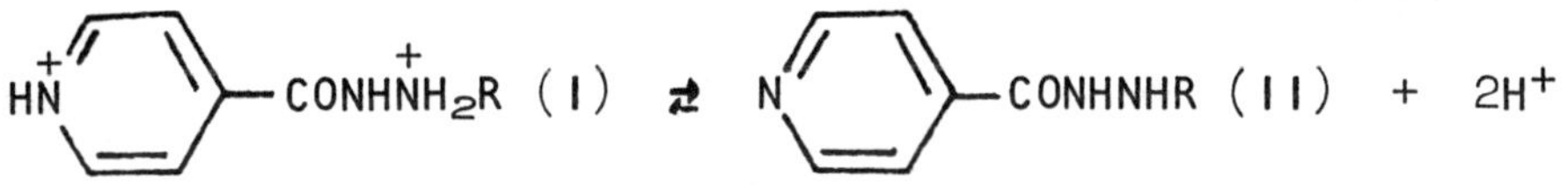

$$\overset{+}{H}N\!\!\bigcirc\!\!-CONH\overset{+}{N}H_2R \;(I) \;\rightleftarrows\; N\!\!\bigcirc\!\!-CONHNHR \;(II) \;+\; 2H^+ \qquad 387\text{-}1$$

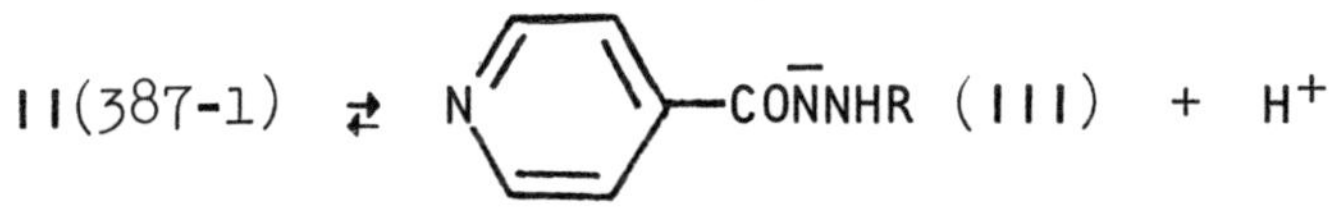

$$II(387\text{-}1) \;\rightleftarrows\; N\!\!\bigcirc\!\!-CO\bar{N}NHR \;(III) \;+\; H^+ \qquad 387\text{-}2$$

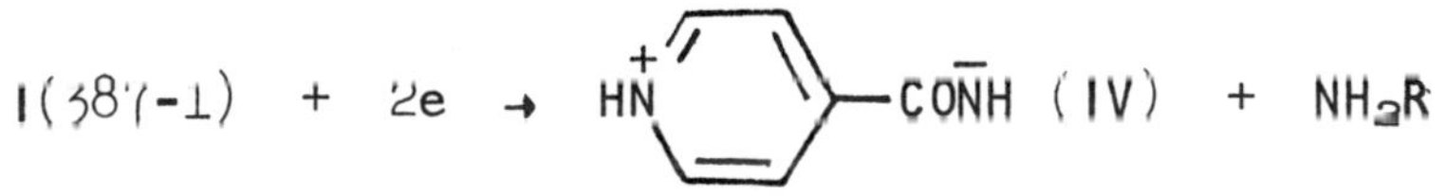

$$I(387\text{-}1) \;+\; 2e \;\rightarrow\; \overset{+}{H}N\!\!\bigcirc\!\!-CO\bar{N}H \;(IV) \;+\; NH_2R \qquad 387\text{-}3$$

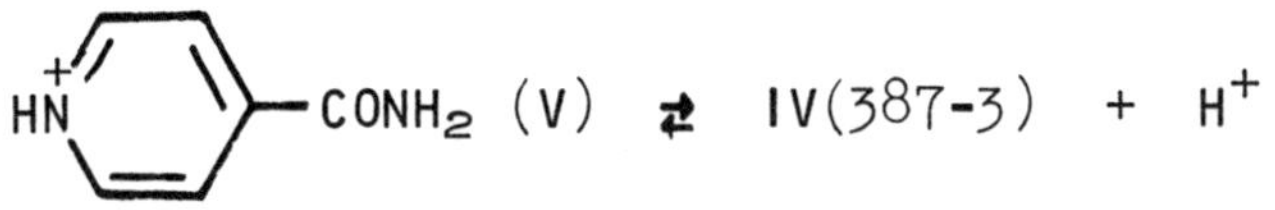

$$\overset{+}{H}N\!\!\bigcirc\!\!-CONH_2 \;(V) \;\rightleftarrows\; IV(387\text{-}3) \;+\; H^+ \qquad 387\text{-}4$$

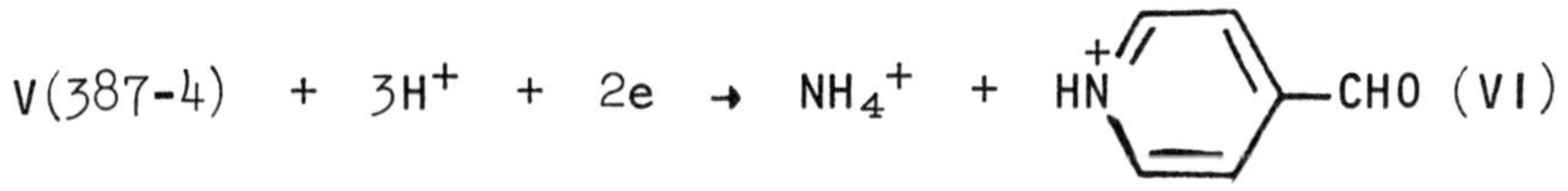

$$V(387\text{-}4) \;+\; 3H^+ \;+\; 2e \;\rightarrow\; NH_4^+ \;+\; \overset{+}{H}N\!\!\bigcirc\!\!-CHO \;(VI) \qquad 387\text{-}5$$

In 387-6, R = H

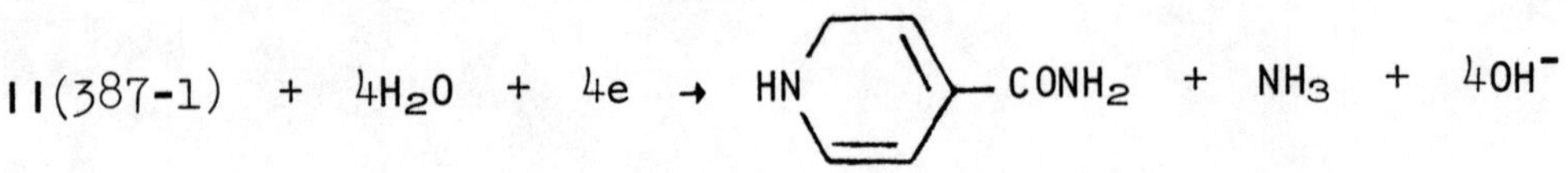

II(387-1) + 4H₂O + 4e → [HN-pyridine-CONH₂] + NH₃ + 4OH⁻   387-6

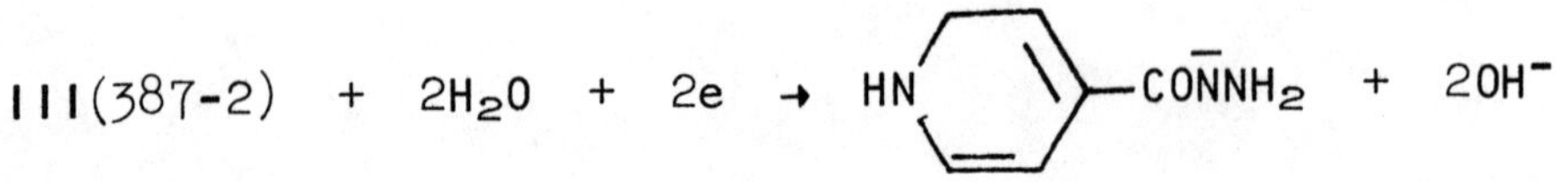

III(387-2) + 2H₂O + 2e → [HN-pyridine-CON̄NH₂] + 2OH⁻   387-7

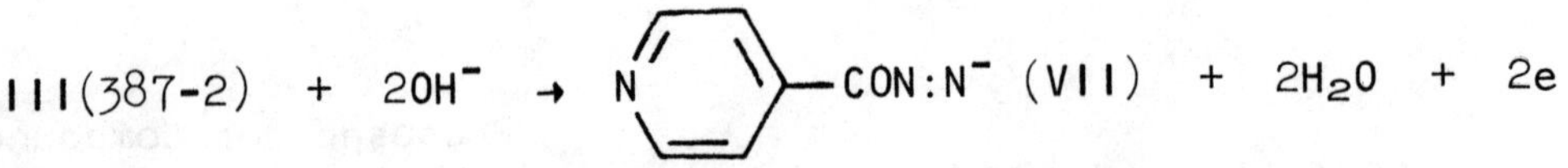

III(387-2) + 2OH⁻ → [N-pyridine-CON:N⁻] (VII) + 2H₂O + 2e   387-8

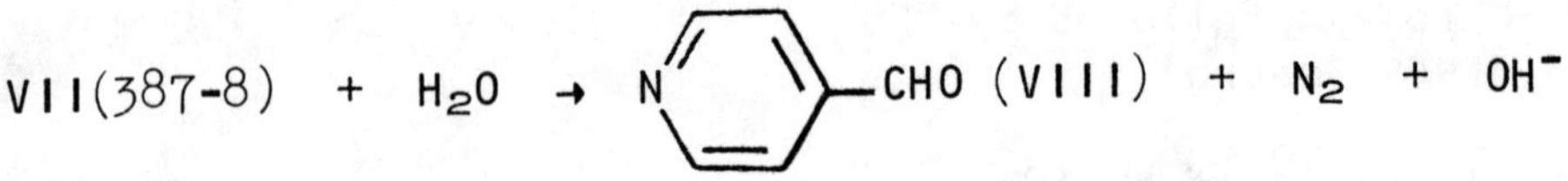

VII(387-8) + H₂O → [N-pyridine-CHO] (VIII) + N₂ + OH⁻   387-9

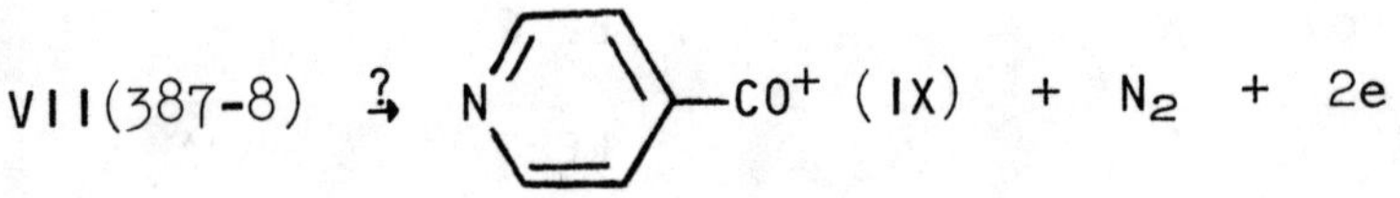

VII(387-8) →? [N-pyridine-CO⁺] (IX) + N₂ + 2e   387-10

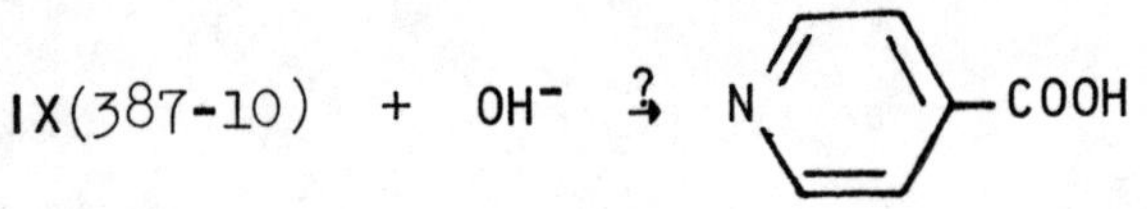

IX(387-10) + OH⁻ →? [N-pyridine-COOH]   387-11

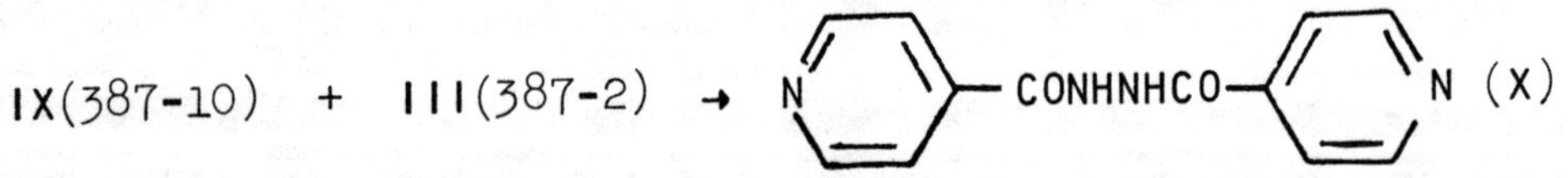

IX(387-10) + III(387-2) → [N-pyridine-CONHNHCO-pyridine-N] (X)   387-12

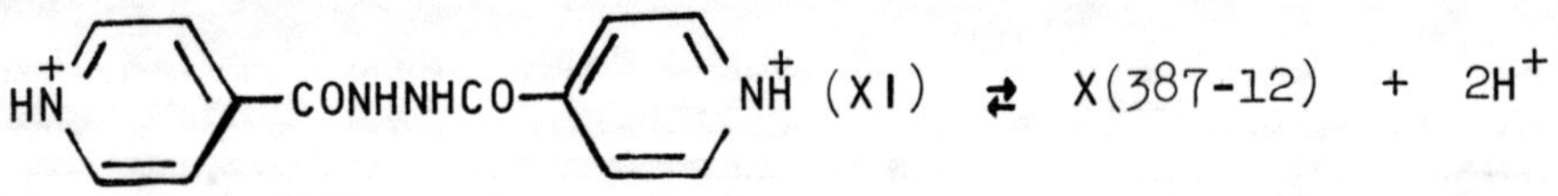

[HN⁺-pyridine-CONHNHCO-pyridine-NH⁺] (XI) ⇌ X(387-12) + 2H⁺   387-13

XI(387-13) + 8H⁺ + 6e → 2VI(387-5) + 2NH₄⁺   387-14

R = H or C₆H₅

TABLE III. Courses and Mechanisms of Half-Reactions

388

Mechanism:
388a: 388-1, and -4 to -6 (acid)

Proposed for Compound:
EA30, EA38, EA52, EA63, EA74

388b: 388-2, -3, -7, and -8 (base)

EA30, EA38, EA52, EA63, EA74

$RC(:\overset{+}{N}HNHCONH_2)COOH \rightleftarrows RC(:NNHCONH_2)COOH + H^+$	388-1
$RC(:NNHCONH_2)COOH \rightleftarrows RC(:NNHCONH_2)COO^- + H^+$	388-2
$RC(:\overset{+}{N}HNHCONH_2)COOH \rightleftarrows RC(:\overset{+}{N}HNHCONH_2)COO^- + H^+$	388-3
$RC(:\overset{+}{N}HNHCONH_2)COOH + H^+ + 2e \rightleftarrows RCH(NHNHCONH_2)COOH$	388-4
$RCH(NHNHCONH_2)COOH + H^+ \overset{k_5}{\rightarrow} RC(:NH_2^+)COOH + NH_2CONH_2$	388-5
$RC(:NH_2^+)COOH + H^+ + 2e \rightarrow RCH(NH_2)COOH$	388-6
$RC(:NNHCONH_2)COO^- + 2H^+ + 2e \rightarrow RCH(NHNHCONH_2)COO^-$	388-7
$RCH(NHNHCONH_2)COO^- + H^+ \overset{k_8}{\rightarrow} RC(:NH_2^+)COO^- + NH_2CONH_2$	388-8

389

Mechanism:
389a: 389-1 to -3

Proposed for Compound:
EC57

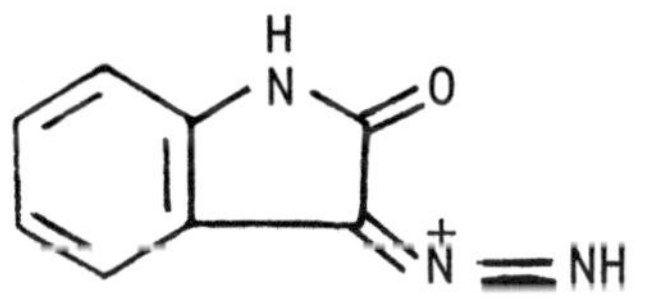

(I) $\rightleftarrows$ (II) $+ H^+$	389-1
I(389-1) $+ 2e \rightarrow$ (III)	389-2

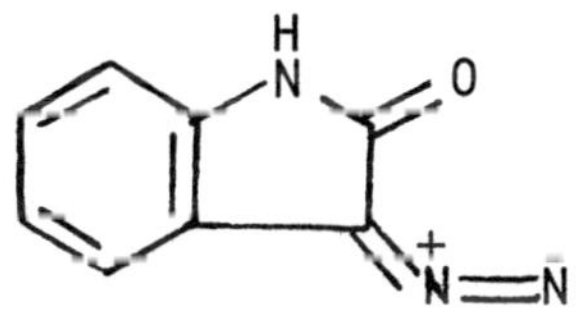

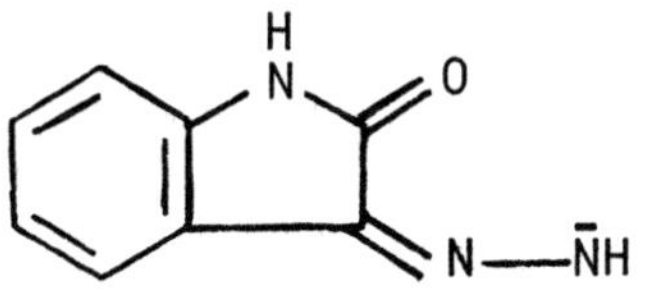

337

[indolin-3-imine-2-one structure] ⇌ III(389-2) + H⁺   389-3

390

Mechanism:

390a: 390-1 to -4, followed by 283b (pH < 4)   Proposed for Compound: EC65

390b: 390-2, -5, and -4, followed by 283b   EC65
(4 < pH < 10)

390c: 390-6 and -7, followed by 283d (pH > 10)   EC65

$[ArCOCH:NNH_3^+]H^+ \rightleftarrows ArCOCH:NNH_3^+ + H^+$ 390-1

$ArCOCH:NNH_3^+ \rightleftarrows ArCOCH:NNH_2 + H^+$ 390-2

$[ArCOCH:NNH_3^+]H^+ + H^+ + 2e \rightarrow ArCOCH_2NHNH_3^+$ 390-3

$ArCOCH_2NHNH_3^+ + 2H^+ + 2e \rightarrow ArCOCH_2NH_3^+ + NH_3$ 390-4

$ArCOCH:NNH_3^+ + 2H^+ + 2e \rightarrow ArCOCH_2NHNH_3^+$ 390-5

$ArCOCH:NNH_2 + 2H^+ + 2e \rightarrow ArCOCH_2NHNH_2$ 390-6

$ArCOCH_2NHNH_2 + 2H^+ + 2e \rightarrow ArCOCH_2NH_2 + NH_3$ 390-7

Ar = C_6H_5

TABLE III. Courses and Mechanisms of Half-Reactions

391

Mechanism: Proposed for Compound:

391a: 391-1, -3, -4, -5, and -10 (pH < 9.5)   EC12

391b: 391-2, and -6 to -10 (pH > 9.5)   EC12

(I) ⇌ (II) + H$^+$	391-1
II(391-1) ⇌ (III) + H$^+$	391-2
I(391-1) + 4H$^+$ + 4e → (IV) + H$_2$O	391-3
[structure]H$^+$ (V) ⇌ IV(391-3) + H$^+$	391-4
V(391-4) + 2e → (VI) + NH$_3$	391-5
II(391-1) + 4H$^+$ + 4e → (VII) + H$_2$O	391-6
IV(391-3) ⇌ VII(391-6) + H$^+$	391-7

IV(391-3) + 2e → [structure] (VIII) + NH₃   391-8

VI(391-5) ⇌ VIII(391-8) + H⁺   391-9

VI(391-5) + 2H⁺ + 2e → [structure]   391-10

Code No.	pK_1	pK_2
EC12	<1	8.6

392

Mechanism: Proposed for Compound:

392a: 392-1, and -3 to -8 (pH < 8)   EC25

392b: 392-1, -2, and -9 (8.5 < pH < 10.5)   EC25

392c: 392-2 and -10 (pH > 10.5)   EC25

[structure] (I) ⇌ [structure] (II) + H⁺   392-1

TABLE III. Courses and Mechanisms of Half-Reactions

II(392-1) ⇌ [pyrrolizine with =NOH and =NO⁻ substituents] (III) + H⁺ 392-2

I(392-1) + 6e + 5H⁺ → [pyrrolizine with two NH₂ groups] (IV) + 2H₂O 392-3

IV(392-3) + H⁺ ⇌ [pyrrolizine with NH₂, H, and =NH₂⁺] (V) 392-4

V(392-4) + H⁺ + 2e → [pyrrolizine with NH₂] (VI) + NH₃ 392-5

VI(392-5) + H⁺ → [pyrrolizidine with NH₃⁺] (VII) 392-6

VII(392-6) + nH⁺ + 2e(?) → products 392-7

IV(392-3) + H₂O → [pyrrolizinone with NH₂] + NH₃ 392-8

II(392-1) + 6H⁺ + 6e → IV(392-3)(?) + 2H₂O   392-9

III(392-2) + 7H⁺ + 6e → IV(392-3)(?) + 2H₂O   392-10

Code No.	pK_1	pK_1'	pK_2	pK_4'	pK_6'
EC25	<1.9	≈9.7	11.2	≈7	≈7

i_3 at pH < 3 probably accompanied by catalytic H-evolution.

393

Mechanism:
393a: 393-1, -4, and -5

393b: 393-2, -4, and -5

393c: 393-3 to -5

Proposed for Compound:
ED47, ED48, ED49, ED73, ED76, EE26, EE61

ED46, ED77, EE25

ED52, ED78

[adamantane-X] → [adamantane-X with +]   (I) + H⁺ + 2e   393-1

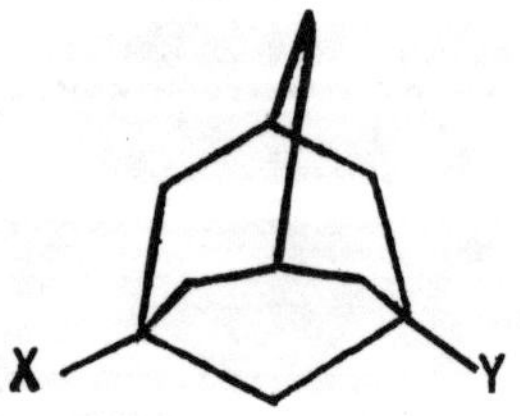

[adamantane with X and Y] → I(393-1) + Y⁺ + 2e   393-2

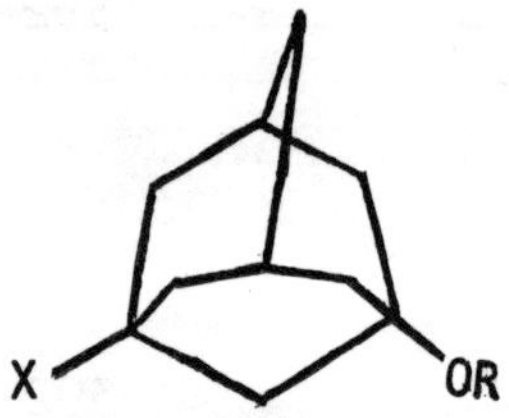

[adamantane with X and OR] + H⁺ → I(393-1) + ROH   393-3

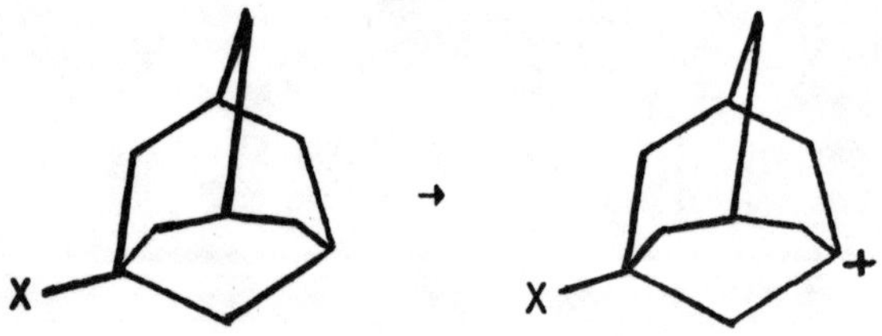

I(393-1) + CH₃CN → [adamantane with X and $\overset{+}{N}\equiv CCH_3$]   (II)   393-4

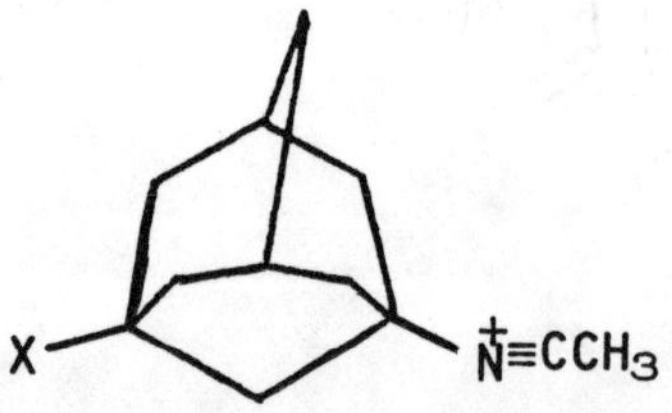

II(393-4) + H$_2$O → [adamantane structure with X and NHCOCH$_3$ substituents] + H$^+$ 393-5

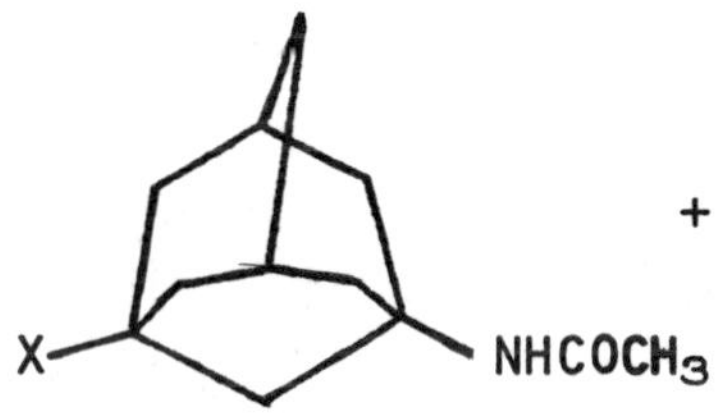

X = H, Cl, F, CH$_3$, CH$_2$OCCH$_3$ (O=), —C(=O)—OCH$_3$, or CN

Y = Br, CH$_2$OH, or COCH$_3$

R = H or CH$_3$

394

Mechanism:
394a: 394-1 to -3

394b: 394-1, -2, -4, and -5

Proposed for Compound:
EC95

EC55

[isatin structure with R on N] (I) + e → [isatin radical anion]$^{\bullet-}$ (II)   394-1

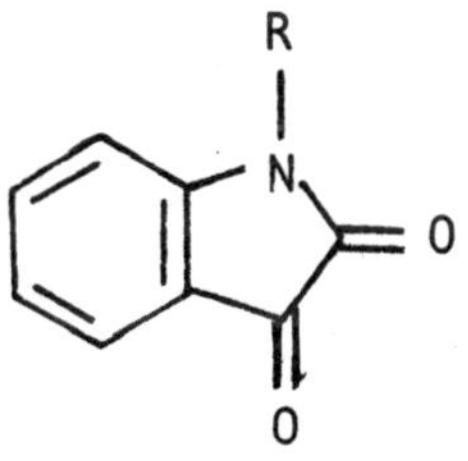

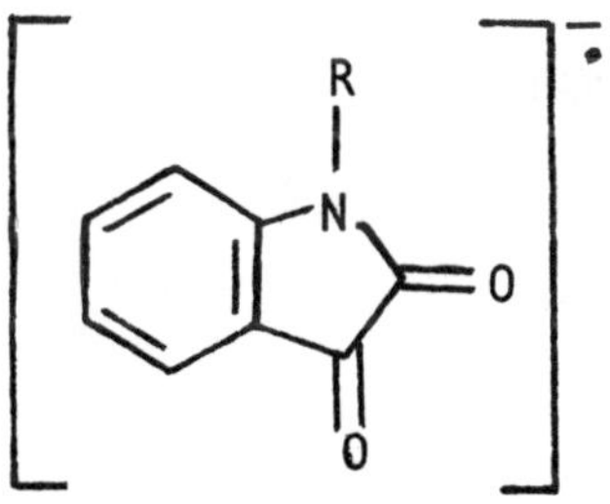

II(394-1) + e → [isatin dianion]$^{2-}$ (III)   394-2

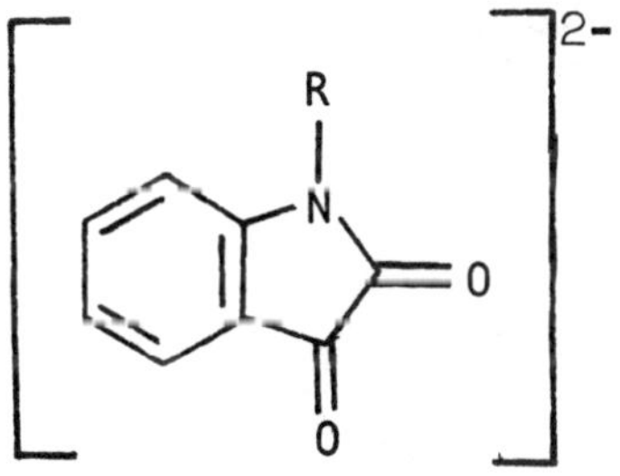

2 II(394-1) ⇌ I(394-1) + III(394-2)   394-3

In 394-4 and -5, R = H

3 I(394-1) + 2e ⇄ [indolin-2-one with 3-OH] + 2 [isatin] (IV)   394-4

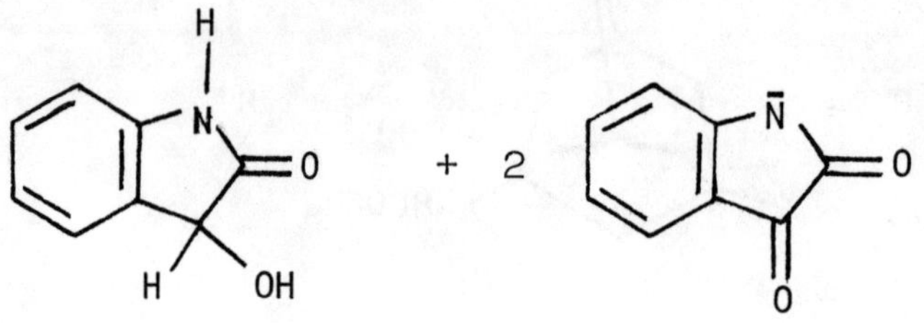

IV(394-4) + e → [isatin radical anion]⁻·   394-5

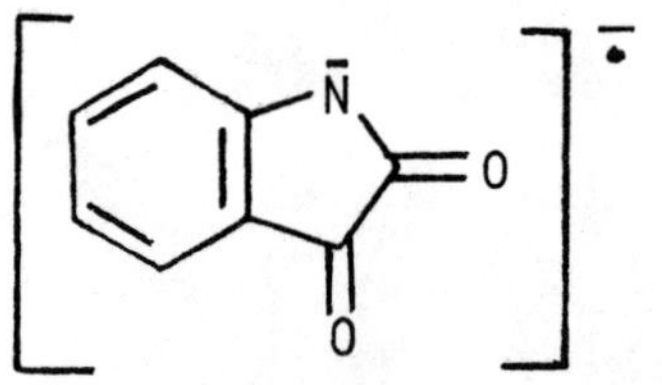

R = H or CH₃

395

Mechanism: Proposed for Compound:
395a: 395-1 to -3 (acidic)   ED29

395b: 395-4 to -7 (basic)   ED29

[pyridinium-cyclopentadiene] (I) ⇄ [pyridinium-cyclopentadienide] (II) + H⁺   395-1

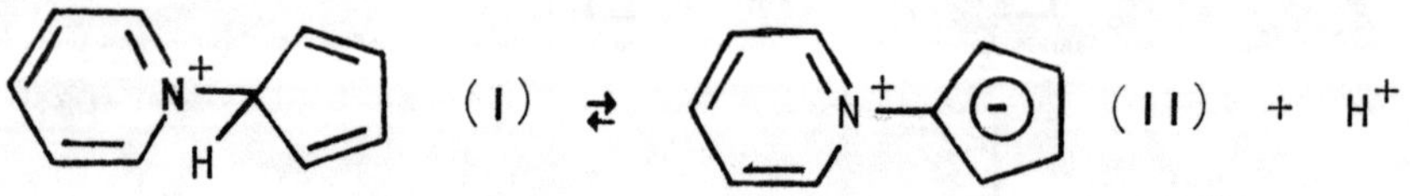

[I cannot be oxidixed.]

I(395-1) + e ⇄ [dihydropyridinyl radical] (III)   395-2

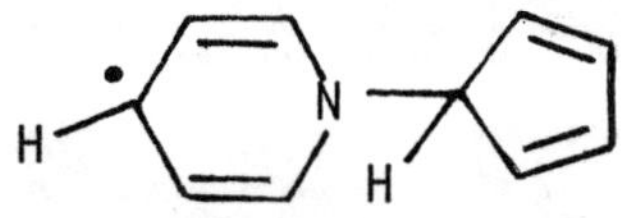

2 III(395-2) → [dimer structure]   395-3

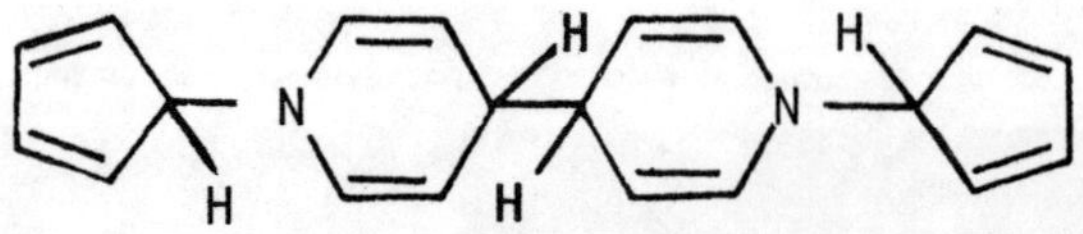

TABLE III. Courses and Mechanisms of Half-Reactions

II(395-1) + e ⇌ [structure IV: pyridinyl radical with cyclopentadienide] (IV)   395-4

2 IV(395-4) → [dimer structure]   395-5

II(395-1) ⇌ [structure V: pyridinium with cyclopentadienyl radical] (V) + e   395-6

2 V(395-6) → [dimer structure]   395-7

396

Mechanism: Proposed for Compound:
396a: 396-1 to -4   EE35

[acridine structure] (I) ⇌ [protonated acridine radical] (II) + e   396-1

II(396-1) + I(396-1) $\xrightarrow{k_2}$ [structure] (?) (III)   396-2

III(396-2) $\rightleftarrows$ [structure] (IV) + H^+ + e   396-3

II(396-1) + III(396-2) $\rightleftarrows$ IV(396-3) + I(396-1) + H^+ 396-4

Code No.	k_2, dm^{3}mol^{-1}s^{-1}
EE35	1.3×10^5

397

Mechanism:
397a: 397-1 to -4

Proposed for Compound:
EF22, EF23, EF24

[structure] (I) $\rightleftarrows$ [structure] (II)

                                   + H^+ 397-1

II(397-1) $\rightleftarrows$ [structure] (III) + e   397-2

TABLE III. Courses and Mechanisms of Half-Reactions

III(397-2) ⇌ [structure: Ar¹, Ar² substituted imidazole connected to quinoid =NR₂⁺] + e   397-3

2 III(397-2) → dimer   397-4

Code No.	R	Ar¹, Ar²	$pK = \frac{\alpha nF}{2.3RT}(E_{\frac{1}{2}}^1 - E_{\frac{1}{2}}^2)$
EF22	CH_3	$4\text{-}ClC_6H_4$	7.32
EF23	CH_3	[2,2'-dimethylbiphenyl structure]	6.57
EF24	CH_3	C_6H_5	8.05

398

Mechanism: Proposed for Compound:

398a: 398-1, and -3 to -6 (pH > 7.1)   EB18

398b: 398-1, -2, and -7 to -20, followed by 19b   EB18

398c: 398-11, and -17 to 20, followed by 19b   EB20

398d: 398-20, followed by 19b   EA42

[Pteridine structure with OH and H] (I) ⇌(slow)⇌ [Pteridine structure] (II) + H_2O 398-1

I(398-1) ⇌ (III) + H⁺   398-2

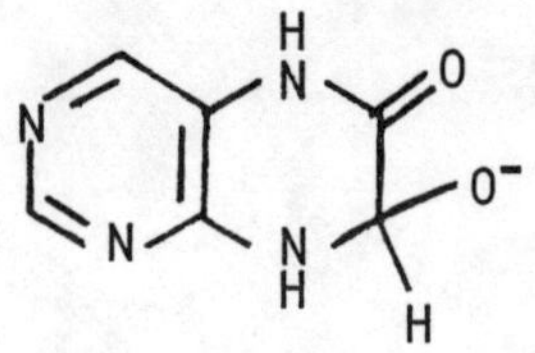

II(398-1) ⇌(fast) (IV) + H⁺   398-3

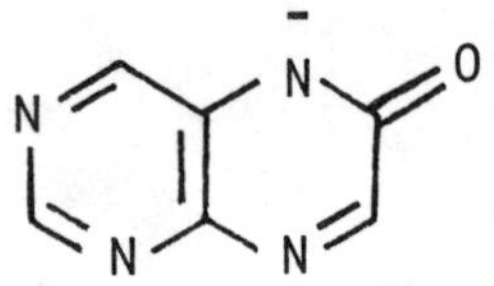

IV(398-3) ↔ (V)   398-4

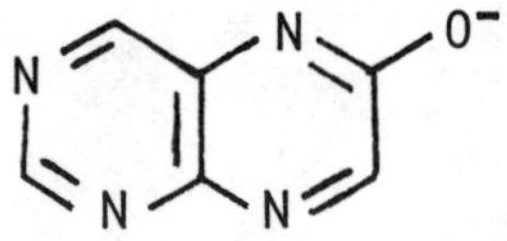

II(398-1) + 2H⁺ + 2e → [(VI)]$_{ads}$ 398-5

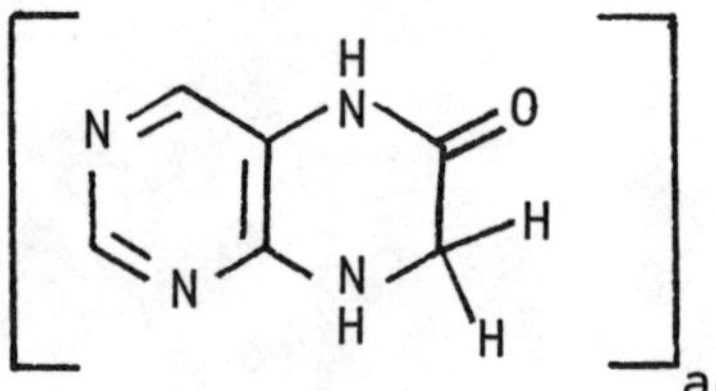

n VI(398-5) → polymeric products   398-6

III(398-2) → (VII) + H⁺ + 2e   398-7

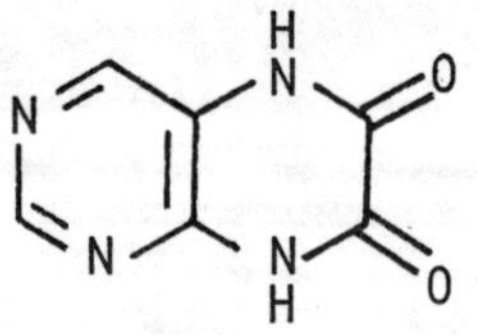

VII(398-7) ⇌ (VIII)   398-8

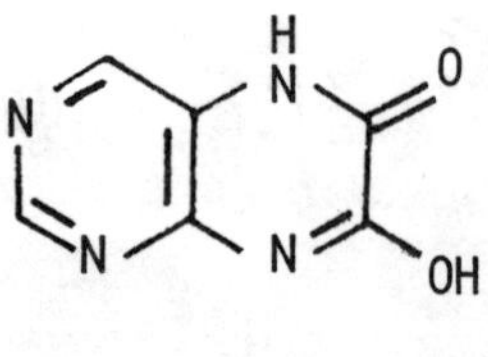

TABLE III. Courses and Mechanisms of Half-Reactions

VII(398-7) → [structure: pteridine-dione] (IX) + 2e + 2H$^+$ 398-9

IX(398-9) + 2H$_2$O → [structure: dihydroxy-dihydropteridine-dione] (X)   398-10

X(398-10) + H$_2$O → [structure: trioxo-hydroxy pteridine derivative] (XI) + 2H$^+$ + 2e   398-11

X(398-10) → [structure: pyrazine-dione] (XII) + H$_2$NCOCONH$_2$ 398-12

XII(398-12) + 2H$_2$O → HC(=NH)(NH$_2$) + HCOCOCOOH   398-13

HC(=NH)(NH$_2$) + H$_2$O → HCONH$_2$ + NH$_3$ 398-14

HCONH$_2$ + H$_2$O → HCOOH + NH$_3$ 398-15

HCOCOCOOH + H₂O → H₂CO + HOOCCOOH   398-16

XI(398-11) + H₂O → (H₂N)₂CO + [structure with OH, CHO substituents on diketopiperazine ring] (XIII)   398-17

XIII(398-17) + H₂O → [structure with OH, COOH substituents on diketopiperazine ring] (XIV) + 2H⁺ + 2e   398-18

XIV(398-18) → [tetraketopiperazine structure] (XV) + CO₂ + 2H⁺ + 2e   398-19

XV(398-19) + H₂O → H₂NCOCOOH + 0.5 H₂NCOCONH₂ + 0.5 HOOCCOOH   398-20

Code No.	pK₂'	pK₃	pK₃'
EB18	4	6.7	≈9

399

Mechanism: Proposed for Compound:
399a: 399-1, and -4 to -6 (pH < 4)   EE33

399b: 399-1, -4, -7, and -8 (4 < pH < 6)   EE33

399c: 399-2, -3, -9, and -10 (6 < pH < 9)   EE33

TABLE III. Courses and Mechanisms of Half-Reactions

399d: 399-2, -11, and -12 (pH > 9)   EE33

[Structure I] (I) ⇌ [Structure II] (II) + H⁺   399-1

II(399-1) ⇌ [Structure III] (III) + H⁺   399-2

[Structure IV] (IV) ⇌ III(399-2) + H⁺   399-3

I(399-1) + e ⇌ [Structure V] (V)   399-4

[Structure VI] (VI) ⇌ V(399-4) + H⁺   399-5

VI(399-5) + e ⇌ [Structure VII] (VII)   399-6

V(399-4) + e ⇌ [structure VIII: acenaphthene with Cl-CH and CH-COOH bridge] (VIII)   399-7

VII(399-6) ⇌ VIII(399-7) + H⁺   399-8

IV(399-3) + 2e → [structure IX: acenaphthene with Cl-CH and CH-COO⁻ bridge] (IX)   399-9

[structure X: naphthalene with Cl-CH₂ and CH-COO⁻ substituents] (X) ⇌ IX(399-10) + H⁺   399-10

III(399-2) + 2e → [structure XI: acenaphthylene with Cl and COO⁻ on the bridge double bond] (XI)   399-11

X(399-10) ⇌ XI(399-11) + 2H⁺   399-12

TABLE III. Courses and Mechanisms of Half-Reactions

400

Mechanism:
400a: 400-1 to -3 (pH < 4.5)

Proposed for Compound:
EA89, EA90, EA91, EB72, EB73, EC45, EC46, EC47, ED17

400b: 400-4 to -7; or 400-4, -8, -9, and -6 (pH > 5)

EA89, EA90, EA91, EB72, EB73, EC45, EC46, EC47, ED17

$$\text{Het}\underset{(CH_2)_y}{\overset{(CH_2)_x}{\diagup\!\!\!\diagdown}}C=OH^+ \;(I) \rightleftarrows \text{Het}\underset{(CH_2)_y}{\overset{(CH_2)_x}{\diagup\!\!\!\diagdown}}C=O \;(II) + H^+ \qquad 400\text{-}1$$

$$I(400\text{-}1) + e \rightarrow \text{Het}\underset{(CH_2)_y}{\overset{(CH_2)_x}{\diagup\!\!\!\diagdown}}\overset{\bullet}{C}-OH \;(III) \qquad 400\text{-}2$$

$$2\;III(400\text{-}2) \rightarrow \text{dimer} \qquad 400\text{-}3$$

$$II(400\text{-}1) + e \rightarrow \left[\text{Het}\underset{(CH_2)_y}{\overset{(CH_2)_x}{\diagup\!\!\!\diagdown}}C=O\right]^{\overline{\bullet}} \;(IV) \qquad 400\text{-}4$$

$$IV(400\text{-}4) + e \rightarrow \left[\text{Het}\underset{(CH_2)_y}{\overset{(CH_2)_x}{\diagup\!\!\!\diagdown}}C=O\right]^{2-} \;(V) \qquad 400\text{-}5$$

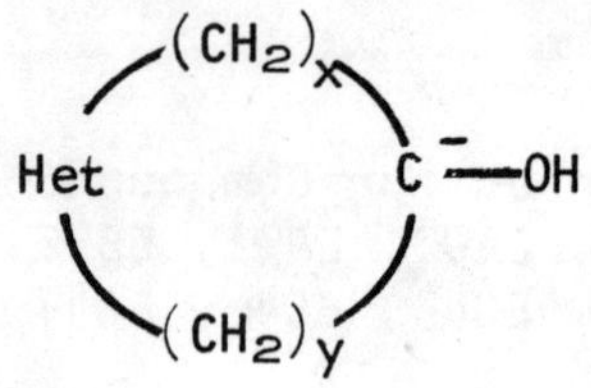

$$\text{Het}\underset{(CH_2)_y}{\overset{(CH_2)_x}{\diagup\!\!\!\diagdown}} \overset{-}{C}\!\!-\!\!OH \quad (VI) \;\rightleftarrows\; V(400\text{-}5) \;+\; H^+ \qquad\qquad 400\text{-}6$$

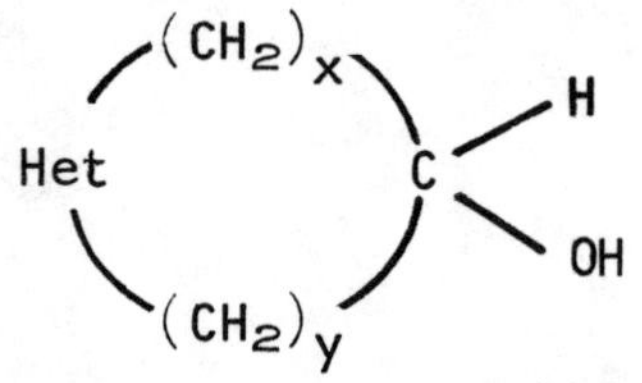

$$\text{Het}\underset{(CH_2)_y}{\overset{(CH_2)_x}{\diagup\!\!\!\diagdown}} C\!\!\underset{OH}{\overset{H}{\diagdown\!\!\!\diagup}} \quad \rightleftarrows\; VI(400\text{-}6) \;+\; H^+ \qquad\qquad 400\text{-}7$$

$$III(400\text{-}2) \;\rightleftarrows\; IV(400\text{-}4) \;+\; H^+ \qquad\qquad 400\text{-}8$$

$$III(400\text{-}2) \;+\; e \;\rightarrow\; VI(400\text{-}6) \qquad\qquad 400\text{-}9$$

Het = S or O

x = 1, 2, or 3

y = 2, 3, or 4

Code No.	Het	x	y	
EA89	S	3	1	
EA90	S	2	2	
EA91	O	2	2	
EB72	S	1	4	
EB73	S	2	3	
EC45	S	2	3	CH_3 on C-2
EC46	S	1	4	
EC47	S	3	3	
ED17	S	2	2	

TABLE III. Courses and Mechanisms of Half-Reactions

401

Mechanism:
401a: 401-1, -2, -3, and -5
401b: 401-1, -2, -4, and -5

Proposed for Compound:
EC29
EC77

$ArCH_2OH \rightarrow ArCH_2OH^{+\cdot} + e$	401-1
$n\ ArCH_2OH^{+\cdot} \rightarrow$ polymeric products	401-2
$ArCH_2OH^{+\cdot} \rightarrow Ar\dot{C}HOH + H^+$	401-3
$ArCH_2OH^{+\cdot} + B \rightarrow Ar\dot{C}HOH + BH^+$	401-4
$Ar\dot{C}HOH \rightarrow ArCHO + H^+ + e$	401-5

$Ar = C_6H_5$ or $4\text{-}CH_3OC_6H_4$

B = pyridine

402

Mechanism:
402a: 402-1 to -4

Proposed for Compound:
EA33

$CH_3COCH_3 + OH^- \overset{k_1}{\rightleftarrows} CH_3COCH_2^- + H_2O$	402-1
$Hg \rightarrow Hg^{2+} + 2e$	402-2
$CH_3COCH_2^- + Hg^{2+} \rightleftarrows CH_3COCH_2Hg^+$	402-3
$CH_3COCH_2Hg^+ + OH^- \rightleftarrows CH_3COCH_2HgOH$	402-4

Code No.	k_1, dm^{3}mol^{-1}s^{-1}
EA33	7.6×10^{-2}

403

Mechanism:
403a: 403-1

Proposed for Compound:
ED31

$$[Co(III)(\eta-C_5H_5)_2]^+ + e \rightleftarrows [Co(II)(\eta-C_5H_5)_2] \qquad 403-1$$

404

Mechanism:
404a: 404-1 to -3

Proposed for Compound:
EB23, EB24, EB26, EB27, EB34, EB35, EB48, EC33

$$ArSeO_2H_2^+ \xrightleftharpoons{fast} ArSeO_2H + H^+ \qquad 404-1$$

$$ArSeO_2H_2^+ + e \rightarrow ArSeO + H_2O \qquad 404-2$$

$$2ArSeO + 4H^+ + 4e \rightarrow ArSeSeAr + 2H_2O \qquad 404-3$$

Ar = C_6H_5, 3-BrC_6H_4, 4-BrC_6H_4, 3-ClC_6H_4, 4-ClC_6H_4, 3-$O_2NC_6H_4$, 4-$O_2NC_6H_4$, or 4-$CH_3C_6H_4$

405

Mechanism:
405a: 405-1 to -5

Proposed for Compound:
EE55

$$ArSCH_2SAr \rightarrow ArS\cdot + ArSCH_2^+ + e \qquad 405-1$$

$$ArSCH_2^+ + H_2O \rightarrow ArSCH_2OH + H^+ \qquad 405-2$$

$$ArSCH_2OH \rightarrow ArS\cdot + \overset{+}{C}H_2OH + e \qquad 405-3$$

$$2ArS\cdot \rightarrow ArSSAr \qquad 405-4$$

$$\overset{+}{C}H_2OH \rightarrow HCHO + H^+ \qquad 405-5$$

Ar = C_6H_5

TABLE III. Courses and Mechanisms of Half-Reactions

406

Mechanism:
406a: 406-1 to -6 (MeCN)

Proposed for Compound:
ED70

$ArCHS + e \rightleftarrows ArCHS^-$	406-1
$ArCHS^- + H^+ \rightleftarrows ArCHSH·$	406-2
$ArCHSH· + e \rightleftarrows ArCHSH^-$	406-3
$ArCHSH^- + H^+ + 2e \rightarrow ArCH_2^- + SH^-$	406-4
$ArCHS \rightleftarrows ArCHS^+ + e$	406-5
$2ArCHS^+ \rightarrow (ArCHS)_2^{2+}$	406-6

Ar = [1,2,3-trimethylindolizine structure]

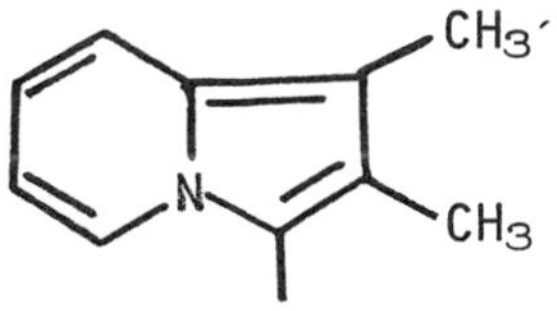

407

Mechanism:
407a: 407-1 to -3, and -5 (MeCN)

407b: 407-1 to -4 (MeCN + C_6H_6)

Proposed for Compound:
EE13

EE13

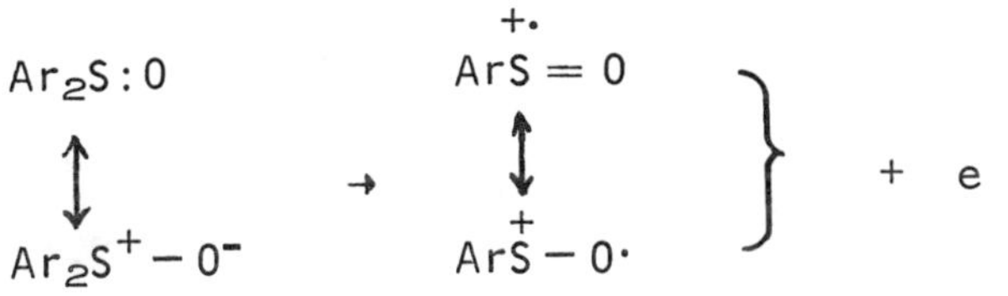

$$Ar_2S:O \leftrightarrow Ar_2S^+-O^- \rightarrow \left\{ ArS\overset{+·}{=}O \leftrightarrow Ar\overset{+}{S}-O· \right\} + e \quad\quad 407\text{-}1$$

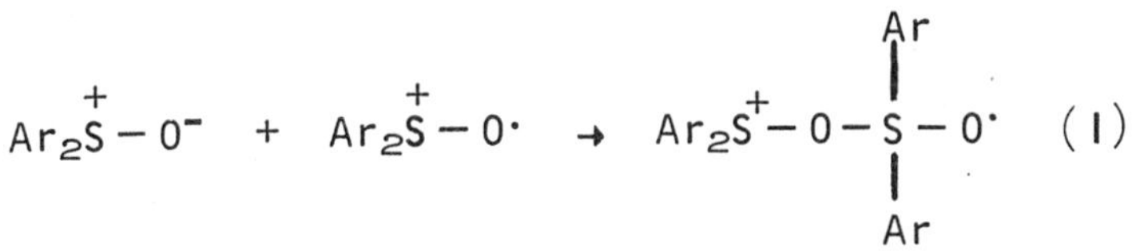

$$Ar_2\overset{+}{S}-O^- + Ar_2\overset{+}{S}-O· \rightarrow Ar_2\overset{+}{S}-O-\underset{Ar}{\overset{Ar}{\underset{|}{\overset{|}{S}}}}-O· \quad (I) \quad\quad 407\text{-}2$$

$I(407-1) \rightarrow Ar_2S^{+\cdot} + Ar_2S\begin{smallmatrix}O\\O\end{smallmatrix}$ 407-3

$Ar_2S^{+\cdot} + C_6H_6 \rightarrow Ar_2S^+C_6H_5 + H^+ + e$ 407-4

$Ar_2S^{+\cdot} + CH_3CN \rightarrow Ar_2\dot{S}-N=\overset{+}{C}-CH_3$ 407-5

$Ar_2\dot{S}-N=\overset{+}{C}-CH_3 \rightarrow$ further oxidation   407-6

408

Mechanism: Proposed for Compound:
408a: 408-1, -2, and -3 or -4 (Pt electrode)   EA75

408b: 408-5 to -10 (Cu electrode)   EA75

$Pt + H_2O \rightleftarrows Pt-OH + H^+ + e$ 408-1

$Pt-OH + (RO)_2P(S)S^- \rightleftarrows (RO)_2P(S)S-Pt + OH^-$ 408-2

$(RO)_2P(S)S-Pt + (RO)_2P(S)S^- \rightleftarrows [(RO)_2P(S)S]_2 + Pt + e$ 408-3

$2(RO)_2P(S)S-Pt \rightleftarrows [(RO)_2P(S)S]_2 + 2Pt$ 408-4

$(RO)_2P(S)S^- \rightleftarrows (RO)_2P(S)S^{\cdot} + e$ 408-5

$2(RO)_2P(S)S^{\cdot} \rightleftarrows [(RO)_2P(S)S]_2$ 408-6

$(RO)_2P(S)S^{\cdot} + (RO)_2P(S)S^- \rightleftarrows [(RO)_2P(S)S]_2 + e$ 408-7

$Cu + (RO)_2P(S)S^{\cdot} \rightleftarrows (RO)_2P(S)SCu^{\cdot}$ 408-8

$(RO)_2P(S)SCu^{\cdot} + (RO)_2P(S)S^{\cdot} \rightleftarrows [(RO)_2P(S)S]_2Cu$ 408-9

$Cu + (RO)_2P(S)S^{\cdot} + (RO)_2P(S)S^- \rightleftarrows [(RO)_2P(S)S]_2Cu + e$ 408-10

409

Mechanism: Proposed for Compound:
409a: 409-1 to -3; or 409-1, -4, and -5; or   EE16
409-1, -6, and -7; followed in any
case by -8 and -9 (Pt electrode, MeCN)

TABLE III. Courses and Mechanisms of Half-Reactions

$$C_6H_5SC_6H_5 \rightarrow C_6H_5\overset{+\cdot}{S}C_6H_5 + e \qquad \qquad 409\text{-}1$$

$$C_6H_5\overset{+\cdot}{S}C_6H_5 \rightleftarrows C_6H_5S\text{-}(C_6H_4)\cdot \; (I) + H^+ \qquad \qquad 409\text{-}2$$

$$C_6H_5\overset{+\cdot}{S}C_6H_5 + I(409\text{-}2) \rightarrow (C_6H_5)_2\overset{+}{S}\text{-}C_6H_4\text{-}SC_6H_5 \; (II) \qquad \qquad 409\text{-}3$$

$$C_6H_5\overset{+\cdot}{S}C_6H_5 \rightarrow C_6H_5S\text{-}(C_6H_4)^+ \; (III) + H^+ + e \qquad \qquad 409\text{-}4$$

$$C_6H_5SC_6H_5 + III(409\text{-}4) \rightarrow II(409\text{-}3) \qquad \qquad 409\text{-}5$$

$$C_6H_5SC_6H_5 + C_6H_5\overset{+\cdot}{S}C_6H_5 \rightarrow (C_6H_4)_2\overset{+\cdot}{S}\text{-}C_6H_4\text{-}SC_6H_5 \; (IV) \qquad \qquad 409\text{-}6$$

$$IV(409\text{-}6) \rightarrow II(409\text{-}3) + H^+ + e \qquad \qquad 409\text{-}7$$

$$2\,II(409\text{-}3) \overset{?}{\rightarrow} (C_6H_5)_2S^+\text{-}C_6H_4\text{-}\underset{C_6H_5}{\overset{+}{S}}\text{-}C_6H_4\text{-}S\text{-}C_6H_4\text{-}\overset{+\cdot}{S}(C_6H_5)_2 \; (V)$$

$$+ H^+ + 2e \qquad \qquad 409\text{-}8$$

V(409-8) + H₂O $\xrightarrow{?}$ (C₆H₅)₂$\overset{+}{S}$—⟨C₆H₄⟩—$\overset{\overset{+}{|}}{\underset{C_6H_5}{S}}$—⟨C₆H₄⟩—$\overset{\overset{+}{|}}{\underset{O}{S}}$—⟨C₆H₄⟩—$\overset{+}{S}$(C₆H₅)₂

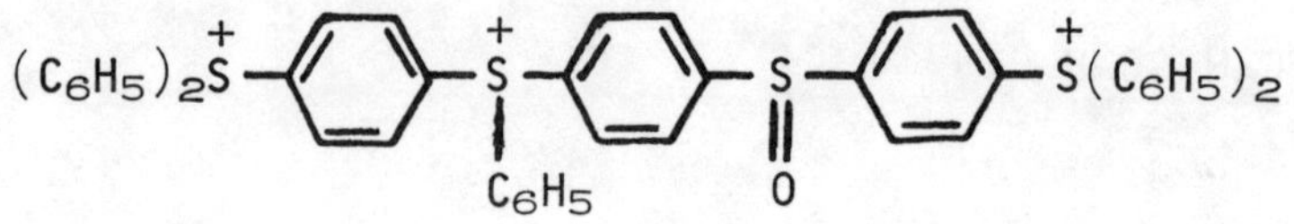

+ 2H⁺ + 2e   409-9

410

Mechanism:	Proposed for Compound:
410a: 410-1 to -8 (Pt electrode, C₆H₅I added)	EE17
410b: 410-1, -2, and -7 to -10 (Pt electrode, MeCN)	EE17

$ArSSAr \rightarrow Ar\overset{+\cdot}{S}SAr + e$ 410-1

$Ar\overset{+}{S}SAr \rightarrow 2ArS^+ + e$ 410-2

$ArS^+ + C_6H_5I \rightarrow ArS\overset{+}{I}C_6H_5$ 410-3

$Ar\overset{+\cdot}{S}SAr + C_6H_5I \rightarrow Ar\overset{\cdot}{S}SAr$ (I)   410-4
$\phantom{Ar\overset{+\cdot}{S}SAr + C_6H_5I \rightarrow Ar\overset{\cdot}{S}}|$
$\phantom{Ar\overset{+\cdot}{S}SAr + C_6H_5I \rightarrow Ar\overset{\cdot}{S}}I^+$
$\phantom{Ar\overset{+\cdot}{S}SAr + C_6H_5I \rightarrow Ar\overset{\cdot}{S}}|$
$\phantom{Ar\overset{+\cdot}{S}SAr + C_6H_5I \rightarrow Ar\overset{\cdot}{S}}C_6H_5$

I(410-4) + C₆H₅I → 2ArSI⁺C₆H₅ + e   410-5

ArSI⁺C₆H₅ → ArSC₆H₄I + H⁺   410-6

ArSSAr ⇌ 2ArS·   410-7

ArS· → ArS⁺ + e   410-8

$ArS^+ + CH_3CN \rightarrow ArS-\overset{+}{N}\equiv CCH_3$ 410-9

$ArS-\overset{+}{N}\equiv CCH_3 \rightarrow$ further reactions   410-10

TABLE III. Courses and Mechanisms of Half-Reactions

411

Mechanism:

411a: 411-1; -2, -3, and -4 as one step; and -6

411b: 411-1 to -4 (0 < pH < 4)

411c: 411-5 (pH > 4)

Proposed for Compound:

C157

EC96, ED30, ED69, EE65, EE71

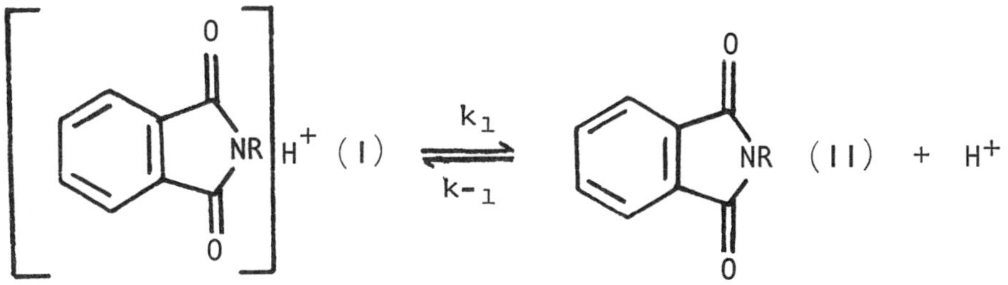

EC96, ED30, ED69, EE65, EE71

[phthalimide with NH]·H⁺ (I) ⇌ (k₁/k₋₁) phthalimide NR (II) + H⁺   411-1

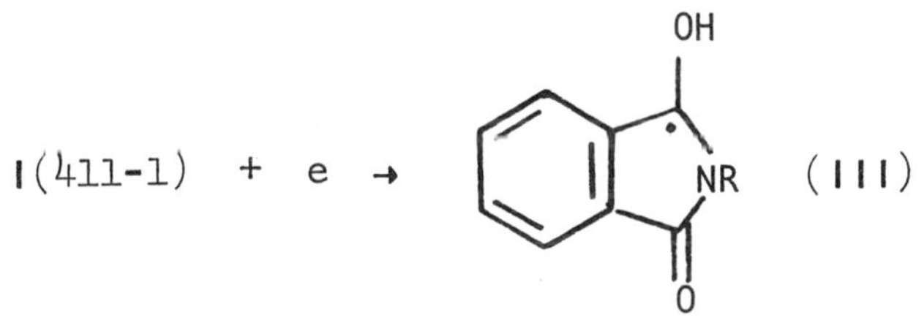

I(411-1) + e → [3-hydroxy-isoindolinone radical] (III)   411-2

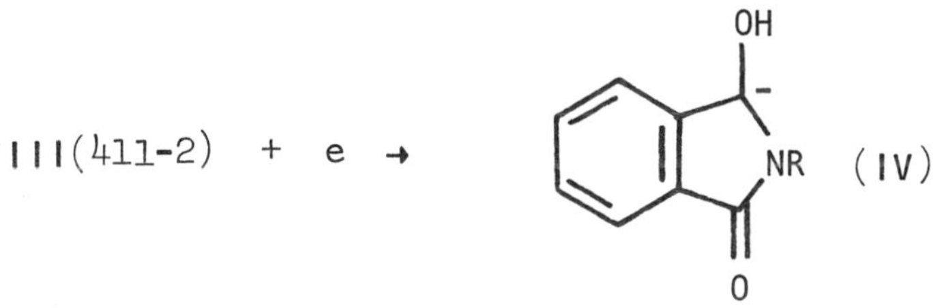

III(411-2) + e → [3-hydroxy-isoindolinone anion] (IV)   411-3

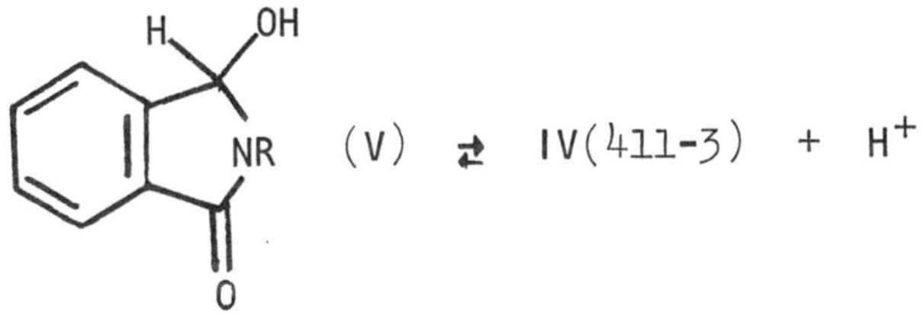

[3-hydroxy-3H-isoindolinone] (V) ⇌ IV(411-3) + H⁺   411-4

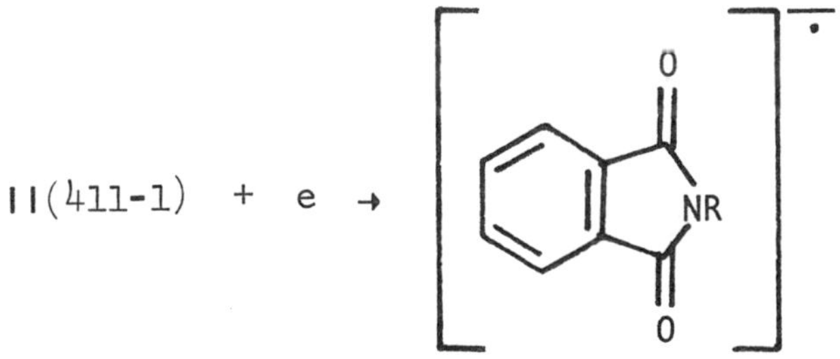

II(411-1) + e → [phthalimide radical anion]·⁻   411-5

In 411-6, R = NH₆H₅

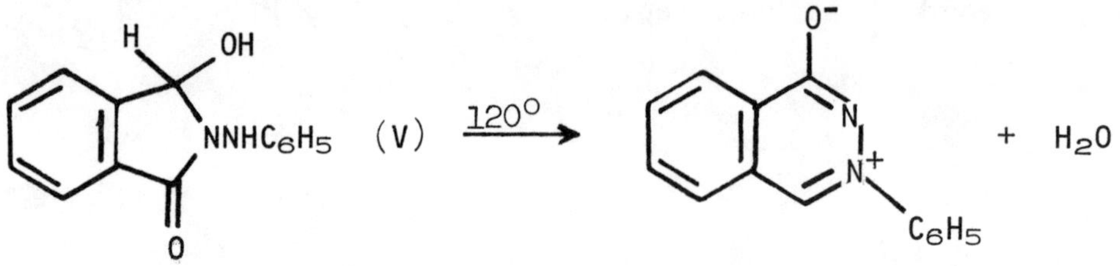

[structure V with OH, NHC₆H₅, carbonyl on isoindolinone] →(120°) [phthalazinium N-oxide with N⁺-C₆H₅] + H₂O   411-6

R = NHC₆H₅, CH₃, C₂H₅, CH₂CH₂CH₃, C₆H₅, or cyclohexyl

412

Mechanism:	Proposed for Compound:
412a: 412-1 and -2	CC00, CC01, CC06, CC55, CC57, CC58, CC70, CC72, CC74, CC75, CD23, CD40, CD57
412b: 412-3 to -7	EB44, EC28
412c: 412-5	EE76

In 412-1 and -2, R = H

$ArSO_2NHR + H^+ + 2e \rightarrow ArH + SO_2NHR^-$ 412-1

$SO_2NHR^- + H_2O \rightleftarrows HSO_3^- + RNH_2$ 412-2

In 412-3 through -7, R ≠ H

$ArSO_2NHR + e \rightleftarrows ArSO_2NHR^{\cdot -}$ 412-3

$ArSO_2NHR^{\cdot -} \xrightarrow{slow} ArSO_2\bar{N}R + H\cdot$ 412-4

$ArSO_2\bar{N}R + e \rightleftarrows [ArSO_2\bar{N}R]^{\cdot -}$ 412-5

$ArSO_2NHR^{\cdot -} + e \rightleftarrows ArSO_2NHR^{2-}$ 412-6

$ArSO_2NHR^{2-} \xrightarrow{fast} [ArSO_2\bar{N}R]^{\cdot -} + H\cdot$ 412-7

R = H or CH₃
Ar = substituted aromatic ring

TABLE III. Courses and Mechanisms of Half-Reactions

413

Mechanism:
413a: 413-1 to -9

Proposed for Compound:
EA45

[pyridazine diprotonated] (I) ⇌ [pyridazine] (II) + 2H⁺   413-1

I(413-1) + 2e ⇌ [dihydropyridazine] (III)   413-2

III(413-2) ⇌ [tetrahydropyridazine] (IV)   413-3

IV(413-4) + H_2O → $H_2NNHCH:CHCH_2CHO$	413-4
$H_2NNHCH:CHCH_2CHO$ ⇌ $H_2NN:CHCH_2CH_2CHO$	413-5
$H_2NN:CHCH_2CH_2CHO$ + $2H^+$ + 2e → $H_2NNHCH_2CH_2CH_2CHO$	413-6
$H_2NN:CHCH_2CH_2CHO$ + H_2O → $OHCCH_2CH_2CHO$ + H_2NNH_2	413-7
$OHCCH_2CH_2CHO$ + ne → products	413-8
$2H_2NN:CHCH_2CH_2CHO$ → $H_2NNHCH:CHCH_2CH:NNHCH:CHCH_2CHO$ + H_2O	413-9

Code No.	pH	[Pyridazine], mol dm^{-3}	k_3, s^{-1}
EA45	0	37	56
	0	47.1	60
	1.39	3.6	68
	3.01	2.5	84
	4.66	3.6	1.20
	6.82	3.6	33

414

Mechanism:
414a: 414-1 followed by 283d (pH < 10)

Proposed for Compound:
EC64

414b: 414-1 followed by 283b (pH > 10)

EC64

$$\underset{R}{ArCOC}=NOH + 4H^+ + 4e \rightarrow \underset{R}{ArCOCHNH_2} + H_2O \qquad 414\text{-}1$$

415

Mechanism:
415a: 415-1 and -2

Proposed for Compound:
EA95

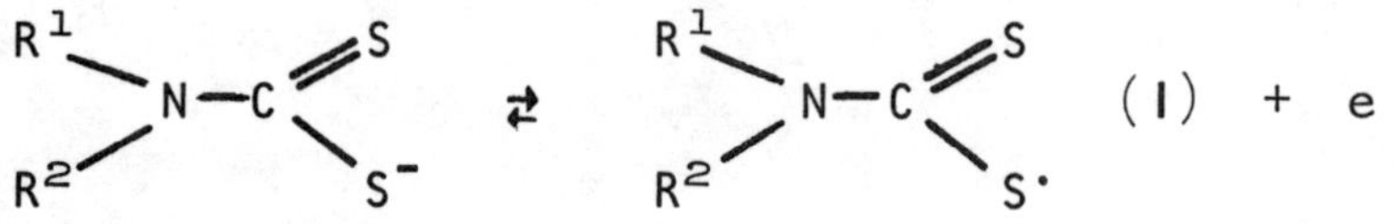

$$\underset{R^2}{\overset{R^1}{N}}-C\underset{S^-}{\overset{S}{=}} \rightleftarrows \underset{R^2}{\overset{R^1}{N}}-C\underset{S^\cdot}{\overset{S}{=}} \; (I) + e \qquad 415\text{-}1$$

$$2\,I(415\text{-}1) \xrightarrow{k_2} \left[\underset{R^2}{\overset{R^1}{N}}-C\underset{S}{\overset{S}{=}}\right]_2 \qquad 415\text{-}2$$

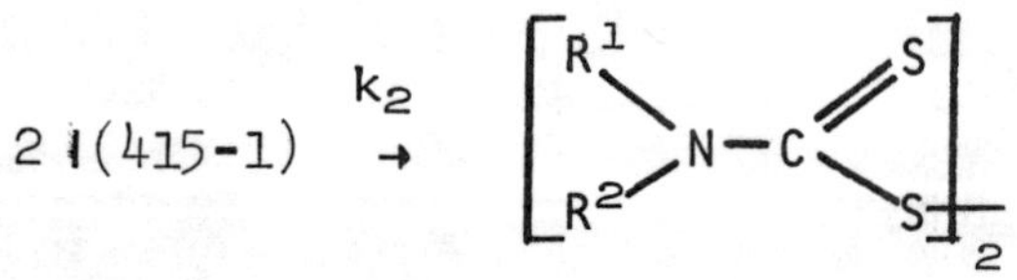

Code No.	k_2, mol^{-1}dm^{3}s^{-1}
EA95	2×10^5

TABLE III. Courses and Mechanisms of Half-Reactions

416

Mechanism:
416a: 416-1 and -2 (pH = 1)

Proposed for Compound:
AD62

(I) + 6H$^+$ + 6e → (II) + 2H$_2$O   416-1

II(416-1) + 4e → products   416-2

417

Mechanism:
417a: 417-1 to -3

Proposed for Compound:
EC11

(I) → (II) + e   417-1

I(417-1) + II(417-1) → (III) + H$^+$ 417-2

III(417-2) → + H$^+$ + S + e   417-3

TABLE IV.
COMPOUNDS INCLUDED IN TABLE I

This index is a list of the compounds that appear in Table I. It contains the names that are used in Table I and also, for most of the compounds, one or more synonyms as well as a few common trivial names. All these names are arranged in alphabetical order, and each is followed by the code number assigned to the compound in Table I.

TABLE IV. Compounds Included in Table I

Acenaphthenequinone, ED79
Acenaphthenon-2-carboxaldehyde, EE34
Acenaphthoquinone, ED79
Acenaphthylene, ED80
Acenaphthylenedione, ED79
Acetaldehyde, EA16
Acetaldehyde oxime, EA18
Acetaldoxime, EA18
Acetamide oxime, EA20
Acetamidoxime, EA20
Acetol semicarbazone, EA98
Acetone, EA33
Acetone oxime, EA34
Acetonitrile, EA13
Acetonoxime, EA34
Acetophenone pinacol, EE95
2-Acetoxybenzoic acid, ED01
β-Acetylacrylic acid, EA85
1-Acetyladamantane, EE25
N-Acetyl-1-aminoadamantane, EE28
Acetylsalicylic acid, ED01
Acridine, EE35
Acrolein, EA24
Acrylaldehyde, EA24
Adamantane, ED49
1-Adamantanecarboxylic acid methyl ester, EE26
1-Adamantanol, ED52
N-(1-Adamantyl)acetamide, EE28
1-Adamantyl bromide, ED46
1-Adamantylcarbinyl acetate, EE61
1-Adamantylcarbonitrile, ED73
1-Adamantyl chloride, ED47
1-Adamantyl fluoride, ED48
1-Adamantylmethanol, ED77
1-Adamantyl methyl ether, ED78
1-Adamantyl methyl ketone, EE25
Adenine, EA82
Adenosine, ED41
Adenosine 5'-diphosphate trisodium salt, ED36
Adenosine cyclo-2',3'-hydrogen phosphate, ED37

Adenosine cyclo-3',5'-hydrogen phosphate, ED38
Adenosine 5'-monophosphate, ED45
Adenosine 2',3'-phosphoric acid, ED37
Adenosine 5'-triphosphate disodium salt, ED43
Adenosine 5'-(trisodiumdiphosphate), ED36
5'-Adenylic acid, ED45
Adenylyl (3'→5') adenosine, EF17
Adenylyl (3'→5') adenylyl (3'→5') adenosine, EF26
Adenylyl (3'→5') adenylyl (3'→5') adenylyl (3'→5') adenosine, EF27
Adenylyl (3'→5') adenylyl (3'→5') adenylyl (3'→5') adenylyl (3'→5') adenylyl (3'→5') adenosine, EF28
Adenylyl (3'→5') cytidine, EF11
Adenylyl (3'→5') guanosine, EF18
Adenylyl (3'→5') uridine, EF10
L-Alanine, EA35
L-α-Alanine, EA35
Aldoxime, EA18
π-Allylbromotricarbonylruthenium, EB25
π-Allyltricarbonylruthenium bromide, EB25
2-Amino-8-aza-6-hydroxypurine, EA49
2-Amino-8-azahypoxanthine, EA49
Aminobenzene, EB50
2-Aminobenzenesulfonic acid, EB57
3-Aminobenzenesulfonic acid, EB56
4-Aminobenzenesulfonic acid, EB58
[2-(Aminocarbonyl)hydrazino]acetic acid, EA38
2-[(Aminocarbonyl)hydrazino]-1,3-dihydroxyethane, EA72
3-[(Aminocarbonyl)hydrazino]-1,2-dihydroxyethane, EA73
2-[(Aminocarbonyl)hydrazino]-1,3-ethanediol, EA72
3-[(Aminocarbonyl)hydrazino]-1,2-ethanediol, EA73
2-[(2-(Aminocarbonyl)hydrazino]propanoic acid, EA74
[(Aminocarbonyl)hydrazono] acetic acid, EA30

3-[(Aminocarbonyl)hydrazono]-2-butanol, EA98

3-[(Aminocarbonyl)hydrazono]-2-hydroxybutane, EA98

1-[(Aminocarbonyl)hydrazono]-1-phenyl-2-ethanol, ED08

2-[(Aminocarbonyl)hydrazono]propanedioic acid, EA52

2-[(Aminocarbonyl)hydrazono]propanoic acid, EA63

5-Amino-1,4-dihydro-7H-1,2,3-triazolo-[4,5-d]pyrimidine-7-one, EA49

2-Amino-4,6-dihydroxy-7-methylpteridine, EC27

α-Amino-β-(3,4-dihydroxyphenyl)propionic acid, ED07

4-Aminodiphenylamine-2-sulfonic acid, EE21

(1-Aminoethylidene)azanol, EA20

2-Aminohexanoic acid, EB86

2-Aminohydroxybenzene, EB51

3-Aminohydroxybenzene, EB52

4-Aminohydroxybenzene, EB53

4-Amino-2-hydroxybenzene-1-azo-2-pyridine, ED68

4-(2-Amino-1-hydroxyethyl)-1,2-benzenediol, EC82

2-Amino-4-hydroxy-6-methylpteridine, EC26

2-Amino-3-mercaptopropanoic acid, EA36

2-Amino-3-mercaptopropionic acid, EA36

4-Amino-4'-methoxydiphenylamine, EE57

1-Amino-2-methylbenzene, EC36

1-Amino-3-methylbenzene, EC37

1-Amino-4-methylbenzene, EC38

α-(Aminomethyl)-3,4-dihydroxybenzyl alcohol, EC82

α-(Aminomethyl)-4-hydroxy-3-methoxybenzenemethanol, ED12

2-Amino-3-methylpentanoic acid, EB84

2-Amino-4-methylpentanoic acid, EB85

2-Amino-7-methyl-4,6-pteridinediol, EC27

2-Amino-6-methyl-4-pteridinol, EC26

2-Amino-6-methyl-4(1H)-pteridinone, EC26

2-Amino-3-methylvaleric acid, EB84

2-Amino-4-methylvaleric acid, EB85

2-Amino-1-nitrobenzene, EB40

3-Amino-1-nitrobenzene, EB41

4-Amino-1-nitrobenzene, EB42

4'-Amino-2-nitrodiphenyl sulfide, EE08

2-Aminophenol, EB51

3-Aminophenol, EB52

4-Aminophenol, EB53

5-Amino-2-(phenylamino)benzenesulfonic acid, EE21

1-Amino-2-phenylethenylideneazenol, ED02

4-Aminophenyl 2-nitrophenyl sulfide, EE08

L-2-Aminopropanoic acid, EA35

L-α-Aminopropionic acid, EA35

2-Aminopteridine, EB36

6-Aminopurine, EA82

5-Amino-2-(2-pyridylazo)phenol, ED68

6-Amino-9-β-D-ribofuranosyl-9H-purine, ED41

2-Amino-9-β-D-ribofuranosyl-9H-purine-6(1H)-one-5'-triphosphate disodium salt, ED44

4-Amino-2-sulfodiphenylamine, EE21

2-Amino-1,3,5,8-tetraazonaphthalene, EB36

α-Amino-β-thiolpropionic acid, EA36

2-Aminotoluene, EC36

3-Aminotoluene, EC37

4-Aminotoluene, EC38

Aniline, EB50

4'-Anilino-2-nitrophenylthioether, EE08

2-Anisaldehyde, EC68

Anisyl alcohol, EC77

9,10-Anthracenedione, EE62

Anthraquinone, EE62

9,10-Anthraquinone, EE62

ApA, EF17

(Ap)$_2$A, EF26

(Ap)$_3$A, EF27

(Ap)$_5$A, EF28

TABLE IV. Compounds Included in Table I

ApC, EF11
ApG, EF18
ApU, EF10
Ascorbic acid, EB65
L-Ascorbic acid, EB65
Aspirin, ED01
9-Azaanthracene, EE35
7-Azabenz[j]anthrancene, EE97
8-Azaguanine, EA49
1-Azanaphthalene, EC93
1-Azaphenanthrene, EE37
4-Azaphenanthrene, EE38
5-Azauracil, EA23
6-Azauracil riboside, EC84
6-Azauridine, EC84
Azobenzene, ED99
2,2'-Azodiphenol, EE03
3,3'-Azodiphenol, EE05
4,4'-Azodiphenol, EE06
Azoxybenzene, EE00

Benz[c]acridine, EE97
3,4-Benzacridine, EE97
Benzaldehyde, EC13
Benzamide, EC21
Benzeneacetaldehyde, EC66
1,2-Benzenedicarboxaldehyde, EC59
1,3-Benzenedicarboxaldehyde, EC58
1,4-Benzenedicarboxaldehyde, EC60
1,4-Benzenediol, EB47
Benzenemethanol, EC29
Benzenepropanal, FD04
Benzeneseleninic acid, EB48
Benzenesulfo-4-nitranilide, EE11
Benzenethiol, EB49
Benzo[c]isoquinoline, EE36
3,4-Benzoisoquinoline, EE36
Benzonitrile, EC02
Benzophenone, EE44
Benzo[b]pyridine, EC93
Benzo[f]quinoline, EE37

Benzo[h]quinoline, EE38
p-Benzoquinone, EB22
1,4-Benzoquinone, EB22
Benzoquinone chloride, EB07
1-Benzosuberone, ED71
Benzothiazole-2-thiol, EC11
Benzoylamide, EC21
4-Benzoylbiphenyl, EF06
Benzoylcarbinol semicarbazone, ED08
Benzoyl formaldoxime, EC64
Benzoylmethylideneazenol, EC64
(Benzoyl)methylidenediazane, EC65
Benzyl alcohol, EC29
N-Benzyl-1,4-dihydronicotinamide, EE58
N-Benzylmaleimide, ED60
Bi(4-amino-3-methylphenyl), EE73
Bibenzyl, EE68
Biotin, ED50
Biphenyl-4,4'-diol, EE14
Bis(2-amino-2-carboxyethyl) diselenide, EB81
Bis(2-amino-2-carboxyethyl) disulfide, EB80
p-Bis(azobenzene), EF01
4,5-Bis(4-chlorophenyl)-2-(4-dimethylaminophenyl)imidazole, EF22
Bis(η-cyclopentadienyl)cobalt(III) hexafluorophosphate, ED32
Bis(η-cyclopentadienyl)cobalt(III) perchlorate, ED31
Biscyclopentadienyliron(II), ED33
Bis[7-11-η-7,8-dicarbaundecaborane-(II)]cobaltate(1$^-$) tetrabutylammonium salt, EF21
Bis[π-(3)-1,2-dicarbollyl]nickel(III), FA79
Bis[π-(3)-1,7-dicarbollyl]nickel(III), EA78
2,6-Bis(1,1-dimethylethyl)-2,5-cyclohexadiene-1,4-dione, EE74
Bis(1,3-dithiolylidene), EB66
1,2-Bisisonicotinoylhydrazine, EE12
1,2-Bis(4-nitrophenyl)ethene, EE66
1,2-Bis(4-nitrophenyl)ethylene, EE66
1,4-Bis(phenylazo)benzene, EF01
Bis(phenylthio)methane, EE55

Bis-2-pyridoyl, ED88

1,2-Bis(2-pyridyl)-1,2-dihydroxyethane N,N'-dioxide, EE22

2,2'-Bispyridylethyleneglycol bis-N-oxide, EE22

1,2-Bis(2-pyridyl)-2-oxoethanol, EE07

1,2-Bis(2-pyridyl)-2-oxoethanol N,N'-dioxide, EE10

Bis[(7,8,9,10,11-η)-undecahydro-1,7-dicarbaundecaborato(11)$^{2-}$]nickelate-(1-), EA78

Bis[(7,8,9,10,11-η)-undecahydro-7,8-dicarbaundecaborato(11)$^{2-}$]nickelate-(1-), EA79

$\Delta^{2,2'}$-Bi-1,3-thiolane, EB66

1-Bromoadamantane, ED46

3-Bromobenzeneseleninic acid, EB23

4-Bromobenzeneseleninic acid, EB24

1-Bromofluorenone, EE31

1-Bromo-9-fluorenone, EE31

3-Bromofluorenone, EE32

3-Bromo-9-fluorenone, EE32

4'-Bromo-2-nitrodiphenyl sulfone, ED81

4'-Bromo-4-nitrodiphenyl sulfone, ED82

4-Bromophenyl 2-nitrophenyl sulfone, ED81

4-Bromophenyl 4-nitrophenyl sulfone, ED82

2-(4-Bromophenylsulfonyl)nitrobenzene, ED81

4-(4-Bromophenylsulfonyl)nitrobenzene, ED82

5-Bromopyridine-2-azo-(2',4'-dihydroxybenzene), ED59

2-(5-Bromo-2-pyridylazo)-5-(diethylamino)phenol, EE82

4-[(5-Bromo-2-pyridyl)azo]-2',4'-dihydroxybenzene, ED59

2-[(5-Bromo-2-pyridyl)azo]-5-dimethylaminophenol, EE56

4-[(5-Bromo-2-pyridyl)azo]resorcinol, ED59

5-Bromo-1-β-D-ribofuranosyl-2,4(1H,3H)-pyrimidinedione, ED05

5-Bromo-1-β-D-ribofuranosyluracil, ED05

Bromotricarbonyl(η^3-2-propenyl)ruthenium, EB25

5-Bromouridine, ED05

2,3-Butandione dioxime, EA67

trans-2-Butenal, EA56

1-Butene, EA64

2-Butene, EA65

3-Buten-2-one, EA58

5-tert-Butyl-6-azauracil, EC43

O-Butylcarbonodithioate potassium salt, EA94

2-tert-Butyl-2,3-diazeniumbicyclo-[2.2.1]hept-2-ene tetrafluoroborate, ED19

2-tert-Butyl-2,3-diazeniumbicyclo-[2.2.2]oct-2-ene tetrafluoroborate, ED55

N-Butylmaleimide, EC81

N-Butylmaleinimide, EC81

2-tert-Butyl-3-methyl-2,3-diazabicyclo[2.2.1]heptane, ED56

2-tert-Butyl-3-methyl-2,3-diazanorbornane, ED56

1-Butyl-3-pyrroline-2,5-dione, EC81

6-tert-Butyl-1,2,4-triazacyclohex-6-ene-3,5-dione, EC43

6-tert-Butyl-1,2,4-triazine-3,5-(2H,4H)-dione, EC43

Butylxanthic acid potassium salt, EA94

5-Carbamoyl-2,4-dimethylpyrimidine, EC41

2[(Carbamoyl)hydrazino]ethanol, EA37

1-[(Carbamoyl)hydrazono]-2-hydroxy-1-phenylethane, ED08

3-Carbamoyl-1-methylpyridinium iodide, EC34

3-Carbamoylpyridine, EB39

4-Carbamoylpyridine, EB38

5-Carbethoxy-2,4-dimethylpyrimidine, ED09

1-Carbmethoxyadamantane, EE26

2-Carboxybenzaldehyde, EC61

4-Carboxybenzaldehyde, EC62

2-Carboxyphenyl acetate, ED01

Chloranil, EB01

Chloranilic acid, EB06

TABLE IV. Compounds Included in Table I

1-Chloroacenaphthylene-2-carboxylic acid, EE33
2-Chloro-1-acenaphthylenecarboxylic acid, EE33
Chloroacetic acid, EA09
1-Chloroadamantane, ED47
6-Chloro-1-azanaphthalene, EC91
3-Chlorobenzaldehyde, EB99
4-Chlorobenzaldehyde, EC00
6-Chloro-1-benzazine, EC91
3-Chlorobenzeneseleninic acid, EB26
4-Chlorobenzeneseleninic acid, EB27
6-Chlorobenzo[b]pyridine, EC91
Chlorobenzoquinone, EB07
(E)-Chloro-2-butenedioic acid, EA43
(Z)-Chloro-2-butenedioic acid, EA44
6-Chloro-2-cyclohexyl-2,3-dihydro-5-sulfamyl-1,3-dioxo-1H-isoindole, EE71
5-Chloro-N-cyclohexyl-4-sulfamyl-phthalimide, EE71
Chloroethanoic acid, EA09
cis-Chloro-1,2-ethylenedicarboxylic acid, EA44
trans-Chloro-1,2-ethylenedicarboxylic acid, EA43
Chlorofumaric acid, EA43
2-Chloro-5-hydroxybenzoquinone, EB08
5-Chloro-1H-isoindole-1,3(2H)-dione, EC53
Chloromaleic acid, EA44
2-Chloro-6-methyl-1,4-benzoquinone, EC01
1-Chloro-4-(methylsulfinyl)benzene, EC20
2-Chloro-4'-nitrodiphenyl sulfide, ED83
2'-Chloro-4-nitrodiphenyl sulfide, ED83
2'-Chloro-2-nitrodiphenyl sulfone, ED84
2'-Chloro-4-nitrodiphenyl sulfone, ED85
3'-Chloro-2-nitrodiphenyl sulfone, ED86
3'-Chloro-4-nitrodiphenyl sulfone, ED87
4-Chlorophenyl methyl sulfoxide, EC20
2-Chlorophenyl 4-nitrophenyl sulfide, ED83
2-Chlorophenyl 2-nitrophenyl sulfone, ED84
2-Chlorophenyl 4-nitrophenyl sulfone, ED85
3-Chlorophenyl 2-nitrophenyl sulfone, ED86
3-Chlorophenyl 4-nitrophenyl sulfone, ED87
4-Chlorophthalimide, EC53
2-(5-Chloro-2-pyridylazo)-5-(diethylamino)phenol, EE83
6-Chloroquinoline, EC91
2-Chloroquinone, EB07
6-Chlorotoluquinone, EC01
Cinchomeronic acid, EC09
Cinnamaldehyde, EC99
Cinnamic acid amidoxime, ED02
Cobaltocenium hexafluorophosphate, ED32
Cobaltocenium perchlorate, ED31
CpA, EF12
CpC, EF04
CpG, EF13
CpU, EF03
Crotonaldehyde, EA56
1-Cyanoadamantane, ED73
Cyanobenzene, EC02
2-Cyanofluorobenzene, EB96
3-Cyanofluorobenzene, EB97
4-Cyanofluorobenzene, EB98
Cyanomethane, EA13
2-Cyanopyridine, EB10
3-Cyanopyridine, EB11
Cycloheptanone, EC44
Cyclohexane, EB78
Cyclohexene, EB68
O-Cyclohexylcarbonodithioate potassium salt, EC42
1-Cyclohexyl-1H-isoindole-1,3(2H)-dione, ED40
N-Cyclohexylmaleimide, ED40
O-Cyclohexyl potassium dithiocarbonate, ED42
Cyclooctanone, EC86
η-Cyclopentadienyl-η-(3)-1,2-dicarbolly-cobalt(III), EC52
(η-Cyclopentadienyl)[7-11-η-undecarborane(11)$^{2-}$]cobalt (1-), EC52

(η^5-2,4-Cyclopentadien-1-yl)[(7,8,9,10,11-η)-undecahydro-7,8-dicarbaundecaborato(2-)]cobaltate(1-), EC52

Cysteine, EA36

Cystine, EB80

Cytidine-5'-triphosphate disodium salt, ED13

Cytidylyl (3'→5') adenosine, EF12

Cytidylyl (3'→5') cytidine, EF04

Cytidylyl (3'→5') guanosine, EF13

Cytidylyl (3'→5') uridine, EF03

Decane, ED57

n-Decane, ED57

2,3-Diacetyldioxime, EA67

1,4-Diaminobenzene, EB61

1,4-Diaminobutan-N,N,N',N'-tetraacetic acid, EE29

1,2-Di[(aminocarbonyl)hydrazono]propane, EA96

2,3-Di[(aminocarbonyl)hydrazono]-1-propanol, EA97

2,3-Diamino-1,4-dihydroxy-1,4-diaza-1,3-butadiene, EA21

1,2-Diamino-1,2-di(hydroxyimino)-ethane, EA21

4,4'-Diamino-3,3'-dimethylbiphenyl, EE73

1,2-Diazaspiro[2.5]octane, EB79

1,2-Diazine, EA45

1,3-Diazine, EA46

3-Diazo-1,3-dihydro-2H-indol-2-one, EC57

3-Diazoindol-2(1H)-one, EC57

3-Diazooxindole, EC57

Dibenzo[a,c]-9,10-diazaanthracene, EF14

Dibenzo[a,c]-9,10-phenazine, EF14

Dibenzo(b,e)pyridine, EE35

Dibenzo[f,h]quinoxaline, EE87

5,6,7,8-Dibenzoquinoxaline, EE87

Dibenzothiophene, ED89

Dibenzo[b,d]thiophene, ED89

Dibenzoylacetylene, EE88

1,4-Dibenzoylbutadiyne, EE98

Dibenzoyldiacetylene, EE98

trans-Dibenzoylethene, EE94

trans-Dibenzoylethylene, EE94

1,4-Dibromobutane, EA66

1,3-Dibromopropane, EA32

2,6-Di(tert-butyl)-1,4-benzoquinone, EE74

2,6-Di-tert-butyl-2,5-cyclohexadiene-1,4-dione, EE74

2,3-Dicarboxypyridine, EC05

2,4-Dicarboxypyridine, EC06

2,5-Dicarboxypyridine, EC07

2,6-Dicarboxypyridine, EC08

3,4-Dicarboxypyridine, EC09

3,5-Dicarboxypyridine, EC10

Dichloroacetic acid, EA06

Dichloroacetic acid ethyl ester, EA53

1,4-Dichlorobenzene, EB09

2,6-Dichlorobenzoquinone, EB04

(E)-Dichlorobutenedioic acid, EA40

(Z)-Dichlorobutenedioic acid, EA41

2,5-Dichloro-3,6-dihydroxy-1,4-benzoquinone, EB06

3,6-Dichloro-2,5-dihydroxybenzoquinone, EB06

Dichloroethanoic acid, EA06

cis-1,2-Dichloro-1,2-ethylenedicarboxylic acid, EA41

trans-1,2-Dichloro-1,2-ethylenedicarboxylic acid, EA40

Dichlorofumaric acid, EA40

3,5-Dichloro-2-hydroxybenzoquinone, EB05

Dichloromaleic acid, EA41

2,6-Dichloroquinone, EB04

Dicyclohexenoannulenedione, EF25

Dicyclohexeno[d,j]cyclooctadeca-2,4,10,12-tetraene-6,8,15,17-tetrayne-1,14-dione, EF25

Dicyclopentadienylcobalt(III) hexafluorophosphate, ED32

Dicyclopentadienylcobalt(III) perchlorate, ED31

Dicyclopentadienyliron(II), ED33

2,3-Dicyanopyridine, EB90

2,4-Dicyanopyridine, EB91

2,5-Dicyanopyridine, EB92

TABLE IV. Compounds Included in Table I

2,6-Dicyanopyridine, EB93
3,4-Dicyanopyridine, EB94
3,5-Dicyanopyridine, EB95
5-(Diethylamino)-2-[(5-iodo-2-pyridyl)-azo]phenol, EE84
5-(Diethylamino)-2-(2-pyridylazo)-phenol, EE85
5-(Diethylamino)-2-(2-quinolylazo)-phenol, EF07
5-(Diethylamino)-2-(2-thiazolylazo)-phenol, EE60
Diethyl 2-[(carbamoyl)hydrazono]propandioate, EC85
Diethyl mesoxalate semicarbazone, EC85
O,O-Diethylphosphorodithioate potassium salt, EA75
Difluoroacetic acid, EA07
Difluoroacetic acid ethyl ester, EA54
Difluoroethanoic acid, EA07
1,2-Diformylbenzene, EC59
1,3-Diformylbenzene, EC58
1,4-Diformylbenzene, EC60
5,6-Dihydro-6-azauracil, EA29
1,4-Dihydro-N-benzyl-3-pyridinecarboxamide, EE58
1,6-Dihydro-3,5-dihydroxy-as-triazine, EA29
5,6-Dihydro-2,4-dioxopyrimidine, EA55
1,3-Dihydro-3-hydroxy-2H-indol-2-one, EC63
5,6-Dihydro-7-hydroxypteridine, EB45
7,8-Dihydro-6-hydroxypteridine, EB46
3,4-Dihydro-1(2H)-naphthalenone, ED34
5,8-Dihydro-6,7-pteridinedione, EB20
5,8-Dihydro-2,4,6,7(1H,3H)-pteridinetetrone, EB21
1,7-Dihydro-6(5H)-pteridinone, EB46
5,8-Dihydro-7(6H)-pteridinone, EB45
1,7-Dihydro-6H-purine-6-thione, EA80
5,6-Dihydro-2,4-(1H,3H)-pyrimidinedione, EA55
1,2-Dihydropyrrolizine-1,2-dione-dioxime, EC25
Dihydro-2H-thiopyran-3(4H)-one, EA89
Dihydro-1,2,4-triazine-3,5(2H,4H)-dione, EA29

5,6-Dihydrouracil, EA55
2,4-Dihydroxypyrimidine, EA48
Dihydroxyacetone semicarbazone, EA72
2,2'-Dihydroxyazobenzene, EE03
2,4-Dihydroxyazobenzene, EE04
3,3'-Dihydroxyazobenzene, EE05
4,4'-Dihydroxyazobenzene, EE06
1,4-Dihydroxybenzene, EB47
2,4-Dihydroxybenzene-1-azobenzene, EE04
2,3-Dihydroxybenzene-1-azo-2-hydroxybenzene, EE09
2,4-Dihydroxybenzene-1-azo-2'-pyridine, ED65
2,4-Dihydroxybenzene-1-azo-3'-pyridine, ED66
2,4-Dihydroxybenzene-1-azo-4'-pyridine, ED67
2,5-Dihydroxybenzoic acid, EC17
4,4'-Dihydroxybiphenyl, EE14
2,5-Dihydroxy-3,6-dichloroquinone, EB06
1,3-Dihydroxy-2,4-dinitrosobenzene, EB15
N,N''-Dihydroxyethanediimidamide, EA21
2,2-Dihydroxy-1,3-indanedione, EC92
2,2-Dihydroxy-1H-indene-2,3(2H)-dione, EC92
1,4-Dihydroxynaphthalene, ED28
3-(3,4-Dihydroxyphenyl)alanine, ED07
3-(3,4-Dihydroxyphenyl)-2-aminopropanoic acid, ED07
2,3-Dihydroxypropanal semicarbazone, EA73
1,3-Dihydroxy-2-propanone semicarbazone, EA72
6,7-Dihydroxypteridine, EB20
4,5-Dihydroxypyrimidine, EA47
6,7-Dihydroxy-1 3,5,8-tetraazanaphthalene, EB20
2,5-Dihydroxytoluene, EC32
2,4-Dihydroxy-s-triazine, EA23
2,4-Dihydroxy-1,3,5-triazine, EA23
1,2-Diisonicotinoylhydrazine, EE12
1,4-Dimethoxybenzene, EC78

4-Dimethylaminobenzaldehyde, ED06

4-Dimethylaminobenzenecarbonal, ED06

2-(Dimethylamino)ethanethiol hydrochloride, EA77

2-(4-Dimethylamino-2-hydroxyphenylazo)-5-nitrothiazole, EE59

2-(4-Dimethylaminophenyl)-4,5-(2,2'-biphenylene)imidazole sodium, EF23

2-(4-Dimethylaminophenyl)-4,5-di-(4-chlorophenyl)imidazole sodium, EF22

2-(4-Dimethylaminophenyl)-4,5-diphenylimidazole sodium, EF24

4-[(Dimethylamino)sulfonyl]-1-nitrobenzene, EC75

2-Dimethylammoniumethanethiol chloride, EA77

N,N-Dimethylaniline, EC80

3,5-Dimethyl-6-azauracil, EA86

3,3'-Dimethylbenzidine, EE73

2,5-Dimethyl-1,4-benzoquinone, EC69

2,4-Dimethyl-5-carbethoxypyrimidine, ED09

5,14-Dimethylcyclooctadeca-2,4,11,13-tetraen-6,8,15,17-tetryn-1,10-dione, EF15

10,15-Dimethylcyclooctadeca-7,9,15,17-tetraen-2,4,11,13-tetryn-1,6-dione, EF16

2,3-Dimethyl-2,3-diazabicyclo[2.2.1]-heptane, EC50

2,3-Dimethyl-2,3-diazabicyclo[2.2.2]-octane, EC89

2-(1,1-Dimethylethyl)-3-aza-2-azoniabicyclo[2.2.1]hept-2-ene tetrafluoroborate, ED19

2-(1,1-Dimethylethyl)-3-aza-2-azoniabicyclo[2.2.2]oct-2-ene tetrafluoroborate, ED55

sym-Dimethylethylene, EA65

Dimethylglyoxime, EA67

1,1-Dimethylhydrazine, EA22

1,2-Dimethylindolizine-3-thial, ED70

Dimethyl ketone, EA33

Dimethylketoxime, EA34

Dimethyl(4-methylphenyl)amine, ED11

N,N-Dimethyl-4-nitrobenzenesulfonamide, EC75

3,7-Dimethyl-2,6-octadienal, ED53

Dimethylphenylamine, EC80

2,4-Dimethylpyrimidine-5-carboxamide, EC41

1,2-Dimethylthio-3-indolizinecarboxaldehyde, ED70

N,N-Dimethyl-4-toluidine, ED11

Dimethyl(p-tolyl)amine, ED11

4,6-Dimethyl-1,2,4-triazacyclohex-6-en-3,5-dione, EA86

4,6-Dimethyl-3H,5H-triazin-3,5-dione, EA86

4,6-Dimethyl-1,2,4-triazin-3,5(3H,5H)-dione, EA86

Dinicotinic acid, EC10

1,2-Dinitrobenzene, EB12

1,3-Dinitrobenzene, EB13

1,4-Dinitrobenzene, EB14

2,4-Dinitrosoresorcinol, EB15

4,4'-Dinitrostilbene, EE66

Dioxindole, EC63

9,10-Dioxoanthracene, EE62

1,2-Dioxo-1H,2,3-dihydroindolizine dioxime, EC25

Diphenylacetaldehyde, EE67

Diphenylaminobenzene, EF02

N,N-Diphenylbenzenamine, EF02

2,3-Diphenyl-2,3-butanediol, EE95

trans-1,4-Diphenyl-3-buten-1,4-dione, EE94

1,4-Diphenyl-2-butyn-1,4-dione, EE88

Diphenyldiazene, ED99

Diphenyldiazene oxide, EE00

trans-1,4-Diphenyl-1,4-dioxo-3-butene, EE94

Diphenyl disulfide, EE17

Diphenylene disulfide, ED90

2,2-Diphenylethanal, EE67

1,2-Diphenylethane, EE68

1,6-Diphenyl-2,4-hexadiyn-1,6-dione, EE98

Diphenyl ketone, EE44

Diphenylmethanone, EE44

Diphenyl sulfide, EE16

TABLE IV. Compounds Included in Table I

Diphenyl sulfone, EE15
Diphenyl sulfoxide, EE13
Diphenylthallium(III) nitrate, ED98
Dipicolinic acid, EC08
Dipotassium 1,4-naphthohydro-
  quinone, ED21
Di-2-pyridinylethanedione, ED88
1,2-Di(2-pyridyl)-1,2-ethanedione,
  ED88
3,3'-Diselenobis(alanine), EB81
3,3'-Dithiobis(2-aminopropanoic acid),
  EB80
(Dithiodiethylene)bis(trimethyl-
  ammonium)dication, ED58
1,3-Dithiolane-2-thione, EA25
Dithiooxamide, EA15
Duroquinone, ED39

EDTA, ED51
Eicosane, EF20
Ergadenylic acid, ED45
D-Erythrose, EA69
L-Erythrulose, EA70
Ethanal, EA16
Ethanediamide, EA14
Ethanedioic acid, EA08
Ethanedithioamide, EA15
Ethanenitrile, EA13
Ethanethioamide, EA19
2-(2-Ethoxy-2-methyl-1-propyl)-1,4-
  benzohydroquinone, EE27
2-Ethyl 3-aza-2-azoniabicyclo[2.2.1]-
  hept-2-ene tetrafluoroborate, EC49
2-Ethylbutanal, EB82
2-Ethyl-2-butenal, EB69
2-Ethylbutyraldehyde, EB82
Ethyl 2[(carbamoyl)hydrazono]pro-
  panoate, EB77
Ethylcarbonodithioate potassium salt,
  EA28
Ethyl chloroacetate, EA59
Ethyl chloroethanoate, EA59

Ethyl cinnamate, ED72
2-Ethyl-2,3-diazeniumbicyclo[2.2.1]-
  hept-2-ene tetrafluoroborate, EC49
Ethyl dichloroacetate, EA53
Ethyl dichloroethanoate, EA53
Ethyl difluoroacetate, EA54
Ethyl difluoroethanoate, EA54
Ethyl 1,4-dihydro-2-methyl-4-oxo-5-
  pyrimidinecarboxylate, EC73
Ethyl 2,4-dimethylpyrimidine-5-car-
  boxylate, ED09
Ethylene bis(3-mercaptopropionate),
  EC88
Ethylenediaminetetraacetic acid,
  ED51
(Ethylenedinitrilo)tetraacetic acid,
  ED51
Ethylene trithiocarbonate, EA25
Ethylethylene, EA64
Ethyl fluoroacetate, EA61
Ethyl fluoroethanoate, EA61
2-Ethyl-2-hexenal, EC87
Ethylideneazenol, EA18
1-Ethyl-1H-isoindole-1,3(2H)-dione,
  ED30
N-Ethylmaleimide, EB55
3-Ethyl-3-methyldiaziridine, EA76
Ethyl 2-methyl-4-hydroxypyrimidine-
  5-carboxylate, EC73
1-Ethyl-4-(methylsulfinyl)benzene, ED10
Ethyl 2-methylthio-4-hydroxypyri-
  midine-5-carboxylate, EC74
Ethyl β-phenylacrylate, ED72
4-Ethylphenyl methyl sulfoxide, ED10
Ethyl 3-phenylpropenoate, ED72
N-Ethylphthalimide, ED30
Ethyl pyruvate semicarbazone, EB77
2-Ethylthio-1,3-dithianium tetrafluoroborate, EB76
2-(Ethylthio)-1,3-dithian-2-ylium
  tetrafluoroborate(1-), EB76
2-Ethylthio-1,3-dithiolanium tetrafluoroborate, EA92
2-(Ethylthio)-1,3-dithiolan-2-ylium
  tetrafluoroborate, EA92

2-Ethylthio-4-phenyl-1,3-dithiolanium methyl sulfate, EE24

2-(Ethylthio)-4-phenyl-1,3-dithiolan-2-ylium methyl sulfate, EE24

Ethyl trichloroacetate, EA50

Ethyl trichloroethanoate, EA50

Ethyl trifluoroacetate, EA51

Ethyl trifluoroethanoate, EA51

Ethylxanthic acid potassium salt, EA28

Ferrocene, ED33

Fluoroacetic acid, EA11

1-Fluoroadamantane, ED48

2-Fluorobenzonitrile, EB96

3-Fluorobenzonitrile, EB97

4-Fluorobenzonitrile, EB98

Fluoroethanoic acid, EA11

Formaladehyde, EA00

Formaldehyde-d_2, EA01

Formaldehyde oxime, EA03

Formaldoxime, EA03

Formic acid, EA02

Formylbenzene, EC13

2-Formylbenzoic acid, EC61

4-Formylbenzoic acid, EC62

1-Formyl-3-chlorobenzene, EB99

1-Formyl-4-chlorobenzene, EC00

4-Formyl-N,N-dimethylaniline, ED06

2-Formyl-1-hydroxybenzene, EC16

1-Formyl-2-methoxybenzene, EC68

1-Formyl-3-methylbenzene, EC67

1-Formyl-3-nitrobenzene, EC03

1-Formyl-4-nitrobenzene, EC04

2-Formylphenol, EC16

4-Formylphenol, EC14

3-Formyltoluene, EC67

D-Glucose, EB83

μ-L-Glutamyl-L-cysteinylglycine, ED54

Glutathione, ED54

Glyceraldehyde semicarbazone, EA73

L-Glycero-tetrulose, EA70

Glycolaldehyde, EA17

Glycolaldehyde semicarbazone, EA37

Glycol dimercaptopropionate, EC88

Glycolic aldehyde, EA17

Glyoxylic acid semicarbazone, EA30

GpA, EF19

Guajazulene, EE72

Guanosine-5'-triphosphate disodium salt, ED44

Guanylyl (3'→5') adenosine, EF19

Heptane, EC51

n-Heptane, EC51

Hexadecane, EE96

2,4-Hexadienal, EB64

Hexahydro-2-oxo-1H-thieno[3,4-d]-imidazole-4-pentanoic acid, ED50

1,2,4,5,7,8-Hexamethylacridine, EF08

Hexane, EB87

5-(Hydrazinocarbonyl)-4-hydroxy-2-methylpyrimidine, EB63

4-Hydrazinonitrobenzene, EB60

Hydroquinol, EB47

Hydroquinone, EB47

Hydroquinone dimethyl ether, EC78

Hydroxomercurioacetone, EA27

2-Hydroxy-1-acenaphthylenecarboxaldehyde, EE34

Hydroxyacetaldehyde EA17

N^2-Hydroxyacetamidine, EA20

2-Hydroxyacetophenone semicarbazone, ED08

1-Hydroxyadamantane, ED52

Hydroxyaminobenzene, EB54

4-Hydroxyaminodiazoaminobenzene, EE23

TABLE IV. Compounds Included in Table I

1-[4-Hydroxyamino)phenyl]-3-phenyl-1-triazene, EE23

2-Hydroxyaniline, EB51

3-Hydroxyaniline, EB52

4-Hydroxyaniline, EB53

4-Hydroxyanisole, EC31

8-Hydroxy-1-azanaphthalene, EC94

2-Hydroxyazobenzene, EE01

4-Hydroxyazobenzene, EE02

2-Hydroxyazylene-1,2-dihydropyrrolizin-1-one, EC12

4-Hydroxyazyl-1-nitrobenzene, EB43

2-Hydroxybenzaldehyde, EC16

4-Hydroxybenzaldehyde, EC14

8-Hydroxy-1-benzazine, EC94

8-Hydroxybenzo[b]pyrazine, EC94

3-Hydroxy-2-butanone semicarbazone, EA98

5-Hydroxy-2-chloroquinone, EB08

N^2-Hydroxycinnamamidine, ED02

3-Hydroxy-1,2-di[(carbamoyl)hydrazono]-propane, EA97

3-Hydroxy-2,6-dichloroquinone, EB05

2-(2-Hydroxy-4-dimethylaminophenyl-azo)-5-nitrothiazole, EE59

4-Hydroxydiphenylamine, EE18

4-Hydroxydiphenyldiazene, EE02

2-Hydroxy-1,2-di-2-pyridinylethanone, EE07

2-Hydroxyethanal semicarbazone, EA37

N^2-Hydroxyethanamidine, EA20

N-Hydroxyethanimidamine, EA20

3-Hydroxy-5-(hydroxymethyl)-2-methyl-isonicotinaldehyde, EC71

3-Hydroxy-5-(hydroxymethyl)-2-methyl-isonicotinaldehyde 5-phosphate, EC72

3-Hydroxy-1-(2-hydroxyphenyl)-3-(2-pyridyl)-1-propanone oxime, EE69

3-Hydroxy-1-(2-hydroxyphenyl)-3-(4-pyridyl)-1-propanone oxime, EE70

2-Hydroxyimino-1,2-dihydropyrrolizin-1-one, EC12

3-Hydroxy-2-indolinone, EC63

4-(Hydroxylamino)nitrobenzene, EB43

4-Hydroxy-3-methoxybenzaldehyde, EC70

1-(4-Hydroxy-3-methoxyphenyl)-2-aminoethanol, ED12

3-Hydroxy methylglyoxal bissemicarbazone, EA97

2-(2-Hydroxy-1-methylpropylidene)-hydrazine carboxamide, EA98

2-Hydroxynaphthalene-1-azobenzene, EE91

4-Hydroxynaphthalene-1-azobenzene, EE92

2-Hydroxynaphthalene-1-azo-(2'-nitrobenzene), EE89

2-Hydroxynaphthalene-1-azo-(4'-nitrobenzene), EE90

[(2-Hydroxynaphthyl)azo-2-nitrobenzene, EE77

2-(2-Hydroxy-1-naphthylazo)thiazole, EE39

4-Hydroxy-4'-nitroazobenzene, ED97

N-Hydroxy-4-nitrobenzamine, EB43

2-Hydroxynitrobenzene, EB31

3-Hydroxynitrobenzene, EB32

4-Hydroxynitrobenzene, EB33

1-Hydroxy-4-nitrosobenzene, EB30

2-Hydroxyphenylazobenzene, EE01

4-Hydroxyphenylazobenzene, EE02

1-(2-Hydroxyphenylazo)-2-naphthol, EE93

2-(2'-Hydroxyphenylazo)pyridine, ED61

2-(4'-Hydroxyphenylazo)pyridine, ED62

3-(4'-Hydroxyphenylazo)pyridine, ED63

4-(4'-Hydroxyphenylazo)pyridine, ED64

1-(4-Hydroxyphenyl)-2-phenyldiazene, EE02

N-Hydroxy-3-phenyl-2-propenimidamide, ED02

2-Hydroxy-1-phenylsemicarbazonoethane, ED08

N-Hydroxy-4-(3-phenyl-1-triazenyl)benzenamine, EE23

4-Hydroxypteridine, EB17

6-Hydroxypteridine, EB18

7-Hydroxypteridine, EB19

Hydroxypyruvaldehyde semicarbazone, EA97

8-Hydroxyquinoline, EC94

4-Hydroxy-1,3,5,8-tetraazanaphthalene, EB17

6-Hydroxy-1,3,5,8-tetraazanaphthalene, EB18

7-Hydroxy-1,3,5,8-tetraazanaphthalene, EB19

6-Hydroxy-2,3,5-trichloro-1,4-benzoquinone, EB03

3-Hydroxytyrosine, ED07

1-Indanone, ED00

1,2,3-Indantrione 2-monohydrate, EC92

1H-Indole-2,3-dione, EC55

2,3-Indolinedione, EC55

INH, EB59

2-(5-Iodo-2-pyridylazo)-5-(diethylamino)phenol, EE84

Isatin, EC55

1,3-Isobenzofurandione, EC54

4-Isobutylphenyl methyl sulfoxide, ED74

Isobutyraldehyde, EA68

Isocinchomeronic acid, EC07

1H-Isoindol-1,3(2H)dione, EC56

1,3-Isoindoledione, EC56

Isoleucine, EB84

Isoniazid, EB59

Isonicotinamide, EB38

Isonicotinic acid amide, EB38

Isonicotinic acid hydrazide, EB59

Isonicotinohydrazide, EB59

N-Isonicotinoylhydrazine, EB59

1-Isonicotinoyl-2-phenylhydrazine, EE20

Isophthalaldehyde, EC58

Isopropenyl methyl ketone, EA88

N-Isopropylmaleimide, EC39

N-Isopropylmaleinimide, EC39

Leucine, EB85

Lutidinic acid, EC06

β-Mercaptoalanine, EA36

2-Mercaptobenzothiazole, EC11

2-Mercaptoethyl-N,N-dimethylammonium chloride, EA77

(β-Mercaptoethyl)trimethylammonium iodide, EB00

6-Mercaptopurine, EA80

2-Mercapto-N,N,N-trimethylethanaminium iodide, EB00

Mercuriacetone nitrate, EA27

Mesityl oxide, EB71

Mesoxalic acid semicarbazone, EA52

Metadiazine, EA46

Metanilic acid, EB56

Methacrolein, EA57

Methanal, EA00

Methanal-d_2, EA01

Methanoic acid, EA02

1-Methoxyadamantane, ED78

4-Methoxy-4'-aminodiphenylamine, EE57

2-Methoxybenzaldehyde, EC68

4-Methoxybenzenemethanol, EC77

p-Methoxybenzyl alcohol, EC77

(2-Methoxy-2-methylpropyl)hydroquinone, ED75

2-Methoxynitrobenzene, EC22

3-Methoxynitrobenzene EC23

4-Methoxynitrobenzene, EC24

4-Methoxy-2'-nitrodiphenyl sulfide, EE50

4-Methoxy-4'-nitrodiphenyl sulfone, EE54

4-Methoxyphenol, EC31

4-Methoxyphenyl methyl sulfoxide, EC79

N-(4-Methoxyphenyl)-4-phenylenediamine, EE57

4-[(4-Methoxyphenyl)sulfonyl]nitrobenzene, EE54

β-Methylacrolein, EA56

1-Methyladamantane, ED76

Methyl 1-adamantanecarboxylate, EE26

(Methylamino)benzene, EC35

4-[(Methylamino)sulfonyl]-1-nitrobenzene, EC28

TABLE IV. Compounds Included in Table I

N-Methylaniline, EC35
2-Methylaniline, EC36
3-Methylaniline, EC37
4-Methylaniline, EC38
3-Methylbenzaldehyde, EC67
α-Methylbenzeneacetaldehyde, ED03
4-Methylbenzeneseleninic acid, EC33
2-Methyl-1,4-benzoquinone, EC15
2-Methyl-2-butenal, EA87
3-Methyl-3-buten-2-one, EA88
2-Methyl-6-chloro-1,4-benzoquinone, EC01
2-Methylcrotonaldehyde, EA87
Methyl cyanide, EA13
Methyldiazane, EA04
4-Methyl-N,N-dimethylaniline, ED11
1-Methyl-2,5-dioxo-1-azacyclopenta-3-ene, EA81
N-Methyl-2,5-dioxo-3-pyrroline, EA81
5-Methyl-2,4-dithiouracil, EA84
Methylene oxide, EA00
2-(2-Methyl-2-ethoxypropyl)-1,4-hydroquinone, EE27
O-(1-Methylethyl)carbonodithioate potassium salt, EA62
Methylethyldiaziridine, EA76
1-Methylethylideneazanol, EA34
7-(1-Methylethyl)-1-methylazulene, EE72
1-(1-Methylethyl)-1H-pyrrole-2,5-dione, EC39
N-(1-Methylethyl)-3-pyrrolin-2,5-dione, EC39
Methylglyoxal bissemicarbazone, EA96
Methylhydrazine, EA04
Methylhydroquinone, EC32
3-Methyl-3-hydroxy-4-formyl-5-hydroxymethylpyridine, EC71
2-Methyl-4-hydroxypyrimidine-5-carbohydrazide, EB63
Methylideneazenol, EA03
1-Methylindole-2,3-dione, EC95
N-Methylisatin, EC95
N-Methyl-1,3-isoindoledione, EC96
1-Methyl-7-isopropylazulene, EE72

N-Methylmaleimide, EA81
N-Methylmaleinimide, EA81
2-(2-Methyl-2-methoxypropyl)hydroquinone, ED75
1-Methyl-4-(methylsulfinyl)benzene, EC76
1-Methyl-4-methylthiouracil, EB62
N-Methylnicotinamide iodide, EC34
N-Methyl-4-nitrobenzenesulfonamide, EC28
2-Methyl-2'-nitrodiphenyl sulfide, EE45
2-Methyl-4'-nitrodiphenyl sulfide, EE46
3-Methyl-2'-nitrodiphenyl sulfide, EE47
3-Methyl-4'-nitrodiphenyl sulfide, EE48
4-Methyl-2'-nitrodiphenyl sulfide, EE49
2-Methyl-2'-nitrodiphenyl sulfone, EE51
3-Methyl-2'-nitrodiphenyl sulfone, EE52
3-Methyl-4'-nitrodiphenyl sulfone, EE53
Methyl 2-nitrophenyl ether, EC22
Methyl 3-nitrophenyl ether, EC23
Methyl 4-nitrophenyl ether, EC24
3-Methylpentane, EB88
2-Methyl-2-pentenal, EB70
4-Methyl-3-penten-2-one, EB71
4-Methylphenyl methyl sulfoxide, EC76
2-[(2-Methylphenyl)sulfonyl]nitrobenzene, EE51
2-[(3-Methylphenyl)sulfonyl]nitrobenzene, EE52
4-[(3-Methylphenyl)sulfonyl]nitrobenzene, EE53
Methyl phenyl sulfoxide, EC30
1-[(4-Methylphenyl)thio]-2-nitrobenzene, EE49
N-Methylphthalimide, EC96
2-Methylpropanal, EA68
2-Methyl-2-propenal, EA57

2-Methylpropionaldehyde, EA68

4-(2-Methylpropyl)-1-methylsulfinyl-benzene, ED74

6-Methylpterin, EC26

1-Methylpyridinium-3-carbamoyl iodide, EC34

5-Methyl-2,4(1H,3H)-pyrimidine-dithione, EA84

1-Methyl-3-pyrroline-2,5-dione, EA81

Methylquinone, EC15

p-(Methylsulfinyl)anisole, EC79

Methylsulfinylbenzene, EC30

4-(Methylsulfinyl)-1-methoxybenzene, EC79

4-(Methylsulfinyl)methylbenzene, EC76

p-(Methylsulfinyl)toluene, EC76

3-Methyl-4-thiacycloheptanone, EC45

4-Methyl-3-thiacycloheptanone, EC46

2-Methyl-4-thiepanone, EC45

7-Methyl-3-thiepanone, EC46

2-Methylthio-1,3-dithianium perchlorate, EA93

2-(Methylthio)-1,3-dithian-2-ylium perchlorate, EA93

2-Methylthio-1,3-dithietanium fluorosulfonate, EA26

2-(Methylthio)-1,3-dithietan-2-ylium, fluorosulfate, EA26

2-Methylthio-1,3-dithiolanium perchlorate, EA60

2-(Methylthio)-1,3-dithiolan-2-ylium perchlorate, EA60

4-Methylthio-1-methyl-2(1H,2H)-pyrimidinone, EB62

4-Methylthio-2-oxopyrimidine, EA83

2-(Methylthio)-4-phenyl-1,3-dithiolan-2-ylium fluorosulfonate, ED35

4-Methylthio-2(1H)-pyrimidinone, EA83

S-Methyl-4-thio-1-β-D-ribofuranosyluracil, ED42

4-(Methylthio)-1-β-D-ribofuranosyluracil, ED42

S-Methyl-4-thiouracil, EA83

4-Methylthiouracil, EA83

S-Methyl-4-thiouridine, ED42

4-Methylthiouridine, ED42

2-Methyl-3,5,6-trichloro-1,4-benzoquinone, EB89

1-Methyltricyclo[3.3.1.1^{3,7}]decane, ED76

Methyl vinyl ketone, EA58

Methylxanthic acid potassium salt, EA12

7-Methylxanthopterin, EC27

Miazine, EA46

Monochloroacetic acid, EA09

Naphthalene, ED27

1,4-Naphthalenediol, ED28

1,4-Naphthalenediol dipotassium salt, ED21

1,2-Naphthalenedione, ED22

1,4-Naphthalenedione, ED23

1,4-Naphthaquinol, ED28

1,4-Naphthohydroquinone, ED28

1,2-Naphthoquinone, ED22

1,4-Naphthoquinone, ED23

Niacinamide, EB39

Nicotinamide, EB39

Nicotinic acid amide, EB39

Nicotinic acid nitrile, EB11

Nicotinonitrile, EB11

Ninhydrin, EC92

2-Nitroaniline, EB40

3-Nitroaniline, EB41

4-Nitroaniline, EB42

2-Nitroanisole, EC22

3-Nitroanisole, EC23

4-Nitroanisole, EC24

9-Nitroanthracene, EE63

4-Nitroazobenzene, ED96

3-Nitrobenzaldehyde, EC03

4-Nitrobenzaldehyde, EC04

7-Nitrobenz[a]anthracene, EE99

7-Nitro-1,2-benzanthracene, EE99

Nitrobenzene, EB29

TABLE IV. Compounds Included in Table I

3-Nitrobenzeneseleninic acid, EB34
4-Nitrobenzeneseleninic acid, EB35
4-Nitrobenzenesulfinic acid, EE75
4-Nitrobenzenesulfonamide, EB44, EE76
1-Nitrobenzo[def]phenanthrene, EE86
2-Nitrobiphenyl, ED91
4-Nitrobiphenyl, ED92
6-Nitrochrysene, EF00
4-Nitrodiphenyldiazene, ED96
2-Nitrodiphenyl sulfide, ED93
2-Nitrodiphenyl sulfone, ED94
4-Nitrodiphenyl sulfone, ED95
2-Nitro-2'-methyldiphenyl sulfide, EE45
1-Nitronaphthalene, ED24
2-Nitronaphthalene, ED25
9-Nitrophenanthrene, EE64
2-Nitrophenol, EB31
3-Nitrophenol, EB32
4-Nitrophenol, EB33
1-[(2-Nitrophenyl)azo]-2-naphthol, EE89
1-[(4-Nitrophenyl)azo]-2-naphthol, EE90
N-(4-Nitrophenyl)benzenesulfonamide, EE11
4-Nitrophenylhydrazine, EB60
4-Nitrophenylhydroxylamine, EB43
4-Nitrophenylphenyldiazene, ED96
2-Nitrophenyl phenyl sulfide, ED93
2-Nitrophenyl phenyl sulfone, ED94
4-Nitrophenyl phenyl sulfone, ED95
4-(2-Nitrophenylsulfonyl)bromobenzene, ED81
4-(4-Nitrophenylsulfonyl)bromobenzene, ED82
2-(2-Nitrophenylsulfonyl)chlorobenzene, ED84
2-(4-Nitrophenylsulfonyl)chlorobenzene, ED85
3-(2-Nitrophenylsulfonyl)chlorobenzene, ED86
3-(4-Nitrophenylsulfonyl)chlorobenzene, ED87
4-(2-Nitrophenylthio)aniline, EE08
1-Nitropyrene, EE86
Nitrosobenzene, EB28

4-Nitrosophenol, FB30
2-[(5-Nitro-2-thiazolyl)azo]-5-diethylaminophenol, EE59
2-[2-(5-Nitrothiazolyl)azo]-5-diethylaminophenol, EE59
Norepinephrine, EC82
Norleucine, EB86
Normetanephrine, ED12

Octafluoronaphthalene, ED20
Octahydro-4H-1-benzothipyrane 1,1-dioxide, ED15
Octahydro-4H-benzothipyran-4-one, ED14
Octane, EC90
Oizine, EA45
ONB, ED91
Orthanilic acid, EB57
Orthodiazine, EA45
4-Oxacyclohexanone, EA91
Oxalamide, EA14
Oxalic acid, EA08
Oxalic acid bisamidoxime, EA21
Oxalic acid diamide, EA14
Oxamide, EA14
Oxamide bisoxime, EA21
Oxan-4-one, EA91
Oxine, EC94
Oxocycloheptane, EC44
1-Oxo-2-hydroxyimino-1H-2,3-dihydroindolizine, EC12
4-Oxooxacyclohexane, EA91
4-Oxooxane, EA91
4-Oxo-2-pentenoic acid, EA85
Oxopropanedioic acid semicarbazone, EA52
2-Oxopropanoic acid semicarbazone, EA63
2-Oxopropylmercury nitrate EA27
1-Oxo-1,2,3,4-tetrahydronaphthalene, ED34
3-Oxothiacyclohexane, EA89
4-Oxothiacyclohexane, EA90

3-Oxothiane, EA89
4-Oxothiane, EA90
5-Oxothiocane, EC47
5-Oxothiocane 1,1-dioxide, EC48

3,3-Pentamethylene diaziridine, EB79
Pentane, EA99
Perfluoronaphthalene, ED20
N-(Phenethyl)maleimide, EE19
Phenylacetaldehyde, EC66
9-Phenylacridine, EF05
β-Phenylacrolein, EC99
Phenylamine, EB50
N-Phenyl-4-aminophenol, EE18
4-(Phenylamino)phenol, EE18
Phenylazanol, EB54
4-(Phenylazo)azobenzene, EF01
3-(Phenylazo)catechol, EE04
1-(Phenylazo)-2-naphthol, EE91
4-(Phenylazo)-1-naphthol, EE92
2-Phenylazophenol, EE01
4-Phenylazophenol, EE02
4-Phenylbenzophenone, EF06
Phenyl cyanide, EC02
4-(2-Phenyl-1-diazenyl)phenyl-1-phenyldiazene, EF01
1-Phenyl-3,3-dimethyltriazene, EC83
Phenyl disulfide, EE17
(Phenyldithio)benzene, EE17
4-Phenylenediamine, EB61
1-(2-Phenylethyl)-2,5-dioxo-3-pyrroline, EE19
1-(2-Phenylethyl)-1H-pyrrole-2,5-dione, EE19
Phenylglyoxal aldoxime, EC64
Phenylglyoxal hydrazone, EC65
Phenylglyoxal 2-hydrazone, EC65
2-Phenylhydrazide-2-pyridinecarboxylic acid, EE20
4-(Phenylhydrazinocarbonyl)pyridine, EE20
Phenylhydroxylamine, EB54

2'-Phenylisonicotinohydrazide, EE20
N-Phenylmaleimide, ED26
Phenylmercaptan, EB49
Phenylmethanol, EC29
1-(Phenylmethyl)-3-pyrroline-2,5-dione, ED60
2-Phenylnitrobenzene, ED91
4-Phenylnitrobenzene, ED92
N-Phenylphthalimide, EE65
2-Phenylpropanal, ED03
3-Phenylpropanal, ED04
3-Phenylpropenal, EC99
2-Phenylpropionaldehyde, ED03
3-Phenylpropionaldehyde, ED04
1-Phenyl-1H-pyrrole-2,5-dione, ED26
1-Phenyl-3-pyrrolin-2,5-dione, ED26
Phenyl sulfide, EE16
(Phenylsulfinyl)benzene, EE13
Phenyl sulfone, EE15
4-[(Phenylsulfonyl)amino]nitrobenzene, EE11
(Phenylsulfonyl)benzene, EE15
2-(Phenylsulfonyl)nitrobenzene, ED94
4-(Phenylsulfonyl)nitrobenzene, ED95
Phenyl sulfoxide, EE13
(Phenylthio)benzene, EE16
(2-Phenylthio)nitrobenzene, ED93
S-Phenylthiophenol, EE16
Phenyl xenyl ketone, EF06
Phthalaldehyde, EC59
Phthalic anhydride, EC54
Phthalimide, EC56
Picolinic acid nitrile, EB10
Picolinonitrile, EB10
Piperazinetetrone, EA42
Potassium O-butyldithiocarbonate, EA94
Potassium O-butyl xanthate, EA94
Potassium O-butyl xanthogenate, EA94
Potassium O-cyclohexyldithiocarbonate, EC42
Potassium O-cyclohexylxanthate, EC42
Potassium cyclohexylxanthogenate, EC42
Potassium O,O-diethyl dithiophosphate, EA75

TABLE IV. Compounds Included in Table I

Potassium ethyldithiocarbonate, EA28
Potassium ethylxanthate, EA28
Potassium ethylxanthogenate, EA28
Potassium O-isopropyl dithiocarbonate, EA62
Potassium O-isopropylxanthate, EA62
Potassium isopropylxanthogenate, EA62
Potassium methyldithiocarbonate, EA12
Potassium 1-methylethyldithiocarbonate, EA62
Potassium 1-methylethylxanthogenate, EA62
Potassium O-methylxanthate, EA12
Potassium methylxanthogenate, EA12
2-Propanone, EA33
2-Propanonemercury nitrate, EA27
2-Propanone oxime, EA34
2-Propenal, EA24
Propene, EA31
Propylene, EA31
1-Propyl-1H-isoindol-1,3(2H)-dione, ED69
N-(1-Propyl)maleimide, EC40
N-Propylmaleinimide, EC40
N-Propylphthalimide, ED69
1-Propyl-1H-pyrrole-2,5-dione, EC40
N-Propyl-3-pyrrolin-2,5-dione, EC40
Protocatechuic aldehyde, EC70
Pteridine, EB16
2,4,6,7-Pteridinetetrol, EB21
4-Pteridinol, EB17
4(1H)-Pteridinone, EB17
6(5H)-Pteridinone, EB18
7(1H)-Pteridinone, EB19
1H-Purin-6-amine, EA82
6-Purinethiol, EA80
Pyridazine, EA45
2,2'-Pyridil, ED88
Pyridine-2-azo-(2'-hydroxybenzene), ED61
Pyridine-2-azo-(4'-hydroxybenzene), ED62
Pyridine-3-azo-(4'-hydroxybenzene), ED63
Pyridine-4-azo(4'-hydroxybenzene), ED64

2-Pyridinecarbonitrile, EB10
3-Pyridinecarbonitrile, EB11
3-Pyridinecarboxamide, EB39
4-Pyridinecarboxamide, EB38
Pyridinecyclopentadienylide, ED29
2,3-Pyridinedicarbonitrile, EB90
2,4-Pyridinedicarbonitrile, EB91
2,5-Pyridinedicarbonitrile, EB92
2,6-Pyridinedicarbonitrile, EB93
3,4-Pyridinedicarbonitrile, EB94
3,5-Pyridinedicarbonitrile, EB95
2,3-Pyridinedicarboxylic acid, EC05
2,4-Pyridinedicarboxylic acid, EC06
2,5-Pyridinedicarboxylic acid, EC07
2,6-Pyridinedicarboxylic acid, EC08
3,4-Pyridinedicarboxylic acid, EC09
3,5-Pyridinedicarboxylic acid, EC10
(1-Pyridinium)cyclopentadienide, ED29
2,2'-Pyridoin, EE07
2,2'-Pyridoin bis-N-oxide, EE10
Pyridoxal, EC71
Pyridoxal-5-monophosphoric acid ester, EC72
Pyridoxal 5-phosphate, EC72
4-Pyridoylhydrazine, EB59
4-(2-Pyridylazo)-1,3-dihydroxybenzene, ED65
4-(3-Pyridylazo)-1,3-dihydroxybenzene, ED66
4-(4-Pyridylazo)-1,3-dihydroxybenzene, ED67
2-(2-Pyridylazo)hydroxybenzene, ED61
4-(2-Pyridylazo)hydroxybenzene, ED62
4-(3-Pyridylazo)hydroxybenzene, ED63
4-(4-Pyridylazo)hydroxybenzene, ED64
1-(2-Pyridylazo)-2-naphthol, EE77
1-(3-Pyridylazo)-2-naphthol, EE78
1-(4-Pyridylazo)-2-naphthol, EE79
4-(2-Pyridylazo)-1-naphthol, EE80
4-(2-Pyridylazo)-1-naphthylamine, EE81
2-(2'-Pyridylazo)phenol, ED61
4-(2'-Pyridylazo)phenol, ED62
4-(3'-Pyridylazo)phenol, ED63

4-(4'-Pyridylazo)phenol, ED64

4-(2-Pyridylazo)resorcinol, ED65

4-(3-Pyridylazo)resorcinol, ED66

4-(4-Pyridylazo)resorcinol, ED67

1-(2-Pyridyl)-3-(3-pyridyl)-2-propen-1-one, EE40

1-(3-Pyridyl)-3-(2-pyridyl)-2-propen-1-one, EE41

1-(3-Pyridyl)-3-(3-pyridyl)-2-propen-1-one, EE42

1-(3-Pyridyl)-3-(4-pyridyl)-2-propen-1-one, EE43

Pyrimidine, EA46

4,5-Pyrimidinediol, EA47

2,4(1H,3H)-Pyrimidinedione, EA48

1H-Pyrrolizine-1,2(3H)-dione dioxime, EC25

1H-Pyrrolizine-1,2(3H)-dione 2-oxime, EC12

Pyruvaldehyde bissemicarbazone, EA96

Pyruvic acid semicarbazone, EA63

Quinoline, EC93

Quinolinic acid, EC05

8-Quinolinol, EC94

Quinone, EB22

Quinone oximine, EB30

9-β-D-Ribofuranosidoadenine, ED41

2-β-D-Ribofuranosyl-as-triazine-3,5-(2H,4H)-dione, EC84

Rubeanic acid, EA15

Salicyladehyde, EC16

Selenocystine, EB81

Semicarbazidoacetic acid, EA38

2-Semicarbazidopropionic acid, EA74

Sodium benzenethiolate, EB37

Sodium diethylcarbamodithioate, EA95

Sodium diethyldithiocarbamate, EA95

Sodium diethyldithiocarboxylate, EA95

Sodium 2-(4-dimethylaminophenyl)-4,5-(2,2'-biphenylene)imidazol-1-ide, EF23

Sodium 2[4-(dimethylamino)phenyl]-4,5-di(4-chlorophenyl)imidazol-1-ide, EF22

Sodium 2-(4-dimethylaminophenyl)-4,5-diphenylimidazol-1-ide, EF24

Sodium sulfidobenzene, EB37

Sodium thiophenolate, EB37

Sorbaldehyde, EB64

Sorbicaldehyde, EB64

Spergon, EB01

Spiro(cyclohexane-1,3'-diaziridine), EB79

4-Sulfamoyl-1-nitrobenzene, EB44 (see also EE76)

Sulfanilic acid, EB58

2-Sulfoaniline, EB57

3-Sulfoaniline, EB56

4-Sulfoaniline, EB58

Terephthalaldehyde, EC60

1,3,5,8-Tetraazanaphthalene, EB16

Tetrabutylammonium bis[π-(3)-1,2-dicarbollyl]cobaltate(III), EF21

Tetrabutylammonium bis[(7,8,9,10,11-η)-undecahydro-7,8-dicarbaundecaborato(1-)cobalt(1-), EF21

Tetrachloro-1,4-benzoquinone, EB01

2,3,5,6-Tetrachloro-2,5-cyclohexadiene-1,4-dione, EB01

Tetrachloroquinone, EB01

Tetraethylammonium 4-nitrobenzenesulfinate, EE75

Tetraethylammonium 4-nitrobenzenesulfonamidate, EE76 (see also EB14)

6,7,8,9-Tetrahydro-5H-benzocyclohepten-5-one, ED71

TABLE IV. Compounds Included in Table I

1,2,3,3a-Tetrahydro-1,2-di[(aminocarbonyl)hydrazino]-3a-azapentalene, EC25

5'-(Tetrahydrogen triphosphate)-adenosine disodium salt, ED43

5'-(Tetrahydrogen triphosphate)-guanosine disodium salt, ED44

cis-Tetrahydro-2-oxothieno[3,4-d]-imidazoline-4-valeric acid, ED50

Tetrahydropyran-4-one, EA91

Tetrahydro-2,2,6,6-tetramethyl-4H-thiopyran-4-one, ED16

Tetrahydro-3,3,5,5-tetramethyl-4H-thiopyran-4-one, ED17

Tetrahydro-2,2,6,6-tetramethyl-4H-thiopyran-4-one 1,1-dioxide, ED18

Tetrahydrothiopyran-3-one, EA89

Tetrahydrothiopyran-4-one, EA90

2,4,6,7-Tetrahydroxypteridine, EB21

2,4,6,7-Tetrahydroxy-1,3,5,8-tetraazanaphthalene, EB21

Tetraketopiperazine, EA42

1-Tetralone, ED34

2,3,5,6-Tetramethyl-1,4-benzoquinone, ED39

Tetramethylene dibromide, EA66

(Tetramethylenedinitrilo)tetraacetic acid, EE29

2,2,6,6,Tetramethyl-4-oxothiane, ED16

3,3,5,5-Tetramethyl-4-oxothiane, ED17

2,2,6,6-Tetramethyl-4-oxothiane 1,1-dioxide, ED18

2,2,6,6-Tetramethylthiacyclohexan-4-one, ED16

2,2,6,6-Tetramethyl-4-thiacyclohexanone, ED17

3,3,5,5-Tetramethyl-4-thiacyclohexanone 1,1-dioxide, ED18

2,2,6,6-Tetramethylthian-4-one, ED16

2,2,6,6-Tetramethylthian-4-one 1,1-dioxide, ED18

Tetraoxopiperazine, EA42

2,2',5,5'-Tetrathiabicyclopentylidene, EB66

2-Thiabicyclo[4.4.0]decan-5-one, ED14

2-Thiabicyclo[4.4.0]decan-5-one 2,2-dioxide, ED15

Thiacycloheptan-3-one, EB72

Thiacycloheptan-4-one, EB73

Thiacycloheptan-3-one 1,1-dioxide, EB74

Thiacycloheptan-4-one 1,1-dioxide, EB75

3-Thiacyclohexanone, EA89

4-Thiacyclohexanone, EA90

Thiacyclooctan-5-one, EC47

5-Thiacyclooctanone, EC47

5-Thiacyclooctanone 5,5-dioxide, EC48

Thian-3-one, EA89

Thian-4-one, EA90

Thianthracene, ED90

Thianthrene, ED90

1-(2-Thiazolylazo)-2-naphthol, EE39

4-(2-Thiazolylazo)phenol, EC97

4-(2-Thiazolylazo)resorcinol, EC98

3-Thiepanone, EB72

4-Thiepanone, EB73

3-Thiepanone 1,1-dioxide, EB74

4-Thiepanone 1,1-dioxide, EB75

Thioacetamide, EA19

5-Thiocanone, EC47

5-Thiocanone 1,1-dioxide, EC48

Thiocholine disulfide, ED58

Thiocholine iodide, EB00

Thiophenol, EB49

6-Thiolpurine, EA80

D-Threose, EA71

o-Tolidine, EE73

2-Tolidine, EE73

3-Tolualdehyde, EC67

p-Tolueneselenic acid, EC33

2-Toluidine, EC36

3-Toluidine, EC37

4-Toluidine, EC38

2-Toluquinone, EC15

1,3,5-Triazine-2,4(1H,3H)-dione, EA23

Tributyl phosphate, EE30

Trichloroacetamidoxime, EA10

Trichloro-1,4-benzoquinone, EB02

2,3,5-Trichloro-6-hydroxybenzoquinone, EB03

2,2,2-Trichloro-N-hydroxyethanimid-
   amide, EA10
2,3,6-Trichloro-5-hydroxyquinone,
   EB03
2,3,5-Trichloro-6-hydroxybenzoquinone,
   EB03
2,3,5-Trichloro-6-methylbenzoquinone,
   EB89
2,3,5-Trichloro-6-methyl-2,5-cyclohexa-
   diene-1,4-dione, EB89
2,3,5-Trichloroquinone, EB02
3,5,6-Trichlorotoluquinone, EB89
Tricyclo[3.3.1.1^{3,7}]decane, ED49
Tricyclo[3.3.1.1^{3,7}]decane-1-methanol
   acetate, EE61
Tricyclo[3.3.1.1^{3,7}]decyl bromide,
   ED46
Tricyclo[3.3.1.1^{3,7}]decyl chloride,
   ED47
Tricyclo[3.3.1.1^{3,7}]decyl fluoride,
   ED48
Trifluoroacetic acid, EA05
Trifluoroethanoic acid, EA05
2,2',3-Trihydroxyazobenzene, EE09
2,3,6-Trihydroxybenzoic acid, EC18
2,4,5-Trihydroxybenzoic acid, EC19
(R)-2,3,4-Trihydroxybutanal, EA69
(S)-1,3,4-Trihydroxy-2-butanone, EA70
1,3,5-Trimethyl-6-azauracil, EB67
Trimethylene dibromide, EA32
Trimethylhydrazine, EA39
2,4,6-Trimethyl-1,2,4-triazacyclo-
   hex-6-ene-3,5-dione, EB67
2,4,6-Trimethyl-1,2,4-triazine-3,5-
   (2H,4H)-dione, EB67
Triphenylamine, EF02

UpA, EF09
Uracil, EA48
Uridylyl (3'→5') adenosine, EF09
Uridylyl (5'→3') adenosine, EF10

Vanillin, EC70
Vitamin C, EB65
Vitamin H, ED50

p-Xyloquinone, EC69

TABLE V.
FUNCTIONAL-GROUP INDEX

This index divides the 528 compounds that appear in Table I into 107 groups and subgroups of chemically related compounds. It is preceded by a list of these groups which begins on the following page.

So that none of the final categories would contain more than about 50 compounds, many main groups have been divided and some have been subdivided. Frequently the division was made, first according to the molecular frame (e.g., alicyclic, aliphatic, aromatic or benzenoid, and heterocyclic compounds) and then according to the number of substituents present, but for halogenated compounds the first division was according to the kind of halogen atom, and organometallic ones were divided according to the metallic atom.

In each category there appear, in alphabetical order, the names of the compounds, together with the code numbers assigned to them in Table I, in which the functional group named is, or may be, electroactive. Compounds in which a given functional group is present but is believed to be electroinactive are omitted from the entry for that functional group. Compounds for which the products of the half-reaction are unknown are listed under each of the groups they contain.

TABLE V. Functional-Group Index

ACETYLENES, see HYDROCARBONS and UNSATURATED COMPOUNDS

ACIDS, CARBOXYLIC (see also KETONES and UNSATURATED COMPOUNDS)
  Aliphatic
  Aromatic
  Heterocyclic and Others

ACID SALTS, see ACIDS

ACIDS, SULFONIC, see SULFUR COMPOUNDS

ALCOHOLS, see HYDROXY COMPOUNDS

ALDEHYDES
  Aliphatic
  Aromatic
  Heterocyclic

AMIDES
  Aliphatic
  Aromatic
  Uracils
  Other Heterocyclic
  Sulfonamides
  Others

AMINES (see also NITROSO COMPOUNDS)
  Aliphatic
  Anilines
    N-Unsubstituted
    N-Substituted
  Other Aromatic
  Heterocyclic
  Quaternary

AMINE SALTS, see AMINES

AMINO ACIDS, see ACIDS, CARBOXYLIC

ANHYDRIDES, CARBOXYLIC

AZINES, see IMINES

AZO AND AZOXY COMPOUNDS

BENZOQUINONES, see QUINONES

CARBAMATES, see ESTERS, NON-CARBOXYLIC

CARBAZIDES, CARBAZONES, SEMI-CARBAZIDES, SEMICARBAZONES, AND THIO ANALOGS

CARBAZONES, see CARBAZIDES

CARBOHYDRATES AND SUGARS

CARBOXYLIC ACIDS, see ACIDS, CARBOXYLIC

DEUTERATED COMPOUNDS, see ISOTOPES

DIAZO COMPOUNDS

ESTERS, CARBOXYLIC

ESTERS, NONCARBOXYLIC
  Carbamates, Thiocarbamates, and Xanthates
  Phosphates

ETHERS AND EPOXIDES

HALOGEN COMPOUNDS
  Aliphatic
    Bromo
    Chloro
    Fluoro
  Alicyclic
    Bromo
  Aromatic and Quinoid
    Bromo
    Chloro
    Fluoro
  Heterocyclic
  Organometallic

HETEROCYCLIC COMPOUNDS
  Nitrogen Heterocycles
    Pyridines
    One Ring-Nitrogen Atom, Monocyclic (Except Pyridines)
    One Ring-Nitrogen Atom, Polycyclic
    Two Ring-Nitrogen Atoms, Monocyclic
    Two Ring-Nitrogen Atoms, Polycyclic
    Three Ring-Nitrogen Atoms
    Four Ring-Nitrogen Atoms
    Five and More Ring-Nitrogen Atoms
  Oxygen Heterocycles (see also ANHYDRIDES, CARBOXYLIC; CARBOHYDRATES AND SUGARS; and LACTONES)
  Sulfur Heterocycles (see also SULFUR COMPOUNDS)

HYDRAZIDES, see HYDRAZINES

HYDRAZINES, HYDRAZIDES, AND HYDRAZONES (see also CARBAZIDES)

HYDRAZONES, see HYDRAZINES

HYDROCARBONS
  Aliphatic and Alicyclic
  Aromatic

HYDROXY COMPOUNDS (see also QUINONES)
  Aliphatic and Alicyclic
  Aromatic
    Phenols, Unsubstituted and Monosubstituted

    Phenols, Polysubstituted
    Two or more Rings
  Heterocyclic (see also CARBOHYDRATES)
  Hydroxylamines

HYDROXY SALTS, see HYDROXY COMPOUNDS

IMIDES (see also HETEROCYCLIC COMPOUNDS)

IMINES AND OTHER AZOMETHINES (see also CARBAZIDES; HYDRAZINES; and OXIMES)

ISOTOPES

KETONES (see also HETEROCYCLIC COMPOUNDS; IMIDES; and UNSATURATED COMPOUNDS)
  Dialkyl Ketones
  Aryl-Alkyl Ketones
  Diaryl Ketones
  Ketones Containing Heterocyclic Rings and Other Groups
  Alicyclic Ketones
  Other Carbocyclic Ketones
  Heterocyclic Ketones

LACTAMS, see LACTONES

LACTONES AND LACTAMS (see also HETEROCYCLIC COMPOUNDS)

METHOXIMES, see OXIMES

NITRILES

NITRO COMPOUNDS
  Nitrobenzenes
  Nitroaromatics with Condensed Rings
  Nitro Derivatives of Heterocycles

NITROSO AND NITROSYL COMPOUNDS

ORGANOMETALLICS
  B
  Co
  Ferrocenes
  Hg
  Ni
  Ru
  Tl

N-OXIDES

OXIMES AND ALKOXIMES
  Aldoximes
  Ketoximes and Amidoximes

PHENOLS, see HYDROXY COMPOUNDS

QUARTERNARY COMPOUNDS, see AMINES

QUINONES AND HYDROQUINONES (see also HYDROXY COMPOUNDS)
  Benzoquinones
  Naphthoquinones
  Anthraquinones
  Polycyclic Quinones

SELENIUM AND TELLURIUM COMPOUNDS

SEMICARBAZIDES AND SEMICARBAZONES, see CARBAZIDES

SUGARS, see CARBOHYDRATES AND SUGARS

SULFIDES, see SULFUR COMPOUNDS

SULFONAMIDES, see SULFUR COMPOUNDS

SULFONATES, see SULFUR COMPOUNDS

SULFONES, see SULFUR COMPOUNDS

SULFONIC ACIDS, see SULFUR COMPOUNDS

SULFUR COMPOUNDS (see also ESTERS, NONCARBOXYLIC; HETEROCYCLIC COMPOUNDS; and ORGANOMETALLICS)
  Thiols
  Sulfides, Thioethers, and Sulfonium Ions
  Disulfides
  Thiocarbamates, Dithiocarbamates, and Xanthates
  Thiones, Thioamides, Thiobarbiturates, Thiopyrimidines, and Thiopurines
  Sulfoxides, Sulfones, and S-dioxides
  Sulfuric and Thiosulfuric Acid Derivatives, Sulfonates, Sulfinates, and Sulfonamides

TELLURIUM, see SELENIUM

THIOLS, see SULFUR COMPOUNDS

THIOSEMICARBAZIDES AND THIOSEMICARBAZONES, see CARBAZIDES

UNSATURATED COMPOUNDS
  Ethylenic
    Double Bond Isolated from Nonhydrocarbon Functional Groups
    α,β-Unsaturated Aldehydes
    α,β-Unsaturated Ketones
    α,β-Unsaturated Acids, Thioacids, Esters, and Amides
    Unsaturated Amines, Halogens, Alcohols, Ethers, and Nitrocompounds
    Cyclopentadienyl Compounds
  Acetylenic

TABLE V. Functional-Group Index

ACETYLENES, see HYDROCARBONS and UNSATURATED COMPOUNDS

ACIDS, CARBOXYLIC (see also KETONES and UNSATURATED COMPOUNDS)

Aliphatic

L-alanine, EA35
L-ascorbic acid, EB65
biotin, ED50
chloroacetic acid, EA09
chlorofumaric acid, EA43
chloromaleic acid, EA44
cysteine, EA36
cystine, EB80
1,4-diaminobutan-N,N,N',N'-tetraacetic acid, EE29
dichloroacetic acid, EA06
dichlorofumaric acid, EA40
dichloromaleic acid, EA41
difluoroacetic acid, EA07
3-(3,4-dihydroxyphenyl)alanine, ED07
ethylenediaminetetraacetic acid, ED51
fluoroacetic acid, EA11
formic acid, EA02
glutathione, ED54
glyoxylic acid semicarbazone, EA30
isoleucine, EB84
leucine, EB85
mesoxalic acid semicarbazone, EA52
norleucine, EB86
oxalic acid, EA08
4-oxo-2-pentenoic acid, EA85
pyruvic acid semicarbazone, FA63
selenocystine, EB81
semicarbazidoacetic acid, EA38
2-semicarbazidopropionic acid, EA74
trifluoroacetic acid, EA05

Aromatic

acetylsalicylic acid, ED01
2-chloro-1-acenaphthylenecarboxylic acid, EE33
2,5-dihydroxybenzoic acid, EC17
2-formylbenzoic acid, EC61
4-formylbenzoic acid, EC62
2,3,6-trihydroxybenzoic acid, EC18
2,4,5-trihydroxybenzoic acid, EC19

Heterocyclic and Others

2,3-pyridinedicarboxylic acid, EC05
2,4-pyridinedicarboxylic acid, EC06
2,5-pyridinedicarboxylic acid, EC07
2,6-pyridinedicarboxylic acid, EC08
3,4-pyridinedicarboxylic acid, EC09
3,5-pyridinedicarboxylic acid, EC10

ACID SALTS, see ACIDS

ACIDS, SULFONIC, see SULFUR COMPOUNDS

ALCOHOLS, see HYDROXY COMPOUNDS

ALDEHYDES

Aliphatic

acetaldehyde, EA16
acrolein, EA24
cinnamaldehyde, EC99
crotonaldehyde, EA56
3,7-dimethyl-2,6-octadienal, ED53
diphenylacetaldehyde, EE67
D-erythrose, EA69
2-ethylbutanal, EB82
2-ethyl-2-butenal, EB69
2-ethyl-2-hexenal, EC87
formaldehyde, EA00
formaldehyde-d_2, EA01
D-glucose, EB83
glycolaldehyde, EA17
2,4-hexadienal, EB64
methacrolein, EA57

ALDEHYDES (cont.)

  Aliphatic (cont.)

    2-methyl-2-butenal, EA87

    2-methyl-2-pentenal, EB70

    2-methylpropionaldehyde, EA68

    phenylacetaldehyde, EC66

    2-phenylpropanal, ED03

    3-phenylpropanal, ED04

    D-threose, EA71

  Aromatic

    2-anisaldehyde, EC68

    benzaldehyde, EC13

    3-chlorobenzaldehyde, EB99

    4-chlorobenzaldehyde, EC00

    4-dimethylaminobenzaldehyde, ED06

    2-formylbenzoic acid, EC61

    4-formylbenzoic acid, EC62

    2-hydroxy-1-acenaphthylene-carboxaldehyde, EE34

    4-hydroxybenzaldehyde, EC14

    isophthalaldehyde, EC58

    3-nitrobenzaldehyde, EC03

    4-nitrobenzaldehyde, EC04

    phthalaldehyde, EC59

    salicylaldehyde, EC16

    terephthalaldehyde, EC60

    3-tolualdehyde, EC67

    vanillin, EC70

  Heterocyclic

    pyridoxal, EC71

    pyridoxal 5-phosphate, EC72

AMIDES

  Aliphatic

    glutathione, ED54

    N^2-hydroxycinnamamidine, ED02

    oxamide, EA14

    trichloroacetamidoxime, EA10

  Aromatic

    benzamide, EC21

Uracils

  adenylyl (3'→5') cytidine, EF11

  adenylyl (3'→5') guanosine, EF18

  adenylyl (3'→5') uridine, EF10

  6-azauridine, EC84

  5-bromouridine, ED05

  5-tert-butyl-6-azauracil, EC43

  cytidine-5'-triphosphate disodium salt, ED13

  cytidylyl (3'→5') adenosine, EF12

  cytidylyl (3'→5') cytidine, EF04

  cytidylyl (3'→5') guanosine, EF13

  cytidylyl (3'→5') uridine, EF03

  3,5-dimethyl-6-azauracil, EA86

  guanylyl (3'→5') adenosine, EF19

  1-methyl-4-methylthiouracil, EB62

  4-methylthiouracil, EA83

  5-methyl-4-thiouridine, ED42

  1,3,5-trimethyl-6-azauracil, EB67

  uridylyl (3'→5') adenosine, EF09

Other Heterocyclic

  2-amino-4,6-dihydroxy-7-methyl-pteridine, EC27

  N-benzylmaleimide, ED60

  biotin, ED50

  5-chloro-N-cyclohexyl-4-sulfamyl-phthalimide, EE71

  4-chlorophthalimide, EC53

  N-cyclohexylmaleimide, ED40

  3-diazo-1,3-dihydro-2H-indol-2-one, EC57

  1,3-dihydro-3-hydroxy-2H-indol-2-one, EC63

  5,6-dihydro-7-hydroxypteridine, EB45

  7,8-dihydro-6-hydroxypteridine, EB46

  1,2-diisonicotinoylhydrazine, EE12

TABLE V. Functional-Group Index

AMIDES (cont.)

  Other Heterocyclic (cont.)

    2,4-dimethylpyrimidine-5-carboxamide, EC41

    N-ethylphthalimide, ED30

    isatin, EC55

    isonicotinamide, EB38

    isonicotinic acid hydrazide, EB59

    1-isonicotinoyl-2-phenylhydrazine, EE20

    N-isopropylmaleimide, EC39

    2-methyl-4-hydroxypyrimidine-5-carbohydrazide, EB63

    N-methylisatin, EC95

    N-methylphthalimide, EC96

    6-methylpterin, EC26

    1-methylpyridinium-3-carbamoyl iodide, EC34

    nicotinamide, EB39

    N-(phenethyl)maleimide, EE19

    N-phenylmaleimide, ED26

    N-phenylphthalimide, EE65

    phthalimide, EC56

    N-(1-propyl)maleimide, EC40

    N-propylphthalimide, ED69

    tetraoxopiperazine, EA42

  Sulfonamides

    N,N-dimethyl-4-nitrobenzenesulfonamide, EC75

    N-methyl-4-nitrobenzenesulfonamide, EC28

    4-nitrobenzenesulfonamide, EB44

    N-(4-nitrophenyl)benzenesulfonamide, EE11

  Others

    N-(1-adamantyl)acetamide, EE28

    dithiooxamide, EA15

AMINES (see also NITROSO COMPOUNDS)

  Aliphatic

    L-alanine, EA35

    cysteine, EA36

    cystine, EB80

    1,4-diaminobutan-N,N,N',N'-tetraacetic acid, EE29

    3-(3,4-dihydroxyphenyl)alanine, ED07

    ethylenediaminetetraacetic acid, ED51

    glutathione, ED54

    1-(4-hydroxy-3-methoxyphenyl)-2-aminoethanol, ED12

    isoleucine, EB84

    leucine, EB85

    2-mercaptoethyl-N,N-dimethylammonium chloride, EA77

    norepinephrine, EC82

    norleucine, EB86

    selenocystine, EB81

    semicarbazidoacetic acid, EA38

  Anilines

    N-Unsubstituted

      4-aminodiphenylamine-2-sulfonic acid, EE21

      4-amino-2-hydroxybenzene-1-azo-2-pyridine, ED68

      4-amino-4'-methoxydiphenylamine, EE57

      4'-amino-2-nitrodiphenyl sulfide, EE08

      2-aminophenol, EB51

      3-aminophenol, EB52

      4-aminophenol, EB53

      aniline, EB50

      1,4-diaminobenzene, EB61

      metanilic acid, EB56

      2-nitroaniline, EB40

      3-nitroaniline, EB41

      4-nitroaniline, EB42

      orthanilic acid, EB57

      sulfanilic acid, EB58

      o-tolidine, EE73

      2-toluidine, EC36

      3-toluidine, EC37

      4-toluidine, EC38

AMINES (see also NITROSO COMPOUNDS) (cont.)
  Anilines (cont.)
    N-Substituted

      4-aminodiphenylamine-2-sulfonic acid, EE21

      4-amino-4'-methoxydiphenylamine, EE57

      4,5-bis(4-chlorophenyl)-2-(4-dimethylaminophenyl)imidazole sodium salt, EF22

      2-(5-bromo-2-pyridylazo)-5-(diethylamino)phenol, EE82

      2-(5-bromo-2-pyridylazo)-5-(dimethylamino)phenol, EE56

      2-(5-chloro-2-pyridylazo)-5-(diethylamino)phenol, EE83

      5-(diethylamino)-2-(2-pyridylazo)-phenol, EE85

      5-(diethylamino)-2-(2-quinolylazo)-phenol, EF07

      5-(diethylamino)-2-(2-thiazolylazo)phenol, EE60

      4-dimethylaminobenzaldehyde, ED06

      2-(4-dimethylaminophenyl)-4,5-(2,2-biphenylene)imidazole sodium, EF23

      2-(4-dimethylaminophenyl)-4,5-diphenylimidazole sodium, EF24

      N,N-dimethylaniline, EC80

      N,N-dimethyl-4-toluidine, ED11

      2-(5-iodo-2-pyridylazo)-5-(diethylamino)phenol, EE84

      N-methylaniline, EC35

      2-[(5-nitro-2-thiazolyl)azo]-5-(diethylamino)phenol, EE59

      4-(phenylamino)phenol, EE18

      1-phenyl-3,3-dimethyltriazene, EC83

      triphenylamine, EF02

  Other Aromatic

    4-(2-pyridylazo)-1-naphthylamine, EE81

  Heterocyclic

    adenine, EA82
    adenosine, ED41

    adenosine 5'-diphosphate trisodium salt, ED36

    adenosine cyclo-2',3'-hydrogen phosphate, ED37

    adenosine cyclo-3',5'-hydrogen phosphate, ED38

    adenosine 5'-triphosphate disodium salt, ED43

    5'-adenylic acid, ED45

    adenylyl (3'→5') adenosine, EF17

    adenylyl (3'→5') adenylyl (3'→5') adenosine, EF26

    adenylyl (3'→5') adenylyl (3'→5') adenylyl (3'→5') adenosine, EF27

    adenylyl (3'→5') adenylyl (3'→5') adenylyl (3'→5') adenylyl (3'→5') adenylyl (3'→5') adenosine, EF28

    adenylyl (3'→5') cytidine, EF11

    adenylyl (3'→5') guanosine, EF18

    adenylyl (3'→5') uridine, EF10

    2-amino-4,6-dihydroxy-7-methyl-pteridine, EC27

    2-aminopteridine, EB36

    8-azaguanine, EA49

    cytidine-5'-triphosphate disodium salt, ED13

    cytidylyl (3'→5') adenosine, EF12

    cytidylyl (3'→5') cytidine, EF04

    cytidylyl (3'→5') guanosine, EF13

    cytidylyl (3'→5') uridine, EF03

    guanosine-5'-triphosphate disodium salt, ED44

    guanylyl (3'→5') adenosine, EF19

    6-methylpterin, EC26

    uridylyl (3'→5') adenosine, EF09

  Quaternary

    2-(1,1-dimethylethyl)-3-aza-2-azoniabicyclo[2.2.1]hept-2-ene tetrafluoroborate, ED19

    2-(1,1-dimethylethyl)-3-aza-2-azoniabicyclo[2.2.2]oct-2-ene tetrafluoroborate, ED55

TABLE V. Functional-Group Index

AMINES (see also NITROSO COMPOUNDS) (cont.)

  Quaternary (cont.)

    2-ethyl-2,3-diazeniumbicyclo-[2.2.1]hept-2-ene tetrafluoroborate, EC49

    1-methylpyridinium-3-carbamoyl iodide, EC34

    thiocholine disulfide, ED58

    thiocholine iodide, EB00

AMINE SALTS, see AMINES

AMINO ACIDS, see ACIDS, CARBOXYLIC

ANHYDRIDES, CARBOXYLIC

    phthalic anhydride, EC54

AZINES, see IMINES

AZO AND AZOXY COMPOUNDS

    4-amino-2-hydroxybenzene-1-azo-2-pyridine, ED68

    azobenzene, ED99

    azoxybenzene, EE00

    5-bromopyridine-2-azo-(2',4'-dihydroxybenzene), ED59

    2-(5-bromo-2-pyridylazo)-5-(diethylamino)phenol, EE82

    2-(5-bromo-2-pyridylazo)-5-(dimethylamino)phenol, EE56

    2-(5-chloro-2-pyridylazo)-5-(dimethylamino)phenol, EE83

    5-(diethylamino)-2-(2-pyridylazo)-phenol, EE85

    5-(diethylamino)-2-(2-quinolylazo)phenol, EF07

    5-(diethylamino)-2-(2-thiazolylazo)phenol, EE60

    2,2'-dihydroxyazobenzene, EE03

    2,4-dihydroxyazobenzene, EE04

    3,3'-dihydroxyazobenzene, EE05

    4,4'-dihydroxyazobenzene, EE06

    2,4-dihydroxybenzene-1-azo-2'-pyridine, ED65

    2,4-dihydroxybenzene-1-azo-3'-pyridine, ED66

    2,4-dihydroxybenzene-1-azo-4'-pyridine, ED67

    2-(1,1-dimethylethyl)-3-aza-2-azoniabicyclo[2.2.1]hept-2-ene tetrafluoroborate, ED19

    2-(1,1-dimethylethyl)-3-aza-2-azoniabicyclo[2.2.2]oct-2-ene tetrafluoroborate, ED55

    2-ethyl-2,3-diazeniumbicyclo-[2.2.1]hept-2-ene tetrafluoroborate, EC49

    4-hydroxyaminodiazoaminobenzene, EE23

    2-hydroxyazobenzene, EE01

    4-hydroxyazobenzene, EE02

    2-hydroxynaphthalene-1-azobenzene, EE91

    4-hydroxynaphthalene-1-azobenzene, EE92

    2-hydroxynaphthalene-1-azo-(2'-nitrobenzene), EE89

    2-hydroxynaphthalene-1-azo-(4'-nitrobenzene), EE90

    4-hydroxy-4'-nitroazobenzene, ED97

    1-(2-hydroxyphenylazo)-2-naphthol, EE93

    2-(2'-hydroxyphenylazo)pyridine, ED61

    2-(4'-hydroxyphenylazo)pyridine, ED62

    3-(4'-hydroxyphenylazo)pyridine, ED63

    4-(4'-hydroxyphenylazo)pyridine, ED64

    2-(5-iodo-2-pyridylazo)-5-(diethylamino)phenol, EE84

    4-nitroazobenzene, ED96

    2-[(5-nitro-2-thiazolyl)azo]-5-(diethylamino)phenol, EE59

    4-(phenylazo)azobenzene, EF01

    1-phenyl-3,3-dimethyltriazene, EC83

    1-(2-pyridylazo)-2-naphthol, EE77

    1-(3-pyridylazo)-2-naphthol, EE78

    1-(4-pyridylazo)-2-naphthol, EE79

    4-(2-pyridylazo)-1-naphthol, EE80

AZO AND AZOXY COMPOUNDS (cont.)

- 4-(2-pyridylazo)-1-naphthylamine, EE81
- 1-(2-thiazoylazo)-2-naphthol, EE39
- 4-(2-thiazolylazo)phenol, EC97
- 4-(2-thiazolylazo)resorcinol, EC98
- 2,2',3-trihydroxyazobenzene, EE09

BENZOQUINONES, see QUINONES

CARBAMATES, see ESTERS, NON-CARBOXYLIC

CARBAZIDES, CARBAZONES, SEMICARBAZIDES, SEMICARBAZONES, AND THIO ANALOGS

- diethyl mesoxalate semicarbazone, EC85
- dihydroxyacetone semicarbazone, EA72
- ethyl pyruvate semicarbazone, EB77
- glyceraldehyde semicarbazone, EA73
- glycolaldehyde semicarbazone, EA37
- glyoxylic acid semicarbazone, EA30
- 2-hydroxyacetophenone semicarbazone, ED08
- 3-hydroxy-2-butanone semicarbazone, EA98
- hydroxypyruvaldehyde semicarbazone, EA97
- mesoxalic acid semicarbazone, EA52
- methylglyoxal bissemicarbazone, EA96
- pyruvic acid semicarbazone, EA63
- semicarbazidoacetic acid, EA38
- 2-semicarbazidopropionic acid, EA74

CARBAZONES, see CARBAZIDES

CARBOHYDRATES AND SUGARS

- adenosine, ED41
- adenosine 5'-diphosphate trisodium salt, ED36
- adenosine cyclo-2',3'-hydrogen phosphate, ED37
- adenosine cyclo-3',5'-hydrogen phosphate, ED38
- adenosine 5'-triphosphate disodium salt, ED43
- 5'-adenylic acid, ED45
- adenylyl (3'→5') adenosine, EF17
- adenylyl (3'→5') adenylyl (3'→5') adenosine, EF26
- adenylyl (3'→5') adenylyl (3'→5') adenylyl (3'→5') adenosine, EF27
- adenylyl (3'→5') adenylyl (3'→5') adenylyl (3'→5') adenylyl (3'→5') adenylyl (3'→5') adenosine, EF28
- adenylyl (3'→5') cytidine, EF11
- adenylyl (3'→5') guanosine, EF18
- adenylyl (3'→5') uridine, EF10
- L-ascorbic acid, EB65
- 6-azauridine, EC84
- 5-bromouridine, ED05
- cytidine-5'-triphosphate disodium salt, ED13
- cytidylyl (3'→5') adenosine, EF12
- cytidylyl (3'→5') cytidine, EF04
- cytidylyl (3'→5') guanosine, EF13
- cytidylyl (3'→5') uridine, EF03
- D-erythrose, EA69
- D-glucose, EB83
- L-glycero-tetrulose, EA70
- guanosine-5'-triphosphate disodium salt, ED44
- guanylyl (3'→5') adenosine, EF19
- s-methyl-4-thiouridine, ED42
- D-threose, EA71
- uridylyl (3'→5') adenosine, EF09

CARBOXYLIC ACIDS, see ACIDS, CARBOXYLIC

DEUTERATED COMPOUNDS, see ISOTOPES

DIAZO COMPOUNDS

- 3-diazo-1,3-dihydro-2H-indol-2-one, EC57
- 2,3-dimethyl-2,3-diazabicyclo[2.2.2]octane, EC89
- 2-(1,1-dimethylethyl)-3-aza-2-azoniabicyclo[2.2.1]hept-2-ene tetrafluoroborate, ED19

DIAZO COMPOUNDS (cont.)

    2-(1,1-dimethylethyl)-3-aza-2-azoniabicyclo[2.2.2]oct-2-ene tetrafluoroborate, ED55

    2-ethyl-3-aza-2-azoniabicyclo-[2.2.1]hept-2-ene tetrafluoroborate, EC49

    4-hydroxyaminodiazoaminobenzene, EE23

    1-phenyl-3,3-dimethyltriazene, EC83

ESTERS, CARBOXYLIC

    acetylsalicylic acid, ED01

    1-adamantylcarbinyl acetate, EE61

    diethyl mesoxalate semicarbazone, EC85

    ethyl chloroacetate, FA59

    ethyl cinnamate, ED72

    ethyl dichloroacetate, EA53

    ethyl difluoroacetate, EA54

    ethyl 2,4-dimethylpyrimidine-5-carboxylate, ED09

    ethyl fluoroacetate, EA61

    ethyl 2-methyl-4-hydroxypyrimidine-5-carboxylate, EC73

    ethyl 2-methylthio-4-hydroxypyrimidine-5-carboxylate, EC74

    ethyl pyruvate semicarbazone, EB77

    ethyl trichloroacetate, EA50

    ethyl trifluoroacetate, EA51

    glycol dimercaptopropionate, EC88

    methyl 1-adamantanecarboxylate, EE26

    phthalic anhydride, EC54

ESTERS, NONCARBOXYLIC

    Carbamates, Thiocarbamates, and Xanthates

        potassium O-butyldithiocarbonate, EA94

        potassium O-cyclohexyldithiocarbonate, EC42

        potassium ethyldithiocarbonate, EA28

        potassium O-isopropyldithiocarbonate, EA62

        potassium methyldithiocarbonate, EA12

        sodium diethyldithiocarbamate, EA95

Phosphates

    adenosine 5'-diphosphate trisodium salt, ED36

    adenosine cyclo-2',3'-hydrogen phosphate, ED37

    adenosine cyclo-3',5'-hydrogen phosphate, ED38

    adenosine 5'-triphosphate disodium salt, ED43

    5'-adenylic acid, ED45

    adenylyl (3'→5') adenosine, EF17

    adenylyl (3'→5') adenylyl (3'→5') adenosine, EF26

    adenylyl (3'→5') adenylyl (3'→5') adenylyl (3'→5') adenosine, EF27

    adenylyl (3'→5') adenylyl (3'→5') adenylyl (3'→5') adenylyl (3'→5') adenosine, EF28

    adenylyl (3'→5') cytidine, EF11

    adenylyl (3'→5') guanosine, EF18

    adenylyl (3'→5') uridine, EF10

    cytidine 5'-triphosphate disodium salt, ED13

    cytidylyl (3'→5') adenosine, EF12

    cytidylyl (3'→5') cytidine, EF04

    cytidylyl (3'→5') guanosine, EF13

    cytidylyl (3'→5') uridine, EF03

    guanosine 5'-triphosphate disodium salt, ED44

    guanylyl (3'→5') adenosine, EF19

    potassium O,O-diethyl dithiophosphate, EA75

    pyridoxal 5-phosphate, EC72

    tributylphosphate, EE30

    uridylyl (3'→5') adenosine, EF09

ETHERS AND EPOXIDES

    4-amino-4'-methoxydiphenylamine, EE57

    2-anisaldehyde, EC68

    2-anisyl alcohol, EC77

    1,4-dimethoxybenzene, EC78

    2-(2-ethoxy-2-methyl-1-propyl)-1,4-benzohydroquinone, EE27

    1-(4-hydroxy-3-methoxyphenyl)-2-aminoethanol, ED12

    1-methoxyadamantane, ED78

    4-methoxy-2'-nitrodiphenyl sulfide, EE50

    4-methoxy-4'-nitrodiphenyl sulfone, EE54

    4-methoxyphenol, EC31

    4-methoxyphenyl methyl sulfoxide, EC79

    2-(2-methyl-2-methoxypropyl)hydroquinone, ED75

    2-nitroanisole, EC22

    3-nitroanisole, EC23

    4-nitroanisole, EC24

    vanillin, EC70

HALOGEN COMPOUNDS

  Aliphatic

    Bromo

      1,4-dibromobutane, EA66

      1,3-dibromopropane, EA32

    Chloro

      chloroacetic acid, EA09

      chlorofumaric acid, EA43

      chloromaleic acid, EA44

      dichloroacetic acid, EA06

      dichlorofumaric acid, EA40

      dichloromaleic acid, EA41

      ethyl chloroacetate, EA59

      ethyl dichloroacetate, EA53

      ethyl trichloroacetate, EA50

      trichloroacetamidoxime, EA10

    Fluoro

      difluoroacetic acid, EA07

      ethyl difluoroacetate, EA54

      ethyl fluoroacetate, EA61

      ethyl trifluoroacetate, EA51

      fluoroacetic acid, EA11

      trifluoroacetic acid, EA05

  Alicyclic

    1-bromoadamantane, ED46

    1-chloroadamantane, ED47

    1-fluoroadamantane, ED48

  Aromatic and Quinoid

    Bromo

      3-bromobenzeneseleninic acid, EB23

      4-bromobenzeneseleninic acid, EB24

      1-bromo-9-fluorenone, EE31

      3-bromo-9-fluorenone, EE32

      4'-bromo-2-nitrodiphenyl sulfone, ED81

      4'-bromo-4-nitrodiphenyl sulfone, ED82

    Chloro

      4,5-bis(4-chlorophenyl)-2-(4-dimethylaminophenyl)imidazole sodium salt, EF22

      chloranil, EB01

      2-chloro-1-acenaphthalenecarboxylic acid, EE33

      3-chlorobenzaldehyde, EB99

      4-chlorobenzaldehyde, EC00

      3-chlorobenzeneseleninic acid, EB26

      4-chlorobenzeneseleninic acid, EB27

      chlorobenzoquinone, EB07

      5-chloro-N-cyclohexyl-4-sulfamylphthalimide, EE71

      2-chloro-5-hydroxybenzoquinone, EB08

TABLE V. Functional-Group Index

HALOGEN COMPOUNDS (cont.)

Aromatic and Quinoid

Chloro (cont.)

2-chloro-6-methyl-1,4-benzoquinone, EC01

2'-chloro-4-nitrodiphenyl sulfide, ED83

2'-chloro-2-nitrodiphenyl sulfone, ED84

2'-chloro-4-nitrodiphenyl sulfone, ED85

3'-chloro-2-nitrodiphenyl sulfone, ED86

3'-chloro-4-nitrodiphenyl sulfone, ED87

4-chlorophenyl methyl sulfoxide, EC20

4-chlorophthalimide, EC53

6-chloroquinoline, EC91

1,4-dichlorobenzene, EB09

2,6-dichlorobenzoquinone, EB04

3,6-dichloro-2,5-dihydroxybenzoquinone, EB06

3,5-dichloro-2-hydroxybenzoquinone, EB05

2,3,5-trichlorobenzoquinone, EB02

2,3,5-trichloro-6-hydroxybenzoquinone, EB03

2,3,5-trichloro-6-methylbenzoquinone, EB89

Fluoro

2-fluorobenzonitrile, EB96

3-fluorobenzonitrile, EB97

4-fluorobenzonitrile, EB98

octafluoronaphthalene, ED20

Heterocyclic

5-bromopyridine-2-azo-(2',4'-dihydroxybenzene), ED59

2-(5-bromo-2-pyridylazo)-5-(diethylamino)phenol, EE82

2-(5-bromo-2-pyridylazo)-5-(dimethylamino)phenol, EE56

5-bromouridine, ED05

2-(5-chloro-2-pyridylazo)-5-(diethylamino)phenol, EE83

2-(5-iodo-2-pyridylazo)-5-(diethylamino)phenol, EE84

Organometallic

bromotricarbonyl-η^3-2-propenylruthenium, EB25

HETEROCYCLIC COMPOUNDS

Nitrogen Heterocycles

Pyridines

4-amino-2-hydroxybenzene-1-azo-2-pyridine, ED68

1,2-bis(2-pyridyl)-1,2-dihydroxyethane N,N'-dioxide, EE22

5-bromopyridine-2-azo-(2',4'-dihydroxybenzene), ED59

2-(5-bromo-2-pyridylazo)-5-(diethylamino)phenol, EE82

2-(5-bromo-2-pyridylazo)-5-(dimethylamino)phenol, EE56

2-(5-chloro-2-pyridylazo)-5-(diethylamino)phenol, EE83

2-cyanopyridine, EB10

3-cyanopyridine, EB11

2,3-dicyanopyridine, EB90

2,4-dicyanopyridine, EB91

2,5-dicyanopyridine, EB92

2,6-dicyanopyridine, EB93

3,4-dicyanopyridine, EB94

3,5-dicyanopyridine, ED95

5-(diethylamino)-2-(2-pyridylazo)phenol, EE85

1,4-dihydro-N-(benzyl)-3-pyridinecarboxamide, EE58

2,4-dihydroxybenzene-1-azo-2'-pyridine, ED65

2,4-dihydroxybenzene-1-azo-3'-pyridine, ED66

2,4-dihydroxybenzene-1-azo-4'-pyridine, ED67

1,2-diisonicotinoylhydrazine, EE12

HETEROCYCLIC COMPOUNDS (cont.)

Nitrogen Heterocycles

Pyridines (cont.)

3-hydroxy-1-(2-hydroxyphenyl)-3-(2-pyridyl)-1-propanone oxime, EE69

3-hydroxy-1-(2-hydroxyphenyl)-3-(4-pyridyl)-1-propanone oxime, EE70

2-(2'-hydroxyphenylazo)pyridine, ED61

2-(4'-hydroxyphenylazo)pyridine, ED62

3-(4'-hydroxyphenylazo)pyridine, ED63

4-(4'-hydroxyphenylazo)pyridine, ED64

2-(5-iodo-2-pyridylazo)-5-(diethylamino)phenol, EE84

isonicotinamide, EB38

isonicotinic acid hydrazide, EB59

1-isonicotinoyl-2-phenylhydrazine, EE20

1-methylpyridinium-3-carbamoyl iodide, EC34

nicotinamide, EB39

2,2'-pyridil, ED88

2,3-pyridinedicarboxylic acid, EC05

2,4-pyridinedicarboxylic acid, EC06

2,5-pyridinedicarboxylic acid, EC07

2,6-pyridinedicarboxylic acid, EC08

3,4-pyridinedicarboxylic acid, EC09

3,5-pyridinedicarboxylic acid, EC10

(1-pyridinium)cyclopentadienide, ED29

2,2'-pyridoin, EE07

2,2'-pyridoin bis-N-oxide, EE10

pyridoxal, EC71

pyridoxal 5-phosphate, EC72

1-(3-pyridylazo)-2-naphthol, EE78

1-(4-pyridylazo)-2-naphthol, EE79

4-(2-pyridylazo)-1-naphthol, EE80

4-(2-pyridylazo)-1-naphthylamine, EE81

1-(2-pyridyl)-3-(3-pyridyl)-2-propen-1-one, EE40

1-(3-pyridyl)-3-(2-pyridyl)-2-propen-1-one, EE41

1-(3-pyridyl)-3-(3-pyridyl)-2-propen-1-one, EE42

1-(3-pyridyl)-3-(4-pyridyl)-2-propen-1-one, EE43

One Ring-Nitrogen Atom, Monocyclic (Except Pyridines)

N-benzylmaleimide, ED60

N-butylmaleimide, EC81

N-cyclohexylmaleimide, ED40

5-(diethylamino)-2-(2-thiazolylazo)phenol, EE60

N-ethylmaleimide, EB55

N-isopropylmaleimide, EC39

N-methylmaleimide, EA81

2-[(5-nitro-2-thiazolyl)azo]-5-(diethylamino)phenol, EE59

N-(phenethyl)maleimide, EE19

N-phenylmaleimide, ED26

N-(1-propyl)maleimide, EC40

1-(2-thiazoylazo)-2-naphthol, EE39

4-(2-thiazolylazo)phenol, EC97

4-(2-thiazolylazo)resorcinol, EC98

One Ring-Nitrogen Atom, Polycyclic

acridine, EE35

benz[c]acridine, EE97

benzo[c]isoquinoline, EE36

benzo[f]quinoline, EE37

benzo[h]quinoline, EE38

benzothiazole-2-thiol, EC11

biotin, ED50

4-chlorophthalimide, EC53

6-chloroquinoline, EC91

3-diazo-1,3-dihydro-2H-indol-2-one, EC57

5-(diethylamino)-2-(2-quinolylazo)phenol, EF07

1,3-dihydro-3-hydroxy-2H-indol-2-one, EC63

1,2-dimethylindolizine-3-thial, ED70

TABLE V. Functional-Group Index

HETEROCYCLIC COMPOUNDS (cont.)

Nitrogen Heterocycles

One Ring-Nitrogen Atom, Polycyclic (cont.)

1,2-dioxo-1H,2,3-dihydroindolizine dioxime, EC25

N-ethylphthalimide, ED30

1,2,4,5,7,8-hexamethylacridine, EF08

8-hydroxyquinoline, EC94

isatin, EC55

N-methylisatin, EC95

N-methylphthalimide, EC96

1-oxo-2-hydroxyimino-1H-2,3-dihydroindolizine, EC12

9-phenylacridine, EF05

phthalimide, EC56

N-propylphthalimide, ED69

quinoline, EC93

Two Ring-Nitrogen Atoms, Monocyclic

adenylyl (3'→5') cytidine, EF11

adenylyl (3'→5') uridine, EF10

4,5-bis(4-chlorophenyl)-2-(4-dimethylaminophenyl)imidazole sodium salt, EF22

5-bromouridine, ED05

cytidine 5'-triphosphate disodium salt, ED13

cytidylyl (3'→5') adenosine, EF12

cytidylyl (3'→5') cytidine, EF04

cytidylyl (3'→5') guanosine, EF13

cytidylyl (3'→5') uridine, EF03

1,2-diazaspiro[2.5]octane, EB79

5,6-dihydrouracil, EA55

4,5-dihydroxypyrimidine, EA47

2-(4-dimethylaminophenyl)-4,5-(2,2'-biphenylene)imidazole sodium, EF23

2-(4-dimethylaminophenyl)-4,5-diphenylimidazole sodium, EF24

2,4-dimethylpyrimidine-5-carboxamide, EC41

ethyl 2,4-dimethylpyrimidine-5-carboxylate, ED09

3-ethyl-3-methyldiaziridine, EA76

ethyl 2-methyl-4-hydroxypyrimidine-4-carboxylate, EC73

ethyl 2-methylthio-4-hydroxy-pyrimidine-5-carboxylate, EC74

5-methyl-2,4-dithiouracil, EA84

2-methyl-4-hydroxypyrimidine-5-carbohydrazide, EB63

1-methyl-4-methylthiouracil, EB62

4-methylthiouracil, EA83

S-methyl-4-thiouridine, ED42

pyridazine, EA45

pyrimidine, EA46

tetraoxopiperazine, EA42

uracil, EA48

uridylyl (3'→5') adenosine, EF09

Two Ring-Nitrogen Atoms, Polycyclic

2-tert-butyl-3-methyl-2,3-diazanorbornane, ED56

dibenzo[a,c]phenazine, EF14

dibenzo[f,h]quinoxaline, EE87

2,3-dimethyl-2,3-diazabicyclo[2.2.1]heptane, EC50

2,3-dimethyl-2,3-diazabicyclo[2.2.2]octane, EC89

2-(1,1-dimethylethyl)-3-aza-2-azoniabicyclo[2.2.1]hept-2-ene tetrafluoroborate, ED19

2-(1,1-dimethylethyl)-3-aza-2-azoniabicyclo[2.2.2]oct-2-ene tetrafluoroborate, ED55

2-ethyl-2,3-diazeniumbicyclo[2.2.1]hept-2-ene tetrafluoroborate, EC49

Three Ring-Nitrogen Atoms

5-azauracil, EA23

6-azauridine, EC84

5-tert-butyl-6-azauracil, EC43

5,6-dihydro-6-azauracil, EA29

3,5-dimethyl-6-azauracil, EA86

1,3,5-trimethyl-6-azauracil, EB67

HETEROCYCLIC COMPOUNDS (cont.)

Nitrogen Heterocycles

Four Ring-Nitrogen Atoms

adenine, EA82

adenosine, ED41

adenosine 5'-diphosphate trisodium salt, ED36

adenosine cyclo-2',3'-hydrogen
  phosphate, ED37

adenosine cyclo-3',5'-hydrogen
  phosphate, ED38

adenosine 5'-triphosphate disodium
  salt, ED43

5'-adenylic acid, ED45

adenylyl (3'→5') adenosine, EF17

adenylyl (3'→5') adenylyl
  (3'→5') adenosine, EF26

adenylyl (3'→5') adenylyl
  (3'→5') adenylyl (3'→5')
  adenosine, EF27

adenylyl (3'→5') adenylyl
  (3'→5') adenylyl (3'→5')
  adenylyl (3'→5') adenylyl
  (3'→5') adenosine, EF28

adenylyl (3'→5') cytidine, EF11

adenylyl (3'→5') guanosine, EF18

adenylyl (3'→5') uridine, EF10

2-amino-4,6-dihydroxy-7-methyl-
  pteridine, EC27

2-aminopteridine, EB36

cytidylyl (3'→5') adenosine, EF12

cytidylyl (3'→5') guanosine, EF13

cytidylyl (3'→5') uridine, EF03

5,6-dihydro-7-hydroxypteridine,
  EB45

7,8-dihydro-6-hydroxypteridine,
  EB46

6,7-dihydroxypteridine, EB20

guanosine 5'-triphosphate disodium
  salt, ED44

guanylyl (3'→5') adenosine, EF19

4-hydroxypteridine, EB17

6-hydroxypteridine, EB18

7-hydroxypteridine, EB19

6-mercaptopurine, EA80

6-methylpterin, EC26

pteridine, EB16

2,4,6,7-tetrahydroxypteridine,
  EB21

uridylyl (3'→5') adenosine, EF09

Five and More Ring-Nitrogen Atoms

8-azaguanine, EA49

Oxygen Heterocycles (see also ANHYDRIDES, CARBOXYLIC; CARBO-HYDRATES AND SUGARS; and LACETONES]

adenosine, ED41

adenosine 5'-diphosphate trisodium salt, ED36

adenosine cyclo-2',3'-hydrogen
  phosphate, ED37

adenosine cyclo-3',5'-hydrogen
  phosphate, ED38

adenosine 5'-triphosphate disodium salt, ED43

5'-adenylic acid, ED45

adenylyl (3'→5') adenosine, EF17

adenylyl (3'→5') adenylyl
  (3'→5') adenosine, EF26

adenylyl (3'→5') adenylyl
  (3'→5') adenylyl (3'→5')
  adenosine, EF27

adenylyl (3'→5') adenylyl
  (3'→5') adenylyl (3'→5')
  adenylyl (3'→5') adenylyl
  (3'→5') adenosine, EF28

adenylyl (3'→5') cytidine, EF11

adenylyl (3'→5') guanosine, EF18

adenylyl (3'→5') uridine, EF10

L-ascorbic acid, EB65

6-azauridine, EC84

5-bromouridine, ED05

cytidine 5'-triphosphate disodium
  salt, ED13

cytidylyl (3'→5') adenosine,
  EF12

cytidylyl (3'→5') cytidine, EF04

cytidylyl (3'→5') guanosine,
  EF13

cytidylyl (3'→5') uridine, EF03

D-erythrose, EA69

L-glycero-tetrulose, EA70

TABLE V. Functional-Group Index

HETEROCYCLIC COMPOUNDS (cont.)

Oxygen Heterocycles (see also ANHYDRIDES, CARBOXYLIC; CARBOHYDRATES AND SUGARS; and LACTONES) (cont.)

D-glucose, EB83

guanosine 5'-triphosphate disodium salt, ED44

guanylyl (3'→5') adenosine, EF19

4-oxooxane, EA91

phthalic anhydride, EC54

D-threose, EA71

uridylyl (3'→5') adenosine, EF09

Sulfur Heterocycles (see also SULFUR COMPOUNDS)

benzothiazole-2-thiol, EC11

biotin, ED50

dibenzothiophene, ED89

5-(diethylamino)-2-(2-thiazolylazo)phenol, EE60

1,3-dithiolane-2-thione, EA25

2-(ethylthio)-1,3-dithian-2-ylium tetrafluoroborate, EB76

2-ethylthio-1,3-dithiolanium tetrafluoroborate, EA92

2-ethylthio-4-phenyl-1,3-dithiolanium methyl sulfate, EE24

2-methyl-4-thiepanone, EC45

7-methyl-3-thiepanone, EC46

2-methylthio-1,3-dithianium perchlorate, EA93

2-methylthio-1,3-dithietanium fluorosulfonate, EA26

2-(methylthio)-1,3-dithiolan-2-ylium perchlorate, EA60

2-(methylthio)-4-phenyl-1,3-dithiolan-2-ylium fluorosulfonate, ED35

S-methyl-4-thiouridine, ED42

2-[(5-nitro-2-thiazolyl)azo]-5-(diethylamino)phenol, EE59

3-oxothiane, EA89

4-oxothiane, EA90

2,2,6,6-tetramethyl-4-oxothiane, ED16

3,3,5,5-tetramethyl-4-oxothiane, ED17

2,2,6,6-tetramethyl-4-oxothiane 1,1-dioxide, ED18

2,2',5,5'-tetrathiabicyclopentylidene, EB66

2-thiabicyclo[4.4.0]decan-5-one, ED14

2-thiabicyclo[4.4.0]decan-5-one 2,2-dioxide, ED15

thianthrene, ED90

4-(2-thiazolylazo)phenol, EC97

4-(2-thiazolylazo)resorcinol, EC98

1-(2-thiazoylazo)-2-naphthol, EE39

3-thiepanone, EB72

4-thiepanone, EB73

3-thiepanone 1,1-dioxide, EB74

4-thiepanone 1,1-dioxide, EB75

5-thiocanone, EC47

5-thiocanone 1,1-dioxide, EC48

HYDRAZIDES, see HYDRAZINES

HYDRAZINES, HYDRAZIDES, AND HYDRAZONES (see also CARBAZIDES)

2-tert-butyl-3-methyl-2,3-diazanorbornane, ED56

1,2-diazaspiro[2.5]octane, EB79

1,2-diisonicotinoylhydrazine, EE12

2,3-dimethyl-2,3-diazabicyclo[2.2.1]heptane, EC50

2,3-dimethyl-2,3-diazabicyclo[2.2.2]octane, EC89

1,1-dimethylhydrazine, EA22

3-ethyl-3-methyldiaziridine, EA76

4-hydroxyaminodiazoaminobenzene, EE23

isonicotinic acid hydrazide, EB59

1-isonicotinoyl-2-phenylhydrazine, EE20

methylhydrazine, EA04

2-methyl-4-hydroxypyrimidine-5-carbohydrazide, EB63

4-nitrophenylhydrazine, EB60

phenylglyoxal 2-hydrazone, EC65

HYDRAZINES, HYDRAZIDES, AND HYDRAZONES (see also CARBAZIDES) (cont.)

    trimethylhydrazine, EA39

HYDRAZONES, see HYDRAZINES

HYDROCARBONS

  Aliphatic and Alicyclic

    adamantane, ED49

    cyclohexane, EB78

    decane, ED57

    eicosane, EF20

    heptane, EC51

    hexadecane, EE96

    hexane, EB87

    1-methyladamantane, ED76

    3-methylpentane, EB88

    octane, EC90

    pentane, EA99

  Aromatic

    acenaphthylene, ED80

    bibenzyl, EE68

    1-methyl-7-isopropylazulene, EE72

    naphthalene, ED27

    (1-pyridinium)cyclopentadienide, ED29

HYDROXY COMPOUNDS (see also QUINONES)

  Aliphatic and Alicyclic

    1-adamantanol, ED52

    1-adamantylmethanol, ED77

    anisyl alcohol, EC77

    L-ascorbic acid, EB65

    benzyl alcohol, EC29

    1,2-bis(2-pyridyl)-1,2-dihydroxy-ethane N,N'-dioxide, EE22

    dihydroxyacetone semicarbazone, EA72

    2,3-diphenyl-2,3-butanediol, EE95

    D-erythrose, EA69

    D-glucose, EB83

    glyceraldehyde semicarbazone, EA73

    L-glycero-tetrulose, EA70

    glycolaldehyde, EA17

    glycolaldehyde semicarbazone, EA37

    2-hydroxyacetophenone semicarbazone, ED08

    3-hydroxy-2-butanone, EA98

    3-hydroxy-1-(2-hydroxyphenyl)-3-(2-pyridyl)-1-propanone oxime, EE69

    3-hydroxy-1-(2-hydroxyphenyl)-3-(4-pyridyl)-1-propanone oxime, EE70

    1-(4-hydroxy-3-methoxyphenyl)-2-aminoethanol, ED12

    hydroxypyruvaldehyde semicarbazone, EA97

    norepinephrine, EC82

    2,2'-pyridoin, EE07

    2,2'-pyridoin bis-N-oxide, EE10

    pyridoxal, EC71

    pyridoxal 5-phosphate, EC72

    D-threose, EA71

  Aromatic

    Phenols, Unsubstituted and Monosubstituted

    2-aminophenol, EB51

    3-aminophenol, EB52

    4-aminophenol, EB53

    2,2'-dihydroxyazobenzene, EE03

    3,3'-dihydroxyazobenzene, EE05

    4,4'-dihydroxyazobenzene, EE06

    4,4'-dihydroxybiphenyl, EE14

    hydroquinone, EB47

    2-hydroxyazobenzene, EE01

    4-hydroxyazobenzene, EE02

    4-hydroxybenzaldehyde, EC14

    3-hydroxy-1-(2-hydroxyphenyl)-3-(2-pyridyl)-1-propanone oxime, EE69

    3-hydroxy-1-(2-hydroxyphenyl)-3-(4-pyridyl)-1-propanone oxime, EE70

    4-hydroxy-4'-nitroazobenzene, ED97

HYDROXY COMPOUNDS (see also QUINONES) (cont.)

Aromatic

Phenols, Unsubstituted and Monosubstituted (cont.)

1-(2-hydroxyphenylazo)-2-naphthol, EE93

2-(2'-hydroxyphenylazo)pyridine, ED61

2-(4'-hydroxyphenylazo)pyridine, ED62

3-(4'-hydroxyphenylazo)pyridine, ED63

4-(4'-hydroxyphenylazo)pyridine, ED64

4-methoxyphenol, EC31

2-nitrophenol, EB31

3-nitrophenol, EB32

4-nitrophenol, EB33

4-nitrosophenol, EB30

4-(phenylamino)phenol, EE18

salicylaldehyde, EC16

4-(2-thiazolylazo)phenol, EC97

Phenols, Polysubstituted

4-amino-2-hydroxybenzene-1-azo-2-pyridine, ED68

5-bromopyridine-2-azo-(2',4'-dihydroxybenzene), ED59

2-(5-bromo-2-pyridylazo)-5-(diethylamino)phenol, EE82

2-(5-bromo-2-pyridylazo)-5-(dimethylamino)phenol, EE56

2-chloro-5-hydroxybenzoquinone, EB08

2-(5-chloro-2-pyridylazo)-5-(diethylamino)phenol, EE83

3,6-dichloro-2,5-dihydroxybenzoquinone, EB06

3,5-dichloro-2-hydroxybenzoquinone, EB05

5-(diethylamino)-2-(2-pyridylazo)phenol, EE85

5-(diethylamino)-2-(2-quinolylazo)phenol, EF07

5-(diethylamino)-2-(2-thiazolylazo)phenol, EE60

2,4-dihydroxyazobenzene, EE04

2,4-dihydroxybenzene-1-azo-2'-pyridine, ED65

2,4-dihydroxybenzene-1-azo-3'-pyridine, ED66

2,4-dihydroxybenzene-1-azo-4'-pyridine, ED67

2,5-dihydroxybenzoic acid, EC17

3-(3,4-dihydroxyphenyl)alanine, ED07

2,4-dinitrosoresorcinol, EB15

2-(2-ethoxy-2-methyl-1-propyl)-1,4-benzohydroquinone, EE27

2-(5-iodo-2-pyridylazo)-5-(diethylamino)phenol, EE84

(2-methoxy-2-methylpropyl)hydroquinone, ED75

methyl hydroquinone, EC32

2-[(5-nitro-2-thiazolyl)azo]-5-(diethylamino)phenol, EE59

norepinephrine, EC82

4-(2-thiazolylazo)resorcinol, EC98

2,2',3-trihydroxyazobenzene, EE09

2,3,6-trihydroxybenzoic acid, EC18

2,4,5-trihydroxybenzoic acid, EC19

vanillin, EC70

Two or More Rings

dipotassium 1,4-naphthohydroquinone, ED21

2-hydroxy-1-acenaphthylenecarboxaldehyde, EE34

2-hydroxynaphthalene-1-azobenzene, EE91

4-hydroxynaphthalene-1-azobenzene, EE92

2-hydroxynaphthalene-1-azo-(2'-nitrobenzene), EE89

2-hydroxynaphthalene-1-azo-(4'-nitrobenzene), EE90

8-hydroxyquinoline, EC94

1,4-naphthohydroquinone, ED28

ninhydrin, EC92

1-(2-pyridylazo)-2-naphthol, EE77

HYDROXY COMPOUNDS (see also QUINONES) (cont.)

Aromatic

Two or More Rings (cont.)

1-(3-pyridylazo)-2-naphthol, EE78
1-(4-pyridylazo)-2-naphthol, EE79
4-(2-pyridylazo)-1-naphthol, EE80
1-(2-thiazoylazo)-2-naphthol, EE39

Heterocyclic (see also CARBOHYDRATES)

2-amino-4,6-dihydroxy-7-methyl-pteridine, EC27
8-azaguanine, EA49
5-azauracil, EA23
6-azauridine, EC84
5-bromouridine, ED05
5-tert-butyl-6-azauracil, EC43
cytidine 5'-triphosphate disodium salt, ED13
cytidylyl (3'→5') adenosine, EF12
cytidylyl (3'→5') cytidine, EF04
cytidylyl (3'→5') guanosine, EF13
cytidylyl (3'→5') uridine, EF03
5,6-dihydro-6-azauracil, EA29
1,3-dihydro-3-hydroxy-2H-indol-2-one, EC63
5,6-dihydro-7-hydroxypteridine, EB45
7,8-dihydro-6-hydroxypteridine, EB46
5,6-dihydrouracil, EA55
6,7-dihydroxypteridine, EB20
4,5-dihydroxypyrimidine, EA47
ethyl 2-methyl-4-hydroxypyrimidine-5-carboxylate, EC73
ethyl 2-methylthio-4-hydroxy-pyrimidine-5-carboxylate, EC74
guanosine 5'-triphosphate disodium salt, ED44
guanylyl (3'→5') adenosine, EF19
4-hydroxypteridine, EB17
6-hydroxypteridine, EB18
7-hydroxypteridine, EB19

2-methyl-4-hydroxypyrimidine-5-carbohydrazide, EB63
6-methylpterin, EC26
S-methyl-4-thiouridine, ED42
pyridoxal, EC71
pyridoxal 5-phosphate, EC72
2,4,6,7-tetrahydroxypteridine, EB21
1,3,5-trimethyl-6-azauracil, EB67
uracil, EA48

Hydroxylamines

4-hydroxyaminodiazoaminobenzene, EE23
4-nitrophenylhydroxylamine, EB43
phenylhydroxylamine, EB54

HYDROXY SALTS, see HYDROXY COMPOUNDS

IMIDES (see also HETEROCYCLIC COMPOUNDS)

adenylyl (3'→5') cytidine, EF11
adenylyl (3'→5') uridine, EF10
6-azauridine, EC84
N-benzylmaleimide, ED60
5-bromouridine, ED05
5-tert-butyl-6-azauracil, EC43
N-butylmaleimide, EC81
5-chloro-N-cyclohexyl-4-sulfamyl-phthalimide, EE71
4-chlorophthalimide, EC53
N-cyclohexylmaleimide, ED40
cytidylyl (3'→5') adenosine, EF12
cytidylyl (3'→5') uridine, EF03
N-ethylmaleimide, EB55
N-ethylphthalimide, ED30
N-isopropylmaleimide, EC39
N-methylmaleimide, EA81
N-methylphthalimide, EC96
S-methyl-4-thiouridine, ED42
N-(phenethyl)maleimide, EE19
N-phenylmaleimide, ED26
N-phenylphthalimide, EE65
N-(1-propyl)maleimide, EC40

TABLE V. Functional-Group Index

IMIDES (see also HETEROCYCLIC COMPOUNDS) (cont.)

   N-propylphthalimide, ED69

   2,4,6,7-tetrahydroxypteridine, EB21

   tetraoxopiperazine, EA42

   1,3,5-trimethyl-6-azauracil, EB67

   uridylyl (3'→5') adenosine, EF09

IMINES AND OTHER AZOMETHINES (see also CARBAZIDES; HYDRAZINES; and OXIMES)

   adenylyl (3'→5') guanosine, EF18

   6-azauridine, EC84

   5-tert-butyl-6-azauracil, EC43

   guanosine 5'-triphosphate disodium salt, ED44

   guanylyl (3'→5') adenosine, EF19

   oxamide bisoxime, EA21

   1,3,5-trimethyl-6-azauracil, EB67

ISOTOPES

   formaldehyde-d_2, EA01

KETONES (see also HETEROCYCLIC COMPOUNDS; IMIDES; and UNSATURATED COMPOUNDS)

   Dialkyl Ketones

      acetone, EA33

      1-acetyladamantane, EE25

      L-glycero-tetrulose, EA70

      3-methyl-3-buten-2-one, EA88

      4-methyl-3-penten-2-one, EB71

      methyl vinyl ketone, EA58

      4-oxo-2-pentenoic acid, EA85

      2-oxopropylmercury nitrate, EA27

   Aryl-Alkyl Ketones

      dibenzoylacetylene, EE88

      1,4-dibenzoylbutadiyne, EE98

      trans-dibenzoylethene, EE94

      phenylglyoxal aldoxime, EC64

      phenylglyoxal 2-hydrazone, EC65

   Diaryl Ketones

      benzophenone, EE44

      4-phenylbenzophenone, EF06

   Ketones Containing Heterocyclic Rings and Other Groups

      2,2'-pyridil, ED88

      2,2'-pyridoin, EE07

      2,2'-pyridoin bis-N-oxide, EE10

      1-(2-pyridyl)-3-(3-pyridyl)-2-propen-1-one, EE40

      1-(3-pyridyl)-3-(2-pyridyl)-2-propen-1-one, EE41

      1-(3-pyridyl)-3-(3-pyridyl)-2-propen-1-one, EE42

      1-(3-pyridyl)-3-(4-pyridyl)-2-propen-1-one, EE43

   Alicyclic Ketones

      cycloheptanone, EC44

      cyclooctanone, EC86

      dicyclohexeno[d,j]cyclooctadeca-2,4,10,12-tetraen-6,8,15,17-tetrayn-1,14-dione, EF25

      5,14-dimethylcyclooctadeca-2,4,11,13-tetraen-6,8,15,17-tetryn-1,10-dione, EF15

      10,15-dimethylcyclooctadeca-7,9,15,17-tetraen-2,4,11,13-tetryn-1,6-dione, EF16

      1-indanone, ED00

      ninhydrin, EC92

      6,7,8,9-tetrahydro-5H-benzocyclohepten-5-one, ED71

      1-tetralone, ED34

   Other Carbocyclic Ketones

      acenaphthenequinone, ED79

      1-bromo-9-fluorenone, EE31

      3-bromo-9-fluorenone, EE32

KETONES (see also HETEROCYCLIC COMPOUNDS, IMIDES, and UNSATURATED COMPOUNDS) (cont.)

Heterocyclic Ketones

- 5-chloro-N-cyclohexyl-4-sulfamyl-phthalimide, EE71
- 3-diazo-1,3-dihydro-2H-indol-2-one, EC57
- 1,3-dihydro-3-hydroxy-2H-indol-2-one, EC63
- N-ethylphthalimide, ED30
- isatin, EC55
- N-methylisatin, EC95
- N-methylphthalimide, EC96
- 2-methyl-4-thiepanone, EC45
- 7-methyl-3-thiepanone, EC46
- 1-oxo-2-hydroxyimino-1H-2,3-dihydroindolizine, EC12
- 4-oxooxane, EA91
- 3-oxothiane, EA89
- 4-oxothiane, EA90
- N-(phenethyl)maleimide, EE19
- N-phenylphthalimide, EE65
- phthalimide, EC56
- N-propylphthalimide, ED69
- 2,2,6,6-tetramethyl-4-oxothiane, ED16
- 3,3,5,5-tetramethyl-4-oxothiane, ED17
- 2,2,6,6-tetramethyl-4-oxothiane 1,1-dioxide, ED18
- 2-thiabicyclo[4.4.0]decan-5-one, ED14
- 2-thiabicyclo[4.4.0]decan-5-one 2,2-dioxide, ED15
- 3-thiepanone, EB72
- 4-thiepanone, EB73
- 3-thiepanone 1,1-dioxide, EB74
- 4-thiepanone 1,1-dioxide, EB75
- 5-thiocanone, EC47
- 5-thiocanone 1,1-dioxide, EC48

LACTAMS, see LACTONES

LACTONES AND LACTAMS (see also HETEROCYCLIC COMPOUNDS)

- adenylyl (3'→5') cytidine, EF11
- adenylyl (3'→5') guanosine, EF18
- adenylyl (3'→5') uridine, EF10
- L-ascorbic acid, EB65
- biotin, ED50
- cytidine 5'-triphosphate disodium salt, ED13
- cytidylyl (3'→5') adenosine, EF12
- cytidylyl (3'→5') cytidine, EF04
- cytidylyl (3'→5') guanosine, EF13
- cytidylyl (3'→5') uridine, EF03
- 3-diazo-1,3-dihydro-2H-indol-2-one, EC57
- 1,3-dihydro-3-hydroxy-2H-indol-2-one, EC63
- 3,5-dimethyl-6-azauracil, EA86
- guanylyl (3'→5') adenosine, EF19
- N-methylmaleimide, EA81
- 1-methyl-4-methylthiouracil, EB62
- 4-methylthiouracil, EA83
- phthalic anhydride, EC54
- phthalimide, EC56
- 2,4,6,7-tetrahydroxypteridine, EB21
- uridylyl (3'→5') adenosine, EF09

METHOXIMES, see OXIMES

NITRILES

- acetonitrile, EA13
- benzonitrile, EC02
- 1-cyanoadamantane, ED73
- 2-cyanopyridine, EB10
- 3-cyanopyridine, EB11
- 2,3-dicyanopyridine, EB90
- 2,4-dicyanopyridine, EB91
- 2,5-dicyanopyridine, EB92
- 2,6-dicyanopyridine, EB93
- 3,4-dicyanopyridine, EB94
- 3,5-dicyanopyridine, EB95
- 2-fluorobenzonitrile, EB96

TABLE V. Functional-Group Index

NITRILES (cont.)

    3-fluorobenzonitrile, EB97

    4-fluorobenzonitrile, EB98

NITRO COMPOUNDS

  Nitrobenzenes

    4'-amino-2-nitrodiphenyl sulfide, EE08

    4'-bromo-2-nitrodiphenyl sulfone, ED81

    4'-bromo-4-nitrodiphenyl sulfone, ED82

    2'-chloro-4-nitrodiphenyl sulfide, ED83

    2'-chloro-2-nitrodiphenyl sulfone, ED84

    2'-chloro-4-nitrodiphenyl sulfone, ED85

    3'-chloro-2-nitrodiphenyl sulfone, ED86

    3'-chloro-4-nitrodiphenyl sulfone, ED87

    N,N-dimethyl-4-nitrobenzenesulfonamide, EC75

    1,2-dinitrobenzene, EB12

    1,3-dinitrobenzene, EB13

    1,4-dinitrobenzene, EB14

    4,4'-dinitrostilbene, EE66

    2-hydroxynaphthalene-1-azo-(2'-nitrobenzene), EE89

    2-hydroxynaphthalene-1-azo-(4'-nitrobenzene), EE90

    4-methoxy-2'-nitrodiphenyl sulfide, EE50

    4-methoxy-4'-nitrodiphenyl sulfone, EE54

    N-methyl-4-nitrobenzenesulfonamide, EC28

    2-methyl-2'-nitrodiphenyl sulfide, EE45

    2-methyl-4'-nitrodiphenyl sulfide, EE46

    3-methyl-2'-nitrodiphenyl sulfide, EE47

    3-methyl-4'-nitrodiphenyl sulfide, EE48

    4-methyl-2'-nitrodiphenyl sulfide, EE49

    2-methyl-2'-nitrodiphenyl sulfone, EE51

    3-methyl-2'-nitrodiphenyl sulfone, EE52

    3-methyl-4'-nitrodiphenyl sulfone, EE53

    2-nitroaniline, EB40

    3-nitroaniline, EB41

    4-nitroaniline, EB42

    2-nitroanisole, EC22

    3-nitroanisole, EC23

    4-nitroanisole, EC24

    4-nitroazobenzene, ED96

    3-nitrobenzaldehyde, EC03

    4-nitrobenzaldehyde, EC04

    nitrobenzene, EB29

    3-nitrobenzeneseleninic acid, EB34

    4-nitrobenzeneseleninic acid, EB35

    4-nitrobenzenesulfonamide, EB44

    2-nitrobiphenyl, ED91

    4-nitrobiphenyl, ED92

    2-nitrodiphenyl sulfide, ED93

    2-nitrodiphenyl sulfone, ED94

    4-nitrodiphenyl sulfone, ED95

    2-nitrophenol, EB31

    3-nitrophenol, EB32

    4-nitrophenol, EB33

    N-(4-nitrophenyl)benzenesulfonamide, EE11

    4-nitrophenylhydrazine, EB60

    4-nitrophenylhydroxylamine, EB43

    tetraethylammonium 4-nitrobenzenesulfinate, EE75

    tetraethylammonium 4-nitrobenzenesulfonamidate, EE76

  Nitroaromatics with Condensed Rings

    9-nitroanthracene, EE63

    7-nitrobenz[a]anthracene, EE99

    6-nitrochrysene, EF00

NITRO COMPOUNDS (cont.)

Nitroaromatics with Condensed Rings (cont.)

1-nitronaphthalene, ED24
2-nitronaphthalene, ED25
9-nitrophenanthrene, EE64
1-nitropyrene, EE86

Nitro Derivatives of Heterocycles

2-[(5-nitro-2-thiazolyl)azo]-5-(diethylamino)phenol, EE59

NITROSO AND NITROSYL COMPOUNDS

2,4-dinitrosoresorcinol, EB15
nitrosobenzene, EB28
4-nitrosophenol, EB30

ORGANOMETALLICS

B

bis[7,11-η-7,8-dicarbaundecaborane(11)]cobaltate(1-) tetrabutylammonium salt, EF21
bis[(7,8,9,10,11-η)-undecahydro-1,7-dicarbaundecaborato(11)$^{2-}$]-nickelate(1-), EA78
bis[(7,8,9,10,11-η)-undecahydro-7,8-dicarbaundecaborato(11)$^{2-}$]-nickelate(1-), EA79

Co

bis(η-cyclopentadienyl)cobalt-(III) hexafluorophosphate, ED32
bis(η-cyclopentadienyl)cobalt-(III) perchlorate, ED31
bis[7,11-η-7,8-dicarbaundecaborane(11)]cobaltate(1-) tetrabutylammonium salt, EF21
(η-cyclopentadienyl)[7,11-η-undecaborane(11)$^{2-}$]cobalt(1-), EC52

Ferrocenes

ferrocene, ED33

Hg

2-oxopropylmercury nitrate, EA27

Ni

bis[(7,8,9,10,11-η-undecahydro-1,7-dicarbaundecaborato(11)$^{2-}$]-nickelate(1-), EA78
bis[7,8,9,10,11-η)-undecahydro-7,8-dicarbaundecaborato(11)$^{2-}$]-nickelate(1-), EA79

Ru

bromotricarbonyl-η^3-2-propenyl-ruthenium, EB25

Tl

diphenylthallium(III) nitrate, ED98

N-OXIDES

azoxybenzene, EE00
1,2-bis(2-pyridyl)-1,2-dihydroxyethane N,N'-dioxide, EE22
2,2'-pyridoin bis-N-oxide, EE10

OXIMES AND ALKOXIMES

Aldoximes

acetaldehyde oxime, EA18
formaldehyde oxime, EA03
phenylglyoxal aldoxime, EC64

Ketoximes and Amidoximes

acetamide oxime, EA20
dimethylglyoxime, EA67
1,2-dioxo-1H,2,3-dihydroindolizine dioxime, EC25
N^2-hydroxycinnamamidine, ED02
3-hydroxy-1-(2-hydroxyphenyl)-3-(2-pyridyl)-1-propanone oxime, EE69
3-hydroxy-1-(2-hydroxyphenyl)-3-(4-pyridyl)-1-propanone oxime, EE70

TABLE V. Functional-Group Index

OXIMES AND ALKOXIMES (cont.)

Ketoximes and Amidoximes (cont.)

oxamide bisoxime, EA21

1-oxo-2-hydroxyimino-1H-2,3-dihydroindolizine, EC12

2-propanone oxime, EA34

trichloroacetamidoxime, EA10

PHENOLS, see HYDROXY COMPOUNDS

QUATERNARY COMPOUNDS, see AMINES

QUINONES AND HYDROQUINONES (see also HYDROXY COMPOUNDS)

Benzoquinones

1,4-benzoquinone, EB22

chloranil, EB01

chlorobenzoquinone, EB07

2-chloro-5-hydroxybenzoquinone, EB08

2-chloro-6-methyl-1,4-benzoquinone, EC01

2,6-di(tert-butyl)-1,4-benzoquinone, EE74

2,6-dichlorobenzoquinone, EB04

3,6-dichloro-2,5-dihydroxybenzoquinone, EB06

3,5-dichloro-2-hydroxybenzoquinone, EB05

2,5-dimethyl-1,4-benzoquinone, EC69

duroquinone, ED39

2-methyl-1,4-benzoquinone, EC15

methyl hydroquinone, EC32

2,3,5-trichlorobenzoquinone, EB02

2,3,5-trichloro-6-hydroxybenzoquinone, EB03

2,3,5-trichloro-6-methylbenzoquinone, EB89

Naphthoquinones

1,2-naphthoquinone, ED22

1,4-naphthoquinone, ED23

Anthraquinones

anthraquinone, EE62

Polycyclic Quinones

acenaphthenequinone, ED79

SELENIUM AND TELLURIUM COMPOUNDS

benzeneseleninic acid, EB48

3-bromobenzeneseleninic acid, EB23

4-bromobenzeneseleninic acid, EB24

3-chlorobenzeneseleninic acid, EB26

4-chlorobenzeneseleninic acid, EB27

4-methylbenzeneseleninic acid, EC33

3-nitrobenzeneseleninic acid, EB34

4-nitrobenzeneseleninic acid, EB35

selenocystine, EB81

SEMICARBAZIDES AND SEMICARBAZONES, see CARBAZIDES

SUGARS, see CARBOHYDRATES AND SUGARS

SULFIDES, see SULFUR COMPOUNDS

SULFONAMIDES, see SULFUR COMPOUNDS

SULFONATES, see SULFUR COMPOUNDS

SULFONES, see SULFUR COMPOUNDS

SULFONIC ACIDS, see SULFUR COMPOUNDS

SULFUR COMPOUNDS (see also ESTERS, NONCARBOXYLIC; HETEROCYCLIC COMPOUNDS; and ORGANOMETALLICS)

Thiols

benzenethiol, EB49

benzothiazole-2-thiol, EC11

cysteine, EA36

glutathione, ED54

SULFUR COMPOUNDS (see also ESTERS, NONCARBOXYLIC; HETEROCYCLIC COMPOUNDS; and ORGANOMETALLICS)(cont.)

Thiols (cont.)

glycol dimercaptopropionate, EC88

2-mercaptoethyl-N,N-dimethyl-ammonium chloride, EA77

6-mercaptopurine, EA80

5-methyl-2,4-dithiouracil, EA84

sodium thiophenolate, EB37

thiocholine iodide, EB00

Sulfides, Thioethers, and Sulfonium Ions

4'-amino-2-nitrodiphenyl sulfide, EE08

bis(phenylthio)methane, EE55

2'-chloro-4-nitrodiphenyl sulfide, ED83

diphenyl sulfide, EE16

ethyl 2-methylthio-4-hydroxy-pyrimidine-5-carboxylate, EC74

2-(ethylthio)-1,3-dithian-2-ylium tetrafluoroborate(1-), EB76

2-ethylthio-1,3-dithiolanium tetrafluoroborate, EA92

2-ethylthio-4-phenyl-1,3-dithiolan-ium methyl sulfate, EE24

4-methoxy-2'-nitrodiphenyl sulfide, EE50

1-methyl-4-methylthiouracil, EB62

2-methyl-2'-nitrodiphenyl sulfide, EE45

2-methyl-4'-nitrodiphenyl sulfide, EE46

3-methyl-2'-nitrodiphenyl sulfide, EE47

3-methyl-4'-nitrodiphenyl sulfide, EE48

4-methyl-2'-nitrodiphenyl sulfide, EE49

2-methylthio-1,3-dithianium perchlorate, EA93

2-methylthio-1,3-dithietanium fluorosulfonate, EA26

2-(methylthio)-1,3-dithiolan-2-ylium perchlorate, EA60

2-(methylthio)-4-phenyl-1,3-dithiolan-2-ylium fluorosulfonate, ED35

4-methylthiouracil, EA83

S-methyl-4-thiouridine, ED42

2-nitrodiphenyl sulfide, ED93

Disulfides

cystine, EB80

diphenyl disulfide, EE17

thiocholine disulfide, ED58

Thiocarbamates, Dithiocarbamates, and Xanthates

potassium O-butyldithiocarbonate, EA94

potassium O-cyclohexyldithiocarbonate, EC42

potassium O,O-diethyldithiophosphate, EA75

potassium ethyldithiocarbonate, EA28

potassium O-isopropyl carbonate, EA62

potassium methyldithiocarbonate, EA12

sodium diethyldithiocarbamate, EA95

Thiones, Thioamides, Thiobarbiturates, Thiopyrimidines, and Thiopurines

1,2-dimethylindolizine-3-thial, ED70

1,3-dithiolane-2-thione, EA25

dithiooxamide, EA15

thioacetamide, EA19

Sulfoxides, Sulfones, and S-dioxides

4'-bromo-2-nitrodiphenyl sulfone, ED81

4'-bromo-4-nitrodiphenyl sulfone, ED82

2'-chloro-2-nitrodiphenyl sulfone, ED84

2'-chloro-4-nitrodiphenyl sulfone, ED85

TABLE V. Functional-Group Index

SULFUR COMPOUNDS (see also ESTERS, NONCARBOXYLIC; HETEROCYCLIC COMPOUNDS; and ORGANOMETALLICS)(cont.)

Sulfoxides, Sulfones, and S-dioxides (cont.)

3'-chloro-2-nitrodiphenyl sulfone, ED86

3'-chloro-4-nitrodiphenyl sulfone, ED87

4-chlorophenyl methyl sulfoxide, EC20

diphenyl sulfone, EE15

diphenyl sulfoxide, EE13

4-ethylphenyl methyl sulfoxide, ED10

4-isobutylphenyl methyl sulfoxide, ED74

4-methoxy-4'-nitrodiphenyl sulfone, EE54

4-methoxyphenyl methyl sulfoxide, EC79

2-methyl-2'-nitrodiphenyl sulfone, EE51

3-methyl-2'-nitrodiphenyl sulfone, EE52

3-methyl-4'-nitrodiphenyl sulfone, EE53

4-methylphenyl methyl sulfoxide, EC76

methyl phenyl sulfoxide, EC30

2-nitrodiphenyl sulfone, ED94

4-nitrodiphenyl sulfone, ED95

2,2,6,6-tetramethyl-4-oxothiane 1,1-dioxide, ED18

2-thiabicyclo[4.4.0]decan-5-one 2,2-dioxide, ED15

3-thiepanone 1,1-dioxide, EB74

4-thiepanone 1,1-dioxide, EB75

5-thiocanone 1,1-dioxide, EC48

Sulfuric and Thiosulfuric Acid Derivatives, Sulfonates, Sulfinates, and Sulfonamides

4-aminodiphenylamine-2-sulfonic acid, EE21

5-chloro-N-cyclohexyl-4-sulfamyl-phthalimide, EE71

N,N-dimethyl-4-nitrobenzene-sulfonamide, EC75

metanilic acid, EB56

N-methyl-4-nitrobenzenesulfonamide, EC28

4-nitrobenzenesulfonamide, EB44

N-(4-nitrophenyl)benzene-sulfonamide, EE11

orthanilic acid, EB57

sulfanilic acid, EB58

tetraethylammonium 4-nitrobenzene-sulfinate, EE75

tetraethylammonium 4-nitrobenzene-sulfonamidate, EE76

TELLURIUM, see SELENIUM

THIOLS, see SULFUR COMPOUNDS

THIOSEMICARBAZIDES AND THIOSEMI-CARBAZONES, see CARBAZIDES

UNSATURATED COMPOUNDS

Ethylenic

Double Bond Isolated from Non-hydrocarbon Functional Groups

bromotricarbonyl-(η^3-2-propenyl)-ruthenium, EB25

1-butene, EA64

2-butene, EA65

cyclohexene, EB68

propylene, EA31

α,β-Unsaturated Aldehydes

acrolein, EA24

cinnamaldehyde, EC99

crotonaldehyde, EA56

3,7-dimethyl-2,6-octadienal, ED53

2-ethyl-2-butenal, ED69

2-ethyl-2-hexenal, EC87

2,4-hexadienal, EB64

2-hydroxy-1-acenaphthylenecarbox-aldehyde, EE34

methacrolein, EA57

2-methyl-2-butenal, EA87

2-methyl-2-pentenal, EB70

UNSATURATED COMPOUNDS (cont.)
Ethylenic

α,β-Unsaturated Ketones

L-ascorbic acid, EB65

6-azauridine, EC84

1,4-benzoquinone, EB22

5-tert-butyl-6-azauracil, EC43

chloranil, EB01

chlorobenzoquinone, EB07

2-chloro-5-hydroxybenzoquinone, EB08

2-chloro-6-methyl-1,4-benzoquinone, EC01

trans-dibenzoylethene, EE94

2,6-di(tert-butyl)-1,4-benzoquinone, EE74

2,6-dichlorobenzoquinone, EB04

3,6-dichloro-2,5-dihydroxybenzoquinone, EB06

3,5-dichloro-2-hydroxybenzoquinone, EB05

dicyclohexeno[d,j]cyclooctadeca-2,4,10,12-tetraen-6,8,15,17-tetrayn-1,14-dione, EF25

5,14-dimethylcyclooctadeca-2,4,11,13-tetraen-6,8,15,17-tetryn-1,10-dione, EF15

10,15-dimethylcyclooctadeca-7,9,15,17-tetraen-2,4,11,13-tetryn-1,6-dione, EF16

2,5-dimethyl-1,4-benzoquinone, EC69

duroquinone, ED39

2-methyl-1,4-benzoquinone, EC15

3-methyl-3-buten-2-one, EA88

4-methyl-3-penten-2-one, EB71

methyl vinyl ketone, EA58

1,2-naphthoquinone, ED22

1,4-naphthoquinone, ED23

4-oxo-2-pentenoic acid, EA85

1-(2-pyridyl)-3-(3-pyridyl)-2-propen-1-one, EE40

1-(3-pyridyl)-3-(2-pyridyl)-2-propen-1-one, EE41

1-(3-pyridyl)-3-(3-pyridyl)-2-propen-1-one, EE42

1-(3-pyridyl)-3-(4-pyridyl)-2-propen-1-one, EE43

2,3,5-trichlorobenzoquinone, EB02

2,3,5-trichloro-6-hydroxybenzoquinone, EB03

2,3,5-trichloro-6-methylbenzoquinone, EB89

1,3,5-trimethyl-6-azauracil, EB67

α,β-Unsaturated Acids, Thioacids, Esters, and Amides

N-benzylmaleimide, ED60

N-butylmaleimide, EC81

2-chloro-1-acenaphthylenecarboxylic acid, EE33

chlorofumaric acid, EA43

chloromaleic acid, EA44

N-cyclohexylmaleimide, ED40

dichlorofumaric acid, EA40

dichloromaleic acid, EA41

ethyl cinnamate, ED72

N-ethylmaleimide, EB55

N-isopropylmaleimide, EC39

N-methylmaleimide, EA81

4-oxo-2-pentenoic acid, EA85

N-(phenethyl)maleimide, EE19

N-phenylmaleimide, ED26

N-(1-propyl)maleimide, EC40

Unsaturated Amines, Halogens, Alcohols, Ethers, and Nitrocompounds

L-ascorbic acid, EB65

chloranil, EB01

chlorobenzoquinone, EB07

chlorofumaric acid, EA43

2-chloro-5-hydroxybenzoquinone, EB08

chloromaleic acid, EA44

2-chloro-6-methyl-1,4-benzoquinone, EC01

2,6-dichlorobenzoquinone, EB04

3,6-dichloro-2,5-dihydroxybenzoquinone, EB06

dichlorofumaric acid, EA40

UNSATURATED COMPOUNDS (cont.)
Ethylenic

Unsaturated Amines, Halogens, Alcohols, Ethers, and Nitrocompounds (cont.)

3,5-dichloro-2-hydroxybenzoquinone, EB05

dichloromaleic acid, EA41

4,4'-dinitrostilbene, EE66

2-hydroxy-1-acenaphthylenecarboxaldehyde, EE34

N^2-hydroxycinnamamidine, ED02

2,2',5,5'-tetrathiabicyclopentylidene, EB66

2,3,5-trichlorobenzoquinone, EB02

2,3,5-trichloro-6-hydroxybenzoquinone, EB03

2,3,5-trichloro-6-methylbenzoquinone, EB89

Cyclopentadienyl Compounds

bis(η-cyclopentadienyl)cobalt(III) hexafluorophosphate, ED32

bis(η-cyclopentadienyl)cobalt(III) perchlorate, ED31

(η-cyclopentadienyl)[7-11-η-undecaborane(11)$^{2-}$]cobalt(1-), EC52

ferrocene, ED33

Acetylenic

dibenzoylacetylene, EE88

dibenzoylbutadiyne, EE98

dicyclohexeno[d,j]cyclooctadeca-2,4,10,12-tetraen-6,8,15,17-tetrayn-1,14-dione, EF25

5,14-dimethylcyclooctadeca-2,4,11,13-tetraen-6,8,15,17-tetryn-1,10-dione, EF15

10,15-dimethylcyclooctadeca-7,9,15,17-tetraen-2,4,11,13-tetryn-1,6-dione, EF16

TABLE VI.
CHEMICAL ABSTRACTS SERVICE REGISTRY NUMBERS

This is a list of the Chemical Abstracts Service Registry Numbers given under the names of the compounds appearing in Table I. Registry Numbers have not been assigned to some of the compounds in Table I, and these compounds are not included in this table. The Registry Numbers are listed in numerical order, and each is followed by the code number assigned to the corresponding compound in Table I.

TABLE VI. Chemical Abstracts Service Registry Numbers

50-00-0, EA00	84-65-1, EE62	100-47-0, EC02
50-44-2, EA80	85-02-9, EE37	100-51-6, EC29
50-78-2, ED01	85-41-6, EC56	100-52-7, EC13
50-81-7, EB65	85-44-9, EC54	100-54-9, EB11
50-99-7, EB83	85-85-8, EE77	100-61-8, EC35
52-90-4, EA36	86-00-0, ED91	100-65-2, EB54
54-25-1, EC84	86-57-7, ED24	100-70-9, EB10
54-47-7, EC72	87-88-7, EB06	101-64-4, EE57
54-85-3, EB59	88-21-1, EB57	102-54-5, ED33
55-21-0, EC21	88-74-4, EB40	103-29-7, EE68
56-41-7, EA35	88-75-5, EB31	103-33-3, ED99
56-89-3, EB80	89-00-9, EC05	103-36-6, ED72
57-14-7, EA22	90-02-8, EC16	104-53-0, ED04
58-61-7, ED41	91-18-9, EB16	104-55-2, EC99
58-85-5, ED50	91-20-3, ED27	104-88-1, EC00
60-00-4, ED51	91-22-5, EC93	104-91-6, EB30
60-34-4, EA04	91-23-6, EC22	105-13-5, EC77
60-92-4, ED38	91-30-5, EE21	105-39-5, EA59
61-19-8, ED45	91-56-5, EC55	106-46-7, EB09
61-71-2, EC63	92-85-3, ED90	106-49-0, EC38
61-90-5, EB85	92-88-6, EE14	106-50-3, EB61
62-53-3, EB50	92-93-3, ED92	106-51-4, EB22
62-55-5, EA19	93-53-8, ED03	106-98-9, EA64
64-18-6, EA02	95-43-2, EA71	107-01-7, EA65
66-22-8, EA48	95-45-4, EA67	107-02-8, EA24
66-72-8, EC71	95-53-4, EC36	107-29-9, EA18
67-64-1, EA33	95-55-6, EB51	108-44-1, EC37
70-18-8, ED54	96-14-0, EB88	108-98-5, EB49
71-33-0, EA23	97-31-4, ED12	109-64-8, EA32
73-24-5, EA82	97-96-1, EB82	109-66-0, EA99
73-32-5, EB84	98-92-0, EB39	110-52-1, EA66
75-05-8, EA13	98-95-3, EB29	110-54-3, EB87
75-07-0, EA16	99-09-2, EB41	110-82-7, EB78
75-17-2, EA03	99-61-6, EC03	110-83-8, EB68
76-05-1, EA05	99-65-0, EB13	111-65-9, EC90
78-84-2, EA68	99-97-8, ED11	112-95-8, EF20
78-85-3, EA57	100-01-6, EB42	115-07-1, EA31
78-94-4, EA58	100-02-7, EB33	118-02-5, EB15
79-11-8, EA09	100-10-7, ED06	118-75-2, EB01
79-40-3, EA15	100-16-3, EB60	119-61-9, EE44
79-43-6, EA06	100-17-4, EC24	119-67-5, EC61
82-86-0, ED79	100-25-4, EB14	121-33-5, EC70
83-33-0, ED00	100-26-5, EC07	121-47-1, EB56

121-57-3, EB58	313-72-4, ED20	602-56-2, EF05
121-69-7, EC80	327-57-1, EB86	602-60-8, EE63
122-37-2, EE18	381-73-7, EA07	603-34-9, EF02
122-78-1, EC66	383-63-1, EA51	608-42-4, EA41
123-08-0, EC14	394-47-8, EB96	610-90-2, EC19
123-30-8, EB53	403-54-3, EB97	612-57-7, EC91
123-31-9, EB47	454-31-9, EA54	617-42-5, EA44
123-73-9, EA56	459-72-3, EA61	617-43-6, EA43
124-18-5, ED57	471-46-5, EA14	619-66-9, EC62
126-73-8, EE30	485-47-2, EC92	620-23-5, EC67
127-06-0, EA34	490-11-9, EC09	623-27-8, EC60
127-63-9, EE15	490-79-9, EC17	623-36-9, EB70
130-15-4, ED23	492-10-4, EC27	625-00-3, EB00
132-65-0, ED89	492-73-9, ED88	626-19-7, EC58
134-58-7, EA49	495-48-7, EE00	634-01-5, ED37
135-02-4, EC68	499-80-9, EC06	634-85-5, EB02
137-18-8, EC69	499-81-0, EC10	643-79-8, EC59
138-07-8, EA38	499-83-2, EC08	645-62-5, EC87
138-65-8, EC82	502-42-1, EC44	695-99-8, EB07
139-66-2, EE16	502-49-8, EC86	697-91-6, EB04
140-89-6, EA28	504-07-4, EA55	700-47-0, EB17
140-92-1, EA62	515-84-4, EA50	700-81-2, EB36
141-46-8, EA17	520-03-6, EE65	708-75-8, EC26
141-79-7, EB71	524-42-5, ED22	711-01-3, EE26
142-82-5, EC51	527-17-3, ED39	719-22-2, EE74
142-83-6, EB64	528-29-0, EB12	768-90-1, ED46
144-49-0, EA11	529-34-0, ED34	768-91-2, ED76
144-62-7, EA08	532-54-7, EC64	768-92-3, ED48
148-18-5, EA95	533-49-3, EA70	768-95-6, ED52
148-24-3, EC94	535-15-9, EA53	770-71-8, ED77
149-30-4, EC11	544-76-3, EE96	778-10-0, EE61
150-76-5, EC31	550-44-7, EC96	814-78-8, EA88
150-78-7, EC78	553-97-9, EC15	822-38-8, EA25
185-79-5, EB79	554-84-7, EB32	826-73-3, ED71
208-96-8, ED80	555-03-3, EC23	842-07-9, EE91
217-68-5, EE87	555-16-8, EC04	871-58-9, EA94
225-51-4, EE97	571-60-8, ED28	877-13-4, EB03
230-27-3, EE38	581-89-5, ED25	882-33-7, EE17
260-94-6, EE35	583-50-6, EA69	917-44-2, EF26
281-23-2, ED49	586-96-9, EB28	928-73-4, EA30
289-80-5, EA45	587-04-2, EB99	930-69-8, EB37
289-95-2, EA46	587-45-1, ED07	930-88-1, EA81

TABLE VI. Chemical Abstracts Service Registry Numbers

934-72-5, EC76	2068-77-1, EF28	4901-75-1, EA76
934-73-6, EC20	2092-65-1, ED36	5022-29-7, ED30
935-56-8, ED47	2128-93-0, EF06	5323-50-2, ED69
941-69-5, ED26	2183-29-1, EA60	5392-40-5, ED53
945-51-7, EE13	2226-86-0, ED09	5522-43-0, EE86
947-91-1, EE67	2246-46-0, EC98	5551-23-5, EA98
954-46-1, EE64	2362-57-4, EE01	6217-61-4, EA84
957-75-5, ED05	2382-64-1, EF03	6221-74-5, ED78
959-28-4, EE94	2382-65-2, EF13	6319-45-5, EC28
987-65-5, ED43	2382-66-3, EF12	6325-93-5, EB44
1072-72-6, EA90	2391-46-0, EF17	6410-09-9, EE89
1073-93-4, EC39	2432-26-0, EB18	6410-10-2, EE90
1087-09-8, EE88	2432-27-1, EB19	6456-44-6, EC34
1115-11-3, EA87	2491-52-3, ED96	6554-00-3, EF19
1123-64-4, EC01	2501-02-2, EE66	6640-54-6, EE45
1141-06-6, EE07	2533-67-7, EA10	6742-34-3, EB45
1141-59-9, ED65	2536-99-4, EF04	6759-49-5, ED63
1144-81-6, EE08	2580-79-2, EA21	6943-90-4, EE19
1146-39-0, ED95	2667-20-1, FA12	6996-92-5, EB48
1147-56-4, EE39	2704-30-5, EA63	7147-90-2, EC53
1161-45-1, EF01	2720-77-6, EC42	7227-91-0, EC83
1193-82-4, EC30	2817-14-3, EB21	7496-02-8, EF00
1194-02-1, EB98	2893-33-6, EB93	7687-22-1, ED62
1195-58-0, EB95	2973-09-3, EC81	10200-47-2, EA96
1435-60-5, ED97	3051-84-1, EF10	10335-29-2, ED61
1453-82-3, EB38	3169-69-5, EE50	10558-42-6, EE60
1464-43-3, EB81	3256-24-4, EF09	11077-17-1, ED31
1533-65-9, EE78	3265-29-0, EC57	11078-84-5, EF21
1631-25-0, ED40	3352-23-6, EF18	12427-42-8, ED32
1633-44-9, EB94	3454-66-8, EA75	13152-82-4, EE33
1636-34-6, EE95	3517-99-5, EC79	13242-44-9, EA77
1664-98-8, EA01	3561-67-9, FF55	13309-06-3, EC41
1689-82-3, EE02	3651-02-3, EE92	13309-07-4, EE42
1741-01-1, EA39	3822-99-9, EE71	13328-57-9, EE43
1798-13-6, EE29	3947-46-4, EB20	13344-54-2, EE40
1823-45-6, EC97	4042-12-9, EF27	14287-89-9, EC50
1829-81-8, EE11	4329-75-3, EE12	14287-92-4, EC89
1962-02-3, ED29	4468-11-5, ED58	14337-52-1, EE85
2041-19-2, EE32	4592-97-6, EB89	14337-53-2, EE82
2050-14-8, EE03	4743-82-2, EA85	14493-15-3, EE84
2050-15-9, EE05	4833-63-0, EF11	14923-66-1, EB77
2058-74-4, EC95	4866-98-2, EE93	

15837-41-9, EA47
16169-16-7, EB43
16219-95-7, EE79
16247-81-7, EE59
16534-78-4, EC18
16566-56-6, ED66
17075-79-5, EE27
17091-08-6, ED59
17132-78-4, EB90
17208-01-4, ED75
17459-03-9, EC75
17539-61-6, ED17
18220-83-2, EE25
18332-84-8, EC25
18802-37-4, EA29
19026-73-4, EE28
19090-03-0, EA89
19780-25-7, EB69
20268-51-3, EE99
20292-75-5, EC65
20701-80-8, EC47
20730-07-8, EB92
20753-52-0, EC33
20753-53-1, EB27
20815-56-9, ED67
20815-66-1, ED64
20825-08-5, EB24
20912-17-8, EE49
21104-13-2, EE58
21205-25-4, EA37
21205-26-5, EA73
21205-27-6, EA72
21675-25-2, EE98
21746-40-7, EC40
22059-22-9, EA20
22072-22-6, EB73
22504-50-3, EC88
22842-41-7, ED16
23074-42-2, ED73
24719-68-4, EB66
25057-68-5, EB72
25144-43-8, EA40

25369-27-1, ED70
26015-51-0, EE83
27255-88-5, EA97
27255-91-0, EA52
28898-50-2, ED18
29181-50-8, EB91
29943-42-8, EA91
30013-72-0, EE34
31515-43-2, ED94
31781-74-5, EB25
31909-31-6, EF24
31909-36-1, EF22
31909-48-5, EC85
33350-63-9, EB26
33350-65-1, EB23
33350-70-8, EB35
33439-12-2, EB34
33667-99-1, ED83
34529-57-2, EE23
35140-51-3, EE80
35551-31-6, EA83
36051-68-0, ED13
36165-01-2, EB74
36165-02-3, EC48
36165-03-4, ED15
36804-63-4, EE31
38481-84-4, ED14
40505-10-0, EF08
41322-56-9, ED19
41511-88-0, EB75
41634-03-1, EC12
41736-65-6, EB52
42085-93-8, ED85
42485-44-9, EF07
42842-99-9, ED56
43008-09-9, EC49
43008-10-2, ED55
49715-78-8, EA42
49844-94-2, EB62
50783-82-9, EE56
51036-16-9, EB46
51159-91-2, EA79

51210-86-7, EE69
51210-87-8, EE70
51348-29-9, ED35
51348-31-3, EE24
51348-33-5, EA93
51606-61-2, EF15
51790-22-8, ED68
51823-96-2, EA92
51823-98-4, EB76
51850-04-5, EC52
51850-05-6, EA78
52236-30-3, EC43
52641-57-3, ED60
53120-15-3, ED10
53135-24-3, EC73
53400-11-6, EB67
53554-30-6, EC41
55654-09-6, ED02
55846-35-0, EA74
56001-37-7, ED44
56125-79-2, EA26
57434-97-6, ED21
57437-80-6, ED98
58481-06-4, EE20

TABLE VII.
INDEX OF SOLVENTS EMPLOYED

This table is an index that lists the solvents (except pure water) and mixtures of solvents that appear in Table I. It enables the user to identify all of the entries in that table that contain information pertaining to any particular solvent or mixture of solvents.

The entry

Propylene carbonate, EA31

signifies that data are given in Table I for the compound designated by code number EA31 in nominally pure propylene carbonate as the solvent. We have considered a solvent to be nominally pure whenever there was no deliberate addition of a second solvent or whenever the stated purity of the solvent was at least 99%. This criterion does not enable the user of this table to distinguish between entries for, say, reagent-grade acetonitrile and those for acetonitrile that had been rigorously purified to remove as much as possible of the water and other proton donors.

The entry

Dimethylformamide-ethanol(20:80), EC20

provides references to data obtained in binary mixtures containing 20% of dimethylformamide and 80% of ethanol by volume. Binary (and other) mixtures are listed in the alphabetical order of the names of their constituents, so that mixtures of benzene with benzonitrile precede those of benzene with ethanol. Binary mixtures containing the same constituents in different proportions are listed in the order of decreasing percentage, which is always by volume unless otherwise stated, of the constituent whose name is given first. Thus ethanol-water (90:10) precedes ethanol-water (80:20). Occasionally in this series we decided to include data even though the information given in the original literature left us more or less uncertain about the proportions of the components of a mixture. These cases are represented by entries of the form

Acetone-toluene-water(?:?:?), DF96

which denotes a mixture of acetone, toluene, and water in unstated proportions.

The citations for binary mixtures are given only once, in the entry for the constituent whose name occurs first in the alphabet. A cross-reference is always given under the name of the other constituent and in the following form:

Ethanol
-dioxane, see Dioxane-ethanol

Ternary mixtures are treated similarly. For a mixture of acetonitrile, sulfuric acid, and water the citation is given in the entry for acetonitrile, and in that entry it appears as part of the subentry for mixtures of acetonitrile with sulfuric acid, so that the three constituents are named in alphabetical order:

Acetonitrile
-sulfuric acid
-water(63:5:32), EC53,EC56,EC96,ED30,ED69,EE65,EE71

All ternary mixtures are fully cross-referenced so that they can be located in the entries for all of their constituents.

TABLE VII. Index of Solvents Employed

Acetone, EA12, EA28, EA62, EA94, EC42
Acetonitrile, EA13, EA22, EA25, EA26, EA31, EA32, EA39, EA60, EA64, EA65, EA66,
            EA78, EA79, EA92, EA93, EA95, EB14, EB25, EB28, EB47, EB50, EB51,
            EB52, EB53, EB61, EB68, EB76, EB88, EC11, EC20, EC29, EC30, EC49,
            EC50, EC52, EC76, EC77, EC79, EC89, EC90, ED10, ED19, ED27, ED32,
            ED35, ED46, ED47, ED48, ED49, ED52, ED55, ED56, ED70, ED73, ED74,
            ED76, ED77, ED78, ED89, EE00, EE13, EE16, EE17, EE18, EE24, EE25,
            EE26, EE28, EE35, EE36, EE37, EE38, EE55, EE57, EE61, EE87, EE96,
            EF05, EF08, EF14, EF21, EF22, EF23, EF24
    -water (66.7:33.3), EC53, EC56, EC96, ED30, ED69, EE65, EE71
        (30 :70), EC04, EC60
        (25 :75), EA77
        (20 :80), EC04, EC58, EC60
        (10 :90), EC03, EC04, EC58, EC60
    -sulfuric acid-water (63: 5:32), EC53, EC56, EC96, ED30, ED69, EE65, EE71
Aluminum trichloride-sodium chloride (63:37), EB66, ED90, EF02
                                      (50:50), EC80, ED11

Deuterium oxide, EA01
Dichloromethane, EA31, EA64, EA65, EB68, EB88, EC77, EC90
Dimethoxyethane, ED33, EE15
    -water (97: 3), EE15
Dimethylformamide, EA05, EA06, EA07, EA09, EA11, EA50, EA51, EA53, EA54, EA59,
                EA61, EB12, EB13, EB14, EB22, EB28, EB29, EB44, EB96, EB97,
                EB98, EC02, EC20, EC28, EC54, EC55, EC56, EC59, EC63, EC75,
                EC76, EC79, EC95, ED10, ED22, ED23, ED24, ED25, ED74, ED79,
                ED91, ED92, EE00, EE15, EE31, EE32, EE62, EE63, EE64, EE66,
                EE75, EE76, EE86, EE88, EE94, EE98, EE99, EF00, EF15, EF16,
                EF25
    -water (97: 3), EE15
        (90:10), ED72
        (60:40), ED50
        (20:80), EE69, EE70
    -ethanol (20:80), EC20, EC30, EC76, EC79, ED10, ED74
Dimethylsulfoxide, EB30, EB32, EB33, EB37, EB49, EC22, EC23, EC24, EE15, EE17
    -water (97: 3), EE15
Dioxane-water (75:25), EE40, EE41, EE42, EE43
    -ethanol (20:80), EC20, EC30, EC76, EC79, ED10, ED74

Ethanol, EA68, EC20, EC76, EC79, EC83, ED10, ED29, ED71, ED74, EE72
    -water (90:10), EC44, EC86
        (80:20), EE89, EE90, EF01
        (75:25), EB72, EB73, EB74, EB75, EC44, EC47, EC48, EC86, ED14, ED15,
                 ED16, ED18
        (55:45), EB29
        (50:50), EA67, EA77, EA81, EB01, EB71, EC39, EC40, EC81, EC97, EC98,
                 ED26, ED28, ED40, ED59, ED60, ED61, ED62, ED63, ED64, ED65,
                 ED66, ED67, ED68, ED75, ED80, ED81, ED82, ED83, ED84, ED85,
                 ED86, ED87, ED88, ED93, ED94, ED95, ED96, ED97, ED99, EE01,
                 EE02, EE03, EE04, EE05, EE06, EE07, EE08, EE09, EE10, EE19,
                 EE22, EE23, EE27, EE33, EE34, EE39, EE40, EE41, EE42, EE43,
                 EE45, EE46, EE47, EE48, EE49, EE50, EE51, EE52, EE53, EE54,
                 EE56, EE59, EE60, EE77, EE78, EE79, EE80, EE81, EE82, EE83,
                 EE84, EE85, EE90, EE91, EE92, EE93, EF07
        (50 w/w:50), ED99
        (25 :75), EA10, EA20, EA21, ED02
        (20 :80), EB28
        (15 :85), EC66
        (10 :90), EA67, EB99, EC00, EC03, EC04, EC14, EC16, EC58, EC59,
                 EC60, EC61, EC62, EC67, EC68, EC70, EC88, ED06

(5:95), EA16, EA68, EB82, ED04
(2:98), EA00, EA16, EA68, EB82, EC66, ED00, ED03, ED04, ED34, ED71, EE67
Ethanol-dimethylformamide, see Dimethylformamide-ethanol
Ethanol-dioxane, see Dioxane-ethanol
Ethyl nitrate, EA31, EA64, EA65, EB68, EC90

Hexamethylphosphoramide, EB29
Hydrofluoric acid, EB09, EB47, ED20, ED27, EE14

Methanol, EA88, EB15, EB71
    -water (50:50), EA77, EB55
        (25:75), EA58, EA88, EB71
        (20:80), EA76, EB20
        (10:90), EB15, EB29, EB31, EB32, EB33, EB40, EB41, EB42, EB43, EB54, EB60, EC24, EE11
N-Methylpyrrolidone, EC00, EC91, EC93, EC94, EC99, ED99

Nitromethane, EA31, EA64, EA65, EB68, EB88, EC78, EC90, ED27

α-Picoline, ED31
γ-Picoline, ED31
2-Propanol-water (40:60), EC15, EC69, ED39, EE74
Propylene carbonate, EA31, EA64, EA65, EB68
Pyridine, ED31
Pyrrolidone, ED33

Sodium chloride-aluminum chloride, see Aluminum chloride-sodium chloride
Sulfolane, EA31, EA64, EA65, EB68
Sulfuric acid-acetonitrile-water, see Acetonitrile-sulfuric acid-water

Tetrahydrofuran, ED33
Trifluoroacetic acid, EA99, EB78, EB87, EC51, EC90, ED57, EE96, EF20

Water (see introduction to this table)
    -acetonitrile, see Acetonitrile-water
        -sulfuric acid, see Acetonitrile-sulfuric acid-water
    -dimethoxyethane, see Dimethoxyethane-water
    -dimethylformamide, see Dimethylformamide-water
    -dimethylsulfoxide, see Dimethylsulfoxide-water
    -dioxane, see Dioxane-water
    -ethanol, see Ethanol-water
    -methanol, see Methanol-water
    -2-propanol, see 2-Propanol-water
    -sulfuric acid-acetonitrile, see Acetonitrile-sulfuric acid-water

TABLE VIII.
INDEX OF TECHNIQUES EMPLOYED

This is an index that lists the techniques employed in obtaining the data that appear in Table I. It enables the user to identify all of the entries in that table that contain information obtained by any particular technique.

In the typical entry

$$\text{Chronoamperometry(IR), EB28, EB44, EB96, EB97, EB98, EC28, EC29, EE15}$$

the name is the one recommended by the International Union of Pure and Applied Chemistry (*Pure and Applied Chemistry*, 45, 81, 1976). Cross-references are provided liberally, so that, for example, the entries for polarographic coulometry can be found by users who seek them under "dropping-electrode coulometry", "millicoulometry", or "microcoulometry", and these cross-references also serve to notify the user that the name followed by the citations is recommended in preference to one cross-referenced to it.

Following each recommended name there appears, in parentheses, the two-letter symbol by which the technique is denoted in Column 6 (or, occasionally, in Columns 21 or 26) of Table I. This enables the user to tell, without having to consult the list of abbreviations repeatedly, what part of the information contained in a long entry in Table I has been obtained by the technique in which he is interested.

Finally, for every technique except polarography itself, there appear in alphanumeric order the code numbers for all of the compounds about which information obtained by that technique appears in Table I. There are many entries of the form

$$\text{Alternating-current chronopotentiometry(EF)}$$

in which the name of a technique is followed by a two-letter symbol that identifies it as a recommended name, but in which no code numbers follow the symbol. Such an entry signifies that the technique is included in our sphere of activity but has produced no information that is given in this volume. At the other extreme, the entry for polarography is of this same form but for a very different reason: as was stated previously, a list of the code numbers of the compounds for which polarographic data are given would be too long to be of any use. We estimate that it would have contained about 350 citations. If this figure is compared with the total of 295 citations given below, it becomes apparent that, during the period covered by these volumes, the contribution made by polarography to our fundamental knowledge of the electrochemical behaviors of organic, biochemical, and organometallic substances appreciably exceeded that of all the other techniques combined.

TABLE VIII. Index of Techniques Employed

Ac polarography (PV), EA83, EA94, EB18, EC26, EC27, ED31, ED42, EE30, EF03, EF04, EF09, EF10, EF11, EF12, EF13, EF17, EF18, EF19, EF27, EF28
    higher-harmonic, see Higher-harmonic ac polarography
    in-phase (PVI), EA78, EA79, EA82, EB71, EC52, ED32, ED41, ED45, EF09, EF10, EF11, EF12, EF17, EF21
Ac voltammetry (VV), ED89
Alternating-current chronopotentiometry (EF)
Alternating-voltage chronopotentiometry (EK)
Amperometry (IO)
    differential (ID)
    with two indicator electrodes (IB)
Av polarography (PI)

Cathode-ray polarography, see Single-sweep polarography
Chronoamperometry (IR), EB28, EB44, EB96, EB97, EB98, EC28, EC29, EE15
    convective, see Convective chronoamperometry
    double potential-step, see Double potential-step chronoamperometry
    dropping-electrode, with linear potential sweep, see Single-sweep polarography
    polarographic, see Polarographic chronoamperometry
    rotating-disc-electrode, see Convective chronoamperometry
    stirred-pool-electrode, see Convective chronoamperometry
    thin-layer (TI)
    with linear potential sweep, (IL), EA08, EA35, EA47, EA99, EB14, EB18, EB19, EB20, EB21, EB29, EB50, EB56, EB57, EB58, EB65, EB78, EB87, EC36, EC37, EC38, EC51, EC58, EC90, ED46, ED47, ED48, ED49, ED52, ED57, ED73, ED76, ED77, ED78, ED98, EE13, EE25, EE26, EE28, EE35, EE36, EE37, EE38, EE55, EE61, EE66, EE96, EE97, EF04, EF05, EF08, EF09, EF10, EF11, EF12, EF17, EF18, EF19, EF20, EF27, EF28
        second-derivative (XT)
        semi-integral (XT), EB12, EB13, EB14
    with non-linear potential sweep (IM)
Chronocoulometry (QR)
    convective, see Convective chronocoulometry
    double potential-step, see Double potential-step chronocoulometry
    potential-step, see Chronocoulometry
    rotating-disc-electrode, see Convective chronocoulometry
    stirred-pool-electrode, see Convective chronocoulometry
    thin-layer (TQ)
Chronopotentiometry (ER), EA19
    alternating-current (EF)
    alternating-voltage (EK)
    current-cessation, see Current-step chronopotentiometry
    current-reversal, see Current-step chronopotentiometry
    current-step, see Current-step chronopotentiometry
    cyclic, see Cyclic chronopotentiometry
    derivative, see Derivative chronopotentiometry
    programmed-current (EC)
    with linear current sweep (EL)
    with superimposed ac (EV)
Controlled-current coulometry
    with a reagent precursor, see Coulometric titration
    without a reagent precursor, see Coulometry, controlled-current (without a reagent precursor)
Controlled-potential coulometry (QE), EC32, ED49
Controlled-potential electrolysis (CP), EA13, EA25, EA66, EB84, EB85, EB86, EC11, EC29, ED01
Convective chronoamperometry (CA)
Convective chronocoulometry (CC)
Convective triangular-wave voltammetry (CT), EA32, EC29, EC77
Coulometric titration (QT)
    controlled potential, see Controlled-potential coulometry

Coulometry
    controlled-current
        with a reagent precursor, see Coulometric titration
        without a reagent precursor (QP)
    controlled-potential, see Controlled-potential coulometry
    dropping-electrode, see Polarographic coulometry
    polarographic, see Polarographic coulometry
Current-cessation chronopotentiometry, see Current-step chronopotentiometry
Current-reversal chronopotentiometry, see Current-step chronopotentiometry
Current-scanning polarography (PC)
Current-step chronopotentiometry (EE), EA19, EC31, EC56
Cyclic chronopotentiometry (EY)
Cyclic triangular-wave polarography (PR), EB80
Cyclic triangular-wave voltammetry (VR), EA66, EB18, EB19, EB22, EB28, EB44, EB45, EB46, EB47, EB51, EB52, EB53, EB56, EB61, EB96, EB97, EB98, EC02, EC17, EC31, EC36, EC37, EC49, EC50, EC80, EC89, ED07, ED12, ED19, ED21, ED23, ED25, ED55, ED56, ED70, EE00, EE13, EE17, EE35, EE62, EE63, EE88, EE94, EE98, EF15, EF16, EF24, EF25

Dc polarography, see Polarography
Demodulation polarography (FP)
Derivative chronopotentiometry (EM)
Derivative polarography (PD)
Derivative pulse polarography (DP)
Derivative voltammetry (VD)
Differential amperometry, see Amperometry, differential
Differential linear-sweep polarography, see Linear-sweep differential polarography
Differential polarography (PF)
Differential pulse polarography (DI), EA95
Differential voltammetry (VF)
Double potential-step chronoamperometry (IU), EB96
Double potential-step chronocoulometry (QU), ED99
Double-tone polarography (PX)
Dropping-electrode coulometry, see Polarographic coulometry
Dynamic capacity, measurement of (NF)

Electrochemiluminescence (HN)
Electrogravimetry (QW)

Faradaic rectification, high-level (HL)

High-level faradaic rectification, see Faradaic rectification, high-level
Higher-harmonic ac polarography (PH)
    with phase-sensitive rectification (PB)
Hydrodynamic voltammetry (VY), EA00, EA04, EA22, EA31, EA39, EA46, EA64, EA65, EA95, EB09, EB23, EB24, EB26, EB27, EB34, EB35, EB48, EB50, EB53, EB56, EB57, EB58, EB65, EB68, EB88, EC00, EC33, EC35, EE36, EC37, EC38, EC53, EC56, EC77, EC78, EC80, EC90, EC91, EC93, EC94, EC96, EC99, ED07, ED27, ED30, ED33, ED69, ED99, EE18, EE57, EE58, EE65, EE71, EE73, EF22, EF23, EF24

Incremental-charge polarography (DQ)
Intermodulation polarography (PL)

Kalousek polarography (PK)

TABLE VIII. Index of Techniques Employed

Linear-sweep differential polarography (XT)
Linear-sweep voltammetry, see Chronoamperometry with linear potential sweep

Microcoulometry, see Polarographic coulometry
Millicoulometry, see Polarographic coulometry
Modulation polarography (MM)
Multisweep polarography (PM)
Multisweep voltammetry (VM)

Non-faradaic admittance, see Dynamic capacity

Oscillographic polarography, see Oscillopolarography
    single-sweep, see Single-sweep polarography
Oscillopolarography (PO), EA23, EA29, EA48, EA49, EA55, EA80, EA82, EA84, EC21,
                EC84, ED05, ED88, EE68, EE95
Oscillovoltammetry (VO)

Polarographic chronoamperometry (IV)
Polarographic coulometry (PQ)
Polarography (PY)
    ac, see Ac polarography
    av, see Av polarography
    cathode-ray, see Single-sweep polarography
    current scanning, see Current-scanning polarography
    cyclic triangular wave, see Cyclic triangular-wave polarography
    demodulation, see Demodulation polarography
    derivative, see Derivative polarography
    derivative pulse, see Derivative pulse polarography
    differential, see Differential polarography
    differential pulse, see Differential pulse polarography
    double-tone, see Double-tone polarography
    higher-harmonic ac, see Higher-harmonic ac polarography
    incremental-charge, see Incremental-charge polarography
    in-phase ac, see Ac polarography, in-phase
    intermodulation, see Intermodulation polarography
    Kalousek, see Kalousek polarography
    linear-sweep differential, see Linear-sweep differential polarography
    modulation, see Modulation polarography
    multisweep, see Multisweep polarography
    oscillographic, see Oscillopolarography
    pulse, see Pulse polarography
    rf, see Rf polarography
    single-sweep, see Single-sweep polarography
    square-wave, see Square-wave polarography
    staircase, see Staircase polarography
    Tast, see Tast polarography
    triangular-wave, see Triangular-wave polarography
        cyclic, see Cyclic triangular-wave polarography
Programmed-current chronopotentiometry (EC)
Pulse polarography (PP), EB83
Pulse voltammetry (XT)

Rf polarography (RP)

Single-sweep polarography (PW), EA33, EB71, EB80, EC13, ED99, EE44, EF06
Square-wave polarography (PG)
Staircase polarography (PS)

Stirred-pool-electrode chronoamperometry, <u>see</u> Convective chronoamperometry

Tast polarography (PT)
Thin-layer chronoamperometry, <u>see</u> Chronoamperometry, thin-layer
Thin-layer chronocoulometry, <u>see</u> Chronocoulometry, thin-layer
Titration, controlled-potential coulometric, <u>see</u> Controlled-potential coulometry
    coulometric, <u>see</u> Coulometric titration
Triangular wave polarography (PA), EA00, EA01, EB80, EC95, ED99
    cyclic, <u>see</u> Cyclic triangular-wave polarography
Triangular-wave voltammetry (VA), EA00, EA02, EA04, EA22, EA26, EA28, EA45, EA60, EA75, EA83, EA92, EA93, EB18, EB22, EB28, EB33, EB44, EB47, EB62, EB66, EB76, EB78, EC18, EC19, EC34, EC56, EC75, EC80, EC82, ED07, ED11, ED20, ED33, ED35, ED42, ED51, ED70, ED72, ED89, ED90, ED98, EE00, EE14, EE15, EE16, EE17, EE18, EE21, EE24, EE31, EE32, EE34, EE57, EE75, EE76, EE87, EF02, EF14
    convective, <u>see</u> Convective triangular-wave voltammetry
    cyclic, see Cyclic triangular-wave voltammetry

Voltammetry (VY)
    ac, <u>see</u> Ac voltammetry
    cyclic triangular-wave, see Cyclic triangular-wave voltammetry
    derivative, <u>see</u> Derivative voltammetry
    differential, <u>see</u> Differential voltammetry
    hydrodynamic, <u>see</u> Hydrodynamic voltammetry
    linear-sweep, <u>see</u> Chronoamperometry with linear potential sweep
    multisweep, <u>see</u> Multisweep voltammetry
    oscillographic, <u>see</u> Oscillovoltammetry
    pulse, <u>see</u> Pulse voltammetry
    triangular-wave, <u>see</u> Triangular-wave voltammetry
        convective, <u>see</u> Convective triangular-wave voltammetry
        cyclic, <u>see</u> Cyclic triangular-wave voltammetry

TABLE IX.
INDEX OF INDICATOR ELECTRODES EMPLOYED

This is an index that lists the indicator and working electrodes employed in obtaining the data that appear in Table I. It enables the user to identify all of the entries in that table that contain information obtained with any particular indicator or working electrode.

In the typical entry

Carbon, glassy, disc, rotating (RDGC), EA46,EE18,EE58

the initial portion identifies the electrode used and gives its configuration whenever possible and, in parentheses, the abbreviation used to denote the electrode in Column 11 of Table I. This enables the user to tell, without having to consult the list of abbreviations repeatedly, what part of the information contained in a long entry in Table I has obtained with the electrode and configuration in which he is interested. Following the abbreviation, there appear the code numbers of the compounds that have been studied with the electrode. Finally, liberal cross-references are provided to ensure that no entry that might be of interest will be overlooked.

Precious though space in the original literature certainly is, it does not seem to us to be valuable enough to justify the frequency with which sentences like "The graphite indicator electrode was described previously" are used by authors meticulous in specifying the other circumstances under which their data were obtained. As it is well known that electrochemical properties depend on the configuration of the indicator electrode as well as on the material from which it is contructed, our original intention was to provide a complete index of both configurations and materials. A few cases in which the reference number appended to a sentence like the one quoted led only to an exactly identical sentence in another article convinced us, however, that this intention was unlikely to be achievable, and we have therefore confined ourselves to transcribing the information provided. We recognize that this complicates the use of this table, even with the cross-references provided, and urge the authors of future papers to describe their electrodes as carefully as they do the other facets of their work.

There is one null entry in this table, and that is the one for the dropping mercury electrode. This reflects the same fact that led to the omission of polarography from Table VIII. There are some compounds in Table I for which no information obtained with a dropping mercury electrode is given in this volume. The great majority of these are compounds that undergo oxidation at relatively positive potentials and that were therefore studied with platinum, graphite, carbon-paste, or other indicator electrodes. Many such compounds are not amenable to investigation with dropping mercury electrodes, but there are many others that are. Some of these can also be reduced within the range of potentials accessible with a dropping mercury electrode while some undergo oxidation within that range under certain conditions but not under others; for some there may be polarographic data in the prior literature, while for others the possibility of obtaining a polarographic response may not have been investigated. In any event, an entry documenting the use of the dropping mercury electrode in obtaining the data included in this volume would have contained 369 entries and would have been too long to be of any practical use.

TABLE IX. Index Indicator Electrodes Employed

Bead electrodes, <u>see</u> Rhodium; Tungsten
Bead electrodes, spherical, <u>see</u> Platinum
Button electrodes, <u>see</u> Tungsten

Carbon (C), EC32; <u>see also</u> Graphite
    disc, rotating, (C RDE), EC56
    glassy (GCE), EB28, EB29, EB47, EC34, ED20, EE14, EE18
        disc, rotating (RDGC), EA46, ED27, EE18, EE58
    paste (CPE), EB50, EB56, EB57, EB58, EC17, EC18, EC19, EC31, EC36, EC37, EC38,
        EC82, ED12, EE21
        1-bromonaphthalene, EB47
        silicone fluid (MS200), EB53
Copper (Cu), EA35, EA75, EB84, EB85, EB86
    disc, mercury-coated (MPCuD), EC56
        rotating (RMCud), EA00, EC56
Copper sulfide (CuS), EA28

Disc electrodes, <u>see</u> Mercury-coated copper; Platinum
Disc electrodes, rotating, <u>see</u> Carbon; Carbon, glassy; Copper; Copper, Mercury-
            coated; Gold; Gold, mercury-coated; Lead; Platinum;
            Platinum, mercury-coated; Tin
Displacement of solution, <u>see</u> Platinum with periodic displacement of solution
Drop electrodes, see Mercury, hanging drop

Gauze electrodes, see Platinum
Gold (Au), EA04, EB22
    disc, rotating (RGDE), EE58
        mercury-coated, <u>see</u> Mercury-coated gold, rotating disc
    mercury-coated (MPAu), EB22
    rotating (RauE), EA04
Graphite (GE), ED07; <u>see also</u> Carbon
    pyrolytic (PGE), EA08, EA47, EB18, EB19, EB20, EB21, EB45, EB46

Iridium (Ir), EB22

Lead disc, rotating (Pb RDE), EC53, EC55, EC96, ED30, ED69, EE65, EE71

Mercury (Hg), EA66
    -coated copper
        disc (MPCuD), EC56
            rotating (RMCuD), EA00, EC56
    -coated gold (MPAu), EB22
        disc, rotating (RMAuD), EC56
    -coated platinum (PMC), EA66
        disc, rotating (RMPD), EA32
    dropping (DME), <u>see</u> introduction to this table
    hanging drop (HMDE), EA33, EA45, EA83, EB12, EB13, EB14, EB28, EB62, EC58,
            ED23, ED25, ED42, ED51, ED70, ED72, EE00, EE15, EE34,
            EE63, EE66, EF04, EF09, EF10, EF11, EF12, EF17, EF18,
            EF19, EF27, EF28

Mercury (Hg)
    pool (MP), EA19, EC00, EC91, EC93, EC94, EC99
    streaming (SME), EA29, EA48, EA49, EA55, EA80, EA82, EA84, EC84, ED05

Palladium (Pd), EB22
    wire (PdWE), EA02
Periodic displacement of solution, see Platinum with periodic displacement of solution
Platinum (Pt), EA02, EA25, EA26, EA60, EA75, EA92, EA93, EA99, EB22, EB28, EB44, EB65, EB66, EB76, EB78, EB87, EB96, EB97, EB98, EC02, EC11, EC28, EC29, EC51, EC75, EC90, ED20, ED23, ED35, ED46, ED47, ED48, ED49, ED52, ED57, ED73, ED76, ED77, ED78, ED89, ED90, ED98, EE18, EE24, EE25, EE26, EE28, EE35, EE36, EE37, EE38, EE55, EE61, EE76, EE87, EE96, EE97, EF05, EF08, EF14, EF20
    disc (PDE), EB22, EB47, EB51, EB52, EB53, EC29, ED21, ED23, ED33, EE57, EE62, EE88, EE94, EE98, EF15, EF16, EF24, EF25
        rotating (RPDE), EA22, EA31, EA39, EA64, EA65, EA95, EB09, EB50, EB65, EB68, EB88, EC29, EC77, EC78, EC90, ED01, ED27, ED33, ED57, ED58, EF22, EF23, EF24
    gauze (Pt gauze), ED49
    rotating (RPE), ED07, EE18
    silver-amalgam-coated (XE1), EC56
    spherical bead (PBE), EA02, EB28, ED89, EE13, EE16, EE17, EE31, EE32
    thin-layer (Pt TLE), EB53
    with periodic displacement of solution (Pt PDS), EB23, EB24, EB26, EB27, EB34, EB35, EB48, EB56, EB57, EB58, EC33, EC35, EC36, EC37, EC38, EC80, ED98, EE73
Pool electrodes, see Mercury

Rhodium (Rh), EB22
    bead (RhBE), EA02
Rotating electrodes, see Gold; Platinum
Rotating disc electrodes, see Carbon; Carbon, glassy; Copper; Copper, mercury-coated; Gold; Gold, mercury-coated, Lead; Platinum; Platinum, mercury-coated; Tin

Silver amalgam, as coating on platinum (XE1), EC56
Spherical bead electrodes, see Platinum

Thin-layer electrodes, see Platinum, thin-layer
Tin (Sn), EA13
    disc, rotating (Sn RDE), EC56
Tungsten (W), EB66, ED90
    bead (WBE), EC80, ED11
    Button (W button), EF02

Wire electrodes, see Palladium

TABLE X.
KEY TO LITERATURE CITATIONS

This table is provided to permit decoding the literature references that appear in Column 14 of Table I and obtaining full citations from them.

The form and significance of those references are described in the introduction of Table I. To illustrate the use of this table we shall suppose that a full citation is wanted for the information given in this volume for dipotassium 1,4-naphthohydroquinone. In Table I this compound appears under code number ED21, and the literature reference is given as JA092-4139 in Column 14 on the first of the lines dealing with this compound.

As was discussed in the introduction to Table I, the letters "JA" denote the journal. The next three digits give the number of the volume, in this case 92. As that number contains only two digits, it is preceded by a zero (if the volume number contained only one digit, it would be preceded by two zeros). The last four digits give the page number. Here this is 4139. Occasionally one, two, or three zeros are needed to give the total of four digits. Hence the reference is to volume 92, page 4139, of the journal denoted by the two-letter code "JA".

To identify the journal that is denoted by this abbreviation, this table must be inspected. The journals are arranged in the alphabetical order of the two-letter codes assigned to them, and each such code is followed in parentheses by the full title of the journal to which it corresponds. It can be found that the two-letter code "JA" denotes the *Journal of the American Chemical Society*. Hence the reference is to the *Journal of the American Chemical Society*, volume 92, page 4139, and this table provides the further information that the paper was published in 1970.

Names of some journals published in other languages than English are given in transliteration of the original title. When data were obtained from English translations (e.g., of Russian journals) the English translation of the name of the journal is given.

Beneath the abbreviation and the name of the journal, volume numbers are given consecutively along with the corresponding years. Under each volume number and year there are references to all of the publications from which data were taken and are given in Table I. These references are given in exactly the same form as is used in Table I, and are followed by the names of the authors. The references to each volume of each journal are listed in the order of increasing page numbers.

For example, the reference "CO027-0525" is the third one which appears underneath volume 27(1962) of the *Collection of Czechoslovak Chemical Communications*. It gives the author's name as R. Zahradník. Consequently the full citation in the usual form is R. Zahradník, *Collection Czech. Chem. Communs.*, 27, 525 (1962).

Data for more than one compound were often obtained from a single publication. In such a case the code numbers of all the compounds are given in alphanumeric order. Usually these additional code numbers denote compounds that are closely related to the one that was originally of interest, or represent data obtained by similar techniques under similar conditions, and in either case provide cross-references that may further illuminate the information already obtained. It should not be inferred that these are complete lists of the compounds about which information is given in the original references, for some of these may have been omitted from this volume as having been studied in insufficient detail or for other reasons.

TABLE X. Key to Literature Citations

AS (ACTA CHEMICA SCANDANAVICA)

Vol. 17 (1963)
AS017-1077 Lund, H., EB38, EB59, EE12, EE20

CO (COLLECTION OF CZECHOSLOVAK CHEMICAL COMMUNICATIONS)

Vol. 25 (1960)
C0025-3292 Holubek, J., and Volke, J., ED88, EE07, EE10, EE22

Vol. 26 (1961)
C0026-0052 Bartušek, M., and Okáč, A., EA03, EA18, EA34
C0026-0370 Zhdanov, S.I., and Mirkin, L.S., ED29, EE72
C0026-1733 Zahradník, R., and Boček, K., EB29
C0026-2438 Mollin, J., and Kašpárek, F., EA10, EA20, EA21, ED02
C0026-2749 Němečková, A., Maturová M., Pergál, M., and Šantavý, F., EA81, EB55, EC39, EC40, EC81, ED26, ED40, ED60, EE19

Vol. 27 (1962)
C0027-0199 Komenda, J., and Laskafeld, D., EB16, EB17, EB36
C0027-0486 Manoušek, O., and Zuman, P., EC71, EC72
C0027-0525 Zahradník, R., ED81, ED82, ED83, ED84, ED85, ED86, ED87, ED93, ED94, ED95, EE08, EE45, EE46, EE47, EE48, EE49, EE50, EE51, EE52, EE53, EE54
C0027-0546 Krupička, J., and Gut, J., EA86, EB67, EC43
C0027-0693 Fišerova-Bergerová, V., EB00, ED54, ED58
C0027-1997 Stankoviansky, S., and Königstein, J., EE29
C0027-2717 Holeček, V., and Horák, V., EA85

Vol. 28 (1963)
C0028-0026 Heyrovský, M., EA27, EA33
C0028-0838 Kalvoda, R., and Budnikov, G., EC21, EE68, EE95
C0028-2163 Eliášek, J., and Jungwirt, A., EB01, EB02, EB03, EB04, EB05, EB06, EB07, EB08, EB22, EB89, EC01

Vol. 29 (1964)
C0029-0182 Humlová, A., EA23, EA29, EA48, EA49, EA55, EA80, EA82, EA84, EC84, ED05

Vol. 30 (1965)
C0030-2460 Kišová, L., Komenda, J., and Hlávatý, J., EC26, EC27
C0030-4024 Holleck, L., and Lehmann, O., EC92
C0030-4178 Kitaev, Yu. P., and Budnikov, G.K., EA76, EB79
C0030-4192 Berg, H., and Gollmick, F.A., EC15, EC69, ED39, EE74
C0030-4219 Warner, C.R., and Elving, P.J., EC00, EC67

Vol. 32 (1967)
C0032-2140 Petránek, J., Ryba, O., and Doskočilová, D., ED75, EE27

Vol. 34 (1969)
C0034-1413 Matrka, M., Zvěrina, V., Ságner, Z., and Marhold, J., EC83

Vol. 36 (1971)
C0036-0114 Fedoroňko, M., Fülleová, E., and Linek, K., EA69, EA70, EA71
C0036-0331 Fleury, D., and Fleury, M.B., EA30, EA37, EA38, EA52, EA63, EA72, EA73, EA74, EA96, EA97, EA98, EB77, EC85, ED08
C0036-0575 Sümmermann, W., and Bäumgartel, H., EF22, EF23, EF24
C0036-1644 Knobloch, P., EB47, EB51, EB52, EB53, EB61

JA (JOURNAL OF THE AMERICAN CHEMICAL SOCIETY)

Vol. 92 (1970)
JA092-4139 Breslow, R., Grubbs, R., and Murahashi, S., ED21, ED23, ED28

Vol. 94 (1972)
JA094-6812 Mayeda, F.A., Miller, L.L. and Wolf, J.F., EC29, EC77

JA (JOURNAL OF THE AMERICAN CHEMICAL SOCIETY)(cont.)

Vol. 95 (1973)
JA095-6033 Houser, K.J., Bartak, D.E., and Hawley, M.D., EB96, EB97, EB98, EC02
JA095-6454 Nelsen, S.F., and Landis, R.T., EC49, EC50, EC89, ED19, ED55, ED56
JA095-6688 Breslow, R., Murayama, D.R., Murahashi, S., and Grubbs, R., ED21, ED23
JA095-8495 Webb, J.W., Janík, B., and Elving, P.J., EA82, ED41, ED45, EF03, EF04, EF09, EF10, EF11, EF12, EF13, EF17, EF18, EF19, EF26, EF27, EF28
JA095-8631 Koch, V.R., and Miller, L.L., ED46, ED47, ED48, ED49, ED52, ED73, ED76, ED77, ED78, EE25, EE26, EE28, EE61

Vol. 96 (1974)
JA096-0249 Breslow, R., Murayama, D., Drury, R., and Sondheimer, F., EB22, ED23, EE62, EE88, EE94, EE98, EF15, EF25

JE (JOURNAL OF ELECTROANALYTICAL CHEMISTRY AND INTERFACIAL ELECTROCHEMISTRY)

Vol. 33 (1971)
JE033-0061 Krznarić, D., Ćosović, B., and Branica, M., EE30

Vol. 34 (1972)
JE034-001A Cauquis, G., and Serve, D., EE18
JE034-0081 Eisner, U., and Zemer, Y., EA04
JE034-0091 Hampson, N.A., and MacDonald, K.I., EA35, EB84, EB85, EB86
JE034-0283 Van Duyne, R.P., Ridgway, T.H., and Reilley, C.N., ED99
JE034-0439 Sinicki, C., Porteix, M., and Zandanel, A., ED33
JE034-0505 Brown, O.R., Chandra, S., and Harrison, J.A., EC77
JE034-0521 Aylward, G.H., Watton, E.C., Buchanan, G.S., and Lee, R.W., ED31
JE034-0543 Cardinali, M.E., Carelli, I., and Trazza, A., EC57

Vol. 35 (1972)
JE035-0219 Sohr, H., and Wienhold, K., ED13, ED36, ED37, ED38, ED41, ED43, ED44, ED45
JE035-0363 Remane, H., Herzschuh, R., Hoa, L.C., and Borsdorf, R., EB72, EB73, EB74, EB75, EC44, EC47, EC48, EC86, ED14, ED15, ED16, ED18
JE035-0369 Holleck, L., and Kazenifard, G., EE23
JE035-0381 Holleck, L., and Mahapatra, S., EA58, EA88, EB71

Vol. 36 (1972)
JE036-0109 Clarke, S., and Harrison, J.A., EA00
JE036-0117 Zecchin, S., and Pilloni, G., ED98
JE036-0137 Clark, D., Fleischmann, M., and Pletcher, D., EA31
JE036-0157 Miller, I.R., and Teva, J., EB80
JE036-0223 Heyrovský, M., and Vavřička, S., EB29, EB31, EB32, EB33, EB40, EB41, EB42, EB43, EB60, EC24, EE11
JE036-0249 Mittal, M.L., and Pandey, A.V., EA77
JE036-0383 Volke, J., and Skála, V., EB10, EB11, EB90, EB91, EB92, EB93, EB94, EB95
JE036-0389 Magno, F., and Bontempelli, G., EE16
JE036-0479 Fonds, A.W., and Los, J.M., EB83
JE036-0515 Saxena, R.S., and Chaturvedi, U.S., EC88

Vol. 38 (1972)
JE038-0127 Farsang, G., Kovács, M., Saber, T.M.H., and Ladányi, L., EE57
JE038-0185 Brown, O.R., Chandra, S., and Harrison, J.A., EC29
JE038-0191 Kirowa-Eisner, E., and Gileadi, E., EA17
JE038-0245 Chambers, J.A., Moses, P.R., Shelton, R.N., and Coffen, D.L., EA25, EC11
JE038-0381 Eisner, U., and Zemer, Y. AE04
JE038-0389 Papouchado, L., Petrie, G., and Adams, R.N., EC17, EC18, EC19, EC32
JE038-0403 Annino, R., Boczkowski, R.J., Bolton, D.J., Geiger, W.E., Jr., Jackson, D.T., Jr., and Mahler, J., EA40, EA41, EA43, EA44

JE (JOURNAL OF ELECTROANALYTICAL CHEMISTRY AND INTERFACIAL ELECTROCHEMISTRY) (cont.)

Vol. 38 (1972) (cont.)
JE038-0476 Jones, H.L., Baxall, L.G., and Osteryoung, R.A., EC80, EF02

Vol. 39 (1972)
JE039-0195 Herzschuh, R., Hoa, L.C., Borsdorf, R., Remane, H., and Mühlstädt, M., EA89, EA90, EA91, EB72, EB73, EC44, EC45, EC46, EC47, EC86, ED17

JE039-0385 Bauer, D., and Foucault, A., EC78, ED27

JE039-0395 Krygowski, T.M., Stencel, M., and Galus, Z., EB29, ED24, ED25, ED91, ED92, EE63, EE64, EE86, EE99, EF00

JE039-0419 Butkiewicz, K., EE40, EE41, EE42, EE43

JE039-0447 Thévenot, D., and Buvet, R., EC34

Vol. 40 (1972)
JE040-006A Cauquis, G., Chabaud, B., and Genies, M., EA22, EA39

JE040-0013 Březina, M., Koryta, J., Loučka, T., Maršíková, D., and Pradáč, J., EB65

JE040-0069 Caillet, A., and Demange-Guerin, G., ED33

JE040-0197 Thévenot, D., and Buvet, R., EB39

JE040-0295 King, D.M., Kolby, N.I., and Price, J.W., EA19

JE040-0339 Venturini, M., Secco, F., and De Filippo, D., EB23, EB24, EB26, EB27, EB34, EB35, EB48, EC33

JE040-0345 Breant, M., and Sue, J.L., EC00, EC91, EC93, EC94, EC99, ED99

JE040-0448 Saxena, R.S., and Chaturvedi, U.S., EC88

Vol. 41 (1973)
JE041-0105 Shams El Din, A.M., Sabri, H., and Saber, T.M.H., EB15

JE041-0127 Florence, T.M., and Farrar, Y.J., EB65

JE041-0405 Avaca, L.A., and Bewick, A., EB29

JE041-0411 Klatt, L.N., and Rouseff, R.L., EA45

Vol. 42 (1973)
JE042-0049 Cardinall, M.E. Carelli, I., Ceccaroni, G., and Trazza, A., EC12

JE042-0057 Bontempelli, G., Magno, F., and Mazzocchin, G.-A., EE17

JE042-0133 Clark, D.B., Fleischmann, M., and Pletcher, D., EA31, EA64, EA65, EB68, EB88, EC90

JE042-0151 Vanden Born, H.W., and Harris, W.E., EB25

JE042-0189 Lamy, E., Nadjo, L., and Savéant, J.M., ED72

JE042-0253 Lasia, A., EC54, EC59

JE042-0261 Krygowski, T.M., Lipsztajn, M., and Galus, Z., EB29

JE042-0397 Petek, M., Bruckenstein, S., Feinberg, B., and Adams, R.N., EC82, ED12

JE042-0415 Laviron, É., ED99

Vol. 43 (1973)
JE043-0095 Guidelli, R., Foresti, M.L., and Pezzatini, G., EA67

JE043-0135 Jovanović, M.M., and Rekalić, V.J., EA14, EA15

JE043-0205 Cauquis, G., and Lachenal, D., EA95

JE043-0215 Brown, O.R., and Gonzalez, E.R., EA32, EA66

JE043-0257 Cǎluşaru, A., and Voicu, V., EB81

JE043-0267 Breitenbach, M., and Heckner, K.H., EB50

JE043-0308 Fleischmann, M., Mengoli, G., and Pletcher, D., EA13

JE043-0311 Heyrovský, M., and Vavřička, S., EB54

JE043-0318 Koch, V.R. Miller, L.L., Clark, D.B., Fleischmann, M., Joslin, I., and Pletcher, D., EA31

JE043-0349 Floet, B., Jee, R.D., and Little, C.J., EB71

JE043-0365 Manoušek, O., and Volke, J., EB99, EC00, EC03, EC04, EC14, EC16, EC58, EC59, EC60, EC61, EC68, EC70, ED06

JE043-0377 Bontempelli, G., Magno, F., Mazzocchin, G.-A., and Zecchin, S., ED89

JE043-0387 Desideri, P.G., Lepri, L., and Heimler, D., EB50, EB56, EB57, EB58, EE21

JE (JOURNAL OF ELECTROANALYTICAL
    CHEMISTRY AND INTERFACIAL
    ELECTROCHEMISTRY)(cont.)

Vol. 44 (1973)
  JE044-0025 Inesi, A., and Rampazzo,
             L., EA06, EA09, EA50,
             EA53, EA59
  JE044-0239 Capon, A., and Parsons, R.,
             EA02
  JE044-0275 Plichon, V., and Faure,
             G., EB53
  JE044-0291 Canterford, D.R., and
             Buchanan, A.S. EA95

Vol. 45 (1973)
  JE045-0156 Serna, A., Vera, J., and
             Marin, D., ED50
  JE045-0205 Capon, A., and Parsons,
             R., EA02
  JE045-0397 Farnia, G., Capobianco,
             G., and Romanin, A.,
             EC55, EC63, EC95
  JE045-0467 Leedy, D.W., EC31

Vol. 46 (1973)
  JE046-0051 Căluşaru, A., Crişan, I.,
             and Kůta, J., EA00, EA02
  JE046-0077 Ghe, A.M., and Valcher, J.,
             ED80, EE34
  JE046-0089 Thévenot, D., EA46
  JE046-0215 Capon, A., and Parsons,
             R., EB22
  JE046-0323 Barnes, D., and Zuman,
             P., EA00, EA16, EA68,
             EB82, EC66, ED04
  JE046-0343 Barnes, D., and Zuman,
             P., EC66, ED03,
             EE67
  JE046-0411 Kowal, A., and Pomianowski,
             A., EA28
  JE046-0421 Laviron, É., and Vallat,
             A., EB28

Vol. 47 (1973)
  JE047-0081 Fung, K.W., Chambers, J.
             Q., and Mamantov, G.,
             EB66, ED90
  JE047-0190 Căluşaru, A., and Bănică,
             F.G., EA36
  JE047-0215 Ammar, F., and Savéant,
             J.M., EB14, EE66
  JE047-0335 Cardinali, M.E., Carelli,
             I., and Andruzzi, R.,
             EC64, EC65
  JE047-0479 McAllister, D.L., and
             Dryhurst, G., EB18,
             EB19, EB45, EB46
  JE047-0499 Tissier, C., and Agoutin,
             M., EC05, EC06, EC07,
             EC08, EC09, EC10
  JE047-0543 Leduc, P., and Thévenot,
             D., EE58

Vol. 48 (1973)
  JE048-0071 Bond, A.M., Casey, A.T.,
             and Thackeray, J.R.,
             EA12, EA28, EA62, EA94,
             EC42
  JE048-0146 Ammar, F., Nadjo, L., and
             Savéant, J.M., EC13,
             EE44, EF06
  JE048-0167 Shoesmith, D.W., Kutowy,
             O., and Barradas, R.G.,
             ED01
  JE048-0277 Cardinali, M.E., Carelli,
             I., and Trazza, A., EC25
  JE048-0297 Butkiewicz, K., EE69, EE70
  JE048-0433 Wrona, M., and Czochralska,
             B., EA83, EB62, ED42

Vol. 49 (1974)
  JE049-0017 Lipsztajn, M., Krygowski,
             T.M., and Galus, Z.,
             EB29
  JE049-0027 Niki, K., Suzuki, K.,
             Satô, G.P., and Mori,
             N., ED51
  JE049-0085 Inesi, A., and Rompazzo,
             L., EA05, EA07, EA11,
             EA51, EA54, EA61
  JE049-0105 Moses, P.R., and Chambers,
             J.Q., EA26, EA60, EA92,
             EA93, EB76, ED35, EE24
  JE049-0111 Marcoux, L., and Adams,
             R.N., EE35, EE36, EE37,
             EE38, EE87, EE97, EF05,
             EF08, EF14
  JE049-0281 Jones, H.L., and Oster-
             young, R.A., EC80,
             ED11
  JE049-0287 Brun, A., and Rosset, R.,
             ED07
  JE049-0433 Vadaszy, R., and Cover,
             R.E., EA24, EA56, EA57,
             EA87, EB64, EB69, EB70,
             EC87, ED53

Vol. 50 (1974)
  JE050-0031 Geiger, W.E., Jr., and
             Smith, D.E., EA78, EA79,
             EC52, ED32, EF21
  JE050-0073 Farnia, G., Mengoli, G.,
             and Vianello, E., EC23
  JE050-0113 Florence, T.M., Johnson,
             D.A., and Batley, G.E.,
             EC97, EC98, ED59, ED61,
             ED62, ED63, ED64, ED65,
             ED66, ED67, ED68, EE39,
             EE56, EE59, EE60, EE77,
             EE78, EE79, EE80, EE81,
             EE82, EE83, EE84, EE85,
             EF07

Vol. 51 (1974)
  JE051-0075 Cox, J.A., and Ozment, C.
             L., EE15

JE (JOURNAL OF ELECTROANALYTICAL
 CHEMISTRY AND INTERFACIAL
 ELECTROCHEMISTRY) (cont.)

Vol. 51 (1974) (cont.)
  JE051-0456 Doughty, A.G.,
    Fleischmann, M., and
    Pletcher, D., EB47,
    ED20, EE14

Vol. 52 (1974)
  JE052-0037 Lindquist, J., EB47
  JE052-0093 Desideri, P.G., Lepri,
    L., and Heimler, D.,
    EC35, EC36, EC37,
    EC80
  JE052-0105 Desideri, P.G., Lepri,
    L., and Heimler, D.,
    EC38, EE73
  JE052-0115 Florence, T.M., ED96,
    ED97, ED99, EE01, EE02,
    EE03, EE04, EE05, EE06,
    EE09, EE89, EE90, EE91,
    EE92, EE93, EF01
  JE052-0229 Lasia, A., EC56

Vol. 53 (1974)
  JE053-0293 Asirvatham, M.R., and
    Hawley, M.D., EB44,
    EC28, EC75, EE75, EE76

Vol. 54 (1974)
  JE054-0181 Fritz, H.P., and
    Würminghausen, T., EA99,
    EB78, EB87, EC51, EC90,
    ED57, EE96, EF20
  JE054-0221 Grandi, G., Benedetti, L.,
    Andreoli, R., and Gavioli,
    G.B., EC20, EC30, EC76,
    EC79, ED10, ED74
  JE054-0232 Masson, J.P., Derynck, J.,
    and Tremillon, B., EB09,
    ED27
  JE054-0313 Lipsztajn, M., Krygowski,
    T.M., Laren, E., and
    Galus, Z., EE00

Vol. 55 (1974)
  JE055-0069 McAllister, D.L., and
    Dryhurst, G., EA08,
    EA42, EB18, EB19, EB20,
    EB21
  JE055-0109 Bontempelli, G., Magno,
    F., Mazzocchin, G.-A.,
    and Seeber, R., EE13
  JE055-0157 Călușaru, A., and Bănică,
    F.G., EA36
  JE055-0277 Kalinowski, M.K., and
    Tenderende-Guminska, B.,
    EB22, ED22, ED23, ED79
  JE055-0417 Ghe, A.M., and Valcher,
    S., EE33
  JE055-0445 Malik, W.U., Mahesh, V.
    K., Raisinghani, M.,
    and Goyal, R.N., EB63,
    EC41, EC73, EC74, ED09

Vol. 56 (1974)
  JE056-0217 Chander, S., and
    Fuerstenau, D.W., EA75
  JE056-0285 Toure, V., Levy, M., and
    Zuman, P., ED00, ED34,
    ED71
  JE056-0373 Persson, B., and Nygård,
    B., EB37, EB49, EE17
  JE056-0427 Cape, J.N., and Vincent,
    C.A., ED70
  JE056-0443 Grimshaw, J., and Trocha-
    Grimshaw, J., EE31,
    EE32
  JE056-0459 Canfield, N.D., and
    Chambers, J.Q., EE55

Vol. 57 (1974)
  JE057-0027 Andrieux, C.P., and
    Savéant, J.M., EB12,
    EB13, EB14
  JE057-0179 Asirvatham, M.R., and
    Hawley, M.D., EB28,
    EE00
  JE051-0191 Farnia, G., Da Silva,
    A. R., and Vianello,
    E., EB30, EB32, EB33,
    EC23, EC24
  JE057-0339 Lipsztajn, M., Krygowski,
    T.M., Laren, E., and
    Galus, Z., EB28
  JE057-0351 Brown, O.R., Fletcher,
    S., and Harrison, J.A.,
    EC53, EC56, EC96, ED30,
    ED69, EE65, EE71

TABLE XI.
AUTHOR INDEX

This is an index to Table X. It lists, in alphabetical order, the names of all the authors cited in Table X and gives, for each author, the number of each page on which a citation of his or her name appears in Table X. Each reference given on the cited page should be inspected, for there may be two or more references on the page that includes the name sought, and such situations are not identified here.

TABLE XI. Author Index

Adams, R.N., 442-44
Agoutin, M., 444
Ammar, F., 444
Andreoli, R., 445
Andrieux, C.P., 445
Andruzzi, R., 444
Annino, R., 442
Asirvatham, M.R., 445
Avaca, L.A., 443
Aylward, G.H., 442

Bănică, F.G., 444-45
Barnes, D., 444
Barradas, R.G., 444
Bartak, D.E., 442
Bartušek, M., 441
Batley, G.E., 444
Bauer, D., 443
Bäumgartel, H., 441
Baxall, L.G., 443
Benedetti, L., 445
Berg, H., 441
Bewick, A., 443
Boček, K., 441
Boczkowski, R.J., 442
Bolton, D.J., 442
Bond, A.M., 444
Bontempelli, G., 442-45
Borsdorf, R., 442-43
Branica, M., 442
Breant, M., 443
Breitenbach, M., 443
Breslow, R., 441-42
Březina, M., 443
Brown, O.R., 442-43, 445
Bruckenstein, S., 443
Brun, A., 444
Buchanan, A.S., 444
Buchanan, G.S., 442
Budnikov, G.K., 441
Butkiewicz, K., 443-44
Buvet, R., 443

Caillet, A., 443
Călușaru, A., 443-45
Canfield, N.D., 445
Canterford, D.R., 444
Cape, J.N., 445
Capobianco, G., 444
Capon, A., 444
Cardinali, M.E., 442-44
Carelli, I., 442-44
Casey, A.T., 444
Cauquis, G., 442-43
Ceccaroni, G., 443

Chabaud, B., 443
Chambers, J.A., 442, 444
Chander, S., 445
Chandra, S., 442
Chaturvedi, U.S., 442-43
Clark, D., 442
Clark, D.B., 443
Clarke, S., 442
Coffen, D.L., 442
Ćosović, B., 442
Cover, R.E., 444
Cox, J.A., 444
Crișan, I., 444
Czochralska, B., 444

Da Silva, A.R., 445
De Filippo, D., 443
Demange-Guerin, G., 443
Derynck, J., 445
Desideri, P.G., 443, 445
Doskočilová, D., 441
Doughty, A.G., 445
Drury, R., 442
Dryhurst, G., 444-45

Eisner, U., 442
Eliášek, J., 441
Elving, P.J., 441-42

Farnia, G., 444-45
Farrar, Y.J., 443
Farsang, G., 442
Faure, G., 444
Fedoroňko, M., 441
Feinberg, B., 443
Fišerová-Bergerová, V., 441
Fleet, B., 443
Fleischmann, M., 442-43, 445
Fletcher, S., 445
Fleury, D., 441
Fleury, M.B., 441
Florence, T.M., 443-45
Fonds, A.W., 442
Foresti, M.L., 443
Foucault, A., 443
Fritz, H.P., 445
Fuerstenau, D.W., 445
Fülleová, E., 441
Fung, K.W., 444

Galus, Z., 443-45
Gavioli, G.B., 445
Geiger, W.E., Jr., 442, 444
Genies, M., 443
Ghe, A.M., 444-45
Gileadi, E., 442
Gollmick, F.A., 441
Gonzalez, E.R., 443
Goyal, R.N., 445
Grandi, G., 445
Grimshaw, J., 445
Grubbs, R., 441-42
Guidelli, R., 443
Gut, J., 441

Hampson, N.A., 442
Harris, W.E., 443
Harrison, J.A., 442, 445
Hawley, M.D., 442, 445
Heckner, K.H., 443
Heimler, D., 443, 445
Herzschuh, R., 442-43
Heyrovský, M., 441-43
Hlavatý, J., 441
Hoa, L.C., 442-43
Holeček, V., 441
Holleck, L., 441-42
Holubek, J., 441
Horák, V., 441
Houser, K.J., 442
Humlová, A., 441

Inesi, A., 444

Jackson, D.T., Jr., 442
Janík, B., 442
Jee, R.D., 443
Johnson, D.A., 444
Jones, H.L., 443-44
Joslin, T., 443
Jovanović, M.M., 443
Jungwirt, A., 441

Kalinowski, M.K., 445
Kalvoda, R., 441
Kašpárek, F., 441
Kazemifard, G., 442
King, D.M., 443
Kirowa-Eisner, E., 442
Kišová, L., 441
Kitaev, Yu.P., 441
Klatt, L.N., 443
Knobloch, P., 441
Koch, V.R., 442-43
Komenda, J., 441
Königstein, J., 441
Kolby, N.I., 443
Koryta, J., 443
Kovács, M., 442
Kowal, A., 444
Krupička, J., 441
Krygowski, T.M., 443-45
Krznarić, D., 442
Kůta, J., 444
Kutowy, O., 444

Lachenal, D., 443
Ladányi, L., 442
Lamy, E., 443
Landis, R.T., 442
Laren, E., 445
Lasia, A., 443, 445
Laskafeld, D., 441
Laviron, É., 443-44
Leduc, P., 444
Lee, R.W., 442
Leedy, D.W., 444
Lehmann, O., 441
Lepri, L., 443, 445
Levy, M., 445
Lindquist, J., 445
Línek, K., 441
Lipsztajn, M., 443-45
Little, C.J., 443
Los, J.M., 442
Loučka, T., 443
Lund, H., 441

McAllister, D.L., 444-45
MacDonald, K.I., 442
Magno, F., 442-43, 445
Mahapatra, S., 442
Mahesh, V.K., 445
Mahler, J., 442
Malik, W.U., 445

TABLE XI. Author Index

Mamantov, G., 444
Manoušek, O., 441, 443
Marcoux, L., 444
Marhold, J., 441
Marin, D., 444
Maršíková, D., 443
Masson, J.P., 445
Matrka, M., 441
Maturová, M., 441
Mayeda, F.A., 441
Mazzocchin, G.-A., 443, 445
Mengoli, G., 443-44
Miller, I.R., 442
Miller, L.L., 441-43
Mirkin, L.S., 441
Mittal, M.L., 442
Mollin, J., 441
Mori, N., 444
Moses, P.R., 442, 444
Mühlstädt, M., 443
Murahashi, S., 441-42
Murayama, D.R., 442

Nadjo, L., 443-44
Nelsen, S.F., 442
Němečková, A., 441
Niki, K., 444
Nygård, B., 445

Okáč, A., 441
Osteryoung, R.A., 443-44
Ozment, C.L., 444

Pandey, A.V., 442
Papouchado, L., 442
Parsons, R., 444
Pergál, M., 441
Persson, B., 445
Petek, M., 443
Petránek, J., 441
Petrie, G., 442
Pezzatini, G., 443
Pilloni, G., 442
Pletcher, D., 442-43, 445
Plichon, V., 444
Pomianowski, A., 444
Porteix, M., 442
Pradáč, J., 443
Price, J.W., 443

Raisinghani, M., 445
Rampazzo, L., 444
Reilley, C.N., 442
Rekalić, V.J., 443
Remane, H., 442-43
Ridgway, T.H., 442
Romanin, A., 444
Rompazzo, L., 444
Rosset, R., 444
Rouseff, R.L., 443
Ryba, O., 441

Saber, T.M.H., 442-43
Sabri, H., 443
Ságner, Z., 441
Šantavý, F., 441
Satô, G.P., 444
Savéant, J.M., 443-45
Saxena, R.S., 442-43
Secco, F., 443
Seeber, R., 445
Serna, A., 444
Serve, D., 442
Shams El Din, A.M., 443
Shelton, R.N., 442
Shoesmith, D.W., 444
Sinicki, C., 442
Skála, V., 442
Smith, D.E., 444
Sohr, H., 442
Sondheimer, F., 442
Stankoviansky, S., 441
Stencel, M., 443
Sue, J.L., 443
Sümmermann, W., 441
Suzuki, K., 444

Tenderende-Guminska, B., 445
Teva, J., 442
Thackeray, J.R., 444
Thévenot, D., 443-44
Tissier, C., 444
Toure, V., 445
Trazza, A., 442-44
Tremillon, B., 445
Trocha-Grimshaw, J., 445

Vadaszy, R., 444
Valcher, J., 444
Valcher, S., 445
Vallat, A., 444

Vanden Born, H.W., 443
Van Duyne, R.P., 442
Vavřička, S., 442-43
Venturini, M., 443
Vera, J., 444
Vianello, E., 444-45
Vincent, C.A., 445
Voicu, V., 443
Volke, J., 441-43

Warner, C.R., 441
Watton, E.C., 442
Webb, J.W., 442
Wienhold, K., 442
Wolf, J.F., 441
Wrona, M., 444
Würminghausen, T., 445

Zahradník, R., 441
Zandanel, A., 442
Zecchin, S., 442-43
Zemer, Y., 442
Zhdanov, S.I., 441
Zuman, P., 441, 444-45
Zvěrina, V., 441

CORRIGENDA FOR PRECEDING VOLUMES OF THE CRC HANDBOOK SERIES IN ORGANIC ELECTROCHEMISTRY

VOLUME 1
Table 1

Page	Code number	Line number	Column number	For	Read
15	AA04	1,3	26	E	PY,GLC
	AA08	2,3	26	E	PY,GLC
25	AA55	1	27	↑?	(remove)
	AA56	1	27	↑?	(remove)
	AA57	1	27	↑?	(remove)
29	AA71	1	26	E	PY
39	AB05	1-7	26	E	PY
53	AB74	2	26	E	VR
59	AC03	2	23	$\log k^0_{s,h}$	$\log k^0_{f,h}$
63	AC16	1	26	E	VA
95	AD29	1	26	E	PY
101	AD65	2,7,8,10	26	E	PY
122	AE03	14	13	w	ω
123	AE03	6,8,11	26	E	VY
		58,60	27	E	(i)
145	AE65	7	23	ΔE_w	$\Delta E_{su/2}$
147	AE73	2	26	E	PY
171	AF83	3,4	26	E	VR
181	AG08	2,4	26	E	PY
200	AG52	2-3	13	$i=2-0.4(u,F)$	$i(u)=2-0.4F$
		5-6	13	$i=20-1(u,F)$	$i(u)=20-1F$
		8-9	13	$i=50-1.5(u,F)$	$i(u)=50-1.5F$
205	AG64	1	27	$(\log k_2^!)$	$(\log k_1)$ (see mech. 192)
	AG65	1	27	$(\log k_2^I)$	$(\log k_1)$ (see mech. 192)
217	AG94	1,3,5	26	E	IU
		7-13	26	E	VA
251	AI32	2	27	lp	ℓp
265	AJ00	1	26	E	PY
267	AJ13	3-4	19	$i_0 \tau_c^{\frac{1}{2}}/$ $i\tau_d^{\frac{1}{2}}$	$J\tau^{\frac{1}{2}}/$ $J\tau_d^{\frac{1}{2}}$

CORRIGENDA FOR PRECEDING VOLUMES OF THE CRC HANDBOOK SERIES IN ORGANIC ELECTROCHEMISTRY (continued)

VOLUME I
Table I (continued)

Page	Code number	Line number	Column number	For	Read
273	AJ39	1,3	18	SCE(O)	SCE(o)
	AJ40	1,2		SCE(O)	SCE(o)
	AJ41	2,4		SCE(O)	SCE(o)
	AJ53	1,2		SCE(O)	SCE(o)
275	AJ54	1,2		SCE(O)	SCE(o)
	AJ55	1,2		SCE(O)	SCE(o)
282	AJ95	4	11	PDMC	PDME
297	AK67	1	27	$c^{3/2}$	$C^{3/2}$
297	AK70	1	27	τ_a	τ_{sw}
				$c^{3/2}$	$C^{3/2}$
299	AK75	4	19	$i_s(u)$	$i_{su}(u)$
335	AM46	9	27	x	X
337	AM67	1	27	$(\log K_2')$	$(\log k_1)$ (see mech. 116)
348	AN48	3	7	THAM	TRIS
349		2	23	k_{sh}^{app}	$k_{s,h}^{app}$
		2-3	25	E_{pa}	$E_{p,A}$
				E_{pc}	$E_{p,C}$
369	AO61	1	27	M_x	Mx
	AO65	6,12	26	E	PY
	AO66	4,10	26	E	PY
375	AO85	8,12,15	26	E	PY
377	AO97	2-5	13	integrated $\ldots E_p$	(remove)
		3-5	19	$Qp\ldots s^{\frac{1}{2}}$	XI
		2	25	i_p/Qp	$i_p A/Q$
		1	27	$A, \neq, p$	$A, \neq, XI = Q/Av^{\frac{1}{2}}$, Q=charge integrated from -0.024 to E_p, p

CORRIGENDA FOR PRECEDING VOLUMES OF THE CRC HANDBOOK SERIES IN ORGANIC ELECTROCHEMISTRY (continued)

VOLUME I
Table I (continued)

Page	Code number	Line number	Column number	For	Read
379	AP00	2	26	E	PY
433	AR14	1	26	E	PY
458	AS51	1	5	PC	PROC
467	AS96	1	26	E	PY
497	BA25	2	26	E	PY
	BA27	2	26	E	PY
499	BA30	5	26	E	VR
503	BA52	5	26	E	VR
594	BE54	3	5	PC	PROC
596	BE56	3	5	PC	PROC
621	BF58	3	23	$E_{pa} - E_{pc}$	$E_{p,A} - E_{p,C}$
651	BH30	4	19	i_s^n	i_{su}^n
693	BJ40	10	23	ΔE_w	$\Delta E_{p/2}$
697	BJ59	2	19	i_s	i_{su}
713	BK37	3	16	E_s	E_{su}
777	BN17	4-6	27	$\tau/Q \ldots p$	$Q_{ads} = kt^{7/6} C(Ap)$ for $C <$ ca. 0.15, $Q_{ads} = kt^{2/3} \neq f(C)(Ap)$ for $C >$ ca. $0.15, p$
815	BP31	1	16	E°	XE
			27	(F)	XE = "reversible reduction potential"

VOLUME II
Table VI

546   under "Propylene carbonate" add code number BE56

CORRIGENDA FOR PRECEDING VOLUMES OF THE CRC HANDBOOK SERIES IN ORGANIC ELECTROCHEMISTRY (continued)

VOLUME III
Introduction to Table I

Page 9   Under columns 19-21 "Emission", 4th line, for "deficient (D)" read "deficient (De)"

Table I

Page	Code number	Line number	Column number	For	Read
61	CB59	2	13	τ	t_{sw}
		4	19	Q_b/Q_f	Q_A/Q_C
97	CC66	2	27	E_λ	E_T
99	CC76	1	27	ΔE°	$\Delta E_{\frac{1}{2}}^\circ$
145	CE48	1	27	ΔE°	$\Delta E_{\frac{1}{2}}^\circ$
171	CF56	2	27	ΔE°	$\Delta E_{\frac{1}{2}}^\circ$
	CF57	1	27	ΔE°	$\Delta E_{\frac{1}{2}}^\circ$
185	CG10	1	27	ΔE°	$\Delta E_{\frac{1}{2}}^\circ$
	CG11	1	27	ΔE°	$\Delta E_{\frac{1}{2}}^\circ$
	CG12	1	27	ΔE°	$\Delta E_{\frac{1}{2}}^\circ$
217	CH31	1	27	ΔE°	$\Delta E_{\frac{1}{2}}^\circ$
239	CI20	1	27	ΔE°	$\Delta E_{\frac{1}{2}}^\circ$
256	CJ03	5	4	PC	PROC
261	CJ28	1	27	ΔE°	$\Delta E_{\frac{1}{2}}^\circ$
281	CK34	2	26	Q_b/Q_f	Q_A/Q_C
287	CK63	1	27	ΔE°	$\Delta E_{\frac{1}{2}}^\circ$
	CK64	1	27	ΔE°	$\Delta E_{\frac{1}{2}}^\circ$
293	CK97	1	27	ΔE°	$\Delta E_{\frac{1}{2}}^\circ$
303	CL43	1	27	ΔE°	$\Delta E_{\frac{1}{2}}^\circ$
323	CM35	1	27	ΔE°	$\Delta E_{\frac{1}{2}}^\circ$
349	CN08	15, 19	27	(2.46E4/ C/1+75C)	(2.46E4) C/(1+75C),
350	CN16	4	14	PC	PROC
351	CN17	1,3-4, 6-9	21	D	De
352		26	4	PC	PROC
353		1,2,10	21	D	De

CORRIGENDA FOR PRECEDING VOLUMES OF THE CRC HANDBOOK SERIES IN ORGANIC ELECTROCHEMISTRY (continued)

VOLUME III
Table III

Page	Line number	For	Read
383	19	ψ	E
395	1	pK	pK_4
	2	pK_O	pK_5
	4	pK_O and pK	pK_5 and pK_4

Headings of second table:

Page	Col.	For:	Read:
422	2	pK from $pK_{E_{1/2}}$	pK from $E_{1/2}$ vs. pH plot
	3	$\log k_1$ from $pK_{E_{1/2}}, pK'$	$\log k_1$ from pK and pK'
	4	$\log k_{-1}$ from $pK_{E_{1/2}} - pK'$	$\log k_{-1}$ from pK and $\log k_1$

Table VI

632   Under Benzene-propylene carbonate(50:50) add Code number CN17

Table VIII

647   Under Carbon: for "cylinder, ratating" read "cylinder, rotating"
648   Under Platinum: for "(PT)" read "(Pt)"

VOLUME IV
Table I

Page	Code number	Line number	Column number	For	Read
17	DA17	3	3	C.A. 104-89-6	C.A. 140-89-6

Table VI

439–440   DA17   C.A. number 104-89-6 should be deleted from page 439 and number 140-89-6 should be inserted on page 440.

LIST OF ABBREVIATIONS

Handbook Series in Organic Electrochemistry

Vol. V

ALL OTHER (cont.)

n_{app}	an apparent number of electrons transferred	Pr	prewave			
n_b	number of electrons transferred through rate-determining step of an anodic process	PrCN	butyronitrile	RPDE	rotating platinum-disc	
		PRW	Prideaux and Ward			
		PS	staircase polarography	RPE	rotating platinum	
		PSCE	mercury-coated carbon	RPGE	rotating pyrolytic graphite	
N_o	collection efficiency (ring-disc electrode)	PSMC	mercury-coated platinum sphere	RPMC	rotating mercury-coated platinum	
		PSME	mercury-coated platinum sphere	RRDE	ring-disc	
		PT	Tast polarography	RSDE	rotating silver-disc	
O		PV	total ac polarography	RWGE	rotating wax-impregnated graphite	
		PVC	poly(vinyl chloride)			
0	no wave observed	PVI	in-phase ac polarography	rxn	reaction	
(oc)	open circuit			r_d	radius of disc	
oh=	number of hydroxide ions consumed through rate-determining step	PVQ	quadrature polarography	$r_{r,i}$	inner radius of ring	
		PW	single-sweep polarography	$r_{r,o}$	outer radius of ring	
OTE	optically transparent electrode (may be preceded by a chemical symbol)	PX	double-tone polarography			
		PY	polarography			
		PYR	pyridine			

P

		Q		**S**		
P—	merging with previous wave	Q	in characteristic potential column, incision quotient; elsewhere, quantity of charge (μC)	S	merging with succeeding wave	
p	preliminary			satd	saturated	
p=	number of protons consumed through rate-determining step			SCE	saturated calomel electrode	
		QE	controlled-potential coulometry	Se	energy-sufficient	
PA	triangular-wave polarography			SLiE	mercury—mercurous chloride—saturated lithium chloride electrode	
		QI	quasi-reversible			
PB	higher harmonic ac polarography with phase-sensitive rectification	QP	controlled-current coulometry without a reagent precursor			
				SMDE	sessile mercury drop	
		Qp	integrated charge corresponding to the faradaic current (μC)	SME	streaming mercury	
PbDE	lead-disc			SnDE	tin-disc	
PBE	platinum-bead			SnO	tin-oxide electrode	
PbSE	lead sulfide electrode	QR	chronocoulometry	Sr	surface reaction	
PC	current-scanning polarography			SRGE	silicone rubber-graphite	
		QT	controlled-current coulometry with a reagent precursor	SSCE	mercury—mercurous chloride—saturated sodium chloride electrode	
PCA	propylene carbonate					
PD	derivative polarography	QU	double potential-step chronocoulometry	St	slope of tangent $E_{\frac{1}{2}}$	
PDE	platinum-disc			sttd	stated	
PDME	mercury-coated platinum-disc	QW	electrogravimetry	SULF	sulfate	
				SULN	sulfolane	
PDS	with periodic displacement of solution			supp. elect.	supporting electrolyte	
PE	pretreated electrode					
PF	differential polarography					
PG	square-wave polarography	**R**		**T**		
PGE	pyrolytic graphite					
PH	higher harmonic ac polarography	R	reversible			
		r	reliable			
PHEN	phenol	RAuE	rotating gold	T	temperature	
PHOS	phosphate	RCCE	rotating cylindrical carbon	t	time (s)	
PHTH	phthalate			t(c)	controlled time	
PI	alternating-voltage polarography	RCGE	rotating glassy carbon cylinder	$t(E_1)$	time of first step	
				t(oc)	drop time open circuit	
PK	Kalousek polarography	RDCP	rotating carbon-paste	Tc	temperature coefficient of the limiting current (% deg^{-1})	
pK_a	-log(acidic dissociation constant)	RDE	rotating disc			
		RDGC	rotating glassy carbon-disc			
pK'	pH at inflection point of polarographic dissociation curve	RDME	rotating dropping mercury	t(sc)	drop time short circuit	
		RGDE	rotating gold-disc	THF	tetrahydrofuran	
		RhBE	rhodium bead	THFA	tetrahydrofurfuryl alcohol	
PL	intermodulation polarography	RMAuD	rotating mercury-coated gold-disc			
				TLC	thin-layer chromatography	
Pl	plateau			TLE	thin-layer electrode	
PM	multisweep polarography	RMCuD	rotating mercury-coated copper disc	TLIR	thin-layer chronoamperometry	
PMC	mercury-coated platinum					
PO	oscillopolarography	RMDE	rotating mercury-coated ring-disc	TLQE	thin-layer coulometry	
Po	postwave			Tomeš	$E_{\frac{3}{4}} - E_{\frac{1}{4}}$ (mV)	
pos.	positive	RMPD	rotating mercury-coated platinum-disc			
PP	pulse polarography					
ppt	precipitate	rms	root mean square	TQ	thin-layer chronocoulometry	
PQ	polarographic coulometry	R_4NCE	mercury—mercurous chloride—tetraalkyl-ammonium chloride electrode			
PR	cyclic triangular-wave polarography			TRIS	tris(hydroxymethyl)aminomethane	
PROC	propylene carbonate			TX100	Triton X-100	
		RP	radio-frequency polarography			